COMPLETE SOLUTIONS MANUAL

to accompany

LINEAR ALGEBRA

with

APPLICATIONS
4e

by

Gareth Williams

Prepared by
W. J. Mourant   and   Gareth Williams

COMPLETE SOLUTIONS MANUAL

to accompany

LINEAR ALGEBRA

with

APPLICATIONS
4e

by

Gareth Williams

Prepared by
W. J. Mourant   and   Gareth Williams

JONES AND BARTLETT PUBLISHERS

Sudbury, Massachusetts

BOSTON   TORONTO   LONDON   SINGAPORE

World Headquarters
Jones and Bartlett Publishers
40 Tall Pine Drive
Sudbury, MA 01776
978-443-5000
info@jbpub.com
www.jbpub.com

Jones and Bartlett Publishers Canada
2406 Nikanna Road
Mississauga, Ontario
Canada L5C 2W6

Jones and Bartlett Publishers International
Barb House, Barb Mews
London W6 7PA
UK

Copyright © 2001 by Jones and Bartlett Publishers, Inc.

Cover image © Steve Bloom-TCL / MASTERFILE

ISBN: 0-7637-1527-1

All rights reserved. No part of the material protected by this copyright notice may be reproduced or utilized in any form, electronic or mechanical, including photocopying, recording, or any information storage or retrieval system, without written permission from the copyright owner.

Printed in the United States of America
04 03 02 01 00   10 9 8 7 6 5 4 3 2 1

This Complete Solutions Manual contains worked-out solutions to all exercises in the text *Linear Algebra with Applications, Fourth Edition* by Gareth Williams. All references in this manual are to chapters, sections, and exercises in the text.

CONTENTS

Preface

1 Systems of Linear Equations   1
- 1.1 Matrices and Systems of Linear Equations
- 1.2 Gauss-Jordan Elimination
- 1.3 Curve Fitting, Electrical Networks and Traffic Flow
- Chapter 1 Review Exercises

2 Matrices   28
- 2.1 Addition, Scalar Multiplication and Multiplication of Matrices
- 2.2 Properties of Matrix Operations
- 2.3 Symmetric Matrices and an Application in Archaeology
- 2.4 The Inverse of a Matrix
- 2.5 Leontief Input-Output Model in Economics
- 2.6 Stochastic Matrices: A Population Movement Model
- 2.7 Communication Model and Group Relationships
- Chapter 2 Review Exercises

3 Determinants   84
- 3.1 Introduction to Determinants
- 3.2 Properties of Determinants
- 3.3 Numerical Evaluation of a Determinant
- 3.4 Determinants, Matrix Inverses and Systems of Linear Equations
- Chapter 3 Review Exercises

4 The Vector Space R^n   116
- 4.1 Introduction to Vectors
- 4.2 Dot Product, Norm, Angle and Distance
- 4.3 Introduction to Linear Transformations
- 4.4 Matrix Transformations, Computer Graphics and Fractals
- Chapter 4 Review Exercises

5 General Vector Spaces 145
- 5.1 Vector Spaces and Subspaces
- 5.2 Linear Combinations of Vectors
- 5.3 Linear Dependence and Independence
- 5.4 Bases and Dimension
- 5.5 Rank of a Matrix
- 5.6 Orthonormal Vectors, Projections and the Gram-Schmidt Method
- Chapter 5 Review Exercises

6 Eigenvalues and Eigenvectors 204
- 6.1 Eigenvalues, Eigenvectors and Eigenspaces
- 6.2 Demography and Weather Prediction
- 6.3 Diagonalization of Matrices
- 6.4 Quadratic Forms, Difference Equations and Normal Modes
- Chapter 6 Review Exercises

7 Linear Transformations 243
- 7.1 Linear Transformations, Kernel and Range
- 7.2 Transformations and Systems of Linear Equations
- 7.3 Coordinate Vectors
- 7.4 Matrix Representations of Linear Transformations
- Chapter 7 Review Exercises

8 Inner Product Spaces 285
- 8.1 Inner Product Spaces
- 8.2 Non-Euclidean Geometry
- 8.3 Approximation of Functions
- 8.4 Least Squares Curves
- Chapter 8 Review Exercises

9 Numerical Techniques 328
- 9.1 Gaussian Elimination
- 9.2 The Method of LU Decomposition
- 9.3 Practical Difficulties in Solving Systems of Equations
- 9.4 Iterative Methods for Solving Systems of Linear Equations
- 9.5 Eigenvalues by Iteration. Connectivity of Networks
- Chapter 9 Review Exercises

10 Linear Programming   367
    10.1 A Geometrical Introduction to Linear Programming
    10.2 The Simplex Method
    10.3 Geometrical Explanation of the Simplex Method
    Chapter 10 Review Exercises

Appendices
    A Cross Product   392
    B Equations of Lines in Three-Space   397
    C Graphing Calculator Manual   402
    D MATLAB Manual   406

Chapter 1

Exercise Set 1.1

1. (a) 3 x 3   (b) 3 x 2   (c) 2 x 4   (d) 3 x 1   (e) 3 x 5   (f) 1 x 4

2. 1, 4, 9, −1, 3, 8   3. 4, 5, 6, 7, 2, 3

4. $\begin{bmatrix} 1 & 0 & 0 & 0 \\ 0 & 1 & 0 & 0 \\ 0 & 0 & 1 & 0 \\ 0 & 0 & 0 & 1 \end{bmatrix}$

5. (a) $\begin{bmatrix} 1 & 3 \\ 2 & -5 \end{bmatrix}$ and $\begin{bmatrix} 1 & 3 & 7 \\ 2 & -5 & -3 \end{bmatrix}$ (b) $\begin{bmatrix} 5 & 2 & -4 \\ 1 & 3 & 6 \\ 4 & 6 & -9 \end{bmatrix}$ and $\begin{bmatrix} 5 & 2 & -4 & 8 \\ 1 & 3 & 6 & 4 \\ 4 & 6 & -9 & 7 \end{bmatrix}$

   (c) $\begin{bmatrix} -1 & 3 & -5 \\ 2 & -2 & 4 \\ 1 & 3 & 0 \end{bmatrix}$ and $\begin{bmatrix} -1 & 3 & -5 & -3 \\ 2 & -2 & 4 & 8 \\ 1 & 3 & 0 & 6 \end{bmatrix}$ (d) $\begin{bmatrix} 5 & 4 \\ 2 & -8 \\ 1 & 2 \end{bmatrix}$ and $\begin{bmatrix} 5 & 4 & 9 \\ 2 & -8 & -4 \\ 1 & 2 & 3 \end{bmatrix}$

   (e) $\begin{bmatrix} 5 & 2 & -4 \\ 0 & 4 & 3 \\ 1 & 0 & -1 \end{bmatrix}$ and $\begin{bmatrix} 5 & 2 & -4 & 8 \\ 0 & 4 & 3 & 0 \\ 1 & 0 & -1 & 7 \end{bmatrix}$

   (f) $\begin{bmatrix} -1 & 3 & -9 \\ 1 & 0 & -4 \\ 1 & 8 & 0 \end{bmatrix}$ and $\begin{bmatrix} -1 & 3 & -9 & -4 \\ 1 & 0 & -4 & 11 \\ 1 & 8 & 0 & 1 \end{bmatrix}$ (g) $\begin{bmatrix} 1 & 0 & 0 \\ 0 & 1 & 0 \\ 0 & 0 & 1 \end{bmatrix}$ and $\begin{bmatrix} 1 & 0 & 0 & -3 \\ 0 & 1 & 0 & 12 \\ 0 & 0 & 1 & 8 \end{bmatrix}$

   (h) $\begin{bmatrix} -4 & 2 & -9 & 1 \\ 1 & 6 & -8 & -7 \\ 0 & -1 & 3 & -5 \end{bmatrix}$ and $\begin{bmatrix} -4 & 2 & -9 & 1 & -1 \\ 1 & 6 & -8 & -7 & 15 \\ 0 & -1 & 3 & -5 & 0 \end{bmatrix}$

6. (a) $x_1 + 2x_2 = 3$
       $4x_1 + 5x_2 = 6$
   (b) $7x_1 + 9x_2 = 8$
       $6x_1 + 4x_2 = -3$
   (c) $x_1 + 9x_2 = -3$
       $5x_1 = 2$

   (d) $8x_1 + 7x_2 + 5x_3 = -1$
       $4x_1 + 6x_2 + 2x_3 = 4$
       $9x_1 + 3x_2 + 7x_3 = 6$
   (e) $2x_1 - 3x_2 + 6x_3 = 4$
       $7x_1 - 5x_2 - 2x_3 = 3$
       $ 2x_2 + 4x_3 = 0$

1

Section 1.1

(f) $\quad -2x_2 = 4$ $\quad\quad$ (g) $x_1 = 3$ $\quad\quad$ (h) $x_1 + 2x_2 - x_3 = 6$
$\quad\quad 5x_1 + 7x_2 = -3$ $\quad\quad\quad\quad x_2 = 8$ $\quad\quad\quad\quad x_2 + 4x_3 = 5$
$\quad\quad 6x_1 = 8$ $\quad\quad\quad\quad\quad\quad x_3 = 4$ $\quad\quad\quad\quad\quad\quad x_3 = -2$

7. (a) $\begin{bmatrix} 1 & 3 & -2 & 0 \\ 1 & 2 & -3 & 6 \\ 8 & 3 & 2 & 5 \end{bmatrix}$ (b) $\begin{bmatrix} 2 & 7 & 5 & 1 \\ 0 & -8 & 4 & 3 \\ 3 & -5 & 8 & 9 \end{bmatrix}$ (c) $\begin{bmatrix} 1 & 2 & 3 & -1 \\ 0 & 3 & 10 & 0 \\ 0 & -8 & -1 & -1 \end{bmatrix}$

(d) $\begin{bmatrix} 1 & 0 & -1 & -6 \\ 0 & 1 & 2 & 1 \\ 0 & 0 & 11 & -1 \end{bmatrix}$ (e) $\begin{bmatrix} 1 & 0 & 0 & -23 \\ 0 & 1 & 0 & 17 \\ 0 & 0 & 1 & 5 \end{bmatrix}$ (f) $\begin{bmatrix} 1 & 0 & 2 & 7 \\ 0 & 1 & 5 & -3 \\ 0 & 0 & -1 & 4 \end{bmatrix}$

8. (a) Elements in the first column, except for the leading 1, become zero.
x_1 is eliminated from all equations except the first.

(b) The second element in the second row becomes a leading 1.
The coefficient of x_2 in the second equation becomes 1.

(c) The leading 1 in row 2 is moved to the left of the leading nonzero term in row 3.
The second equation now contains x_2 with leading coefficient 1.

(d) Elements in the second column, except for the leading 1, become zero.
x_2 is eliminated from all equations except the second.

9. (a) Elements in the third column, except for the leading 1, become zero.
x_3 is eliminated from all equations except the third.

(b) It is now possible to have the leading 1 in row 1 to the left of leading 1s in other rows. It is now possible to have x_1 in row 1 with leading coefficient 1.

(c) The leading nonzero element in row 3 becomes 1.
Equation 3 is now solved for x_3.

(d) Elements in the third column, except for the leading 1, become zero.
The system of equations is solved: $x_1 = -2$, $x_2 = 5$, $x_3 = -3$.

10. (a) $\begin{bmatrix} 1 & -2 & -8 \\ 2 & -3 & -11 \end{bmatrix} \underset{R2+(-2)R1}{\approx} \begin{bmatrix} 1 & -2 & -8 \\ 0 & 1 & 5 \end{bmatrix} \underset{R1+(2)R2}{\approx} \begin{bmatrix} 1 & 0 & 2 \\ 0 & 1 & 5 \end{bmatrix}$,
so the solution is $x_1 = 2$ and $x_2 = 5$.

(b) $\begin{bmatrix} 2 & 2 & 4 \\ 3 & 2 & 3 \end{bmatrix} \underset{(1/2)R1}{\approx} \begin{bmatrix} 1 & 1 & 2 \\ 3 & 2 & 3 \end{bmatrix} \underset{R2+(-3)R1}{\approx} \begin{bmatrix} 1 & 1 & 2 \\ 0 & -1 & -3 \end{bmatrix} \underset{(-1)R2}{\approx} \begin{bmatrix} 1 & 1 & 2 \\ 0 & 1 & 3 \end{bmatrix}$

Section 1.1

$$\underset{R1+(-1)R2}{\approx} \begin{bmatrix} 1 & 0 & -1 \\ 0 & 1 & 3 \end{bmatrix}, \text{ so the solution is } x_1 = -1 \text{ and } x_2 = 3.$$

(c) $\begin{bmatrix} 1 & 0 & 1 & 3 \\ 0 & 2 & -2 & -4 \\ 0 & 1 & -2 & 5 \end{bmatrix} \underset{(1/2)R2}{\approx} \begin{bmatrix} 1 & 0 & 1 & 3 \\ 0 & 1 & -1 & -2 \\ 0 & 1 & -2 & 5 \end{bmatrix} \underset{R3+(-1)R2}{\approx} \begin{bmatrix} 1 & 0 & 1 & 3 \\ 0 & 1 & -1 & -2 \\ 0 & 0 & -1 & 7 \end{bmatrix}$

$\underset{(-1)R3}{\approx} \begin{bmatrix} 1 & 0 & 1 & 3 \\ 0 & 1 & -1 & -2 \\ 0 & 0 & 1 & -7 \end{bmatrix} \underset{\substack{R1+(-1)R3 \\ R2+R3}}{\approx} \begin{bmatrix} 1 & 0 & 0 & 10 \\ 0 & 1 & 0 & -9 \\ 0 & 0 & 1 & -7 \end{bmatrix},$

so the solution is $x_1 = 10$, $x_2 = -9$, $x_3 = -7$.

(d) $\begin{bmatrix} 1 & 1 & 3 & 6 \\ 1 & 2 & 4 & 9 \\ 2 & 1 & 6 & 11 \end{bmatrix} \underset{\substack{R2+(-1)R1 \\ R3+(-2)R1}}{\approx} \begin{bmatrix} 1 & 1 & 3 & 6 \\ 0 & 1 & 1 & 3 \\ 0 & -1 & 0 & -1 \end{bmatrix} \underset{\substack{R1+(-1)R2 \\ R3+R2}}{\approx} \begin{bmatrix} 1 & 0 & 2 & 3 \\ 0 & 1 & 1 & 3 \\ 0 & 0 & 1 & 2 \end{bmatrix}$

$\underset{\substack{R1+(-2)R3 \\ R2+(-1)R3}}{\approx} \begin{bmatrix} 1 & 0 & 0 & -1 \\ 0 & 1 & 0 & 1 \\ 0 & 0 & 1 & 2 \end{bmatrix}$, so the solution is $x_1 = -1$, $x_2 = 1$, $x_3 = 2$.

(e) $\begin{bmatrix} 1 & -1 & 3 & 3 \\ 2 & -1 & 2 & 2 \\ 3 & 1 & -2 & 2 \end{bmatrix} \underset{\substack{R2+(-2)R1 \\ R3+(-3)R1}}{\approx} \begin{bmatrix} 1 & -1 & 3 & 3 \\ 0 & 1 & -4 & -4 \\ 0 & 4 & -11 & -7 \end{bmatrix} \underset{\substack{R1+R2 \\ R3+(-4)R2}}{\approx} \begin{bmatrix} 1 & 0 & -1 & -1 \\ 0 & 1 & -4 & -4 \\ 0 & 0 & 5 & 9 \end{bmatrix}$

$\underset{(1/5)R3}{\approx} \begin{bmatrix} 1 & 0 & -1 & -1 \\ 0 & 1 & -4 & -4 \\ 0 & 0 & 1 & 9/5 \end{bmatrix} \underset{\substack{R1+R3 \\ R2+(4)R3}}{\approx} \begin{bmatrix} 1 & 0 & 0 & 4/5 \\ 0 & 1 & 0 & 16/5 \\ 0 & 0 & 1 & 9/5 \end{bmatrix},$

so the solution is $x_1 = 4/5$, $x_2 = 16/5$, $x_3 = 9/5$.

(f) $\begin{bmatrix} -1 & 1 & -1 & -2 \\ 3 & 1 & 1 & 10 \\ 4 & 2 & 3 & 14 \end{bmatrix} \underset{(-1)R1}{\approx} \begin{bmatrix} 1 & -1 & 1 & 2 \\ 3 & 1 & 1 & 10 \\ 4 & 2 & 3 & 14 \end{bmatrix} \underset{\substack{R2+(-3)R1 \\ R3+(-4)R1}}{\approx} \begin{bmatrix} 1 & -1 & 1 & 2 \\ 0 & 4 & -2 & 4 \\ 0 & 6 & -1 & 6 \end{bmatrix}$

$\underset{(1/4)R2}{\approx} \begin{bmatrix} 1 & -1 & 1 & 2 \\ 0 & 1 & -1/2 & 1 \\ 0 & 6 & -1 & 6 \end{bmatrix} \underset{\substack{R1+R2 \\ R3+(-6)R1}}{\approx} \begin{bmatrix} 1 & 0 & 1/2 & 3 \\ 0 & 1 & -1/2 & 1 \\ 0 & 0 & 2 & 0 \end{bmatrix} \underset{(1/2)R3}{\approx} \begin{bmatrix} 1 & 0 & 1/2 & 3 \\ 0 & 1 & -1/2 & 1 \\ 0 & 0 & 1 & 0 \end{bmatrix}$

$$\underset{\substack{R1+(-1/2)R3\\R2+(1/2)R3}}{\approx} \begin{bmatrix} 1 & 0 & 0 & 3 \\ 0 & 1 & 0 & 1 \\ 0 & 0 & 1 & 0 \end{bmatrix}, \text{ so the solution is } x_1 = 3, x_2 = 1, x_3 = 0.$$

11. (a) $\begin{bmatrix} 0 & 2 & 4 & 8 \\ 2 & 2 & 0 & 6 \\ 1 & 1 & 1 & 5 \end{bmatrix} \underset{R1 \leftrightarrow R2}{\approx} \begin{bmatrix} 2 & 2 & 0 & 6 \\ 0 & 2 & 4 & 8 \\ 1 & 1 & 1 & 5 \end{bmatrix} \underset{(1/2)R1}{\approx} \begin{bmatrix} 1 & 1 & 0 & 3 \\ 0 & 2 & 4 & 8 \\ 1 & 1 & 1 & 5 \end{bmatrix}$

$\underset{R3+(-1)R1}{\approx} \begin{bmatrix} 1 & 1 & 0 & 3 \\ 0 & 2 & 4 & 8 \\ 0 & 0 & 1 & 2 \end{bmatrix} \underset{(1/2)R2}{\approx} \begin{bmatrix} 1 & 1 & 0 & 3 \\ 0 & 1 & 2 & 4 \\ 0 & 0 & 1 & 2 \end{bmatrix} \underset{R1+(-1)R2}{\approx} \begin{bmatrix} 1 & 0 & -2 & -1 \\ 0 & 1 & 2 & 4 \\ 0 & 0 & 1 & 2 \end{bmatrix}$

$\underset{\substack{R1+(2)R3\\R2+(-2)R3}}{\approx} \begin{bmatrix} 1 & 0 & 0 & 3 \\ 0 & 1 & 0 & 0 \\ 0 & 0 & 1 & 2 \end{bmatrix}, \text{ so the solution is } x_1 = 3, x_2 = 0, x_3 = 2.$

(b) $\begin{bmatrix} 1 & -2 & -4 & -9 \\ 1 & 5 & 10 & 21 \\ 2 & -3 & -5 & -13 \end{bmatrix} \underset{\substack{R2+(-1)R1\\R3+(-2)R1}}{\approx} \begin{bmatrix} 1 & -2 & -4 & -9 \\ 0 & 7 & 14 & 30 \\ 0 & 1 & 3 & 5 \end{bmatrix} \underset{R2 \leftrightarrow R3}{\approx} \begin{bmatrix} 1 & -2 & -4 & -9 \\ 0 & 1 & 3 & 5 \\ 0 & 7 & 14 & 30 \end{bmatrix}$

$\underset{\substack{R1+(2)R2\\R3+(-7)R2}}{\approx} \begin{bmatrix} 1 & 0 & 2 & 1 \\ 0 & 1 & 3 & 5 \\ 0 & 0 & -7 & -5 \end{bmatrix} \underset{(-1/7)R3}{\approx} \begin{bmatrix} 1 & 0 & 2 & 1 \\ 0 & 1 & 3 & 5 \\ 0 & 0 & 1 & 5/7 \end{bmatrix}$

$\underset{\substack{R1+(-2)R3\\R2+(-3)R3}}{\approx} \begin{bmatrix} 1 & 0 & 0 & -3/7 \\ 0 & 1 & 0 & 20/7 \\ 0 & 0 & 1 & 5/7 \end{bmatrix},$

so the solution is $x_1 = -3/7, x_2 = 20/7, x_3 = 5/7$.

(c) $\begin{bmatrix} 1 & 2 & 3 & 14 \\ 2 & 5 & 8 & 36 \\ 1 & -1 & 0 & -4 \end{bmatrix} \underset{\substack{R2+(-2)R1\\R3+(-1)R1}}{\approx} \begin{bmatrix} 1 & 2 & 3 & 14 \\ 0 & 1 & 2 & 8 \\ 0 & -3 & -3 & -18 \end{bmatrix} \underset{\substack{R1+(-2)R2\\R3+(3)R2}}{\approx} \begin{bmatrix} 1 & 0 & -1 & -2 \\ 0 & 1 & 2 & 8 \\ 0 & 0 & 3 & 6 \end{bmatrix}$

$\underset{(1/3)R3}{\approx} \begin{bmatrix} 1 & 0 & -1 & -2 \\ 0 & 1 & 2 & 8 \\ 0 & 0 & 1 & 2 \end{bmatrix} \underset{\substack{R1+R3\\R2+(-2)R3}}{\approx} \begin{bmatrix} 1 & 0 & 0 & 0 \\ 0 & 1 & 0 & 4 \\ 0 & 0 & 1 & 2 \end{bmatrix},$

so the solution is $x_1 = 0, x_2 = 4, x_3 = 2$.

(d) $\begin{bmatrix} 1 & -1 & -1 & -1 \\ -2 & 6 & 10 & 14 \\ 2 & 1 & 6 & 9 \end{bmatrix} \underset{R3+(-2)R1}{\overset{R2+(2)R1}{\approx}} \begin{bmatrix} 1 & -1 & -1 & -1 \\ 0 & 4 & 8 & 12 \\ 0 & 3 & 8 & 11 \end{bmatrix} \underset{(1/4)R2}{\approx} \begin{bmatrix} 1 & -1 & -1 & -1 \\ 0 & 1 & 2 & 3 \\ 0 & 3 & 8 & 11 \end{bmatrix}$

$\underset{R3+(-3)R2}{\overset{R1+R2}{\approx}} \begin{bmatrix} 1 & 0 & 1 & 2 \\ 0 & 1 & 2 & 3 \\ 0 & 0 & 2 & 2 \end{bmatrix} \underset{(1/2)R3}{\approx} \begin{bmatrix} 1 & 0 & 1 & 2 \\ 0 & 1 & 2 & 3 \\ 0 & 0 & 1 & 1 \end{bmatrix} \underset{R2+(-2)R3}{\overset{R1+(-1)R3}{\approx}} \begin{bmatrix} 1 & 0 & 0 & 1 \\ 0 & 1 & 0 & 1 \\ 0 & 0 & 1 & 1 \end{bmatrix}$,

so the solution is $x_1 = 1$, $x_2 = 1$, $x_3 = 1$.

(e) $\begin{bmatrix} 2 & 2 & -4 & 14 \\ 3 & 1 & 1 & 8 \\ 2 & -1 & 2 & -1 \end{bmatrix} \underset{(1/2)R1}{\approx} \begin{bmatrix} 1 & 1 & -2 & 7 \\ 3 & 1 & 1 & 8 \\ 2 & -1 & 2 & -1 \end{bmatrix} \underset{R3+(-2)R1}{\overset{R2+(-3)R1}{\approx}} \begin{bmatrix} 1 & 1 & -2 & 7 \\ 0 & -2 & 7 & -13 \\ 0 & -3 & 6 & -15 \end{bmatrix}$

$\underset{R2 \leftrightarrow R3}{\approx} \begin{bmatrix} 1 & 1 & -2 & 7 \\ 0 & -3 & 6 & -15 \\ 0 & -2 & 7 & -13 \end{bmatrix} \underset{(-1/3)R2}{\approx} \begin{bmatrix} 1 & 1 & -2 & 7 \\ 0 & 1 & -2 & 5 \\ 0 & -2 & 7 & -13 \end{bmatrix}$

$\underset{R3+(2)R2}{\overset{R1+(-1)R2}{\approx}} \begin{bmatrix} 1 & 0 & 0 & 2 \\ 0 & 1 & -2 & 5 \\ 0 & 0 & 3 & -3 \end{bmatrix} \underset{(1/3)R3}{\approx} \begin{bmatrix} 1 & 0 & 0 & 2 \\ 0 & 1 & -2 & 5 \\ 0 & 0 & 1 & -1 \end{bmatrix} \underset{R2+(2)R3}{\approx} \begin{bmatrix} 1 & 0 & 0 & 2 \\ 0 & 1 & 0 & 3 \\ 0 & 0 & 1 & -1 \end{bmatrix}$,

so the solution is $x_1 = 2$, $x_2 = 3$, $x_3 = -1$.

12. (a) $\begin{bmatrix} 3/2 & 0 & 3 & 15 \\ -1 & 7 & -9 & -45 \\ 2 & 0 & 5 & 22 \end{bmatrix} \underset{(2/3)R1}{\approx} \begin{bmatrix} 1 & 0 & 2 & 10 \\ -1 & 7 & -9 & -45 \\ 2 & 0 & 5 & 22 \end{bmatrix} \underset{R3+(-2)R1}{\overset{R2+R1}{\approx}} \begin{bmatrix} 1 & 0 & 2 & 10 \\ 0 & 7 & -7 & -35 \\ 0 & 0 & 1 & 2 \end{bmatrix}$

$\underset{(1/7)R2}{\approx} \begin{bmatrix} 1 & 0 & 2 & 10 \\ 0 & 1 & -1 & -5 \\ 0 & 0 & 1 & 2 \end{bmatrix} \underset{R2+R3}{\overset{R1+(-2)R3}{\approx}} \begin{bmatrix} 1 & 0 & 0 & 6 \\ 0 & 1 & 0 & -3 \\ 0 & 0 & 1 & 2 \end{bmatrix}$,

so the solution is $x_1 = 6$, $x_2 = -3$, $x_3 = 2$.

(b) $\begin{bmatrix} -3 & -6 & -15 & -3 \\ 1 & 3/2 & 9/2 & 1/2 \\ -2 & -7/2 & -17/2 & -2 \end{bmatrix} \underset{(2)R3}{\overset{(-1/3)R1}{\underset{(2)R2}{}}} \begin{bmatrix} 1 & 2 & 5 & 1 \\ 2 & 3 & 9 & 1 \\ -4 & -7 & -17 & -4 \end{bmatrix} \underset{R3+(4)R1}{\overset{R2+(-2)R1}{\approx}} \begin{bmatrix} 1 & 2 & 5 & 1 \\ 0 & -1 & -1 & -1 \\ 0 & 1 & 3 & 0 \end{bmatrix}$

$\underset{(-1)R2}{\approx} \begin{bmatrix} 1 & 2 & 5 & 1 \\ 0 & 1 & 1 & 1 \\ 0 & 1 & 3 & 0 \end{bmatrix} \underset{R3+(-1)R2}{\overset{R1+(-2)R2}{\approx}} \begin{bmatrix} 1 & 0 & 3 & -1 \\ 0 & 1 & 1 & 1 \\ 0 & 0 & 2 & -1 \end{bmatrix} \underset{(1/2)R3}{\approx} \begin{bmatrix} 1 & 0 & 3 & -1 \\ 0 & 1 & 1 & 1 \\ 0 & 0 & 1 & -1/2 \end{bmatrix}$

$$\underset{\substack{R1+(-3)R3\\R2+(-1)R3}}{\approx} \begin{bmatrix} 1 & 0 & 0 & 1/2 \\ 0 & 1 & 0 & 3/2 \\ 0 & 0 & 1 & -1/2 \end{bmatrix}, \text{ so the solution is } x_1 = 1/2,\ x_2 = 3/2,\ x_3 = -1/2.$$

(c) $\begin{bmatrix} 3 & 6 & 0 & -3 & 3 \\ 1 & 3 & -1 & -4 & -12 \\ 1 & -1 & 1 & 2 & 8 \\ 2 & 3 & 0 & 0 & 8 \end{bmatrix} \underset{(1/3)R1}{\approx} \begin{bmatrix} 1 & 2 & 0 & -1 & 1 \\ 1 & 3 & -1 & -4 & -12 \\ 1 & -1 & 1 & 2 & 8 \\ 2 & 3 & 0 & 0 & 8 \end{bmatrix}$

$\underset{\substack{R2+(-1)R1\\R3+(-1)R1\\R4+(-2)R1}}{\approx} \begin{bmatrix} 1 & 2 & 0 & -1 & 1 \\ 0 & 1 & -1 & -3 & -13 \\ 0 & -3 & 1 & 3 & 7 \\ 0 & -1 & 0 & 2 & 6 \end{bmatrix} \underset{\substack{R1+(-2)R2\\R3+(3)R2\\R4+R2}}{\approx} \begin{bmatrix} 1 & 0 & 2 & 5 & 27 \\ 0 & 1 & -1 & -3 & -13 \\ 0 & 0 & -2 & -6 & -32 \\ 0 & 0 & -1 & -1 & -7 \end{bmatrix}$

$\underset{(-1/2)R3}{\approx} \begin{bmatrix} 1 & 0 & 2 & 5 & 27 \\ 0 & 1 & -1 & -3 & -13 \\ 0 & 0 & 1 & 3 & 16 \\ 0 & 0 & -1 & -1 & -7 \end{bmatrix} \underset{\substack{R1+(-2)R3\\R2+R3\\R4+R3}}{\approx} \begin{bmatrix} 1 & 0 & 0 & -1 & -5 \\ 0 & 1 & 0 & 0 & 3 \\ 0 & 0 & 1 & 3 & 16 \\ 0 & 0 & 0 & 2 & 9 \end{bmatrix}$

$\underset{(1/2)R4}{\approx} \begin{bmatrix} 1 & 0 & 0 & -1 & -5 \\ 0 & 1 & 0 & 0 & 3 \\ 0 & 0 & 1 & 3 & 16 \\ 0 & 0 & 0 & 1 & 9/2 \end{bmatrix} \underset{\substack{R1+R4\\R3+(-3)R4}}{\approx} \begin{bmatrix} 1 & 0 & 0 & 0 & -1/2 \\ 0 & 1 & 0 & 0 & 3 \\ 0 & 0 & 1 & 0 & 5/2 \\ 0 & 0 & 0 & 1 & 9/2 \end{bmatrix}$, so the solution is $x_1 = -1/2,\ x_2 = 3,\ x_3 = 5/2,\ x_4 = 9/2$.

(d) $\begin{bmatrix} 1 & 2 & 2 & 5 & 11 \\ 2 & 4 & 2 & 8 & 14 \\ 1 & 3 & 4 & 8 & 19 \\ 1 & -1 & 1 & 0 & 2 \end{bmatrix} \underset{\substack{R2+(-2)R1\\R3+(-1)R1\\R4+(-1)R1}}{\approx} \begin{bmatrix} 1 & 2 & 2 & 5 & 11 \\ 0 & 0 & -2 & -2 & -8 \\ 0 & 1 & 2 & 3 & 8 \\ 0 & -3 & -1 & -5 & -9 \end{bmatrix}$

$\underset{R2 \leftrightarrow R3}{\approx} \begin{bmatrix} 1 & 2 & 2 & 5 & 11 \\ 0 & 1 & 2 & 3 & 8 \\ 0 & 0 & -2 & -2 & -8 \\ 0 & -3 & -1 & -5 & -9 \end{bmatrix} \underset{\substack{R1+(-2)R2\\R4+(3)R2}}{\approx} \begin{bmatrix} 1 & 0 & -2 & -1 & -5 \\ 0 & 1 & 2 & 3 & 8 \\ 0 & 0 & -2 & -2 & -8 \\ 0 & 0 & 5 & 4 & 15 \end{bmatrix}$

$\underset{(-1/2)R3}{\approx} \begin{bmatrix} 1 & 0 & -2 & -1 & -5 \\ 0 & 1 & 2 & 3 & 8 \\ 0 & 0 & 1 & 1 & 4 \\ 0 & 0 & 5 & 4 & 15 \end{bmatrix} \underset{\substack{R1+(2)R3\\R2+(-2)R3\\R4+(-5)R3}}{\approx} \begin{bmatrix} 1 & 0 & 0 & 1 & 3 \\ 0 & 1 & 0 & 1 & 0 \\ 0 & 0 & 1 & 1 & 4 \\ 0 & 0 & 0 & -1 & -5 \end{bmatrix}$

Section 1.1

$$\underset{(-1)R4}{\approx} \begin{bmatrix} 1 & 0 & 0 & 1 & 3 \\ 0 & 1 & 0 & 1 & 0 \\ 0 & 0 & 1 & 1 & 4 \\ 0 & 0 & 0 & 1 & 5 \end{bmatrix} \underset{\substack{R1+(-1)R4 \\ R2+(-1)R4 \\ R3+(-1)R4}}{\approx} \begin{bmatrix} 1 & 0 & 0 & 0 & -2 \\ 0 & 1 & 0 & 0 & -5 \\ 0 & 0 & 1 & 0 & -1 \\ 0 & 0 & 0 & 1 & 5 \end{bmatrix},$$

so the solution is $x_1 = -2$, $x_2 = -5$, $x_3 = -1$, $x_4 = 5$.

(e) $\begin{bmatrix} 1 & 1 & 2 & 6 & 11 \\ 2 & 3 & 6 & 19 & 36 \\ 0 & 3 & 4 & 15 & 28 \\ 1 & -1 & -1 & -6 & -12 \end{bmatrix} \underset{\substack{R2+(-2)R1 \\ R4+(-1)R1}}{\approx} \begin{bmatrix} 1 & 1 & 2 & 6 & 11 \\ 0 & 1 & 2 & 7 & 14 \\ 0 & 3 & 4 & 15 & 28 \\ 0 & -2 & -3 & -12 & -23 \end{bmatrix}$

$\underset{\substack{R1+(-1)R2 \\ R3+(-3)R2 \\ R4+(2)R2}}{\approx} \begin{bmatrix} 1 & 0 & 0 & -1 & -3 \\ 0 & 1 & 2 & 7 & 14 \\ 0 & 0 & -2 & -6 & -14 \\ 0 & 0 & 1 & 2 & 5 \end{bmatrix} \underset{(-1/2)R3}{\approx} \begin{bmatrix} 1 & 0 & 0 & -1 & -3 \\ 0 & 1 & 2 & 7 & 14 \\ 0 & 0 & 1 & 3 & 7 \\ 0 & 0 & 1 & 2 & 5 \end{bmatrix}$

$\underset{\substack{R2+(-2)R3 \\ R4+(-1)R3}}{\approx} \begin{bmatrix} 1 & 0 & 0 & -1 & -3 \\ 0 & 1 & 0 & 1 & 0 \\ 0 & 0 & 1 & 3 & 7 \\ 0 & 0 & 0 & -1 & -2 \end{bmatrix} \underset{(-1)R4}{\approx} \begin{bmatrix} 1 & 0 & 0 & -1 & -3 \\ 0 & 1 & 0 & 1 & 0 \\ 0 & 0 & 1 & 3 & 7 \\ 0 & 0 & 0 & 1 & 2 \end{bmatrix}$

$\underset{\substack{R1+R4 \\ R2+(-1)R4 \\ R3+(-3)R4}}{\approx} \begin{bmatrix} 1 & 0 & 0 & 0 & -1 \\ 0 & 1 & 0 & 0 & -2 \\ 0 & 0 & 1 & 0 & 1 \\ 0 & 0 & 0 & 1 & 2 \end{bmatrix}$, so the solution is $x_1 = -1$, $x_2 = -2$, $x_3 = 1$, $x_4 = 2$.

13. (a) $\begin{bmatrix} 1 & 2 & 3 & 4 & 3 \\ 3 & 5 & 8 & 9 & 7 \end{bmatrix} \underset{R2+(-3)R1}{\approx} \begin{bmatrix} 1 & 2 & 3 & 4 & 3 \\ 0 & -1 & -1 & -3 & -2 \end{bmatrix} \underset{(-1)R2}{\approx} \begin{bmatrix} 1 & 2 & 3 & 4 & 3 \\ 0 & 1 & 1 & 3 & 2 \end{bmatrix}$

$\underset{R1+(-2)R2}{\approx} \begin{bmatrix} 1 & 2 & 1 & -2 & -1 \\ 0 & 1 & 1 & 3 & 2 \end{bmatrix}$, so the solutions are in turn $x_1 = 1$, $x_2 = 1$;

$x_1 = -2$, $x_2 = 3$; and $x_1 = -1$, $x_2 = 2$.

(b) $\begin{bmatrix} 1 & 1 & 0 & 5 & 1 \\ 2 & 3 & 1 & 13 & 2 \end{bmatrix} \underset{R2+(-2)R1}{\approx} \begin{bmatrix} 1 & 1 & 0 & 5 & 1 \\ 0 & 1 & 1 & 3 & 0 \end{bmatrix} \underset{R1+(-1)R2}{\approx} \begin{bmatrix} 1 & 0 & -1 & 2 & 1 \\ 0 & 1 & 1 & 3 & 0 \end{bmatrix}$,

so the solutions are in turn $x_1 = -1$, $x_2 = 1$; $x_1 = 2$, $x_2 = 3$; and $x_1 = 1$, $x_2 = 0$.

(c) $\begin{bmatrix} 1 & -2 & 3 & 6 & -5 & 4 \\ 1 & -1 & 2 & 5 & -3 & 3 \\ 2 & -3 & 6 & 14 & -8 & 9 \end{bmatrix} \underset{\substack{R2+(-1)R1 \\ R3+(-2)R1}}{\approx} \begin{bmatrix} 1 & -2 & 3 & 6 & -5 & 4 \\ 0 & 1 & -1 & -1 & 2 & -1 \\ 0 & 1 & 0 & 2 & 2 & 1 \end{bmatrix}$

7

$$\underset{\substack{R1+(2)R2\\R3+(-1)R2}}{\approx} \begin{bmatrix} 1 & 0 & 1 & 4 & -1 & 2 \\ 0 & 1 & -1 & -1 & 2 & -1 \\ 0 & 0 & 1 & 3 & 0 & 2 \end{bmatrix} \underset{\substack{R1+(-1)R3\\R2+R3}}{\approx} \begin{bmatrix} 1 & 0 & 0 & 1 & -1 & 0 \\ 0 & 1 & 0 & 2 & 2 & 1 \\ 0 & 0 & 1 & 3 & 0 & 2 \end{bmatrix},$$

so the solutions are in turn $x_1 = 1$, $x_2 = 2$, $x_3 = 3$; $x_1 = -1$, $x_2 = 2$, $x_3 = 0$; and $x_1 = 0$, $x_2 = 1$, $x_3 = 2$.

(d) $\begin{bmatrix} 1 & 2 & -1 & -1 & 6 & 0 \\ -1 & -1 & 1 & 1 & -4 & -2 \\ 3 & 7 & -1 & -1 & 18 & -4 \end{bmatrix} \underset{\substack{R2+R1\\R3+(-3)R1}}{\approx} \begin{bmatrix} 1 & 2 & -1 & -1 & 6 & 0 \\ 0 & 1 & 0 & 0 & 2 & -2 \\ 0 & 1 & 2 & 2 & 0 & -4 \end{bmatrix}$

$\underset{\substack{R1+(-2)R2\\R3+(-1)R2}}{\approx} \begin{bmatrix} 1 & 0 & -5 & -5 & 6 & 8 \\ 0 & 1 & 0 & 0 & 2 & -2 \\ 0 & 0 & 2 & 2 & -2 & -2 \end{bmatrix} \underset{(1/2)R3}{\approx} \begin{bmatrix} 1 & 0 & -5 & -5 & 6 & 8 \\ 0 & 1 & 0 & 0 & 2 & -2 \\ 0 & 0 & 1 & 1 & -1 & -1 \end{bmatrix}$

$\underset{R1+(5)R3}{\approx} \begin{bmatrix} 1 & 0 & 0 & 0 & 1 & 3 \\ 0 & 1 & 0 & 0 & 2 & -2 \\ 0 & 0 & 1 & 1 & -1 & -1 \end{bmatrix}$, so the solutions are in turn

$x_1 = 0$, $x_2 = 0$, $x_3 = 1$; $x_1 = 1$, $x_2 = 2$, $x_3 = -1$; and $x_1 = 3$, $x_2 = -2$, $x_3 = -1$.

Exercise Set 1.2

1. (a) Yes. (b) Yes.

   (c) No. The second column contains a leading 1, so other elements in that column should be zero.

   (d) No. The second row does not have 1 as the first nonzero number.

   (e) Yes. (f) Yes. (g) Yes.

   (h) No. The second row does not have 1 as the first nonzero number. (i) Yes.

2. (a) No. The leading 1 in row 2 is not to the right of the leading 1 in row 3.

   (b) Yes. (c) Yes.

   (d) No. The third and fourth columns contain leading 1s, so the other numbers in those columns should be zeros.

   (e) No. The row containing all zeros should be at the bottom of the matrix.

Section 1.2

    (f) Yes.

    (g) No. The leading 1 in row 2 is not to the right of the leading 1 in row 3. Also, since column 3 contains a leading 1, all other numbers in that column should be zero.

    (h) No. The leading 1 in row 3 is not to the right of the leading 1s in rows 1 and 2.

    (i) Yes.

3. (a) $x_1 = 2, x_2 = 4, x_3 = -3$.   (b) $x_1 = 3r + 4, x_2 = -2r + 8, x_3 = r$.

   (c) $x_1 = -3r + 6, x_2 = r, x_3 = -2$.   (d) There is no solution. The last row gives $0 = 1$.

   (e) $x_1 = -5r + 3, x_2 = -6r - 2, x_3 = -2r - 4, x_4 = r$.

   (f) $x_1 = -3r + 2, x_2 = r, x_3 = 4, x_4 = 5$.

4. (a) $x_1 = -2r - 4s + 1, x_2 = 3r - 5s - 6, x_3 = r, x_4 = s$.

   (b) $x_1 = 3r - 2s + 4, x_2 = r, x_3 = s, x_4 = -7$.

   (c) $x_1 = 2r - 3s + 4, x_2 = r, x_3 = -2s + 9, x_4 = s, x_5 = 8$.

   (d) $x_1 = -2r - 3s + 6, x_2 = -5r - 4s + 7, x_3 = r, x_4 = -9s - 3, x_5 = s$.

5. (a) $\begin{bmatrix} 1 & 4 & 3 & 1 \\ 2 & 8 & 11 & 7 \\ 1 & 6 & 7 & 3 \end{bmatrix} \underset{R3+(-1)R1}{\overset{R2+(-2)R1}{\approx}} \begin{bmatrix} 1 & 4 & 3 & 1 \\ 0 & 0 & 5 & 5 \\ 0 & 2 & 4 & 2 \end{bmatrix} \underset{R2 \leftrightarrow R3}{\approx} \begin{bmatrix} 1 & 4 & 3 & 1 \\ 0 & 2 & 4 & 2 \\ 0 & 0 & 5 & 5 \end{bmatrix}$

$\underset{(1/2)R2}{\approx} \begin{bmatrix} 1 & 4 & 3 & 1 \\ 0 & 1 & 2 & 1 \\ 0 & 0 & 5 & 5 \end{bmatrix} \underset{R1+(-4)R2}{\approx} \begin{bmatrix} 1 & 0 & -5 & -3 \\ 0 & 1 & 2 & 1 \\ 0 & 0 & 5 & 5 \end{bmatrix} \underset{(1/5)R3}{\approx} \begin{bmatrix} 1 & 0 & -5 & -3 \\ 0 & 1 & 2 & 1 \\ 0 & 0 & 1 & 1 \end{bmatrix}$

$\underset{R2+(-2)R3}{\overset{R1+(5)R3}{\approx}} \begin{bmatrix} 1 & 0 & 0 & 2 \\ 0 & 1 & 0 & -1 \\ 0 & 0 & 1 & 1 \end{bmatrix}$, so the solution is $x_1 = 2, x_2 = -1, x_3 = 1$.

9

Section 1.2

(b) $\begin{bmatrix} 1 & 2 & 4 & 15 \\ 2 & 4 & 9 & 33 \\ 1 & 3 & 5 & 20 \end{bmatrix} \underset{R3+(-1)R1}{\overset{R2+(-2)R1}{\approx}} \begin{bmatrix} 1 & 2 & 4 & 15 \\ 0 & 0 & 1 & 3 \\ 0 & 1 & 1 & 5 \end{bmatrix} \underset{R2 \leftrightarrow R3}{\approx} \begin{bmatrix} 1 & 2 & 4 & 15 \\ 0 & 1 & 1 & 5 \\ 0 & 0 & 1 & 3 \end{bmatrix}$

$\underset{R1+(-2)R2}{\approx} \begin{bmatrix} 1 & 0 & 2 & 5 \\ 0 & 1 & 1 & 5 \\ 0 & 0 & 1 & 3 \end{bmatrix} \underset{R2+(-1)R3}{\overset{R1+(-2)R3}{\approx}} \begin{bmatrix} 1 & 0 & 0 & -1 \\ 0 & 1 & 0 & 2 \\ 0 & 0 & 1 & 3 \end{bmatrix}$,

so the solution is $x_1 = -1$, $x_2 = 2$, $x_3 = 3$.

(c) $\begin{bmatrix} 1 & 1 & 1 & 7 \\ 2 & 3 & 1 & 18 \\ -1 & 1 & -3 & 1 \end{bmatrix} \underset{R3+R1}{\overset{R2+(-2)R1}{\approx}} \begin{bmatrix} 1 & 1 & 1 & 7 \\ 0 & 1 & -1 & 4 \\ 0 & 2 & -2 & 8 \end{bmatrix} \underset{R3+(-2)R2}{\overset{R1+(-1)R2}{\approx}} \begin{bmatrix} 1 & 0 & 2 & 3 \\ 0 & 1 & -1 & 4 \\ 0 & 0 & 0 & 0 \end{bmatrix}$,

so $x_1 + 2x_3 = 3$ and $x_2 - x_3 = 4$.
Thus the general solution is $x_1 = 3 - 2r$, $x_2 = 4 + r$, $x_3 = r$.

(d) $\begin{bmatrix} 1 & 4 & 1 & 2 \\ 1 & 2 & -1 & 0 \\ 2 & 6 & 0 & 3 \end{bmatrix} \underset{R3+(-2)R1}{\overset{R2+(-1)R1}{\approx}} \begin{bmatrix} 1 & 4 & 1 & 2 \\ 0 & -2 & -2 & -2 \\ 0 & -2 & -2 & -1 \end{bmatrix} \underset{(-1/2)R2}{\approx} \begin{bmatrix} 1 & 4 & 1 & 2 \\ 0 & 1 & 1 & 1 \\ 0 & -2 & -2 & -1 \end{bmatrix}$

$\underset{R3+(2)R2}{\overset{R1+(-4)R2}{\approx}} \begin{bmatrix} 1 & 0 & -3 & -2 \\ 0 & 1 & 1 & 1 \\ 0 & 0 & 0 & 1 \end{bmatrix}$, so there is no solution, since the last row of the

matrix corresponds to the equation $0 = 1$.

(e) $\begin{bmatrix} 1 & -1 & 1 & 3 \\ 2 & -1 & 4 & 7 \\ 3 & -5 & -1 & 7 \end{bmatrix} \underset{R3+(-3)R1}{\overset{R2+(-2)R1}{\approx}} \begin{bmatrix} 1 & -1 & 1 & 3 \\ 0 & 1 & 2 & 1 \\ 0 & -2 & -4 & -2 \end{bmatrix} \underset{R3+(2)R2}{\overset{R1+R2}{\approx}} \begin{bmatrix} 1 & 0 & 3 & 4 \\ 0 & 1 & 2 & 1 \\ 0 & 0 & 0 & 0 \end{bmatrix}$,

so $x_1 + 3x_3 = 4$ and $x_2 + 2x_3 = 1$.
Thus the general solution is $x_1 = 4 - 3r$, $x_2 = 1 - 2r$, $x_3 = r$.

(f) $\begin{bmatrix} 3 & -3 & 9 & 24 \\ 2 & -2 & 7 & 17 \\ -1 & 2 & -4 & -11 \end{bmatrix} \underset{(1/3)R1}{\approx} \begin{bmatrix} 1 & -1 & 3 & 8 \\ 2 & -2 & 7 & 17 \\ -1 & 2 & -4 & -11 \end{bmatrix} \underset{R2+R1}{\overset{R2+(-2)R1}{\approx}} \begin{bmatrix} 1 & -1 & 3 & 8 \\ 0 & 0 & 1 & 1 \\ 0 & 1 & -1 & -3 \end{bmatrix}$

10

$$\underset{R2 \leftrightarrow R3}{\approx} \begin{bmatrix} 1 & -1 & 3 & 8 \\ 0 & 1 & -1 & -3 \\ 0 & 0 & 1 & 1 \end{bmatrix} \underset{R1+R2}{\approx} \begin{bmatrix} 1 & 0 & 2 & 5 \\ 0 & 1 & -1 & -3 \\ 0 & 0 & 1 & 1 \end{bmatrix}$$

$$\underset{\substack{R1+(-2)R3 \\ R2+R3}}{\approx} \begin{bmatrix} 1 & 0 & 0 & 3 \\ 0 & 1 & 0 & -2 \\ 0 & 0 & 1 & 1 \end{bmatrix}, \text{ so the solution is } x_1 = 3,\ x_2 = -2,\ x_3 = 1.$$

6. (a) $\begin{bmatrix} 3 & 6 & -3 & 6 \\ -2 & -4 & -3 & -1 \\ 3 & 6 & -2 & 10 \end{bmatrix} \underset{(1/3)R1}{\approx} \begin{bmatrix} 1 & 2 & -1 & 2 \\ -2 & -4 & -3 & -1 \\ 3 & 6 & -2 & 10 \end{bmatrix} \underset{\substack{R2+(2)R1 \\ R3+(-3)R1}}{\approx} \begin{bmatrix} 1 & 2 & -1 & 2 \\ 0 & 0 & -5 & 3 \\ 0 & 0 & 1 & 4 \end{bmatrix}.$

It is now clear that there is no solution. The last two rows give $-5x_3 = 3$ and $x_3 = 4$.

(b) $\begin{bmatrix} 1 & 2 & 1 & 7 \\ 1 & 2 & 2 & 11 \\ 2 & 4 & 3 & 18 \end{bmatrix} \underset{\substack{R2+(-1)R1 \\ R3+(-2)R1}}{\approx} \begin{bmatrix} 1 & 2 & 1 & 7 \\ 0 & 0 & 1 & 4 \\ 0 & 0 & 1 & 4 \end{bmatrix} \underset{\substack{R1+(-1)R2 \\ R3+(-1)R2}}{\approx} \begin{bmatrix} 1 & 2 & 0 & 3 \\ 0 & 0 & 1 & 4 \\ 0 & 0 & 0 & 0 \end{bmatrix},$

so $x_1 + 2x_2 = 3$ and $x_3 = 4$. Thus the general solution is
$x_1 = 3 - 2r,\ x_2 = r,\ x_3 = 4.$

(c) $\begin{bmatrix} 1 & 2 & -1 & 3 \\ 2 & 4 & -2 & 6 \\ 3 & 6 & 2 & -1 \end{bmatrix} \underset{\substack{R2+(-2)R1 \\ R3+(-3)R1}}{\approx} \begin{bmatrix} 1 & 2 & -1 & 3 \\ 0 & 0 & 0 & 0 \\ 0 & 0 & 5 & -10 \end{bmatrix} \underset{R2 \leftrightarrow R3}{\approx} \begin{bmatrix} 1 & 2 & -1 & 3 \\ 0 & 0 & 5 & -10 \\ 0 & 0 & 0 & 0 \end{bmatrix}$

$\underset{(1/5)R2}{\approx} \begin{bmatrix} 1 & 2 & -1 & 3 \\ 0 & 0 & 1 & -2 \\ 0 & 0 & 0 & 0 \end{bmatrix} \underset{R1+R2}{\approx} \begin{bmatrix} 1 & 2 & 0 & 1 \\ 0 & 0 & 1 & -2 \\ 0 & 0 & 0 & 0 \end{bmatrix}$, so $x_1 + 2x_2 = 1$ and $x_3 = -2$.

Thus the general solution is $x_1 = 1 - 2r,\ x_2 = r,\ x_3 = -2$.

(d) $\begin{bmatrix} 1 & 2 & 3 & 8 \\ 3 & 7 & 9 & 26 \\ 2 & 0 & 6 & 11 \end{bmatrix} \underset{\substack{R2+(-3)R1 \\ R3+(-2)R1}}{\approx} \begin{bmatrix} 1 & 2 & 3 & 8 \\ 0 & 1 & 0 & 2 \\ 0 & -4 & 0 & -5 \end{bmatrix}$, so there is no solution since the

last two rows give $x_2 = 2$ and $-4x_2 = -5$.

(e) $\begin{bmatrix} 0 & 1 & 2 & 5 \\ 1 & 2 & 5 & 13 \\ 1 & 0 & 2 & 4 \end{bmatrix} \underset{R1 \leftrightarrow R2}{\approx} \begin{bmatrix} 1 & 2 & 5 & 13 \\ 0 & 1 & 2 & 5 \\ 1 & 0 & 2 & 4 \end{bmatrix} \underset{R3+(-1)R1}{\approx} \begin{bmatrix} 1 & 2 & 5 & 13 \\ 0 & 1 & 2 & 5 \\ 0 & -2 & -3 & -9 \end{bmatrix}$

Section 1.2

$$\underset{\substack{R1+(-2)R2\\R3+(2)R2}}{\approx} \begin{bmatrix} 1 & 0 & 1 & 3 \\ 0 & 1 & 2 & 5 \\ 0 & 0 & 1 & 1 \end{bmatrix} \underset{\substack{R1+(-1)R3\\R2+(-2)R3}}{\approx} \begin{bmatrix} 1 & 0 & 0 & 2 \\ 0 & 1 & 0 & 3 \\ 0 & 0 & 1 & 1 \end{bmatrix},$$

so the solution is $x_1 = 2$, $x_2 = 3$, $x_3 = 1$.

(f) $\begin{bmatrix} 1 & 2 & 8 & 7 \\ 2 & 4 & 16 & 14 \\ 0 & 1 & 3 & 4 \end{bmatrix} \underset{R2+(-2)R1}{\approx} \begin{bmatrix} 1 & 2 & 8 & 7 \\ 0 & 0 & 0 & 0 \\ 0 & 1 & 3 & 4 \end{bmatrix} \underset{R2 \leftrightarrow R3}{\approx} \begin{bmatrix} 1 & 2 & 8 & 7 \\ 0 & 1 & 3 & 4 \\ 0 & 0 & 0 & 0 \end{bmatrix}$

$\underset{R1+(-2)R2}{\approx} \begin{bmatrix} 1 & 0 & 2 & -1 \\ 0 & 1 & 3 & 4 \\ 0 & 0 & 0 & 0 \end{bmatrix}$, so $x_1 + 2x_3 = -1$ and $x_2 + 3x_3 = 4$.

Thus the general solution is $x_1 = -1 - 2r$, $x_2 = 4 - 3r$, $x_3 = r$.

7. (a) $\begin{bmatrix} 1 & 1 & -3 & 10 \\ -3 & -2 & 4 & -24 \end{bmatrix} \underset{R2+(3)R1}{\approx} \begin{bmatrix} 1 & 1 & -3 & 10 \\ 0 & 1 & -5 & 6 \end{bmatrix} \underset{R1+(-1)R2}{\approx} \begin{bmatrix} 1 & 0 & 2 & 4 \\ 0 & 1 & -5 & 6 \end{bmatrix}$,

so $x_1 + 2x_3 = 4$ and $x_2 - 5x_3 = 6$. Thus the general solution is
$x_1 = 4 - 2r$, $x_2 = 6 + 5r$, $x_3 = r$.

(b) $\begin{bmatrix} 2 & -6 & -14 & 38 \\ -3 & 7 & 15 & -37 \end{bmatrix} \underset{(1/2)R1}{\approx} \begin{bmatrix} 1 & -3 & -7 & 19 \\ -3 & 7 & 15 & -37 \end{bmatrix} \underset{R2+(3)R1}{\approx} \begin{bmatrix} 1 & -3 & -7 & 19 \\ 0 & -2 & -6 & 20 \end{bmatrix}$

$\underset{(-1/2)R2}{\approx} \begin{bmatrix} 1 & -3 & -7 & 19 \\ 0 & 1 & 3 & -10 \end{bmatrix} \underset{R1+(3)R2}{\approx} \begin{bmatrix} 1 & 0 & 2 & -11 \\ 0 & 1 & 3 & -10 \end{bmatrix}$,

so $x_1 + 2x_3 = -11$ and $x_2 + 3x_3 = -10$. Thus the general solution is
$x_1 = -11 - 2r$, $x_2 = -10 - 3r$, $x_3 = r$.

(c) $\begin{bmatrix} 1 & 2 & -1 & -1 & 0 \\ 1 & 2 & 0 & 1 & 4 \\ -1 & -2 & 2 & 4 & 5 \end{bmatrix} \underset{\substack{R2+(-1)R1\\R3+R1}}{\approx} \begin{bmatrix} 1 & 2 & -1 & -1 & 0 \\ 0 & 0 & 1 & 2 & 4 \\ 0 & 0 & 1 & 3 & 5 \end{bmatrix} \underset{\substack{R1+R2\\R3+(-1)R2}}{\approx} \begin{bmatrix} 1 & 2 & 0 & 1 & 4 \\ 0 & 0 & 1 & 2 & 4 \\ 0 & 0 & 0 & 1 & 1 \end{bmatrix}$

$\underset{\substack{R1+(-1)R3\\R2+(-2)R3}}{\approx} \begin{bmatrix} 1 & 2 & 0 & 0 & 3 \\ 0 & 0 & 1 & 0 & 2 \\ 0 & 0 & 0 & 1 & 1 \end{bmatrix}$, so $x_1 + 2x_2 = 3$, $x_3 = 2$, and $x_4 = 1$.

Thus the general solution is $x_1 = 3 - 2r$, $x_2 = r$, $x_3 = 2$, and $x_4 = 1$.

(d) $\begin{bmatrix} 1 & 2 & 0 & 4 & 0 \\ -2 & -4 & 3 & -2 & 0 \end{bmatrix} \underset{R2+(2)R1}{\approx} \begin{bmatrix} 1 & 2 & 0 & 4 & 0 \\ 0 & 0 & 3 & 6 & 0 \end{bmatrix} \underset{(1/3)R2}{\approx} \begin{bmatrix} 1 & 2 & 0 & 4 & 0 \\ 0 & 0 & 1 & 2 & 0 \end{bmatrix}$,

so $x_1 + 2x_2 + 4x_4 = 0$ and $x_3 + 2x_4 = 0$. Thus the general solution is $x_1 = -2r - 4s$, $x_2 = r$, $x_3 = -2s$, $x_4 = s$.

(e) $\begin{bmatrix} 0 & 1 & -3 & 1 & 0 \\ 1 & 1 & -1 & 4 & 0 \\ -2 & -2 & 2 & -8 & 0 \end{bmatrix} \underset{R1 \leftrightarrow R2}{\approx} \begin{bmatrix} 1 & 1 & -1 & 4 & 0 \\ 0 & 1 & -3 & 1 & 0 \\ -2 & -2 & 2 & -8 & 0 \end{bmatrix}$

$\underset{R3+(2)R1}{\approx} \begin{bmatrix} 1 & 1 & -1 & 4 & 0 \\ 0 & 1 & -3 & 1 & 0 \\ 0 & 0 & 0 & 0 & 0 \end{bmatrix} \underset{R1+(-1)R2}{\approx} \begin{bmatrix} 1 & 0 & 2 & 3 & 0 \\ 0 & 1 & -3 & 1 & 0 \\ 0 & 0 & 0 & 0 & 0 \end{bmatrix}$,

so $x_1 + 2x_3 + 3x_4 = 0$ and $x_2 - 3x_3 + x_4 = 0$. Thus the general solution is $x_1 = -2r - 3s$, $x_2 = 3r - s$, $x_3 = r$, $x_4 = s$.

8. (a) $\begin{bmatrix} 1 & 1 & 1 & -1 & -3 \\ 2 & 3 & 1 & -5 & -9 \\ 1 & 3 & -1 & -6 & -7 \\ -1 & -1 & -1 & 0 & 1 \end{bmatrix} \underset{\substack{R2+(-2)R1 \\ R3+(-1)R1 \\ R4+R1}}{\approx} \begin{bmatrix} 1 & 1 & 1 & -1 & -3 \\ 0 & 1 & -1 & -3 & -3 \\ 0 & 2 & -2 & -5 & -4 \\ 0 & 0 & 0 & -1 & -2 \end{bmatrix}$

$\underset{\substack{R1+(-1)R2 \\ R3+(-2)R2}}{\approx} \begin{bmatrix} 1 & 0 & 2 & 2 & 0 \\ 0 & 1 & -1 & -3 & -3 \\ 0 & 0 & 0 & 1 & 2 \\ 0 & 0 & 0 & -1 & -2 \end{bmatrix} \underset{\substack{R1+(-2)R3 \\ R2+(3)R3 \\ R4+R3}}{\approx} \begin{bmatrix} 1 & 0 & 2 & 0 & -4 \\ 0 & 1 & -1 & 0 & 3 \\ 0 & 0 & 0 & 1 & 2 \\ 0 & 0 & 0 & 0 & 0 \end{bmatrix}$,

so $x_1 + 2x_3 = -4$, $x_2 - x_3 = 3$, $x_4 = 2$.

The general solution is $x_1 = -2r - 4$, $x_2 = r + 3$, $x_3 = r$, $x_4 = 2$.

(b) $\begin{bmatrix} 0 & 1 & 2 & 7 \\ 1 & -2 & -6 & -18 \\ -1 & -1 & -2 & -5 \\ 2 & -5 & -15 & -46 \end{bmatrix} \underset{R1 \leftrightarrow R2}{\approx} \begin{bmatrix} 1 & -2 & -6 & -18 \\ 0 & 1 & 2 & 7 \\ -1 & -1 & -2 & -5 \\ 2 & -5 & -15 & -46 \end{bmatrix}$

$\underset{\substack{R3+R1 \\ R4+(-2)R1}}{\approx} \begin{bmatrix} 1 & -2 & -6 & -18 \\ 0 & 1 & 2 & 7 \\ 0 & -3 & -8 & -23 \\ 0 & -1 & -3 & -10 \end{bmatrix} \underset{\substack{R1+(2)R2 \\ R3+(3)R2 \\ R4+R2}}{\approx} \begin{bmatrix} 1 & 0 & -2 & -4 \\ 0 & 1 & 2 & 7 \\ 0 & 0 & -2 & -9 \\ 0 & 0 & -1 & -3 \end{bmatrix}$, and there is no

solution because the last two rows of the matrix give, respectively,

13

$x_3 = 9/2$ and $x_3 = 3$.

(c) $\begin{bmatrix} 2 & -4 & 16 & -14 & 10 \\ -1 & 5 & -17 & 19 & -2 \\ 1 & -3 & 11 & -11 & 4 \\ 3 & -4 & 18 & -13 & 17 \end{bmatrix} \underset{(1/2)R1}{\approx} \begin{bmatrix} 1 & -2 & 8 & -7 & 5 \\ -1 & 5 & -17 & 19 & -2 \\ 1 & -3 & 11 & -11 & 4 \\ 3 & -4 & 18 & -13 & 17 \end{bmatrix}$

$\underset{\substack{R2+R1 \\ R3+(-1)R1 \\ R4+(-3)R1}}{\approx} \begin{bmatrix} 1 & -2 & 8 & -7 & 5 \\ 0 & 3 & -9 & 12 & 3 \\ 0 & -1 & 3 & -4 & -1 \\ 0 & 2 & -6 & 8 & 2 \end{bmatrix} \underset{(1/3)R2}{\approx} \begin{bmatrix} 1 & -2 & 8 & -7 & 5 \\ 0 & 1 & -3 & 4 & 1 \\ 0 & -1 & 3 & -4 & -1 \\ 0 & 2 & -6 & 8 & 2 \end{bmatrix}$

$\underset{\substack{R1+(2)R2 \\ R3+R2 \\ R4+(-2)R2}}{\approx} \begin{bmatrix} 1 & 0 & 2 & 1 & 7 \\ 0 & 1 & -3 & 4 & 1 \\ 0 & 0 & 0 & 0 & 0 \\ 0 & 0 & 0 & 0 & 0 \end{bmatrix}$, so $x_1 + 2x_3 + x_4 = 7$ and $x_2 - 3x_3 + 4x_4 = 1$.

Thus the general solution is $x_1 = 7 - 2r - s$, $x_2 = 1 + 3r - 4s$, $x_3 = r$, $x_4 = s$.

(d) $\begin{bmatrix} 1 & -1 & 2 & 0 & 7 \\ 2 & -2 & 2 & -4 & 12 \\ -1 & 1 & -1 & 2 & -4 \\ -3 & 1 & -8 & -10 & -29 \end{bmatrix} \underset{\substack{R2+(-2)R1 \\ R3+R1 \\ R4+(3)R1}}{\approx} \begin{bmatrix} 1 & -1 & 2 & 0 & 7 \\ 0 & 0 & -2 & -4 & -2 \\ 0 & 0 & 1 & 2 & 3 \\ 0 & -2 & -2 & -10 & -8 \end{bmatrix}$

$\underset{R2 \leftrightarrow R4}{\approx} \begin{bmatrix} 1 & -1 & 2 & 0 & 7 \\ 0 & -2 & -2 & -10 & -8 \\ 0 & 0 & 1 & 2 & 3 \\ 0 & 0 & -2 & -4 & -2 \end{bmatrix} \underset{(-1/2)R2}{\approx} \begin{bmatrix} 1 & -1 & 2 & 0 & 7 \\ 0 & 1 & 1 & 5 & 4 \\ 0 & 0 & 1 & 2 & 3 \\ 0 & 0 & -2 & -4 & -2 \end{bmatrix}$

$\underset{R4+(2)R3}{\approx} \begin{bmatrix} 1 & -1 & 2 & 0 & 7 \\ 0 & 1 & 1 & 5 & 4 \\ 0 & 0 & 1 & 2 & 3 \\ 0 & 0 & 0 & 0 & 4 \end{bmatrix}$. The last row gives $0 = 4$, so there is no solution.

(e) $\begin{bmatrix} 1 & 6 & -1 & -4 & 0 \\ -2 & -12 & 5 & 17 & 0 \\ 3 & 18 & -1 & -6 & 0 \end{bmatrix} \underset{\substack{R2+(2)R1 \\ R3+(-3)R1}}{\approx} \begin{bmatrix} 1 & 6 & -1 & -4 & 0 \\ 0 & 0 & 3 & 9 & 0 \\ 0 & 0 & 2 & 6 & 0 \end{bmatrix}$

$\underset{(1/3)R2}{\approx} \begin{bmatrix} 1 & 6 & -1 & -4 & 0 \\ 0 & 0 & 1 & 3 & 0 \\ 0 & 0 & 2 & 6 & 0 \end{bmatrix} \underset{\substack{R1+R2 \\ R3+(-2)R2}}{\approx} \begin{bmatrix} 1 & 6 & 0 & -1 & 0 \\ 0 & 0 & 1 & 3 & 0 \\ 0 & 0 & 0 & 0 & 0 \end{bmatrix}$,

so $x_1 + 6x_2 - x_4 = 0$ and $x_3 + 3x_4 = 0$.

Thus the general solution is $x_1 = -6r + s$, $x_2 = r$, $x_3 = -3s$, $x_4 = s$.

(f) $\begin{bmatrix} 4 & 8 & -12 & 28 \\ -1 & -2 & 3 & -7 \\ 2 & 4 & -8 & 16 \\ -3 & -6 & 9 & -21 \end{bmatrix} \underset{(1/4)R1}{\approx} \begin{bmatrix} 1 & 2 & -3 & 7 \\ -1 & -2 & 3 & -7 \\ 2 & 4 & -8 & 16 \\ -3 & -6 & 9 & -21 \end{bmatrix}$

$\underset{\substack{R2+R1 \\ R3+(-2)R1 \\ R4+(3)R1}}{\approx} \begin{bmatrix} 1 & 2 & -3 & 7 \\ 0 & 0 & 0 & 0 \\ 0 & 0 & -2 & 2 \\ 0 & 0 & 0 & 0 \end{bmatrix} \underset{R2 \leftrightarrow R3}{\approx} \begin{bmatrix} 1 & 2 & -3 & 7 \\ 0 & 0 & -2 & 2 \\ 0 & 0 & 0 & 0 \\ 0 & 0 & 0 & 0 \end{bmatrix}$

$\underset{(-1/2)R2}{\approx} \begin{bmatrix} 1 & 2 & -3 & 7 \\ 0 & 0 & 1 & -1 \\ 0 & 0 & 0 & 0 \\ 0 & 0 & 0 & 0 \end{bmatrix} \underset{R1+(3)R2}{\approx} \begin{bmatrix} 1 & 2 & 0 & 4 \\ 0 & 0 & 1 & -1 \\ 0 & 0 & 0 & 0 \\ 0 & 0 & 0 & 0 \end{bmatrix}$,

so $x_1 + 2x_2 = 4$ and $x_3 = -1$.

Thus the general solution is $x_1 = 4 - 2r$, $x_2 = r$, $x_3 = -1$.

(g) $\begin{bmatrix} 1 & 1 & 2 \\ 2 & 3 & 3 \\ 1 & 3 & 0 \\ 1 & 2 & 1 \end{bmatrix} \underset{\substack{R2+(-2)R1 \\ R3+(-1)R1 \\ R4+(-1)R1}}{\approx} \begin{bmatrix} 1 & 1 & 2 \\ 0 & 1 & -1 \\ 0 & 2 & -2 \\ 0 & 1 & -1 \end{bmatrix} \underset{\substack{R1+(-1)R2 \\ R3+(-2)R2 \\ R4+(-1)R2}}{\approx} \begin{bmatrix} 1 & 0 & 3 \\ 0 & 1 & -1 \\ 0 & 0 & 0 \\ 0 & 0 & 0 \end{bmatrix}$,

so $x_1 = 3$, $x_2 = -1$.

9. (a) The system of equations

$$3x_1 + 2x_2 - x_3 + x_4 = 4$$
$$3x_1 + 2x_2 - x_3 + x_4 = 1$$

clearly has no solution, since the equations are inconsistent. To make a system that is less obvious, add another equation to the system and replace the second equation by the sum of the second equation and some multiple (2 in the example below) of the third equation:

$$3x_1 + 2x_2 - x_3 + x_4 = 4$$
$$5x_1 + 4x_2 - x_3 - x_4 = 1$$
$$x_1 + x_2 - x_4 = 0$$

Section 1.2

(b) Choose a solution, e.g., $x_1 = 1$, $x_2 = 2$. Now make up equations thinking of x_1 as 1 and x_2 as 2:

$$x_1 + x_2 = 3$$
$$x_1 + 2x_2 = 5$$
$$x_1 - 2x_2 = -3$$

An easy way to ensure that there are no additional solutions is to include $x_1 = 1$ or $x_2 = 2$ as an equation in the system.

10. (a)
$$\begin{bmatrix} 1 & 0 & 0 \\ 0 & 1 & 0 \\ 0 & 0 & 1 \end{bmatrix} \begin{bmatrix} 1 & 0 & * \\ 0 & 1 & * \\ 0 & 0 & 0 \end{bmatrix} \begin{bmatrix} 1 & * & 0 \\ 0 & 0 & 1 \\ 0 & 0 & 0 \end{bmatrix} \begin{bmatrix} 1 & * & * \\ 0 & 0 & 0 \\ 0 & 0 & 0 \end{bmatrix}$$
no solution · unique solution · no solution · many solutions

(b)
$$\begin{bmatrix} 1 & 0 & 0 & * \\ 0 & 1 & 0 & * \\ 0 & 0 & 1 & * \end{bmatrix} \begin{bmatrix} 1 & 0 & * & 0 \\ 0 & 1 & * & 0 \\ 0 & 0 & 0 & 1 \end{bmatrix} \begin{bmatrix} 1 & 0 & * & * \\ 0 & 1 & * & * \\ 0 & 0 & 0 & 0 \end{bmatrix} \begin{bmatrix} 1 & * & 0 & 0 \\ 0 & 0 & 1 & 0 \\ 0 & 0 & 0 & 1 \end{bmatrix}$$
unique solution · no solution · many solutions · no solution

$$\begin{bmatrix} 1 & * & 0 & * \\ 0 & 0 & 1 & * \\ 0 & 0 & 0 & 0 \end{bmatrix} \begin{bmatrix} 1 & * & * & 0 \\ 0 & 0 & 0 & 1 \\ 0 & 0 & 0 & 0 \end{bmatrix} \begin{bmatrix} 1 & * & * & * \\ 0 & 0 & 0 & 0 \\ 0 & 0 & 0 & 0 \end{bmatrix}$$
many solutions · no solution · many solutions

11. (a) $\begin{bmatrix} a & b \\ c & d \end{bmatrix}$ $\underset{\text{if } a \neq 0}{(1/a)R1}$ $\begin{bmatrix} 1 & b/a \\ c & d \end{bmatrix}$ Add $(-c)\tilde{R}1$ to R2 $\begin{bmatrix} 1 & b/a \\ 0 & (da-cb)/a \end{bmatrix}$

$\underset{\text{if } (da-cb) \neq 0}{(a/(da-cb))R2}$ $\begin{bmatrix} 1 & b/a \\ 0 & 1 \end{bmatrix}$ Add $(-b/a)\tilde{R}2$ to R1 $\begin{bmatrix} 1 & 0 \\ 0 & 1 \end{bmatrix}$.

Thus $a \neq 0$, $da-cb \neq 0$ leads to reduced form $\begin{bmatrix} 1 & 0 \\ 0 & 1 \end{bmatrix}$. Result A

16

Section 1.2

<u>Suppose a=0</u> Get

$$\begin{bmatrix} a & b \\ c & d \end{bmatrix} = \begin{bmatrix} 0 & b \\ c & d \end{bmatrix} \xrightarrow{R1 \leftrightarrow R2} \begin{bmatrix} c & d \\ 0 & b \end{bmatrix} \xrightarrow[\text{if } c \neq 0]{(1/c)R1} \begin{bmatrix} 1 & d/c \\ 0 & b \end{bmatrix}$$

$$\xrightarrow[\text{if } b \neq 0]{(1/b)R1} \begin{bmatrix} 1 & d/c \\ 0 & 1 \end{bmatrix} \xrightarrow{\text{Add }(-d/c)R2 \text{ to } R1} \begin{bmatrix} 1 & 0 \\ 0 & 1 \end{bmatrix}.$$

Thus if a=0, b≠0, c≠0, get reduced form $\begin{bmatrix} 1 & 0 \\ 0 & 1 \end{bmatrix}$. <u>Result B</u>

Let us combine results A and B. Assume da-cb≠0. If a≠0 we get result A; if a=0 we get result B.

Thus the reduced form is $\begin{bmatrix} 1 & 0 \\ 0 & 1 \end{bmatrix}$ if da-cb≠0.

(b) $\begin{bmatrix} a & b \\ c & d \end{bmatrix} \xrightarrow[\text{if } a \neq 0]{(1/a)R1} \begin{bmatrix} 1 & b/a \\ c & d \end{bmatrix} \xrightarrow{\text{Add }(-c)R1 \text{ to } R2} \begin{bmatrix} 1 & b/a \\ 0 & (da-cb)/a \end{bmatrix}$

$= \begin{bmatrix} 1 & 0 \\ 0 & 0 \end{bmatrix}$ if b=0 and da-cb = 0.

Thus a≠0, b=0, da-cb=0 leads to reduced form $\begin{bmatrix} 1 & 0 \\ 0 & 0 \end{bmatrix}$. <u>Result C</u>

<u>Suppose a=0</u> Get

$$\begin{bmatrix} a & b \\ c & d \end{bmatrix} = \begin{bmatrix} 0 & b \\ c & d \end{bmatrix} \xrightarrow{R1 \leftrightarrow R2} \begin{bmatrix} c & d \\ 0 & b \end{bmatrix} \xrightarrow[\text{if } c \neq 0]{(1/c)R1} \begin{bmatrix} 1 & d/c \\ 0 & b \end{bmatrix}$$

$= \begin{bmatrix} 1 & 0 \\ 0 & 0 \end{bmatrix}$ if b=0 and d = 0.

Thus a=0, c≠0, b=0, d=0 leads to reduced form $\begin{bmatrix} 1 & 0 \\ 0 & 0 \end{bmatrix}$. <u>Result D</u>

12. (a) If $ax_0 + by_0 = 0$ then $k(ax_0 + by_0) = 0$ so that $a(kx_0) + b(ky_0) = 0$.
Thus $x = kx_0$, $y = ky_0$ is a solution. Likewise for the equation $cx + dy = 0$.

(b) If $ax_0 + by_0 = 0$ and $ax_1 + by_1 = 0$ then $ax_0 + by_0 + ax_1 + by_1 = 0 + 0 = 0$. But $ax_0 + by_0 + ax_1 + by_1 = ax_0 + ax_1 + by_0 + by_1 = a(x_0 + x_1) + b(y_0 + y_1)$ so that $a(x_0 + x_1) + b(y_0 + y_1) = 0$. Thus $x = x_0 + x_1$, $y = y_0 + y_1$ is a solution. Likewise for the equation $cx + dy = 0$.

Section 1.2

13. $a(0) + b(0) = 0$ and $c(0) + d(0) = 0$, so $x = 0, y = 0$ is a solution. The reduced echelon form of the matrix $\begin{bmatrix} a & b & 0 \\ c & d & 0 \end{bmatrix}$ is $\begin{bmatrix} 1 & 0 & 0 \\ 0 & 1 & 0 \end{bmatrix}$ if and only if $x = 0, y = 0$ is the only solution to the system. From Exercise 11, the reduced echelon form is $\begin{bmatrix} 1 & 0 & 0 \\ 0 & 1 & 0 \end{bmatrix}$ if and only if $ad - bc \neq 0$.

14. (a) and (b) If the first system of equations has a unique solution, then the reduced echelon form of the matrix $[A:B_1]$ will be $[I_3:X]$. The reduced echelon form of $[A:B_2]$ must therefore be $[I_3:Y]$. So the second system must also have a unique solution.

    (c) If the first system of equations has many solutions, then at least one row of the reduced echelon form of $[A:B_1]$ will consist entirely of zeros. Therefore the corresponding row(s) of the reduced echelon form of $[A:B_2]$ will have zeros in the first three columns. If any such row has a nonzero number in the fourth column, the system will have no solution.

15. (a) $\begin{bmatrix} 1 & 1 & 5 & 2 & 3 \\ 1 & 2 & 8 & 5 & 2 \\ 2 & 4 & 16 & 10 & 4 \end{bmatrix} \underset{R3+(-2)R1}{\overset{R2+(-1)R1}{\approx}} \begin{bmatrix} 1 & 1 & 5 & 2 & 3 \\ 0 & 1 & 3 & 3 & -1 \\ 0 & 2 & 6 & 6 & -2 \end{bmatrix} \underset{R3+(-2)R2}{\overset{R1+(-1)R2}{\approx}} \begin{bmatrix} 1 & 0 & 2 & -1 & 4 \\ 0 & 1 & 3 & 3 & -1 \\ 0 & 0 & 0 & 0 & 0 \end{bmatrix}$,

    so the general solution to the first system is $x_1 = -1 - 2r, x_2 = 3 - 3r, x_3 = r$, and the general solution to the second system is $x_1 = 4 - 2r, x_2 = -1 - 3r, x_3 = r$.

    (b) $\begin{bmatrix} 1 & 2 & 4 & 8 & 5 \\ 1 & 1 & 2 & 5 & 3 \\ 2 & 3 & 6 & 13 & 11 \end{bmatrix} \underset{R3+(-2)R1}{\overset{R2+(-1)R1}{\approx}} \begin{bmatrix} 1 & 2 & 4 & 8 & 5 \\ 0 & -1 & -2 & -3 & -2 \\ 0 & 1 & 2 & 3 & 5 \end{bmatrix} \overset{(-1)R2}{\approx} \begin{bmatrix} 1 & 2 & 4 & 8 & 5 \\ 0 & 1 & 2 & 3 & 2 \\ 0 & 1 & 2 & 3 & 5 \end{bmatrix}$

    $\underset{R3+(-1)R2}{\overset{R1+(-2)R2}{\approx}} \begin{bmatrix} 1 & 0 & 0 & 2 & 1 \\ 0 & 1 & 2 & 3 & 2 \\ 0 & 0 & 0 & 0 & 3 \end{bmatrix}$, so the general solution to the first system is

    $x_1 = 2, x_2 = 3 - 2r, x_3 = r$, and the second system has no solution.

16. A 3x3 matrix represents the equations of three lines in a plane. In order for there to be a unique solution, the three lines would have to meet in a point. For there to be many solutions, the three lines would all have to be the same. It is far more likely that the lines will meet in pairs (or that one pair will be parallel), i.e., that there will be no solution, the situation represented by the reduced echelon form I_3.

17. A 3x4 matrix represents the equations of three planes. In order for there to be many solutions, the three planes must have at least one line in common. For there to be no solutions, either at least two of the three planes must be parallel or the line of intersection

18

Section 1.3

of two of the planes must lie in a plane that is parallel to the third plane. It is more likely that the three planes will meet in a single point, i.e., that there will be a unique solution. The reduced echelon form therefore will be $[I_3 : X]$.

18. The difference between no solution and at least one solution is the presence of a nonzero number in the last position of a row that otherwise consists entirely of zeros. Round-off error is more likely to produce a nonzero number when there should be a zero than the reverse. Thus the answer is b). Thinking geometrically, a small move by one or more of the linear surfaces (round-off error) may destroy a solution if there is one, but probably won't produce a solution if there is none.

Exercise Set 1.3

Exercises 1, 2, and 3 can be solved simultaneously since the coefficient matrices are the same for all three.

$$a_0 + a_1 + a_2 = b_1$$
$$a_0 + 2a_1 + 4a_2 = b_2$$
$$a_0 + 3a_1 + 9a_2 = b_3$$

, where b_1, b_2, b_3 are the y values 2, 2, 4 in Exercise 1;

14, 22, 32 in Exercise 2; and 5, 7, 9 in Exercise 3.

$$\begin{bmatrix} 1 & 1 & 1 & 2 & 14 & 5 \\ 1 & 2 & 4 & 2 & 22 & 7 \\ 1 & 3 & 9 & 4 & 32 & 9 \end{bmatrix} \underset{R3+(-1)R1}{\overset{R2+(-1)R1}{\approx}} \begin{bmatrix} 1 & 1 & 1 & 2 & 14 & 5 \\ 0 & 1 & 3 & 0 & 8 & 2 \\ 0 & 2 & 8 & 2 & 18 & 4 \end{bmatrix} \underset{R3+(-2)R2}{\overset{R1+(-1)R2}{\approx}} \begin{bmatrix} 1 & 0 & -2 & 2 & 6 & 3 \\ 0 & 1 & 3 & 0 & 8 & 2 \\ 0 & 0 & 2 & 2 & 2 & 0 \end{bmatrix}$$

$$\underset{(1/2)R3}{\approx} \begin{bmatrix} 1 & 0 & -2 & 2 & 6 & 3 \\ 0 & 1 & 3 & 0 & 8 & 2 \\ 0 & 0 & 1 & 1 & 1 & 0 \end{bmatrix} \underset{R2+(-3)R3}{\overset{R1+(2)R3}{\approx}} \begin{bmatrix} 1 & 0 & 0 & 4 & 8 & 3 \\ 0 & 1 & 0 & -3 & 5 & 2 \\ 0 & 0 & 1 & 1 & 1 & 0 \end{bmatrix}$$, so the values of a_0, a_1, a_2

are 4, –3, 1 for Exercise 1; 8, 5, 1 for Exercise 2; and 3, 2, 0 for Exercise 3. Thus the equations of the polynomials are:

1. $4 - 3x + x^2 = y$ 2. $8 + 5x + x^2 = y$ 3. $3 + 2x = y$

4. $a_0 + a_1 + a_2 = 8$
   $a_0 + 3a_1 + 9a_2 = 26$
   $a_0 + 5a_1 + 25a_2 = 60$

$$\begin{bmatrix} 1 & 1 & 1 & 8 \\ 1 & 3 & 9 & 26 \\ 1 & 5 & 25 & 60 \end{bmatrix} \underset{R3+(-1)R1}{\overset{R2+(-1)R1}{\approx}} \begin{bmatrix} 1 & 1 & 1 & 8 \\ 0 & 2 & 8 & 18 \\ 0 & 4 & 24 & 52 \end{bmatrix}$$

$$\underset{(1/2)R2}{\approx} \begin{bmatrix} 1 & 1 & 1 & 8 \\ 0 & 1 & 4 & 9 \\ 0 & 4 & 24 & 52 \end{bmatrix} \underset{R3+(-4)R2}{\overset{R1+(-1)R2}{\approx}} \begin{bmatrix} 1 & 0 & -3 & -1 \\ 0 & 1 & 4 & 9 \\ 0 & 0 & 8 & 16 \end{bmatrix} \underset{(1/8)R3}{\approx} \begin{bmatrix} 1 & 0 & -3 & -1 \\ 0 & 1 & 4 & 9 \\ 0 & 0 & 1 & 2 \end{bmatrix}$$

Section 1.3

$$\underset{\substack{R1+(3)R3 \\ R2+(-4)R3}}{\approx} \begin{bmatrix} 1 & 0 & 0 & 5 \\ 0 & 1 & 0 & 1 \\ 0 & 0 & 1 & 2 \end{bmatrix}, \text{ so } a_0 = 5, a_1 = 1, a_2 = 2, \text{ and the equation is}$$

$5 + x + 2x^2 = y$. When $x = 2$, $y = 5 + 2 + 8 = 15$.

5. $\begin{aligned} a_0 - a_1 + a_2 &= -1 \\ a_0 &= 1 \\ a_0 + a_1 + a_2 &= -3 \end{aligned}$ $\quad \begin{bmatrix} 1 & -1 & 1 & -1 \\ 1 & 0 & 0 & 1 \\ 1 & 1 & 1 & -3 \end{bmatrix} \underset{\substack{R2+(-1)R1 \\ R3+(-1)R1}}{\approx} \begin{bmatrix} 1 & -1 & 1 & -1 \\ 0 & 1 & -1 & 2 \\ 0 & 2 & 0 & -2 \end{bmatrix}$

$\underset{\substack{R1+R2 \\ R3+(-2)R2}}{\approx} \begin{bmatrix} 1 & 0 & 0 & 1 \\ 0 & 1 & -1 & 2 \\ 0 & 0 & 2 & -6 \end{bmatrix} \underset{(1/2)R3}{\approx} \begin{bmatrix} 1 & 0 & 0 & 1 \\ 0 & 1 & -1 & 2 \\ 0 & 0 & 1 & -3 \end{bmatrix} \underset{R2+R3}{\approx} \begin{bmatrix} 1 & 0 & 0 & 1 \\ 0 & 1 & 0 & -1 \\ 0 & 0 & 1 & -3 \end{bmatrix}$,

so $a_0 = 1$, $a_1 = -1$, $a_2 = -2$, and the equation is $1 - x - 3x^2 = y$.
When $x = 3$, $y = 1 - 3 - 27 = -29$.

6. $\begin{aligned} a_0 + a_1 + a_2 + a_3 &= -3 \\ a_0 + 2a_1 + 4a_2 + 8a_3 &= -1 \\ a_0 + 3a_1 + 9a_2 + 27a_3 &= 9 \\ a_0 + 4a_1 + 16a_2 + 64a_3 &= 33 \end{aligned}$ $\quad \begin{bmatrix} 1 & 1 & 1 & 1 & -3 \\ 1 & 2 & 4 & 8 & -1 \\ 1 & 3 & 9 & 27 & 9 \\ 1 & 4 & 16 & 64 & 33 \end{bmatrix}$

$\underset{\substack{R2+(-1)R1 \\ R3+(-1)R1 \\ R4+(-1)R1}}{\approx} \begin{bmatrix} 1 & 1 & 1 & 1 & -3 \\ 0 & 1 & 3 & 7 & 2 \\ 0 & 2 & 8 & 26 & 12 \\ 0 & 3 & 15 & 63 & 36 \end{bmatrix} \underset{\substack{R1+(-1)R2 \\ R3+(-2)R2 \\ R4+(-3)R2}}{\approx} \begin{bmatrix} 1 & 0 & -2 & -6 & -5 \\ 0 & 1 & 3 & 7 & 2 \\ 0 & 0 & 2 & 12 & 8 \\ 0 & 0 & 6 & 42 & 30 \end{bmatrix}$

$\underset{\substack{(1/2)R3 \\ (1/6)R4}}{\approx} \begin{bmatrix} 1 & 0 & -2 & -6 & -5 \\ 0 & 1 & 3 & 7 & 2 \\ 0 & 0 & 1 & 6 & 4 \\ 0 & 0 & 1 & 7 & 5 \end{bmatrix} \underset{\substack{R1+(2)R3 \\ R2+(-3)R3 \\ R4+(-1)R3}}{\approx} \begin{bmatrix} 1 & 0 & 0 & 6 & 3 \\ 0 & 1 & 0 & -11 & -10 \\ 0 & 0 & 1 & 6 & 4 \\ 0 & 0 & 0 & 1 & 1 \end{bmatrix}$

$\underset{\substack{R1+(-6)R4 \\ R2+(11)R4 \\ R3+(-6)R4}}{\approx} \begin{bmatrix} 1 & 0 & 0 & 0 & -3 \\ 0 & 1 & 0 & 0 & 1 \\ 0 & 0 & 1 & 0 & -2 \\ 0 & 0 & 0 & 1 & 1 \end{bmatrix}$,

so $a_0 = -3$, $a_1 = 1$, $a_2 = -2$, $a_3 = 1$ and the equation is $-3 + x - 2x^2 + x^3 = y$.

Section 1.3

7. $I_1 + I_2 - I_3 = 0$
    $2I_1 + 4I_3 = 34$
    $4I_2 + 4I_3 = 28$
    so that $I_1 = 5$, $I_2 = 1$, $I_3 = 6$.

8. $I_1 + I_2 - I_3 = 0$
    $I_1 + 2I_3 = 9$
    $3I_2 + 2I_3 = 17$
    so that $I_1 = 1$, $I_2 = 3$, $I_3 = 4$.

9. $I_1 + I_2 - I_3 = 0$
    $3I_3 = 9$
    $4I_2 + 3I_3 = 13$
    so that $I_1 = 2$, $I_2 = 1$, $I_3 = 3$.

10. $I_1 + I_2 - I_3 = 0$
    $2I_1 + 2I_3 = 4$
    $4I_2 + 2I_3 = 2$
    so that $I_1 = 1$, $I_2 = 0$, $I_3 = 1$.

11. $I_1 - I_2 - I_3 = 0$
    $I_1 + 3I_2 = 31$
    $I_1 + 7I_3 = 31$
    so that $I_1 = 10$, $I_2 = 7$, $I_3 = 3$.

12. $I_1 - I_2 - I_3 = 0$
    $I_3 - I_4 - I_5 = 0$
    $I_2 = 4$
    $I_4 = 4$
    $I_5 = 4$

    so that $I_1 = 12$, $I_2 = 4$, $I_3 = 8$, $I_4 = 4$, $I_5 = 4$.

13. $I_1 - I_2 - I_3 = 0$
    $I_3 - I_4 + I_5 = 0$
    $I_1 + I_2 = 4$
    $I_1 + 2I_4 = 4$
    $2I_4 + 2I_5 = 2$

    so that $I_1 = 7/3$, $I_2 = 5/3$, $I_3 = 2/3$, $I_4 = 5/6$, $I_5 = 1/6$.

14. Assume I_3 flows from A to B.
    $I_1 - I_2 - I_3 = 0$
    $I_1 + 2I_3 = 4$
    $I_2 - 2I_3 = 9$

    gives $I_1 = 6$, $I_2 = 7$, $I_3 = -1$, so the current in AB is 1 amp flowing from B to A.

Section 1.3

15. Let I_1 be the current in the direction from the 16volt battery to A, let I_2 be the current from A to B, and let I_3 be the current in the direction from C to B.

$$\begin{aligned} I_1 - I_2 - I_3 &= 0 : 0 \\ 5I_1 + I_2 &= 16 : 16 \\ -I_2 + 5I_3 &= 9 : 23 \end{aligned}$$

We solve the two systems simultaneously:

$$\begin{bmatrix} 1 & -1 & -1 & 0 & 0 \\ 5 & 1 & 0 & 16 & 16 \\ 0 & -1 & 5 & 9 & 23 \end{bmatrix} \underset{R2+(-5)R1}{\approx} \begin{bmatrix} 1 & -1 & -1 & 0 & 0 \\ 0 & 6 & 5 & 16 & 16 \\ 0 & -1 & 5 & 9 & 23 \end{bmatrix}$$

$$\underset{R2 \leftrightarrow R3}{\approx} \begin{bmatrix} 1 & -1 & -1 & 0 & 0 \\ 0 & -1 & 5 & 9 & 23 \\ 0 & 6 & 5 & 16 & 16 \end{bmatrix} \underset{(-1)R2}{\approx} \begin{bmatrix} 1 & -1 & -1 & 0 & 0 \\ 0 & 1 & -5 & -9 & -23 \\ 0 & 6 & 5 & 16 & 16 \end{bmatrix}$$

$$\underset{\substack{R1+R2 \\ R3+(-6)R2}}{\approx} \begin{bmatrix} 1 & 0 & -6 & -9 & -23 \\ 0 & 1 & -5 & -9 & -23 \\ 0 & 0 & 35 & 70 & 154 \end{bmatrix} \underset{(1/35)R3}{\approx} \begin{bmatrix} 1 & 0 & -6 & -9 & -23 \\ 0 & 1 & -5 & -9 & -23 \\ 0 & 0 & 1 & 2 & 22/5 \end{bmatrix}$$

$$\underset{\substack{R1+(6)R3 \\ R2+(5)R3}}{\approx} \begin{bmatrix} 1 & 0 & 0 & 3 & 17/5 \\ 0 & 1 & 0 & 1 & -1 \\ 0 & 0 & 1 & 2 & 22/5 \end{bmatrix}$$

(a) $I_1 = 3$, $I_2 = 1$, $I_3 = 2$. \qquad (b) $I_1 = 17/5$, $I_2 = -1$, $I_3 = 22/5$.

If $I_2 = 0$, let the voltage at C be denoted by V.

$$\begin{aligned} I_1 - I_3 &= 0 \\ 5I_1 &= 16 \\ 5I_3 &= V \end{aligned}$$

Thus $I_1 = I_3 = 16/5$, so $V = 16$.

Section 1.3

16. A: $x_1 + x_4 = 300$ B: $x_1 + x_2 = 250$

    C: $x_2 + x_3 = 100$ D: $x_3 + x_4 = 150$

Solving these equations simultaneously gives

$$x_1 = -x_4 + 300, \; x_2 = x_4 - 50, \; x_3 = -x_4 + 150.$$

Two solutions are $x_1 = 250, x_2 = 0, x_3 = 100, x_4 = 50$
and $x_1 = 150, x_2 = 100, x_3 = 0, x_4 = 150$.

The minimum value of x_1 comes from taking the maximum value of x_4, which is 150. So the minimum value of x_1 is 150.

17. A: $x_1 - x_4 = 100$ B: $x_1 - x_2 = 200$

    C: $-x_2 + x_3 = 150$ D: $x_3 - x_4 = 50$

Solving these equations simultaneously gives

$$x_1 = x_4 + 100, \; x_2 = x_4 - 100, \; x_3 = x_4 + 50.$$

$x_2 = 0$ is theoretically possible. In that case $x_4 = 100, x_1 = 200, x_3 = 150$.

This flow is not likely to be realized in practice unless branch BC is completely closed.

18. Let $y = a_0 + a_1 x + a_2 x^2$. These polynomials must pass through (1, 2) and (3, 4). Thus

$$\begin{matrix} a_0 + a_1 + a_2 = 2 \\ a_0 + 3a_1 + 9a_2 = 4 \end{matrix} \cdot \begin{bmatrix} 1 & 1 & 1 & 2 \\ 1 & 3 & 9 & 4 \end{bmatrix} \approx \begin{bmatrix} 1 & 0 & -3 & 1 \\ 0 & 1 & 4 & 1 \end{bmatrix}. \; a_0 = 3a_2+1, \; a_1 = -4a_2+1.$$

Let $a_2 = r$. The family of polynomials is $y = (3r+1) + (-4r+1)x + rx^2$. $r=0$ gives the line $y=1+x$ that passes through these points. When $r > 0$ the polynomials open up and when $r < 0$ the polynomials open down.

19. Let $y = a_0 + a_1 x + a_2 x^2 + a_3 x^3$. Polynomials must pass through (1, 2), (3, 4) (4,8). Thus

$$\begin{matrix} a_0 + a_1 + a_2 + a_3 = 2 \\ a_0 + 3a_1 + 9a_2 + 27a_3 = 4 \\ a_0 + 4a_1 + 16a_2 + 64a_3 = 8 \end{matrix} \cdot \begin{bmatrix} 1 & 1 & 1 & 1 & 2 \\ 1 & 3 & 9 & 27 & 4 \\ 1 & 4 & 16 & 64 & 8 \end{bmatrix} \approx \begin{bmatrix} 1 & 0 & 0 & 12 & 4 \\ 0 & 1 & 0 & -19 & -3 \\ 0 & 0 & 1 & 8 & 1 \end{bmatrix}.$$

$a_0 = -12a_3+4, \; a_1 = 19a_3-3, \; a_2 = -8a_3+1$.

Chapter 1 Review Exercises

Let $a_2 = r$. The family of polynomials is $y = (-12r+4) + (19r-3)x + (-8r+1)x^2 + rx^3$.
When $r=1$, $y = -8 + 16x - 7x^2 + x^3$.

Chapter 1 Review Exercises

1. (a) 2×3 (b) 2×2 (c) 1×4 (d) 3×1 (e) 4×6

2. 0, 6, 5, 1, 9

3. $I_5 = \begin{bmatrix} 1 & 0 & 0 & 0 & 0 \\ 0 & 1 & 0 & 0 & 0 \\ 0 & 0 & 1 & 0 & 0 \\ 0 & 0 & 0 & 1 & 0 \\ 0 & 0 & 0 & 0 & 1 \end{bmatrix}$

4. (a) $\begin{bmatrix} 1 & 2 \\ 4 & -3 \end{bmatrix}, \begin{bmatrix} 1 & 2 & 6 \\ 4 & -3 & -1 \end{bmatrix}$

    (b) $\begin{bmatrix} 2 & 1 & -4 \\ 1 & -2 & 8 \\ 3 & 5 & -7 \end{bmatrix}, \begin{bmatrix} 2 & 1 & -4 & 1 \\ 1 & -2 & 8 & 0 \\ 3 & 5 & -7 & -3 \end{bmatrix}$

    (c) $\begin{bmatrix} -1 & 2 & -7 & -2 \\ 3 & -1 & 5 & 3 \\ 4 & 3 & 0 & 5 \end{bmatrix}, \begin{bmatrix} -1 & 2 & -7 \\ 3 & -1 & 5 \\ 4 & 3 & 0 \end{bmatrix}$

    (d) $\begin{bmatrix} 1 & 0 & 0 \\ 0 & 1 & 0 \\ 0 & 0 & 1 \end{bmatrix}, \begin{bmatrix} 1 & 0 & 0 & 1 \\ 0 & 1 & 0 & 5 \\ 0 & 0 & 1 & -3 \end{bmatrix}$

    (e) $\begin{bmatrix} -2 & 3 & -8 & 5 \\ 1 & 5 & 0 & -6 \\ 0 & -1 & 2 & 3 \end{bmatrix}, \begin{bmatrix} -2 & 3 & -8 & 5 & -2 \\ 1 & 5 & 0 & -6 & 0 \\ 0 & -1 & 2 & 3 & 5 \end{bmatrix}$

5. (a) $4x_1 + 2x_2 = 0$
       $-3x_1 + 7x_2 = 8$

    (b) $x_1 + 9x_2 = -3$
       $3x_2 = 2$

    (c) $x_1 + 2x_2 + 3x_3 = 4$
       $5x_1 - 3x_3 = 6$

    (d) $x_1 = 5$
       $x_2 = -8$
       $x_3 = 2$

    (e) $x_1 + 4x_2 - x_3 = 7$
       $x_2 + 3x_3 = 8$
       $x_3 = -5$

6. (a) Yes. (b) Yes.

   (c) No. There is a 2 (a nonzero element) above the leading 1 of row 2.

   (d) Yes.

   (e) No. The leading 1 in row 3 is not positioned to the right of the leading 1 in row 2.

24

Chapter 1 Review Exercises

7. (a) $\begin{bmatrix} 2 & 4 & 2 \\ 3 & 7 & 2 \end{bmatrix} \underset{(1/2)R1}{\approx} \begin{bmatrix} 1 & 2 & 1 \\ 3 & 7 & 2 \end{bmatrix} \underset{R2+(-3)R1}{\approx} \begin{bmatrix} 1 & 2 & 1 \\ 0 & 1 & -1 \end{bmatrix} \underset{R1+(-2)R2}{\approx} \begin{bmatrix} 1 & 0 & 3 \\ 0 & 1 & -1 \end{bmatrix}$,

so the solution is $x_1 = 3$ and $x_2 = -1$.

(b) $\begin{bmatrix} 1 & -2 & -6 & -17 \\ 2 & -6 & -16 & -46 \\ 1 & 2 & -1 & -5 \end{bmatrix} \underset{\substack{R2+(-2)R1 \\ R3+(-1)R1}}{\approx} \begin{bmatrix} 1 & -2 & -6 & -17 \\ 0 & -2 & -4 & -12 \\ 0 & 4 & 5 & 12 \end{bmatrix} \underset{(-1/2)R2}{\approx} \begin{bmatrix} 1 & -2 & -6 & -17 \\ 0 & 1 & 2 & 6 \\ 0 & 4 & 5 & 12 \end{bmatrix}$

$\underset{\substack{R1+(2)R2 \\ R3+(-4)R2}}{\approx} \begin{bmatrix} 1 & 0 & -2 & -5 \\ 0 & 1 & 2 & 6 \\ 0 & 0 & -3 & -12 \end{bmatrix} \underset{(-1/3)R3}{\approx} \begin{bmatrix} 1 & 0 & -2 & -5 \\ 0 & 1 & 2 & 6 \\ 0 & 0 & 1 & 4 \end{bmatrix}$

$\underset{\substack{R1+(2)R3 \\ R2+(-2)R3}}{\approx} \begin{bmatrix} 1 & 0 & 0 & 3 \\ 0 & 1 & 0 & -2 \\ 0 & 0 & 1 & 4 \end{bmatrix}$, so that $x_1 = 3, x_2 = -2, x_3 = 4$.

(c) $\begin{bmatrix} 0 & 1 & 2 & 6 & 21 \\ 1 & -1 & 1 & 5 & 12 \\ 1 & -1 & -1 & -4 & -9 \\ 3 & -2 & 0 & -6 & -4 \end{bmatrix} \underset{R1 \leftrightarrow R2}{\approx} \begin{bmatrix} 1 & -1 & 1 & 5 & 12 \\ 0 & 1 & 2 & 6 & 21 \\ 1 & -1 & -1 & -4 & -9 \\ 3 & -2 & 0 & -6 & -4 \end{bmatrix}$

$\underset{\substack{R3+(-1)R1 \\ R4+(-3)R1}}{\approx} \begin{bmatrix} 1 & -1 & 1 & 5 & 12 \\ 0 & 1 & 2 & 6 & 21 \\ 0 & 0 & -2 & -9 & -21 \\ 0 & 1 & -3 & -21 & -40 \end{bmatrix} \underset{\substack{R1+R2 \\ R4+(-1)R2}}{\approx} \begin{bmatrix} 1 & 0 & 3 & 11 & 33 \\ 0 & 1 & 2 & 6 & 21 \\ 0 & 0 & -2 & -9 & -21 \\ 0 & 0 & -5 & -27 & -61 \end{bmatrix}$

$\underset{(-1/2)R3}{\approx} \begin{bmatrix} 1 & 0 & 3 & 11 & 33 \\ 0 & 1 & 2 & 6 & 21 \\ 0 & 0 & 1 & 9/2 & 21/2 \\ 0 & 0 & -5 & -27 & -61 \end{bmatrix} \underset{\substack{R1+(-3)R3 \\ R2+(-2)R3 \\ R4+(5)R3}}{\approx} \begin{bmatrix} 1 & 0 & 0 & -5/2 & 3/2 \\ 0 & 1 & 0 & -3 & 0 \\ 0 & 0 & 1 & 9/2 & 21/2 \\ 0 & 0 & 0 & -9/2 & -17/2 \end{bmatrix}$

$\underset{(-2/9)R4}{\approx} \begin{bmatrix} 1 & 0 & 0 & -5/2 & 3/2 \\ 0 & 1 & 0 & -3 & 0 \\ 0 & 0 & 1 & 9/2 & 21/2 \\ 0 & 0 & 0 & 1 & 17/9 \end{bmatrix} \underset{\substack{R1+(5/2)R4 \\ R2+(3)R4 \\ R3+(-9/2)R4}}{\approx} \begin{bmatrix} 1 & 0 & 0 & 0 & 56/9 \\ 0 & 1 & 0 & 0 & 17/3 \\ 0 & 0 & 1 & 0 & 2 \\ 0 & 0 & 0 & 1 & 17/9 \end{bmatrix}$,

so that $x_1 = 56/9, x_2 = 17/3, x_3 = 2, x_4 = 17/9$.

Chapter 1 Review Exercises

8. (a) $\begin{bmatrix} 1 & -1 & 1 & 3 \\ -2 & 3 & 1 & -8 \\ 4 & -2 & 10 & 10 \end{bmatrix} \underset{R3+(-4)R1}{\overset{R2+(2)R1}{\approx}} \begin{bmatrix} 1 & -1 & 1 & 3 \\ 0 & 1 & 3 & -2 \\ 0 & 2 & 6 & -2 \end{bmatrix} \underset{R3+(-2)R2}{\overset{R1+R2}{\approx}} \begin{bmatrix} 1 & 0 & 4 & 1 \\ 0 & 1 & 3 & -2 \\ 0 & 0 & 0 & 2 \end{bmatrix}$.

There is no need to continue. The last row gives 0 = 2 so there is no solution.

(b) $\begin{bmatrix} 1 & 3 & 6 & -2 & -7 \\ -2 & -5 & -10 & 3 & 10 \\ 1 & 2 & 4 & 0 & 0 \\ 0 & 1 & 2 & -3 & -10 \end{bmatrix} \underset{R3+(-1)R1}{\overset{R2+(2)R1}{\approx}} \begin{bmatrix} 1 & 3 & 6 & -2 & -7 \\ 0 & 1 & 2 & -1 & -4 \\ 0 & -1 & -2 & 2 & 7 \\ 0 & 1 & 2 & -3 & -10 \end{bmatrix}$

$\underset{\substack{R1+(-3)R2 \\ R3+R2 \\ R4+(-1)R2}}{\approx} \begin{bmatrix} 1 & 0 & 0 & 1 & 5 \\ 0 & 1 & 2 & -1 & -4 \\ 0 & 0 & 0 & 1 & 3 \\ 0 & 0 & 0 & -2 & -6 \end{bmatrix} \underset{\substack{R1+(-1)R3 \\ R2+R3 \\ R4+(2)R3}}{\approx} \begin{bmatrix} 1 & 0 & 0 & 0 & 2 \\ 0 & 1 & 2 & 0 & -1 \\ 0 & 0 & 0 & 1 & 3 \\ 0 & 0 & 0 & 0 & 0 \end{bmatrix}$,

so there are many solutions, and the general solution is

$x_1 = 2, x_2 = -1 - 2r, x_3 = r, x_4 = 3$.

9. If a matrix A is in reduced echelon form, it is clear from the definition that the leading 1 in any row cannot be to the left of the diagonal element in that row. Therefore if $A \neq I_n$, there must be some row that has its leading 1 to the right of the diagonal element in that row. Suppose row j is such a row and the leading 1 is in position (j, k) where $j < k \leq n$. Then if rows $j + 1, j + 2, \ldots, j + (n - k) < n$ all contain nonzero terms, the leading 1 in these rows must be at least as far to the right as columns $k + 1, k + 2, \ldots, k + (n - k) = n$, respectively. The leading 1 in row $j + (n - k) + 1$ must then be to the right of column n. But there is no column to the right of column n, so row $j + (n - k) + 1$ must consist of all zeros.

10. Let E be the reduced echelon form of A. Since B is row equivalent to A, B is also row equivalent to E. But since E is in reduced echelon form, it must be the reduced echelon form of B.

11. The equation is of the form $a_0 + a_1 x + a_2 x^2 = y$, so the system of equations to be solved is

$$a_0 + a_1 + a_2 = 3$$
$$a_0 + 2a_1 + 4a_2 = 6$$
$$a_0 + 3a_1 + 9a_2 = 13$$

$\begin{bmatrix} 1 & 1 & 1 & 3 \\ 1 & 2 & 4 & 6 \\ 1 & 3 & 9 & 13 \end{bmatrix} \underset{R3+(-1)R1}{\overset{R2+(-1)R1}{\approx}} \begin{bmatrix} 1 & 1 & 1 & 3 \\ 0 & 1 & 3 & 3 \\ 0 & 2 & 8 & 10 \end{bmatrix} \underset{R3+(-2)R2}{\overset{R1+(-1)R2}{\approx}} \begin{bmatrix} 1 & 0 & -2 & 0 \\ 0 & 1 & 3 & 3 \\ 0 & 0 & 2 & 4 \end{bmatrix}$

Chapter 1 Review Exercises

$$\underset{(1/2)R3}{\approx} \begin{bmatrix} 1 & 0 & -2 & 0 \\ 0 & 1 & 3 & 3 \\ 0 & 0 & 1 & 2 \end{bmatrix} \underset{R2+(-3)R3}{\overset{R1+(2)R3}{\approx}} \begin{bmatrix} 1 & 0 & 0 & 4 \\ 0 & 1 & 0 & -3 \\ 0 & 0 & 1 & 2 \end{bmatrix}$$, so $a_0 = 4$, $a_1 = -3$, $a_2 = 2$, and the

equation is $4 - 3x + 2x^2 = y$.

12. $\begin{aligned} I_1 - I_2 - I_3 &= 0 \\ 2I_1 + I_3 &= 7 \\ 3I_2 - I_3 &= 5 \end{aligned}$ gives $I_1 = 3$, $I_2 = 2$, $I_3 = 1$.

Chapter 2

Exercise Set 2.1

1. (a) $A + B = \begin{bmatrix} 5-3 & 4+0 \\ -1+4 & 7+2 \\ 9+5 & -3-7 \end{bmatrix} = \begin{bmatrix} 2 & 4 \\ 3 & 9 \\ 14 & -10 \end{bmatrix}$. (b) $2B = \begin{bmatrix} -6 & 0 \\ 8 & 4 \\ 10 & -14 \end{bmatrix}$.

   (c) $-D = \begin{bmatrix} -9 & 5 \\ -3 & 0 \end{bmatrix}$. (d) $C + D = \begin{bmatrix} 10 & -3 \\ 6 & 4 \end{bmatrix}$. (e) $A + D$ does not exist.

   (f) $2A + B = \begin{bmatrix} 10-3 & 8+0 \\ -2+4 & 14+2 \\ 18+5 & -6-7 \end{bmatrix} = \begin{bmatrix} 7 & 8 \\ 2 & 16 \\ 23 & -13 \end{bmatrix}$. (g) $A - B = \begin{bmatrix} 8 & 4 \\ -5 & 5 \\ 4 & 4 \end{bmatrix}$.

2. (a) $A + B$ does not exist. (b) $4B = \begin{bmatrix} 0 & -4 & 16 \\ 24 & -32 & 8 \\ -16 & 20 & 36 \end{bmatrix}$. (c) $-3D = \begin{bmatrix} 9 \\ 0 \\ -6 \end{bmatrix}$.

   (d) $B - 3C = \begin{bmatrix} 0-3 & -1-6 & 4+15 \\ 6+21 & -8-27 & 2-9 \\ -4-15 & 5+12 & 9-0 \end{bmatrix} = \begin{bmatrix} -3 & -7 & 19 \\ 27 & -35 & -7 \\ -19 & 17 & 9 \end{bmatrix}$.

   (e) $-A = \begin{bmatrix} -9 \\ -2 \\ 1 \end{bmatrix}$. (f) $3A + 2D = \begin{bmatrix} 27-6 \\ 6+0 \\ -3+4 \end{bmatrix} = \begin{bmatrix} 21 \\ 6 \\ 1 \end{bmatrix}$. (g) $A + D = \begin{bmatrix} 6 \\ 2 \\ 1 \end{bmatrix}$.

3. (a) $AB = \begin{bmatrix} (1 \times 0)+(0 \times -2) & (1 \times 1)+(0 \times 5) \\ (0 \times 0)+(1 \times -2) & (0 \times 1)+(1 \times 5) \end{bmatrix} = \begin{bmatrix} 0 & 1 \\ -2 & 5 \end{bmatrix} = B$. (b) $BA = B$.

   (c) $AC = C$. (d) CA does not exist. (e) $AD = D$. (f) DC does not exist.

Section 2.1

(g) $BD = \begin{bmatrix} (0\times-1)+(1\times5) & (0\times0)+(1\times7) & (0\times3)+(1\times2) \\ (-2\times-1)+(5\times5) & (-2\times0)+(5\times7) & (-2\times3)+(5\times2) \end{bmatrix} = \begin{bmatrix} 5 & 7 & 2 \\ 27 & 35 & 4 \end{bmatrix}$.   (h) $A^2 = \begin{bmatrix} 1 & 0 \\ 0 & 1 \end{bmatrix}$

4. (a) $BA = \begin{bmatrix} 27 \\ 23 \\ 9 \end{bmatrix}$.   (b) AB does not exist.   (c) $CB = [\,10 \;\; 13 \;\; -5\,]$.

(d) $CA = [27]$.   (e) DA does not exist.   (f) DB does not exist.

(g) $AC = \begin{bmatrix} (-1\times-2) & (-1\times0) & (-1\times5) \\ (2\times-2) & (2\times0) & (2\times5) \\ (5\times-2) & (5\times0) & (5\times5) \end{bmatrix} = \begin{bmatrix} 2 & 0 & -5 \\ -4 & 0 & 10 \\ -10 & 0 & 25 \end{bmatrix}$.

(h) $B^2 = \begin{bmatrix} (0\times0)+(1\times3)+(5\times2) & (0\times1)+(1\times-7)+(5\times3) & (0\times5)+(1\times8)+(5\times1) \\ (3\times0)+(-7\times3)+(8\times2) & (3\times1)+(-7\times-7)+(8\times3) & (3\times5)+(-7\times8)+(8\times1) \\ (2\times0)+(3\times3)+(1\times2) & (2\times1)+(3\times-7)+(1\times3) & (2\times5)+(3\times8)+(1\times1) \end{bmatrix}$

$= \begin{bmatrix} 13 & 8 & 13 \\ -5 & 76 & -33 \\ 11 & -16 & 35 \end{bmatrix}$.

5. (a) $2A - 3(BC) = \begin{bmatrix} -12 & 2 \\ -9 & -24 \\ -20 & -24 \end{bmatrix}$.   (b) AB does not exist.   (c) $AC - BD = \begin{bmatrix} 8 & 2 \\ 4 & 1 \\ 0 & 6 \end{bmatrix}$.

(d) $CD - 2D = \begin{bmatrix} -30 & 0 \\ 15 & 0 \end{bmatrix}$.   (e) BA does not exist.

(f) AD + 2(DC) does not exist.   (g) $C^3 + 2(D^2) = \begin{bmatrix} -14 & 0 \\ 12 & 10 \end{bmatrix}$.

6. (a) 3x2   (b) 4x2   (c) does not exist   (d) 3x2   (e) 4x2

(f) 3x2   (g) does not exist

7. (a) 2x2   (b) 2x3   (c) 2x2   (d) does not exist   (e) 3x2

Section 2.1

(f) does not exist (g) 2x2

8. $A + O_3 = \begin{bmatrix} 1+0 & -8+0 & 4+0 \\ 5+0 & -6+0 & 3+0 \\ 2+0 & 0+0 & -1+0 \end{bmatrix} = \begin{bmatrix} 1 & -8 & 4 \\ 5 & -6 & 3 \\ 2 & 0 & -1 \end{bmatrix} = A.$ In the same manner $O_3 + A = A.$

$BO_3 = \begin{bmatrix} 0\times0+2\times0-3\times0 & 0\times0+2\times0-3\times0 & 0\times0+2\times0-3\times0 \\ 5\times0+6\times0+7\times0 & 5\times0+6\times0+7\times0 & 5\times0+6\times0+7\times0 \\ -1\times0+0\times0+4\times0 & -1\times0+0\times0+4\times0 & -1\times0+0\times0+4\times0 \end{bmatrix} = \begin{bmatrix} 0 & 0 & 0 \\ 0 & 0 & 0 \\ 0 & 0 & 0 \end{bmatrix} = O_3.$

$BI_3 = \begin{bmatrix} 0\times1+2\times0-3\times0 & 0\times0+2\times1-3\times0 & 0\times0+2\times0-3\times1 \\ 5\times1+6\times0+7\times0 & 5\times0+6\times1+7\times0 & 5\times0+6\times0+7\times1 \\ -1\times1+0\times0+4\times0 & -1\times0+0\times1+4\times0 & -1\times0+0\times0+4\times1 \end{bmatrix} = \begin{bmatrix} 0 & 2 & -3 \\ 5 & 6 & 7 \\ -1 & 0 & 4 \end{bmatrix} = B.$

In the same manner $O_3 B = O_3$ and $I_3 B = B.$

9. (a) $c_{31} = (1\times-1)+(0\times5)+(-2\times0) = -1.$ (b) $c_{23} = (2\times-3)+(6\times2)+(3\times6) = 24.$

   (c) $d_{12} = (-1\times3)+(2\times6)+(-3\times0) = 9.$ (d) $d_{22} = (5\times3)+(7\times6)+(2\times0) = 57.$

10. (a) $r_{21} = (4\times0)+(6\times0) = 0.$ (b) $r_{33} = (-1\times3)+(3\times4) = 9.$

    (c) $s_{11} = (0\times1)+(1\times4)+(3\times-1) = 1.$ (d) s_{23} does not exist. S is a 2x2 matrix.

11. (a) $d_{12} = (1\times2)+(-3\times0) + 2\times-4 = -6.$ (b) $d_{23} = (0\times-3)+(4\times-1)+ 2\times0 = -4.$

12. (a) $d_{11} = 2[(1\times1)+(-3\times3)+(0\times-1)] + (2\times2)+(0\times4)+(-2\times1) = -14.$

    (b) $d_{21} = 2[(4\times1)+(5\times3)+(1\times-1)] + (4\times2)+(7\times4)+(-5\times1) = 67.$

    (c) $d_{32} = 2[(3\times1)+(8\times0)+(0\times3)] + (1\times0)+(0\times7)+(-1\times0) = 6.$

30

Section 2.1

13. (a) $\begin{bmatrix} 2 & 3 \\ 3 & -8 \end{bmatrix} \begin{bmatrix} x_1 \\ x_2 \end{bmatrix} = \begin{bmatrix} 4 \\ -1 \end{bmatrix}.$ (b) $\begin{bmatrix} 4 & 7 \\ -2 & 3 \end{bmatrix} \begin{bmatrix} x_1 \\ x_2 \end{bmatrix} = \begin{bmatrix} -2 \\ -4 \end{bmatrix}.$

(c) $\begin{bmatrix} -9 & -3 \\ 6 & -2 \end{bmatrix} \begin{bmatrix} x_1 \\ x_2 \end{bmatrix} = \begin{bmatrix} -4 \\ 7 \end{bmatrix}.$

14. (a) $\begin{bmatrix} 1 & 8 & -2 \\ 4 & -7 & 1 \\ -2 & -5 & -2 \end{bmatrix} \begin{bmatrix} x_1 \\ x_2 \\ x_3 \end{bmatrix} = \begin{bmatrix} 3 \\ -3 \\ 1 \end{bmatrix}.$ (b) $\begin{bmatrix} 5 & 2 \\ 4 & -3 \\ 3 & 1 \end{bmatrix} \begin{bmatrix} x_1 \\ x_2 \end{bmatrix} = \begin{bmatrix} 6 \\ -2 \\ 9 \end{bmatrix}.$

(c) $\begin{bmatrix} 1 & -3 & 6 \\ 7 & 5 & 1 \end{bmatrix} \begin{bmatrix} x_1 \\ x_2 \\ x_3 \end{bmatrix} = \begin{bmatrix} 2 \\ -9 \end{bmatrix}.$ (d) $\begin{bmatrix} 2 & 5 & -3 & 4 \\ 1 & 0 & 9 & 5 \\ 3 & -3 & -8 & 5 \end{bmatrix} \begin{bmatrix} x_1 \\ x_2 \\ x_3 \\ x_4 \end{bmatrix} = \begin{bmatrix} 4 \\ 12 \\ -2 \end{bmatrix}.$

15. The third row of AB is the third row of A times each of the columns of B in turn. Since the third row of A is all zeros, each of the products is zero.

16. The second column of CD is the rows of C in turn multiplied by the second column of D. Since the second column of D is all zeros, each of the products is zero.

17. $c_{ij} = [a_{i1} \; a_{i2} \; \cdots \; a_{ir}] \begin{bmatrix} b_{1j} \\ b_{2j} \\ \vdots \\ b_{rj} \end{bmatrix} = [a_{i1} \; a_{i2} \; \cdots \; a_{ir}] B_j$, i.e., the ith member in the column

matrix C_j is given by the product of the ith row of A and the column matrix B_j. From the definition of matrix multiplication, this means that $C_j = AB_j$.

18. $AB_3 = \begin{bmatrix} 1 & 2 & 3 \\ 0 & 4 & 1 \\ 2 & 5 & 0 \end{bmatrix} \begin{bmatrix} 4 \\ 1 \\ 5 \end{bmatrix} = \begin{bmatrix} 21 \\ 9 \\ 13 \end{bmatrix}$.

19. $A_2 B = \begin{bmatrix} 4 & 0 & 3 \end{bmatrix} \begin{bmatrix} 8 & 1 & 3 \\ 2 & 1 & 0 \\ 4 & 6 & 3 \end{bmatrix} = \begin{bmatrix} 44 & 22 & 21 \end{bmatrix}$.

20. The multiplications are the seven multiplications needed to calculate m_1 through m_7. The additions are $a_{12} - a_{22}$, $a_{11} + a_{22}$, $a_{11} - a_{21}$, $a_{11} + a_{12}$, $a_{21} + a_{22}$, $b_{12} - b_{22}$, $b_{11} + b_{22}$, $b_{21} - b_{11}$, $b_{11} + b_{12}$, $b_{21} + b_{22}$, and the eight additions in $m_1 + m_2 + (-m_4) + m_6$, $m_4 + m_5$, $m_6 + m_7$, $m_2 + (-m_3) + m_5 + (-m_7)$.

21. (a) $m_1 = (a_{12} - a_{22})(b_{21} + b_{22}) = (-3)(4) = -12$. $m_5 = a_{11}(b_{12} - b_{22}) = (1)(-2) = -2$.

    $m_2 = (a_{11} + a_{22})(b_{11} + b_{22}) = (4)(5) = 20$. $m_6 = a_{22}(b_{21} - b_{11}) = (3)(-1) = -3$.

    $m_3 = (a_{11} - a_{21})(b_{11} + b_{12}) = (-1)(3) = -3$. $m_7 = (a_{21} + a_{22})b_{11} = (5)(1) = 5$.

    $m_4 = (a_{11} + a_{12})b_{22} = (1)(4) = 4$.

    $AB = \begin{bmatrix} -12+20-4+(-3) & 4+(-2) \\ -3+5 & 20-(-3)+(-2)-5 \end{bmatrix} = \begin{bmatrix} 1 & 2 \\ 2 & 16 \end{bmatrix}$.

    (b) $m_1 = (-2)(0) = 0$, $m_2 = (5)(-1) = -5$, $m_3 = (3)(5) = 15$, $m_4 = (3)(-1) = -3$, $m_5 = (2)(6) = 12$, $m_6 = (3)(1) = 3$, $m_7 = (2)(0) = 0$.

$$AB = \begin{bmatrix} 0+(-5)-(-3)+3 & -3+12 \\ 3+0 & -5-15+12-0 \end{bmatrix} = \begin{bmatrix} 1 & 9 \\ 3 & -8 \end{bmatrix}.$$

(c) $m_1 = (-1)(3) = -3$, $m_2 = (2)(6) = 12$, $m_3 = (-1)(6) = -6$, $m_4 = (1)(1) = 1$,
$m_5 = (2)(0) = 0$, $m_6 = (0)(-3) = 0$, $m_7 = (3)(5) = 15$.

$$AB = \begin{bmatrix} -3+12-1+0 & 1+0 \\ 0+15 & 12-(-6)+0-15 \end{bmatrix} = \begin{bmatrix} 8 & 1 \\ 15 & 3 \end{bmatrix}.$$

(d) $m_1 = (4)(-1) = -4$, $m_2 = (-1)(1) = -1$, $m_3 = (1)(3) = 3$, $m_4 = (3)(-2) = -6$,
$m_5 = (2)(2) = 4$, $m_6 = (-3)(-2) = 6$, $m_7 = (-2)(3) = -6$.

$$AB = \begin{bmatrix} -4+(-1)-(-6)+6 & -6+4 \\ 6+(-6) & -1-3+4-(-6) \end{bmatrix} = \begin{bmatrix} 7 & -2 \\ 0 & 6 \end{bmatrix}.$$

22. (a) We supplement the rows and columns to make 4x4 matrices.

$$\begin{bmatrix} 1 & -1 & 0 & 0 \\ 2 & 1 & 3 & 0 \\ 0 & 4 & 1 & 0 \\ 0 & 0 & 0 & 0 \end{bmatrix} \begin{bmatrix} 1 & 1 & 2 & 0 \\ 3 & 2 & -1 & 0 \\ 0 & 1 & 1 & 0 \\ 0 & 0 & 0 & 0 \end{bmatrix} = \begin{bmatrix} \begin{bmatrix} 1 & -1 \\ 2 & 1 \end{bmatrix} \begin{bmatrix} 0 & 0 \\ 3 & 0 \end{bmatrix} & \begin{bmatrix} 1 & 1 \\ 3 & 2 \end{bmatrix} \begin{bmatrix} 2 & 0 \\ -1 & 0 \end{bmatrix} \\ \begin{bmatrix} 0 & 4 \\ 0 & 0 \end{bmatrix} \begin{bmatrix} 1 & 0 \\ 0 & 0 \end{bmatrix} & \begin{bmatrix} 0 & 1 \\ 0 & 0 \end{bmatrix} \begin{bmatrix} 1 & 0 \\ 0 & 0 \end{bmatrix} \end{bmatrix}$$

$$= \begin{bmatrix} \begin{bmatrix} 1 & -1 \\ 2 & 1 \end{bmatrix}\begin{bmatrix} 1 & 1 \\ 3 & 2 \end{bmatrix} + \begin{bmatrix} 0 & 0 \\ 3 & 0 \end{bmatrix}\begin{bmatrix} 0 & 1 \\ 0 & 0 \end{bmatrix} & \begin{bmatrix} 1 & -1 \\ 2 & 1 \end{bmatrix}\begin{bmatrix} 2 & 0 \\ -1 & 0 \end{bmatrix} + \begin{bmatrix} 0 & 0 \\ 3 & 0 \end{bmatrix}\begin{bmatrix} 1 & 0 \\ 0 & 0 \end{bmatrix} \\ \begin{bmatrix} 0 & 4 \\ 0 & 0 \end{bmatrix}\begin{bmatrix} 1 & 1 \\ 3 & 2 \end{bmatrix} + \begin{bmatrix} 1 & 0 \\ 0 & 0 \end{bmatrix}\begin{bmatrix} 0 & 1 \\ 0 & 0 \end{bmatrix} & \begin{bmatrix} 0 & 4 \\ 0 & 0 \end{bmatrix}\begin{bmatrix} 2 & 0 \\ -1 & 0 \end{bmatrix} + \begin{bmatrix} 1 & 0 \\ 0 & 0 \end{bmatrix}\begin{bmatrix} 1 & 0 \\ 0 & 0 \end{bmatrix} \end{bmatrix}$$

$$= \begin{bmatrix} \begin{bmatrix} -2 & -1 \\ 5 & 4 \end{bmatrix} + \begin{bmatrix} 0 & 0 \\ 0 & 3 \end{bmatrix} & \begin{bmatrix} 3 & 0 \\ 3 & 0 \end{bmatrix} + \begin{bmatrix} 0 & 0 \\ 3 & 0 \end{bmatrix} \\ \begin{bmatrix} 12 & 8 \\ 0 & 0 \end{bmatrix} + \begin{bmatrix} 0 & 1 \\ 0 & 0 \end{bmatrix} & \begin{bmatrix} -4 & 0 \\ 0 & 0 \end{bmatrix} + \begin{bmatrix} 1 & 0 \\ 0 & 0 \end{bmatrix} \end{bmatrix} = \begin{bmatrix} \begin{bmatrix} -2 & -1 \\ 5 & 7 \end{bmatrix} & \begin{bmatrix} 3 & 0 \\ 6 & 0 \end{bmatrix} \\ \begin{bmatrix} 12 & 9 \\ 0 & 0 \end{bmatrix} & \begin{bmatrix} -3 & 0 \\ 0 & 0 \end{bmatrix} \end{bmatrix}$$

$$= \begin{bmatrix} -2 & -1 & 3 & 0 \\ 5 & 7 & 6 & 0 \\ 12 & 9 & -3 & 0 \\ 0 & 0 & 0 & 0 \end{bmatrix}, \text{ so } AB = \begin{bmatrix} -2 & -1 & 3 \\ 5 & 7 & 6 \\ 12 & 9 & -3 \end{bmatrix}.$$ Of course each of the multiplications of 2x2 matrices is to be executed using the method demonstrated in the solution to Exercise 21.

(b) We supplement the rows and columns to make 4x4 matrices.

$$\begin{bmatrix} 4 & 1 & 2 & 0 \\ 0 & 1 & 3 & 0 \\ 5 & -2 & 1 & 0 \\ 0 & 0 & 0 & 0 \end{bmatrix} \begin{bmatrix} -3 & 0 & 1 & 0 \\ 2 & 5 & -1 & 0 \\ 2 & -2 & 3 & 0 \\ 0 & 0 & 0 & 0 \end{bmatrix} = \begin{bmatrix} \begin{bmatrix} 4 & 1 \\ 0 & 1 \end{bmatrix} \begin{bmatrix} 2 & 0 \\ 3 & 0 \end{bmatrix} \\ \begin{bmatrix} 5 & -2 \\ 0 & 0 \end{bmatrix} \begin{bmatrix} 1 & 0 \\ 0 & 0 \end{bmatrix} \end{bmatrix} \begin{bmatrix} \begin{bmatrix} -3 & 0 \\ 2 & 5 \end{bmatrix} \begin{bmatrix} 1 & 0 \\ -1 & 0 \end{bmatrix} \\ \begin{bmatrix} 2 & -2 \\ 0 & 0 \end{bmatrix} \begin{bmatrix} 3 & 0 \\ 0 & 0 \end{bmatrix} \end{bmatrix}$$

$$= \begin{bmatrix} \begin{bmatrix} 4 & 1 \\ 0 & 1 \end{bmatrix}\begin{bmatrix} -3 & 0 \\ 2 & 5 \end{bmatrix} + \begin{bmatrix} 2 & 0 \\ 3 & 0 \end{bmatrix}\begin{bmatrix} 2 & -2 \\ 0 & 0 \end{bmatrix} & \begin{bmatrix} 4 & 1 \\ 0 & 1 \end{bmatrix}\begin{bmatrix} 1 & 0 \\ -1 & 0 \end{bmatrix} + \begin{bmatrix} 2 & 0 \\ 3 & 0 \end{bmatrix}\begin{bmatrix} 3 & 0 \\ 0 & 0 \end{bmatrix} \\ \begin{bmatrix} 5 & -2 \\ 0 & 0 \end{bmatrix}\begin{bmatrix} -3 & 0 \\ 2 & 5 \end{bmatrix} + \begin{bmatrix} 1 & 0 \\ 0 & 0 \end{bmatrix}\begin{bmatrix} 2 & -2 \\ 0 & 0 \end{bmatrix} & \begin{bmatrix} 5 & -2 \\ 0 & 0 \end{bmatrix}\begin{bmatrix} 1 & 0 \\ -1 & 0 \end{bmatrix} + \begin{bmatrix} 1 & 0 \\ 0 & 0 \end{bmatrix}\begin{bmatrix} 3 & 0 \\ 0 & 0 \end{bmatrix} \end{bmatrix}$$

$$= \begin{bmatrix} \begin{bmatrix} -10 & 5 \\ 2 & 5 \end{bmatrix} + \begin{bmatrix} 4 & -4 \\ 6 & -6 \end{bmatrix} & \begin{bmatrix} 3 & 0 \\ -1 & 0 \end{bmatrix} + \begin{bmatrix} 6 & 0 \\ 9 & 0 \end{bmatrix} \\ \begin{bmatrix} -19 & -10 \\ 0 & 0 \end{bmatrix} + \begin{bmatrix} 2 & -2 \\ 0 & 0 \end{bmatrix} & \begin{bmatrix} 4 & 0 \\ 0 & 0 \end{bmatrix} + \begin{bmatrix} 3 & 0 \\ 0 & 0 \end{bmatrix} \end{bmatrix} = \begin{bmatrix} \begin{bmatrix} -6 & 1 \\ 8 & -1 \end{bmatrix} & \begin{bmatrix} 9 & 0 \\ 8 & 0 \end{bmatrix} \\ \begin{bmatrix} -17 & -12 \\ 0 & 0 \end{bmatrix} & \begin{bmatrix} 10 & 0 \\ 0 & 0 \end{bmatrix} \end{bmatrix}$$

$$= \begin{bmatrix} -6 & 1 & 9 & 0 \\ 8 & -1 & 8 & 0 \\ -17 & -12 & 10 & 0 \\ 0 & 0 & 0 & 0 \end{bmatrix}, \text{ so } AB = \begin{bmatrix} -6 & 1 & 9 \\ 8 & -1 & 8 \\ -17 & -12 & 10 \end{bmatrix}.$$ Of course each of the multiplications of 2x2 matrices is to be executed using the method demonstrated in the solution to Exercise 21.

Exercise Set 2.2

1. (a) $AB = \begin{bmatrix} 4 & 7 & 10 \\ 0 & -5 & -4 \end{bmatrix}.$ BA does not exist.

Section 2.2

(b) $AC = \begin{bmatrix} 14 & 5 \\ -2 & -3 \end{bmatrix}$. $\qquad CA = \begin{bmatrix} -1 & 4 \\ 5 & 12 \end{bmatrix}$.

(c) $AD = \begin{bmatrix} 4 & 4 \\ -2 & 2 \end{bmatrix}$. $\qquad DA = \begin{bmatrix} 4 & 4 \\ -2 & 2 \end{bmatrix}$.

2. $A(BC) = \begin{bmatrix} 1 & 2 \\ -1 & 0 \\ 1 & 1 \end{bmatrix} \begin{bmatrix} 10 \\ 4 \end{bmatrix} = \begin{bmatrix} 18 \\ -10 \\ 14 \end{bmatrix}$. $\qquad (AB)C = \begin{bmatrix} -2 & 10 \\ -2 & -4 \\ 0 & 7 \end{bmatrix} \begin{bmatrix} 1 \\ 2 \end{bmatrix} = \begin{bmatrix} 18 \\ -10 \\ 14 \end{bmatrix}$.

3. $A(BC) = \begin{bmatrix} 1 & 2 \\ -1 & 3 \end{bmatrix} \begin{bmatrix} 8 & 10 \\ 14 & 20 \end{bmatrix} = \begin{bmatrix} 36 & 50 \\ 34 & 50 \end{bmatrix}$.

$(AB)C = \begin{bmatrix} 11 & 7 & 4 \\ 9 & 8 & 1 \end{bmatrix} \begin{bmatrix} 1 & 2 \\ 3 & 4 \\ 1 & 0 \end{bmatrix} = \begin{bmatrix} 36 & 50 \\ 34 & 50 \end{bmatrix}$.

4. (a) $AC^2 = \begin{bmatrix} 1 & 2 \\ 3 & 4 \end{bmatrix} \begin{bmatrix} 4 & 0 \\ 6 & 16 \end{bmatrix} = \begin{bmatrix} 16 & 32 \\ 36 & 64 \end{bmatrix}$.

(b) $A^2 - 3B = \begin{bmatrix} 7 & 10 \\ 15 & 22 \end{bmatrix} - \begin{bmatrix} 6 & -9 \\ 0 & 3 \end{bmatrix} = \begin{bmatrix} 1 & 19 \\ 15 & 19 \end{bmatrix}$.

(c) $BC^2 + B^3 = \begin{bmatrix} 2 & -3 \\ 0 & 1 \end{bmatrix} \begin{bmatrix} 4 & 0 \\ 6 & 16 \end{bmatrix} + \begin{bmatrix} 8 & -21 \\ 0 & 1 \end{bmatrix}$

$= \begin{bmatrix} -10 & -48 \\ 6 & 16 \end{bmatrix} + \begin{bmatrix} 8 & -21 \\ 0 & 1 \end{bmatrix} = \begin{bmatrix} -2 & -69 \\ 6 & 17 \end{bmatrix}$.

(d) $3A^2 + 2A - I_2 = \begin{bmatrix} 21 & 30 \\ 45 & 66 \end{bmatrix} + \begin{bmatrix} 2 & 4 \\ 6 & 8 \end{bmatrix} - \begin{bmatrix} 1 & 0 \\ 0 & 1 \end{bmatrix} = \begin{bmatrix} 22 & 34 \\ 51 & 73 \end{bmatrix}$.

5. (a) $(AB)^2 = \begin{bmatrix} -2 & 2 \\ 11 & 19 \end{bmatrix}^2 = \begin{bmatrix} 26 & 34 \\ 187 & 383 \end{bmatrix}$.

35

(b) $A - 3B^2 = \begin{bmatrix} 2 & 0 \\ -1 & 5 \end{bmatrix} - 3 \begin{bmatrix} 3 & 3 \\ 6 & 18 \end{bmatrix} = \begin{bmatrix} -7 & -9 \\ -19 & -49 \end{bmatrix}$

(c) $A^2B + 2C^3 = \begin{bmatrix} 4 & 0 \\ -7 & 25 \end{bmatrix} \begin{bmatrix} -1 & 1 \\ 2 & 4 \end{bmatrix} + 2 \begin{bmatrix} 27 & 76 \\ 0 & 8 \end{bmatrix}$

$= \begin{bmatrix} -4 & 4 \\ 57 & 93 \end{bmatrix} + \begin{bmatrix} 54 & 152 \\ 0 & 16 \end{bmatrix} = \begin{bmatrix} 50 & 156 \\ 57 & 109 \end{bmatrix}.$

(d) $2A^2 - 2A + 3I_2 = \begin{bmatrix} 8 & 0 \\ -14 & 50 \end{bmatrix} - \begin{bmatrix} 4 & 0 \\ -2 & 10 \end{bmatrix} + \begin{bmatrix} 3 & 0 \\ 0 & 3 \end{bmatrix} = \begin{bmatrix} 7 & 0 \\ -12 & 43 \end{bmatrix}.$

6. (a) does not exist   (b) 4x3   (c) 3x6   (d) 6x6

   (e) does not exist, since A^2 does not exist

7. (a) 3x3   (b) 3x1   (c) does not exist, since PR does not exist

   (d) does not exist, since SP does not exist   (e) 2x3

8. There are mn elements in AB. Each is computed by multiplying a row of A times a column of B, i.e., by doing r multiplications (and r-1 additions). Thus the total number of multiplications is (mn)r or mrn.

   The number of multiplications required to compute AB is mrn and the number required to compute BC is rns. (AB)C is the product of an mxn matrix and an nxs matrix, so the number of multiplications required is mns + mrn. A(BC) is the product of an mxr matrix and an rxs matrix, so the number of multiplications is mrs + rns.

9. (a) 2x3x7 = 42.   (b) 5x2x8 = 80.   (c) 1x9x27 = 243.

   (d) 8x5x12 = 480.

10. To compute A^2, $n \times n \times n = n^3$ multiplications are needed. $A^3 = AA^2$, and A and A^2 are both nxn matrices, so the number of multiplications needed is $n^3 + n^3 = 2n^3$. For each additional matrix A, an additional n^3 multiplications are required. Thus the total number of multiplications required to calculate A^m is $(m-1)n^3$.

11. (a) 2x4x3 + 2x3x1 = 30 for (AB)C.   4x3x1 + 2x4x1 = 20 for A(BC).

    (b) 3x7x5 + 3x5x2 = 135 for (AB)C.   7x5x2 + 3x7x2 = 112 for A(BC).

    (c) 6x2x5 + 6x5x3 = 150 for (AB)C.   2x5x3 + 6x2x3 = 66 for A(BC).

    (d) 3x5x47 + 3x47x5 = 1410 for (AB)C.   5x47x5 + 3x5x5 = 1250 for A(BC).

    (e) 7x97x2 + 7x2x3 = 1400 for (AB)C.   97x2x3 + 7x97x3 = 2619 for A(BC).

12. ((AB)C)D: 5x14x87 + 5x87x3 + 5x3x42 = 8025.

    (A(BC))D: 14x87x3 + 5x14x3 + 5x3x42 = 4494.

    (AB)(CD):   5x14x87 + 87x3x42 + 5x87x42 = 35322.

    A((BC)D):   14x87x3 + 14x3x42 + 5x14x42 = 8358.

    A(B(CD)):   87x3x42 + 14x87x42 + 5x14x42 = 65058.

13. Each of the mn elements of AB is computed by multiplying a row of A times a column of B. This requires adding r terms, i.e., doing r-1 additions. Thus the total number of additions required to compute AB is mn(r-1).

To calculate (AB)C requires mn(r-1) additions for AB and ms(n-1) additions for the second product. Thus the total is mn(r-1) + ms(n-1). To calculate A(BC) requires rs(n-1) additions for BC and ms(r-1) for the second product. Thus the total is rs(n-1) + ms(r-1).

If A, B, and C are 2x2, 2x3, and 3x1, respectively, the number of additions is 2x3x1 + 2x1x2 = 10 for (AB)C and 2x1x2 + 2x1x1 = 6 for A(BC).

14. AB can exist only if the number of rows in B is n, and BA can exist only if the number of columns in B is m. So for both AB and BA to exist, B must be nxm.

Section 2.2

15. (a) The (i,j)th element of A + (B + C) is $a_{ij} + (b_{ij} + c_{ij})$. The (i,j)th element of (A + B) + C is $(a_{ij} + b_{ij}) + c_{ij} = a_{ij} + (b_{ij} + c_{ij})$. Since their elements are the same, A + (B + C) = (A + B) + C.

(b) The (i,j)th element of c(A + B) is $c(a_{ij} + b_{ij})$. The (i,j)th element of cA + cB is $ca_{ij} + cb_{ij} = c(a_{ij} + b_{ij})$. Since their elements are the same, c(A + B) = cA + cB.

(c) The only nonzero element in the jth column of I_n is the 1 in the jth row. Thus the (i,j)th element of AI_n is $a_{ij}(1) = a_{ij}$. Likewise, the only nonzero element in the ith row of I_n is the 1 in column i. Thus the (i,j)th element in I_nA is $(1)a_{ij} = a_{ij}$. Since their elements are the same, $AI_n = I_nA = A$.

16. The only nonzero element in column j of I_n is the 1 in row j. Thus the (i,j)th element of AI_n is $a_{ij}(1) = a_{ij}$. Since their elements are the same, $AI_n = A$.

17. The (i,j)th element of cA is ca_{ij}. If $cA = O_{mn}$, then $ca_{ij} = 0$ for all i and j. So either c = 0 or all $a_{ij} = 0$, in which case $A = O_{mn}$.

18. (a) $A(A - 4B) + 2B(A + B) - A^2 + 7B^2 + 3AB = A^2 - 4AB + 2BA + 2B^2 - A^2 + 7B^2 + 3AB$
$= 9B^2 - AB + 2BA$.

(b) $B(2I_n - BA) + B(4I_n + 5A)B - 3BAB + 7B^2A = 2B - B^2A + 4B + 5BAB - 3BAB + 7B^2A$
$= 6B + 2BAB + 6B^2A$.

(c) $(A - B)(A + B) - (A + B)^2 = A^2 - BA + AB - B^2 - (A^2 + AB + BA + B^2)$
$= -2BA - 2B^2$.

19. (a) $A(A + B) - B(A + B) = A^2 + AB - BA - B^2$.

(b) $A(A - B)B + B2AB - 3A^2 = A^2B - AB^2 + 2BAB - 3A^2$.

38

Section 2.2

(c) $(A+B)^3 - 2A^3 - 3ABA - A3B^2 - B^3$
$= A^3 + A^2B + BA^2 + BAB + ABA + AB^2 + B^2A + B^3 - 2A^3 - 3ABA - 3AB^2 - B^3$
$= -A^3 + A^2B + BA^2 + BAB - 2ABA - 2AB^2 + B^2A.$

20. (a) $\begin{bmatrix} 1 & 0 \\ -1 & 0 \end{bmatrix}\begin{bmatrix} a & b \\ c & d \end{bmatrix} = \begin{bmatrix} a & b \\ -a & -b \end{bmatrix}$ and $\begin{bmatrix} a & b \\ c & d \end{bmatrix}\begin{bmatrix} 1 & 0 \\ -1 & 0 \end{bmatrix} = \begin{bmatrix} a-b & 0 \\ c-d & 0 \end{bmatrix}.$

For equality, it is necessary to have $a = a-b$, $b = 0$, and $-a = c-d$. Thus those matrices that commute with the given matrix are all matrices of the form

$$\begin{bmatrix} a & 0 \\ c & c+a \end{bmatrix}$$

where a and c can take any real values.

(b) $\begin{bmatrix} 1 & 0 \\ 0 & 2 \end{bmatrix}\begin{bmatrix} a & b \\ c & d \end{bmatrix} = \begin{bmatrix} a & b \\ 2c & 2d \end{bmatrix}$ and $\begin{bmatrix} a & b \\ c & d \end{bmatrix}\begin{bmatrix} 1 & 0 \\ 0 & 2 \end{bmatrix} = \begin{bmatrix} a & 2b \\ c & 2d \end{bmatrix}.$

For equality, it is necessary to have $b = 2b$ and $c = 2c$, so that $b = c = 0$. Thus those matrices that commute with the given matrix are all matrices of the form

$$\begin{bmatrix} a & 0 \\ 0 & d \end{bmatrix}$$

where a and d can take any real values.

(c) $\begin{bmatrix} 0 & 1 & 0 \\ 0 & 0 & 1 \\ 0 & 0 & 0 \end{bmatrix}\begin{bmatrix} a & b & c \\ d & e & f \\ g & h & i \end{bmatrix} = \begin{bmatrix} d & e & f \\ g & h & i \\ 0 & 0 & 0 \end{bmatrix}$ and $\begin{bmatrix} a & b & c \\ d & e & f \\ g & h & i \end{bmatrix}\begin{bmatrix} 0 & 1 & 0 \\ 0 & 0 & 1 \\ 0 & 0 & 0 \end{bmatrix} = \begin{bmatrix} 0 & a & b \\ 0 & d & e \\ 0 & g & h \end{bmatrix}.$

For equality, it is necessary to have $d = g = h = 0$, $a = e = i$, and $b = f$. Thus those matrices that commute with the given matrix are all matrices of the form

$$\begin{bmatrix} a & b & c \\ 0 & a & b \\ 0 & 0 & a \end{bmatrix}$$

where a, b, and c can take any real values.

39

Section 2.2

21. $AX_1 = AX_2$ does not imply that $X_1 = X_2$.

22. (a) $A^2 = AA$. Both matrices are nxn, so the number of columns in the first is the same as the number of rows in the second and they can be multiplied. The product matrix will have n rows because the first matrix has n rows, and it will have n columns because the second matrix has n columns.

    (b) $A^2 = AA$. The first matrix is mxn and the second matrix is mxn, so the number of columns in the first matrix is not equal to the number of rows in the second matrix. Therefore they cannot be multiplied.

23. $(A + B)^2 = (A + B)(A + B) = A(A + B) + B(A + B)$ (distributive law)
    $= A^2 + AB + BA + B^2$.

    If $AB = BA$ then $(A + B)^2 = A^2 + 2AB + B^2$.

24. $(AB)^2 = (AB)(AB) = ABAB$. If $AB = BA$ then $ABAB = A(BA)B = A(AB)B = AABB = A^2B^2$.

    Example: Let $A = \begin{bmatrix} 1 & -1 \\ 0 & 2 \end{bmatrix}$, $B = \begin{bmatrix} 1 & 1 \\ 1 & 0 \end{bmatrix}$. $AB = \begin{bmatrix} 0 & 1 \\ 2 & 0 \end{bmatrix}$ and $(AB)^2 = \begin{bmatrix} 2 & 0 \\ 0 & 2 \end{bmatrix}$.

    $A^2 = \begin{bmatrix} 1 & -3 \\ 0 & 4 \end{bmatrix}$, $B^2 = \begin{bmatrix} 2 & 1 \\ 1 & 1 \end{bmatrix}$, and $A^2B^2 = \begin{bmatrix} -1 & -2 \\ 4 & 4 \end{bmatrix} \neq (AB)^2$.

25. $(AB)^n = ABABAB \ldots ABAB = AABABA \ldots BABB = AAABAB \ldots ABBB$
    $= AAAABA \ldots BBBB = \ldots = AAA \ldots AABBB \ldots BB = A^nB^n$.

    The example from Exercise 24 will serve here also.

26. $A^rA^s = A^{r+s} = A^{s+r} = A^sA^r$.

27. (a) The (i,j)th element of $A + B$ is $a_{ij} + b_{ij} = 0 + 0$ if $i \neq j$, so $A + B$ is a diagonal matrix.

Section 2.2

(b) The (i,j)th element of cA is $ca_{ij} = 0$ if $i \neq j$, so cA is a diagonal matrix.

(c) The (i,j)th element of AB is $a_{i1}b_{1j} + a_{i2}b_{2j} + \ldots + a_{in}b_{nj}$, which is zero unless $i = j$, since only a_{ii} and b_{jj} can be nonzero. Thus only the diagonal elements of AB can be nonzero, so AB is a diagonal matrix.

28. The nondiagonal elements of AB and of BA are zero. (See Exercise 27(c).) The diagonal elements of AB are the elements $a_{i1}b_{1i} + a_{i2}b_{2i} + \ldots + a_{in}b_{ni} = a_{ii}b_{ii}$ and the diagonal elements of BA are the elements $b_{i1}a_{1i} + b_{i2}a_{2i} + \ldots + b_{in}a_{ni} = b_{ii}a_{ii} = a_{ii}b_{ii}$. Thus AB = BA.

29. Call the diagonal matrix B. The (i,j)th element of AB is $a_{i1}b_{1j} + a_{i2}b_{2j} + \ldots + a_{in}b_{nj} = a_{ij}b_{jj}$ and the (i,j)th element of BA is $b_{i1}a_{1j} + b_{i2}a_{2j} + \ldots + b_{in}a_{nj} = b_{ii}a_{ij}$. If $b_{ii} \neq b_{jj}$, then a_{ij} must be zero for $i \neq j$. Thus A is a diagonal matrix.

30. (a) Yes, $\begin{bmatrix} 1 & 0 \\ 0 & 1 \end{bmatrix}^2 = \begin{bmatrix} 1 & 0 \\ 0 & 1 \end{bmatrix}$. (b) Yes, $\begin{bmatrix} 1 & 0 \\ 0 & 0 \end{bmatrix}^2 = \begin{bmatrix} 1 & 0 \\ 0 & 0 \end{bmatrix}$.

(c) No, $\begin{bmatrix} 0 & 1 \\ 1 & 0 \end{bmatrix}^2 = \begin{bmatrix} 1 & 0 \\ 0 & 1 \end{bmatrix}$. (d) Yes, $\begin{bmatrix} 3 & -6 \\ 1 & -2 \end{bmatrix}^2 = \begin{bmatrix} 3 & -6 \\ 1 & -2 \end{bmatrix}$.

(e) Yes, $\begin{bmatrix} 1 & 2 & 2 \\ 0 & 0 & -1 \\ 0 & 0 & 1 \end{bmatrix}^2 = \begin{bmatrix} 1 & 2 & 2 \\ 0 & 0 & -1 \\ 0 & 0 & 1 \end{bmatrix}$ (f) No, $\begin{bmatrix} 1 & 3 & 0 \\ 0 & 0 & 1 \\ 0 & 0 & 0 \end{bmatrix}^2 = \begin{bmatrix} 1 & 3 & 3 \\ 0 & 0 & 0 \\ 0 & 0 & 0 \end{bmatrix}$.

31. $\begin{bmatrix} 1 & b \\ c & d \end{bmatrix}^2 = \begin{bmatrix} 1+bc & b+bd \\ c+cd & bc+d^2 \end{bmatrix}$, so the matrix will be idempotent if

$$1 = 1 + bc, \quad b = b + bd, \quad c = c + cd, \quad \text{and} \quad d = bc + d^2.$$

The first equation implies that $bc = 0$ and substituting $bc = 0$ in the last equation gives $d = d^2$. Thus d must be either zero or 1. If one of b and c is nonzero, the second and third equations give $d = 0$. If both b and c are zero, then d can be either zero or 1. Thus,

41

the idempotent matrices of the given form are

$$\begin{bmatrix} 1 & 0 \\ 0 & 0 \end{bmatrix}, \begin{bmatrix} 1 & 0 \\ 0 & 1 \end{bmatrix}, \begin{bmatrix} 1 & b \\ 0 & 0 \end{bmatrix}, \text{ and } \begin{bmatrix} 1 & 0 \\ c & 0 \end{bmatrix}.$$

32. $\begin{bmatrix} a & 0 \\ c & d \end{bmatrix}^2 = \begin{bmatrix} a^2 & 0 \\ ac+cd & d^2 \end{bmatrix}$, so the matrix will be idempotent if

$$a = a^2, \ d = d^2, \text{ and } c = ac + cd = c(a + d).$$

Thus a is zero or 1, d is zero or 1, and if c is nonzero, $a + d = 1$. If $c = 0$, the sum of a and d is unimportant. The idempotent matrices of the given form are

$$\begin{bmatrix} 0 & 0 \\ 0 & 0 \end{bmatrix}, \begin{bmatrix} 1 & 0 \\ 0 & 0 \end{bmatrix}, \begin{bmatrix} 1 & 0 \\ 0 & 1 \end{bmatrix}, \begin{bmatrix} 0 & 0 \\ 0 & 1 \end{bmatrix}, \begin{bmatrix} 1 & 0 \\ c & 0 \end{bmatrix}, \text{ and } \begin{bmatrix} 0 & 0 \\ c & 1 \end{bmatrix}.$$

33. $(AB)^2 = (AB)(AB) = A(BA)B$. If $AB = BA$, then $(AB)^2 = A(AB)B = A^2B^2 = AB$. Thus AB is idempotent.

34. $A^2 = A$, $A^3 = A^2A = AA = A$, $A^4 = A^3A = AA = A$, ..., and $A^n = A^{n-1}A = AA = A$.

35. (a) $\begin{bmatrix} 1 & 1 \\ -1 & -1 \end{bmatrix}^2 = \begin{bmatrix} 0 & 0 \\ 0 & 0 \end{bmatrix}.$ (b) $\begin{bmatrix} -4 & 8 \\ -2 & 4 \end{bmatrix}^2 = \begin{bmatrix} 0 & 0 \\ 0 & 0 \end{bmatrix}.$

(c) $\begin{bmatrix} 3 & -9 \\ 1 & -3 \end{bmatrix}^2 = \begin{bmatrix} 0 & 0 \\ 0 & 0 \end{bmatrix}.$

36. $\begin{bmatrix} 0 & 1 & 0 \\ 0 & 0 & 1 \\ 0 & 0 & 0 \end{bmatrix}^3 = \begin{bmatrix} 0 & 1 & 0 \\ 0 & 0 & 1 \\ 0 & 0 & 0 \end{bmatrix}^2 \begin{bmatrix} 0 & 1 & 0 \\ 0 & 0 & 1 \\ 0 & 0 & 0 \end{bmatrix} = \begin{bmatrix} 0 & 0 & 1 \\ 0 & 0 & 0 \\ 0 & 0 & 0 \end{bmatrix} \begin{bmatrix} 0 & 1 & 0 \\ 0 & 0 & 1 \\ 0 & 0 & 0 \end{bmatrix} = \begin{bmatrix} 0 & 0 & 0 \\ 0 & 0 & 0 \\ 0 & 0 & 0 \end{bmatrix}.$

Section 2.3

Exercise Set 2.3

1. (a) $A^t = \begin{bmatrix} -1 & 2 \\ 2 & -3 \end{bmatrix}$. symmetric (b) $B^t = \begin{bmatrix} 1 & 0 \\ 2 & 3 \end{bmatrix}$. not symmetric

   (c) $C^t = \begin{bmatrix} 3 & 2 \\ -1 & 4 \end{bmatrix}$. not symmetric (d) $D^t = \begin{bmatrix} 4 & -2 & 7 \\ 5 & 3 & 0 \end{bmatrix}$. not symmetric

   (e) $E^t = \begin{bmatrix} 4 & -1 & 0 \\ 5 & 2 & 1 \\ 6 & 3 & 2 \end{bmatrix}$. not symmetric (f) $F^t = \begin{bmatrix} 1 & -1 & 3 \\ -1 & 2 & 0 \\ 3 & 0 & 4 \end{bmatrix}$. symmetric

   (g) $G^t = \begin{bmatrix} -2 & 1 \\ 4 & 0 \\ 5 & 3 \\ 7 & -7 \end{bmatrix}$. not symmetric (h) $H^t = \begin{bmatrix} 1 & 4 & -2 \\ -2 & 5 & 6 \\ 3 & 6 & 7 \end{bmatrix}$. not symmetric

   (i) $K^t = \begin{bmatrix} 7 & 0 & 0 \\ 0 & -3 & 0 \\ 0 & 0 & 9 \end{bmatrix}$. symmetric

2. (a) $\begin{bmatrix} 1 & 2 & 4 \\ 2 & 6 & 5 \\ 4 & 5 & 2 \end{bmatrix}$ (b) $\begin{bmatrix} 3 & 5 & -3 \\ 5 & 8 & 4 \\ -3 & 4 & 3 \end{bmatrix}$ (c) $\begin{bmatrix} -3 & -4 & 8 & 9 \\ -4 & 7 & 2 & 7 \\ 8 & 2 & 6 & 4 \\ 9 & 7 & 4 & 9 \end{bmatrix}$

3. (a) 4x2 (b) 4x3 (c) Does not exist. (d) 4x4 (e) Does not exist.

4. (a) The (i,j)th element of $(A + B)^t$ is the (j,i)th element of A + B, which is $a_{ji} + b_{ji}$.

   The (i,j)th element of A^t is the (j,i)th element of A, which is a_{ji}.

   The (i,j)th element of B^t is the (j,i)th element of B, which is b_{ji}.

   So the (i,j)th element of $A^t + B^t$ is $a_{ji} + b_{ji}$. Hence $(A + B)^t = A^t + B^t$.

   (b) The (i,j)th element of $(cA)^t$ is the (j,i)th element of cA, which is ca_{ji}.

   The (i,j)th element of cA^t is c times the (j,i)th element of A, i.e., ca_{ji}. Hence $(cA)^t = cA^t$.

43

Section 2.3

    (c)   The (i,j)th element of $(A^t)^t$ is the (j,i)th element of A^t, which is the (i,j)th element of A. Thus $(A^t)^t = A$.

5.   (a) $(A + B + C)^t = (A + (B + C))^t = A^t + (B + C)^t = A^t + B^t + C^t$.

    (b) $(ABC)^t = (A(BC))^t = (BC)^t A^t = C^t B^t A^t$.

6.   The (i,j)th element of A^t is the (j,i)th element of A. However, since A is a diagonal matrix, $a_{ji} = a_{ij} = 0$ if $i \ne j$. If $i = j$ then $a_{ij} = a_{ii} = a_{ji}$. Hence $A = A^t$.

7. $(A^n)^t = (\underbrace{AA \ldots A}_{n \text{ terms}})^t = \underbrace{A^t A^t \ldots A^t}_{n \text{ terms}} = (A^t)^n$.

8.   If A is symmetric, then $A = A^t$, so the number of rows in A equals the number of rows in A^t equals the number of columns in A. Thus A is square.

9.   If A is symmetric, then $A = A^t$. From Theorem 2.4 and Exercise 4(c), this means $A^t = A = (A^t)^t$. So A^t is symmetric.

10.   (a) $A = A^t$ and $B = B^t$. Thus $A + B = A^t + B^t = (A + B)^t$ from Theorem 2.4. Thus $A + B$ is symmetric.

    (b) $A = A^t$, so $cA = cA^t = (cA)^t$ from Theorem 2.4. So cA is symmetric.

11.   (a) $\begin{bmatrix} 0 & -1 \\ 1 & 0 \end{bmatrix}$

    (b)   If A is antisymmetric, the number of rows in A equals the number of rows in A^t, which is the number of columns in A. Thus A is square.
If $A = -A^t$, then for diagonal elements $a_{ii} = -a_{ii}$ so that $a_{ii} = 0$.

    (c)   If A and B are antisymmetric, then $A + B = (-A^t) + (-B^t) = -(A^t + B^t) = -(A + B)^t$, so $A + B$ is antisymmetric.

    (d)   If A is antisymmetric, then $A = -A^t$ and

Section 2.3

$cA = c(-A^t) = c((-1)A^t) = c(-1)A^t = -cA^t = -(cA)^t$, so cA is antisymmetric.

12. (a) $(A + A^t)^t = A^t + (A^t)^t = A^t + A = A + A^t$, so $A + A^t$ is symmetric.

    (b) $(A - A^t)^t = A^t - (A^t)^t = A^t - A = -(A - A^t)$, so $A - A^t$ is antisymmetric.

13. $B = (1/2)(A + A^t)$ is symmetric and $C = (1/2)(A - A^t)$ is antisymmetric. $A = B + C$.

14. (a) If A is idempotent, then $A^2 = A$. $(A^t)^2 = (A^2)^t = A^t$, so A^t is idempotent.

    (b) If A^t is idempotent, then $(A^t)^2 = A^t$. Therefore $(A^2)^t = (A^t)^2 = A^t$, so that $A^2 = ((A^2)^t)^t = (A^t)^t = A$, and A is idempotent.

15. (a) $2 + (-4) = -2$. (b) $5 + (-3) + 8 = 10$. (c) $0 + 5 - 7 + 1 = -1$.

16. (a) $\text{tr}(cA) = ca_{11} + ca_{22} + \ldots + ca_{nn} = c(a_{11} + a_{22} + \ldots + a_{nn}) = c\,\text{tr}(A)$.

    (b) The diagonal elements of AB are

    $a_{11}b_{11} + a_{12}b_{21} + \ldots + a_{1n}b_{n1} \quad = \quad b_{11}a_{11} + b_{21}a_{12} + \ldots + b_{n1}a_{1n},$

    $a_{21}b_{12} + a_{22}b_{22} + \ldots + a_{2n}b_{n2} \quad = \quad b_{12}a_{21} + a_{22}b_{22} + \ldots + b_{n2}a_{2n},$

    $\vdots \qquad \qquad \qquad \qquad \qquad \vdots$

    $a_{m1}b_{1m} + a_{m2}b_{2m} + \ldots + a_{mn}b_{nm} \quad = \quad b_{1m}a_{m1} + b_{2m}a_{m2} + \ldots + b_{nm}a_{mn}.$

    The diagonal elements of BA are

    $b_{11}a_{11} + b_{12}a_{21} + \ldots + b_{1m}a_{m1},$

    $b_{21}a_{12} + b_{22}a_{22} + \ldots + b_{2m}a_{m2},$

    $\vdots$

    $b_{n1}a_{1n} + b_{n2}a_{2n} + \ldots + b_{nm}a_{mn}.$

    Notice that the terms that appear in the diagonal elements of AB are the same terms that appear in the diagonal elements of BA, although not necessarily in the same elements. Thus the sum of the diagonal elements of AB will be the same as the sum of the diagonal elements of BA. So $\text{tr}(AB) = \text{tr}(BA)$.

Section 2.3

(c) $tr(A) = a_{11} + a_{22} + \ldots + a_{nn} = tr(A^t)$.

17. $tr(A + B + C) = tr(A + (B + C)) = tr(A) + tr(B + C) = tr(A) + tr(B) + tr(C)$.

18. If $A^t = B^t$, the (i,j)th element of A^t equals the (i,j)th element of B^t. The (i,j)th element of A^t is the (j,i)th element of A and the (i,j)th element of B^t is the (j,i)th element of B, so the (j,i)th element of A equals the (j,i)th element of B and A = B. To show that if A = B then $A^t = B^t$, simply interchange A and A^t and B and B^t everywhere in the above argument.

19. If the number of columns in A is equal to the number of rows in B, then from the definition of matrix multiplication AB exists and its (i,j)th element is obtained by multiplying the corresponding elements of row i of A and column j of B. If AB exists, then from the definition of matrix multiplication the number of columns in A equals the number of rows in B.

20. If $A = O_n$, then for any nxn matrix B the (i,j)th element of AB is $0 \times b_{1j} + 0 \times b_{2j} + \ldots + 0 \times b_{nj} = 0$, so $AB = O_n$. Suppose the (i,j)th element of A is not zero. Let B be the matrix with all elements zero except the (j,k)th element. Then the (i,k)th element of AB is $a_{i1} \times 0 + \ldots + a_{ij-1} \times 0 + a_{ij} \times b_{jk} + a_{ij+1} \times 0 + \ldots + a_{in} \times 0 = a_{ij} \times b_{jk} \neq 0$, so $AB \neq O_n$. Thus the only matrix for which $AB = O_n$ for all nxn matrices B is $A = O_n$.

21. $A + B = \begin{bmatrix} 3+i & 8+i \\ 5+2i & 4-2i \end{bmatrix}$.

$AB = \begin{bmatrix} 5(-2+i)+(3-i)(3-i) & 5(5+2i)+(3-i)(4+3i) \\ (2+3i)(-2+i)+(-5i)(3-i) & (2+3i)(5+2i)+(-5i)(4+3i) \end{bmatrix} = \begin{bmatrix} -2-i & 40+15i \\ -12-19i & 19-i \end{bmatrix}$.

$BA = \begin{bmatrix} (-2+i)5+(5+2i)(2+3i) & (-2+i)(3-i)+(5+2i)(-5i) \\ (3-i)5+(4+3i)(2+3i) & (3-i)(3-i)+(4+3i)(-5i) \end{bmatrix} = \begin{bmatrix} -6+24i & 5-20i \\ 14+13i & 23-26i \end{bmatrix}$.

22. $A + B = \begin{bmatrix} 6+2i & -1-3i \\ 8+2i & 5-6i \end{bmatrix}$.

$AB = \begin{bmatrix} 11 & -19-25i \\ 12+8i & -19-15i \end{bmatrix}$. $BA = \begin{bmatrix} -11 & 4-i \\ 42-20i & 3-15i \end{bmatrix}$.

Section 2.3

23. $\bar{A} = \begin{bmatrix} 2+3i & -5i \\ 2 & 5+4i \end{bmatrix}$ $A^* = \bar{A}^t = \begin{bmatrix} 2+3i & 2 \\ -5i & 5+4i \end{bmatrix}$

   $\bar{B} = \begin{bmatrix} 4 & 5+i \\ 5-i & 6 \end{bmatrix}$ $B^* = \bar{B}^t = \begin{bmatrix} 4 & 5-i \\ 5+i & 6 \end{bmatrix}$ hermitian

   $\bar{C} = \begin{bmatrix} -7i & 4+3i \\ 6-8i & -9 \end{bmatrix}$ $C^* = \bar{C}^t = \begin{bmatrix} -7i & 6-8i \\ 4+3i & -9 \end{bmatrix}$

   $\bar{D} = \begin{bmatrix} -2 & 3+5i \\ 3-5i & 9 \end{bmatrix}$ $D^* = \bar{D}^t = \begin{bmatrix} -2 & 3-5i \\ 3+5i & 9 \end{bmatrix}$ hermitian

24. $\bar{A} = \begin{bmatrix} 3 & 7-2i \\ 7+2i & 5 \end{bmatrix}$ $A^* = \bar{A}^t = \begin{bmatrix} 3 & 7+2i \\ 7-2i & 5 \end{bmatrix}$ hermitian

   $\bar{B} = \begin{bmatrix} 3-5i & 1+2i \\ 1-2i & 5-6i \end{bmatrix}$ $B^* = \bar{B}^t = \begin{bmatrix} 3-5i & 1-2i \\ 1+2i & 5-6i \end{bmatrix}$

   $\bar{C} = \begin{bmatrix} 1 & 2 \\ 2 & 4 \end{bmatrix}$ $C^* = \bar{C}^t = \begin{bmatrix} 1 & 2 \\ 2 & 4 \end{bmatrix}$ hermitian

   $\bar{D} = \begin{bmatrix} 9 & 3i \\ -3i & 8 \end{bmatrix}$ $D^* = \bar{D}^t = \begin{bmatrix} 9 & -3i \\ 3i & 8 \end{bmatrix}$ hermitian

25. (a) The (i,j)th element of $A^* + B^*$ is $\overline{a_{ji}} + \overline{b_{ji}} = \overline{a_{ji} + b_{ji}}$, the (i,j)th element of $(A+B)^*$.

   (b) The (i,j)th element of $(zA)^*$ is $\overline{za_{ji}} = \bar{z}\,\overline{a_{ji}}$, the (i,j)th element of $\bar{z}A^*$.

   (c) The (i,j)th element of $(AB)^*$ is the (j,i)th element of $\overline{AB}$:

   $$\overline{a_{j1}}\,\overline{b_{1i}} + \overline{a_{j2}}\,\overline{b_{2i}} + \ldots + \overline{a_{jn}}\,\overline{b_{ni}} = [\overline{a_{j1}}\ \overline{a_{j2}}\ \ldots\ \overline{a_{jn}}]\begin{bmatrix} \overline{b_{1i}} \\ \overline{b_{2i}} \\ \vdots \\ \overline{b_{ni}} \end{bmatrix}$$

Section 2.3

Row i of B^* is $[\overline{b_{1i}} \ \overline{b_{2i}} \ \ldots \ \overline{b_{ni}}]$, and column j of A^* is $\begin{bmatrix} \overline{a_{j1}} \\ \overline{a_{j2}} \\ \vdots \\ \overline{a_{jn}} \end{bmatrix}$.

Thus the (i,j)th element of B^*A^* is also $\overline{a_{j1}}\,\overline{b_{1i}} + \overline{a_{j2}}\,\overline{b_{2i}} + \ldots + \overline{a_{jn}}\,\overline{b_{ni}}$.

(d) The (i,j)th element of $(A^*)^*$ is the conjugate of the (j,i)th element of A^*, which is the conjugate of the (i,j)th element of A, so the (i,j)th element of $(A^*)^*$ is the conjugate of the conjugate of the (i,j)th element of A. That is, it is the (i,j)th element of A.

26. If $C = C^*$, then both have the same number of rows, but the number of rows in C^* is the same as the number of columns in C. Thus C is square.

If $C = C^*$, then for all the diagonal elements $a_{ii} = \overline{a_{ii}}$. Thus all the diagonal elements are real.

27. (a) $G^t = (AA^t)^t = (A^t)^t A^t = AA^t = G$, and $P^t = (A^tA)^t = A^t(A^t)^t = A^tA = P$.

(b) The number of types of pottery common to graves i and j is the number of types of pottery common to graves j and i, so the (i,j)th element of G equals the (j,i)th element of G which is the (i,j)th element of G^t, and so $G = G^t$. Likewise, the number of graves in which pottery types i and j both appear is the same as the number of graves in which pottery types j and i both appear, so the (i,j)th element of P equals the (j,i)th element of P, which is (i,j)th element of P^t, and so $P = P^t$.

28. (a) $G = \begin{bmatrix} 1 & 0 & 1 \\ 0 & 1 & 1 \\ 1 & 1 & 2 \end{bmatrix}$. $P = \begin{bmatrix} 2 & 1 \\ 1 & 2 \end{bmatrix}$.

$g_{12} = 0$, $g_{13} = 1$, $g_{23} = 1$,
so $1 \to 3 \to 2$ or $2 \to 3 \to 1$.

$p_{12} = 1$, so $1 \to 2$ or $2 \to 1$,
gives no information.

48

(b) $G = \begin{bmatrix} 1 & 0 & 1 & 0 \\ 0 & 2 & 1 & 1 \\ 1 & 1 & 2 & 0 \\ 0 & 1 & 0 & 1 \end{bmatrix}.$ $\qquad P = \begin{bmatrix} 2 & 1 & 1 \\ 1 & 2 & 0 \\ 1 & 0 & 2 \end{bmatrix}.$

$g_{12} = 0, g_{13} = 1, g_{14} = 0,$ $\qquad p_{12} = 1, p_{13} = 1, p_{23} = 0,$
$g_{23} = 1, g_{24} = 1, g_{34} = 0,$ $\qquad$ so $2 \to 1 \to 3$ or $3 \to 1 \to 2$.
so $1 \to 3 \to 2 \to 4$ or $4 \to 2 \to 3 \to 1$.

(c) $G = \begin{bmatrix} 2 & 1 & 2 \\ 1 & 3 & 3 \\ 2 & 3 & 4 \end{bmatrix}.$ $\qquad P = \begin{bmatrix} 2 & 1 & 2 & 1 \\ 1 & 2 & 2 & 2 \\ 2 & 2 & 3 & 2 \\ 1 & 2 & 2 & 2 \end{bmatrix}.$

$g_{12} = 1, g_{13} = 2, g_{23} = 3,$ $\qquad p_{12} = 1, p_{13} = 2, p_{14} = 1,$
so $1 \to 3 \to 2$ or $2 \to 3 \to 1$. $\qquad p_{23} = 2, p_{24} = 2, p_{34} = 2,$
$\qquad\qquad\qquad\qquad\qquad\qquad$ so $1 \to 3 \to \{2,4\}$ or $\{2,4\} \to 3 \to 1$.

(d) $G = \begin{bmatrix} 1 & 0 & 1 & 0 \\ 0 & 2 & 0 & 1 \\ 1 & 0 & 2 & 1 \\ 0 & 1 & 1 & 2 \end{bmatrix}.$ $\qquad P = \begin{bmatrix} 2 & 1 & 1 & 0 \\ 1 & 1 & 0 & 0 \\ 1 & 0 & 2 & 1 \\ 0 & 0 & 1 & 2 \end{bmatrix}.$

$g_{12} = 0, g_{13} = 1, g_{14} = 0,$ $\qquad p_{12} = 1, p_{13} = 1, p_{14} = 0,$
$g_{23} = 0, g_{24} = 1, g_{34} = 1,$ $\qquad p_{23} = 0, p_{24} = 0, p_{34} = 1,$
so $1 \to 3 \to 4 \to 2$ or $2 \to 4 \to 3 \to 1$. $\qquad$ so $2 \to 1 \to 3 \to 4$ or $4 \to 3 \to 1 \to 2$.

(e) $G = \begin{bmatrix} 2 & 1 & 0 & 1 \\ 1 & 1 & 0 & 0 \\ 0 & 0 & 2 & 1 \\ 1 & 0 & 1 & 2 \end{bmatrix}.$ $\qquad P = \begin{bmatrix} 2 & 0 & 1 & 0 \\ 0 & 2 & 1 & 1 \\ 1 & 1 & 2 & 0 \\ 0 & 1 & 0 & 1 \end{bmatrix}.$

$g_{12} = 1, g_{13} = 0, g_{14} = 1,$ $\qquad p_{12} = 0, p_{13} = 1, p_{14} = 0,$
$g_{23} = 0, g_{24} = 0, g_{34} = 1,$ $\qquad p_{23} = 1, p_{24} = 1, p_{34} = 0,$
so $2 \to 1 \to 4 \to 3$ or $3 \to 4 \to 1 \to 2$. $\qquad$ so $1 \to 3 \to 2 \to 4$ or $4 \to 2 \to 3 \to 1$.

(f) $G = \begin{bmatrix} 1 & 0 & 0 & 1 & 1 \\ 0 & 3 & 1 & 0 & 2 \\ 0 & 1 & 1 & 0 & 0 \\ 1 & 0 & 0 & 1 & 1 \\ 1 & 2 & 0 & 1 & 3 \end{bmatrix}$. $P = \begin{bmatrix} 2 & 1 & 1 & 2 \\ 1 & 3 & 0 & 1 \\ 1 & 0 & 2 & 1 \\ 2 & 1 & 1 & 2 \end{bmatrix}$.

$g_{12} = 0$, $g_{13} = 0$, $g_{14} = 1$, $g_{15} = 1$,
$g_{23} = 1$, $g_{24} = 0$, $g_{25} = 2$,
$g_{34} = 0$, $g_{35} = 0$, $g_{45} = 1$,
so 3→2→5→{1´4}
or {1´4}→5→2→3.

$p_{12} = 1$, $p_{13} = 1$, $p_{14} = 2$,
$p_{23} = 0$, $p_{24} = 1$, $p_{34} = 1$,
so 2→{1´4}→3 or 3→{1´4}→2.

29. $p_{ij} = a_{1i}a_{1j} + a_{2i}a_{2j} + \ldots + a_{mi}a_{mj}$, where m is the number of graves. $a_{ki}a_{kj}$ is 1 if grave k contains both the ith and jth types of pottery and is zero otherwise. So the sum of these terms is the number of graves that contain both types of pottery.

30. (a) $f_{ij} = a_{i1}a_{j1} + a_{i2}a_{j2} + \ldots + a_{in}a_{jn}$ as in the graves model. $a_{ik}a_{jk}$ is 1 if both person i and person j are friends of person k and is zero otherwise. So the sum of these terms is the number of friends person i and person j have in common.

(b) The matrix A is symmetric when friendships are mutual and is not symmetric when friendships are not mutual. Define a_{ij} to be 1 if person i considers person j to be his friend and zero otherwise. f_{ij} will then be the number of people both person i and person j consider to be their friends.

Exercise Set 2.4

1. (a) $AB = BA = I_2$, so B is the inverse of A.

   (b) $AB = BA = I_2$, so B is the inverse of A.

   (c) $AB = \begin{bmatrix} 2 & -4 \\ -5 & 3 \end{bmatrix}\begin{bmatrix} 3 & 1 \\ 5 & 2 \end{bmatrix} = \begin{bmatrix} -14 & -6 \\ 0 & 1 \end{bmatrix}$, so B is not the inverse of A.

   (d) $AB = BA = I_2$, so B is the inverse of A.

Section 2.4

2. (a) $AB = BA = I_3$, so B is the inverse of A.

   (b) $AB = BA = I_3$, so B is the inverse of A.

   (c) $AB = \begin{bmatrix} 0 & 1 & -1 \\ 2 & -2 & -1 \\ -1 & 1 & 1 \end{bmatrix} \begin{bmatrix} 1 & 2 & 3 \\ 1 & 1 & 2 \\ 0 & 1 & 1 \end{bmatrix} = \begin{bmatrix} 1 & 0 & 1 \\ 0 & 1 & 1 \\ 0 & 0 & 0 \end{bmatrix}$, so B is not the inverse of A.

3. (a) $\begin{bmatrix} 1 & 0 & 1 & 0 \\ 2 & 1 & 0 & 1 \end{bmatrix} \underset{R2+(-2)R1}{\approx} \begin{bmatrix} 1 & 0 & 1 & 0 \\ 0 & 1 & -2 & 1 \end{bmatrix}$, and the inverse matrix is $\begin{bmatrix} 1 & 0 \\ -2 & 1 \end{bmatrix}$.

   (b) $\begin{bmatrix} 1 & 2 & 1 & 0 \\ 9 & 4 & 0 & 1 \end{bmatrix} \underset{R2+(-9)R1}{\approx} \begin{bmatrix} 1 & 2 & 1 & 0 \\ 0 & -14 & -9 & 1 \end{bmatrix} \underset{(-1/14)R2}{\approx} \begin{bmatrix} 1 & 2 & 1 & 0 \\ 0 & 1 & 9/14 & -1/14 \end{bmatrix}$
   $\underset{(-1/14)R2}{\approx} \begin{bmatrix} 1 & 0 & -2/7 & 1/7 \\ 0 & 1 & 9/14 & -1/14 \end{bmatrix}$, and the inverse matrix is $\begin{bmatrix} -2/7 & 1/7 \\ 9/14 & -1/14 \end{bmatrix}$.

   (c) $\begin{bmatrix} 2 & 1 & 1 & 0 \\ 4 & 3 & 0 & 1 \end{bmatrix} \underset{(1/2)R1}{\approx} \begin{bmatrix} 1 & 1/2 & 1/2 & 0 \\ 4 & 3 & 0 & 1 \end{bmatrix} \underset{R2+(-4)R1}{\approx} \begin{bmatrix} 1 & 1/2 & 1/2 & 0 \\ 0 & 1 & -2 & 1 \end{bmatrix}$
   $\underset{R1+(-1/2)R2}{\approx} \begin{bmatrix} 1 & 0 & 3/2 & -1/2 \\ 0 & 1 & -2 & 1 \end{bmatrix}$, and the inverse is $\begin{bmatrix} 3/2 & -1/2 \\ -2 & 1 \end{bmatrix}$.

   (d) $\begin{bmatrix} 0 & 1 & 1 & 0 \\ 1 & 3 & 0 & 1 \end{bmatrix} \underset{R1 \leftrightarrow R2}{\approx} \begin{bmatrix} 1 & 3 & 0 & 1 \\ 0 & 1 & 1 & 0 \end{bmatrix} \underset{R1+(-3)R2}{\approx} \begin{bmatrix} 1 & 0 & -3 & 1 \\ 0 & 1 & 1 & 0 \end{bmatrix}$,
   and the inverse is $\begin{bmatrix} -3 & 1 \\ 1 & 0 \end{bmatrix}$.

   (e) $\begin{bmatrix} 1 & 2 & 1 & 0 \\ 3 & 6 & 0 & 1 \end{bmatrix} \underset{R2+(-3)R1}{\approx} \begin{bmatrix} 1 & 2 & 1 & 0 \\ 0 & 0 & -3 & 1 \end{bmatrix}$, and the inverse does not exist.

   (f) $\begin{bmatrix} 2 & -3 & 1 & 0 \\ 6 & -7 & 0 & 1 \end{bmatrix} \underset{(1/2)R1}{\approx} \begin{bmatrix} 1 & -3/2 & 1/2 & 0 \\ 6 & -7 & 0 & 1 \end{bmatrix} \underset{R2+(-6)R1}{\approx} \begin{bmatrix} 1 & -3/2 & 1/2 & 0 \\ 0 & 2 & -3 & 1 \end{bmatrix}$
   $\underset{(1/2)R2}{\approx} \begin{bmatrix} 1 & -3/2 & 1/2 & 0 \\ 0 & 1 & -3/2 & 1/2 \end{bmatrix} \underset{R1+(3/2)R2}{\approx} \begin{bmatrix} 1 & 0 & -7/4 & 3/4 \\ 0 & 1 & -3/2 & 1/2 \end{bmatrix}$,
   and the inverse is $\begin{bmatrix} -7/4 & 3/4 \\ -3/2 & 1/2 \end{bmatrix}$.

4. (a) $\begin{bmatrix} 1 & 2 & 3 & 1 & 0 & 0 \\ 0 & 1 & 2 & 0 & 1 & 0 \\ 4 & 5 & 3 & 0 & 0 & 1 \end{bmatrix} \underset{R2+(-4)R1}{\approx} \begin{bmatrix} 1 & 2 & 3 & 1 & 0 & 0 \\ 0 & 1 & 2 & 0 & 1 & 0 \\ 0 & -3 & -9 & -4 & 0 & 1 \end{bmatrix}$

51

Section 2.4

$$\underset{\substack{R1+(-2)R2\\R3+(3)R2}}{\approx} \begin{bmatrix} 1 & 0 & -1 & 1 & -2 & 0 \\ 0 & 1 & 2 & 0 & 1 & 0 \\ 0 & 0 & -3 & -4 & 3 & 1 \end{bmatrix} \underset{(-1/3)R3}{\approx} \begin{bmatrix} 1 & 0 & -1 & 1 & -2 & 0 \\ 0 & 1 & 2 & 0 & 1 & 0 \\ 0 & 0 & 1 & 4/3 & -1 & -1/3 \end{bmatrix}$$

$$\underset{\substack{R1+R3\\R2+(-2)R3}}{\approx} \begin{bmatrix} 1 & 0 & 0 & 7/3 & -3 & -1/3 \\ 0 & 1 & 0 & -8/3 & 3 & 2/3 \\ 0 & 0 & 1 & 4/3 & -1 & -1/3 \end{bmatrix}, \text{ and the inverse is } \begin{bmatrix} 7/3 & -3 & -1/3 \\ -8/3 & 3 & 2/3 \\ 4/3 & -1 & -1/3 \end{bmatrix}.$$

(b) $\begin{bmatrix} 2 & 0 & 4 & 1 & 0 & 0 \\ -1 & 3 & 1 & 0 & 1 & 0 \\ 0 & 1 & 2 & 0 & 0 & 1 \end{bmatrix} \underset{(1/2)R1}{\approx} \begin{bmatrix} 1 & 0 & 2 & 1/2 & 0 & 0 \\ -1 & 3 & 1 & 0 & 1 & 0 \\ 0 & 1 & 2 & 0 & 0 & 1 \end{bmatrix}$

$$\underset{R2+R1}{\approx} \begin{bmatrix} 1 & 0 & 2 & 1/2 & 0 & 0 \\ 0 & 3 & 3 & 1/2 & 1 & 0 \\ 0 & 1 & 2 & 0 & 0 & 1 \end{bmatrix} \underset{(1/3)R2}{\approx} \begin{bmatrix} 1 & 0 & 2 & 1/2 & 0 & 0 \\ 0 & 1 & 1 & 1/6 & 1/3 & 0 \\ 0 & 1 & 2 & 0 & 0 & 1 \end{bmatrix}$$

$$\underset{R3+(-1)R2}{\approx} \begin{bmatrix} 1 & 0 & 2 & 1/2 & 0 & 0 \\ 0 & 1 & 1 & 1/6 & 1/3 & 0 \\ 0 & 0 & 1 & -1/6 & -1/3 & 1 \end{bmatrix} \underset{\substack{R1+(-2)R3\\R2+(-1)R3}}{\approx} \begin{bmatrix} 1 & 0 & 0 & 5/6 & 2/3 & -2 \\ 0 & 1 & 0 & 1/3 & 2/3 & -1 \\ 0 & 0 & 1 & -1/6 & -1/3 & 1 \end{bmatrix},$$

and the inverse is $\begin{bmatrix} 5/6 & 2/3 & -2 \\ 1/3 & 2/3 & -1 \\ -1/6 & -1/3 & 1 \end{bmatrix}$.

(c) $\begin{bmatrix} 1 & 2 & -3 & 1 & 0 & 0 \\ 1 & -2 & 1 & 0 & 1 & 0 \\ 5 & -2 & -3 & 0 & 0 & 1 \end{bmatrix} \underset{\substack{R2+(-1)R1\\R3+(-5)R1}}{\approx} \begin{bmatrix} 1 & 2 & -3 & 1 & 0 & 0 \\ 0 & -4 & 4 & -1 & 1 & 0 \\ 0 & -12 & 12 & -5 & 0 & 1 \end{bmatrix}$

$$\underset{(-1/4)R2}{\approx} \begin{bmatrix} 1 & 2 & -3 & 1 & 0 & 0 \\ 0 & 1 & -1 & 1/4 & -1/4 & 0 \\ 0 & -12 & 12 & -5 & 0 & 1 \end{bmatrix} \underset{\substack{R1+(-2)R2\\R3+(12)R2}}{\approx} \begin{bmatrix} 1 & 0 & -1 & 1/2 & 1/2 & 0 \\ 0 & 1 & -1 & 1/4 & -1/4 & 0 \\ 0 & 0 & 0 & -2 & -3 & 1 \end{bmatrix},$$

so the inverse does not exist.

(d) $\begin{bmatrix} 1 & 2 & -1 & 1 & 0 & 0 \\ 3 & -1 & 0 & 0 & 1 & 0 \\ 2 & -3 & 1 & 0 & 0 & 1 \end{bmatrix} \underset{\substack{R2+(-3)R1\\R3+(-2)R1}}{\approx} \begin{bmatrix} 1 & 2 & -1 & 1 & 0 & 0 \\ 0 & -7 & 3 & -3 & 1 & 0 \\ 0 & -7 & 3 & -2 & 0 & 1 \end{bmatrix}$

$$\underset{(-1/7)R2}{\approx} \begin{bmatrix} 1 & 2 & -1 & 1 & 0 & 0 \\ 0 & 1 & -3/7 & 3/7 & -1/7 & 0 \\ 0 & -7 & 3 & -2 & 0 & 1 \end{bmatrix} \underset{\substack{R1+(-2)R2\\R3+(7)R2}}{\approx} \begin{bmatrix} 1 & 0 & -1/7 & 1/7 & 2/7 & 0 \\ 0 & 1 & -3/7 & 3/7 & -1/7 & 0 \\ 0 & 0 & 0 & 1 & -1 & 1 \end{bmatrix},$$

Section 2.4

and the inverse does not exist.

5. (a) $\begin{bmatrix} 1 & 2 & 3 & 1 & 0 & 0 \\ 2 & -1 & 4 & 0 & 1 & 0 \\ 0 & -1 & 1 & 0 & 0 & 1 \end{bmatrix}$ $\underset{R2+(-2)R1}{\approx}$ $\begin{bmatrix} 1 & 2 & 3 & 1 & 0 & 0 \\ 0 & -5 & -2 & -2 & 1 & 0 \\ 0 & -1 & 1 & 0 & 0 & 1 \end{bmatrix}$

$\underset{R2 \leftrightarrow R3}{\approx}$ $\begin{bmatrix} 1 & 2 & 3 & 1 & 0 & 0 \\ 0 & -1 & 1 & 0 & 0 & 1 \\ 0 & -5 & -2 & -2 & 1 & 0 \end{bmatrix}$ $\underset{(-1)R2}{\approx}$ $\begin{bmatrix} 1 & 2 & 3 & 1 & 0 & 0 \\ 0 & 1 & -1 & 0 & 0 & -1 \\ 0 & -5 & -2 & -2 & 1 & 0 \end{bmatrix}$

$\underset{\substack{R1+(-2)R2 \\ R3+(5)R2}}{\approx}$ $\begin{bmatrix} 1 & 0 & 5 & 1 & 0 & 2 \\ 0 & 1 & -1 & 0 & 0 & -1 \\ 0 & 0 & -7 & -2 & 1 & -5 \end{bmatrix}$ $\underset{(-1/7)R3}{\approx}$ $\begin{bmatrix} 1 & 0 & 5 & 1 & 0 & 2 \\ 0 & 1 & -1 & 0 & 0 & -1 \\ 0 & 0 & 1 & 2/7 & -1/7 & 5/7 \end{bmatrix}$

$\underset{\substack{R1+(-5)R3 \\ R2+R3}}{\approx}$ $\begin{bmatrix} 1 & 0 & 0 & -3/7 & 5/7 & -11/7 \\ 0 & 1 & 0 & 2/7 & -1/7 & -2/7 \\ 0 & 0 & 1 & 2/7 & -1/7 & 5/7 \end{bmatrix}$,

and the inverse is $\begin{bmatrix} -3/7 & 5/7 & -11/7 \\ 2/7 & -1/7 & -2/7 \\ 2/7 & -1/7 & 5/7 \end{bmatrix}$.

(b) $\begin{bmatrix} 1 & 2 & -1 & 1 & 0 & 0 \\ 2 & 4 & -3 & 0 & 1 & 0 \\ 1 & -2 & 0 & 0 & 0 & 1 \end{bmatrix}$ $\underset{\substack{R2+(-2)R1 \\ R3+(-1)R1}}{\approx}$ $\begin{bmatrix} 1 & 2 & -1 & 1 & 0 & 0 \\ 0 & 0 & -1 & -2 & 1 & 0 \\ 0 & -4 & 1 & -1 & 0 & 1 \end{bmatrix}$

$\underset{R2 \leftrightarrow R3}{\approx}$ $\begin{bmatrix} 1 & 2 & -1 & 1 & 0 & 0 \\ 0 & -4 & 1 & -1 & 0 & 1 \\ 0 & 0 & -1 & -2 & 1 & 0 \end{bmatrix}$ $\underset{(-1/4)R2}{\approx}$ $\begin{bmatrix} 1 & 2 & -1 & 1 & 0 & 0 \\ 0 & 1 & -1/4 & 1/4 & 0 & -1/4 \\ 0 & 0 & -1 & -2 & 1 & 0 \end{bmatrix}$

$\underset{R1+(-2)R2}{\approx}$ $\begin{bmatrix} 1 & 0 & -1/2 & 1/2 & 0 & 1/2 \\ 0 & 1 & -1/4 & 1/4 & 0 & -1/4 \\ 0 & 0 & -1 & -2 & 1 & 0 \end{bmatrix}$ $\underset{(-1)R3}{\approx}$ $\begin{bmatrix} 1 & 0 & -1/2 & 1/2 & 0 & 1/2 \\ 0 & 1 & -1/4 & 1/4 & 0 & -1/4 \\ 0 & 0 & 1 & 2 & -1 & 0 \end{bmatrix}$

$\underset{\substack{R1+(1/2)R3 \\ R2+(1/4)R3}}{\approx}$ $\begin{bmatrix} 1 & 0 & 0 & 3/2 & -1/2 & 1/2 \\ 0 & 1 & 0 & 3/4 & -1/4 & -1/4 \\ 0 & 0 & 1 & 2 & -1 & 0 \end{bmatrix}$, and the inverse is $\begin{bmatrix} 3/2 & -1/2 & 1/2 \\ 3/4 & -1/4 & -1/4 \\ 2 & -1 & 0 \end{bmatrix}$.

Section 2.4

(c) $\begin{bmatrix} 1 & -2 & -1 & 1 & 0 & 0 \\ -2 & 4 & 6 & 0 & 1 & 0 \\ 0 & 0 & 5 & 0 & 0 & 1 \end{bmatrix} \underset{R2+(2)R1}{\approx} \begin{bmatrix} 1 & -2 & -1 & 1 & 0 & 0 \\ 0 & 0 & 4 & 2 & 1 & 0 \\ 0 & 0 & 5 & 0 & 0 & 1 \end{bmatrix}$,

and the inverse does not exist.

(d) $\begin{bmatrix} 7 & 0 & 0 & 1 & 0 & 0 \\ 0 & -3 & 0 & 0 & 1 & 0 \\ 0 & 0 & 1/5 & 0 & 0 & 1 \end{bmatrix} \underset{\substack{(1/7)R1 \\ (-1/3)R2 \\ (5)R3}}{\approx} \begin{bmatrix} 1 & 0 & 0 & 1/7 & 0 & 0 \\ 0 & 1 & 0 & 0 & -1/3 & 0 \\ 0 & 0 & 1 & 0 & 0 & 5 \end{bmatrix}$,

and the inverse is $\begin{bmatrix} 1/7 & 0 & 0 \\ 0 & -1/3 & 0 \\ 0 & 0 & 5 \end{bmatrix}$.

6. (a) $\begin{bmatrix} -3 & -1 & 1 & -2 & 1 & 0 & 0 & 0 \\ -1 & 3 & 2 & 1 & 0 & 1 & 0 & 0 \\ 1 & 2 & 3 & -1 & 0 & 0 & 1 & 0 \\ -2 & 1 & -1 & -3 & 0 & 0 & 0 & 1 \end{bmatrix} \underset{R1 \leftrightarrow R3}{\approx} \begin{bmatrix} 1 & 2 & 3 & -1 & 0 & 0 & 1 & 0 \\ -1 & 3 & 2 & 1 & 0 & 1 & 0 & 0 \\ -3 & -1 & 1 & -2 & 1 & 0 & 0 & 0 \\ -2 & 1 & -1 & -3 & 0 & 0 & 0 & 1 \end{bmatrix}$

$\underset{\substack{R2+R1 \\ R3+(3)R1 \\ R4+(2)R1}}{\approx} \begin{bmatrix} 1 & 2 & 3 & -1 & 0 & 0 & 1 & 0 \\ 0 & 5 & 5 & 0 & 0 & 1 & 1 & 0 \\ 0 & 5 & 10 & -5 & 1 & 0 & 3 & 0 \\ 0 & 5 & 5 & -5 & 0 & 0 & 2 & 1 \end{bmatrix} \underset{(1/5)R2}{\approx} \begin{bmatrix} 1 & 2 & 3 & -1 & 0 & 0 & 1 & 0 \\ 0 & 1 & 1 & 0 & 0 & 1/5 & 1/5 & 0 \\ 0 & 5 & 10 & -5 & 1 & 0 & 3 & 0 \\ 0 & 5 & 5 & -5 & 0 & 0 & 2 & 1 \end{bmatrix}$

$\underset{\substack{R1+(-2)R2 \\ R3+(-5)R2 \\ R4+(-5)R2}}{\approx} \begin{bmatrix} 1 & 0 & 1 & -1 & 0 & -2/5 & 3/5 & 0 \\ 0 & 1 & 1 & 0 & 0 & 1/5 & 1/5 & 0 \\ 0 & 0 & 5 & -5 & 1 & -1 & 2 & 0 \\ 0 & 0 & 0 & -5 & 0 & -1 & 1 & 1 \end{bmatrix}$

$\underset{\substack{(1/5)R3 \\ (-1/5)R4}}{\approx} \begin{bmatrix} 1 & 0 & 1 & -1 & 0 & -2/5 & 3/5 & 0 \\ 0 & 1 & 1 & 0 & 0 & 1/5 & 1/5 & 0 \\ 0 & 0 & 1 & -1 & 1/5 & -1/5 & 2/5 & 0 \\ 0 & 0 & 0 & 1 & 0 & 1/5 & -1/5 & -1/5 \end{bmatrix}$

$\underset{\substack{R1+(-1)R3 \\ R2+(-1)R3}}{\approx} \begin{bmatrix} 1 & 0 & 0 & 0 & -1/5 & -1/5 & 1/5 & 0 \\ 0 & 1 & 0 & 1 & -1/5 & 2/5 & -1/5 & 0 \\ 0 & 0 & 1 & -1 & 1/5 & -1/5 & 2/5 & 0 \\ 0 & 0 & 0 & 1 & 0 & 1/5 & -1/5 & -1/5 \end{bmatrix}$

$\underset{\substack{R2+(-1)R4 \\ R3+R4}}{\approx} \begin{bmatrix} 1 & 0 & 0 & 0 & -1/5 & -1/5 & 1/5 & 0 \\ 0 & 1 & 0 & 0 & -1/5 & 1/5 & 0 & 1/5 \\ 0 & 0 & 1 & 0 & 1/5 & 0 & 1/5 & -1/5 \\ 0 & 0 & 0 & 1 & 0 & 1/5 & -1/5 & -1/5 \end{bmatrix}$,

54

so the inverse is $\begin{bmatrix} -1/5 & -1/5 & 1/5 & 0 \\ -1/5 & 1/5 & 0 & 1/5 \\ 1/5 & 0 & 1/5 & -1/5 \\ 0 & 1/5 & -1/5 & -1/5 \end{bmatrix}$.

(b) $\begin{bmatrix} 1 & 1 & 0 & 0 & 1 & 0 & 0 & 0 \\ 0 & 1 & 1 & 0 & 0 & 1 & 0 & 0 \\ 1 & 0 & 0 & 1 & 0 & 0 & 1 & 0 \\ 0 & 0 & 1 & 1 & 0 & 0 & 0 & 1 \end{bmatrix} \underset{R3+(-1)R1}{\approx} \begin{bmatrix} 1 & 1 & 0 & 0 & 1 & 0 & 0 & 0 \\ 0 & 1 & 1 & 0 & 0 & 1 & 0 & 0 \\ 0 & -1 & 0 & 1 & -1 & 0 & 1 & 0 \\ 0 & 0 & 1 & 1 & 0 & 0 & 0 & 1 \end{bmatrix}$

$\underset{\substack{R1+(-1)R2 \\ R3+R2}}{\approx} \begin{bmatrix} 1 & 0 & -1 & 0 & 1 & -1 & 0 & 0 \\ 0 & 1 & 1 & 0 & 0 & 1 & 0 & 0 \\ 0 & 0 & 1 & 1 & -1 & 1 & 1 & 0 \\ 0 & 0 & 1 & 1 & 0 & 0 & 0 & 1 \end{bmatrix}$

$\underset{\substack{R1+R3 \\ R2+(-1)R3 \\ R4+(-1)R3}}{\approx} \begin{bmatrix} 1 & 0 & 0 & 1 & 0 & 0 & 1 & 0 \\ 0 & 1 & 0 & -1 & 1 & 0 & -1 & 0 \\ 0 & 0 & 1 & 1 & -1 & 1 & 1 & 0 \\ 0 & 0 & 0 & 0 & 1 & -1 & -1 & 1 \end{bmatrix}$, so the inverse does not exist.

(c) $\begin{bmatrix} -1 & 0 & -1 & -1 & 1 & 0 & 0 & 0 \\ -3 & -1 & 0 & -1 & 0 & 1 & 0 & 0 \\ 5 & 0 & 4 & 3 & 0 & 0 & 1 & 0 \\ 3 & 0 & 3 & 2 & 0 & 0 & 0 & 1 \end{bmatrix} \underset{(-1)R1}{\approx} \begin{bmatrix} 1 & 0 & 1 & 1 & -1 & 0 & 0 & 0 \\ -3 & -1 & 0 & -1 & 0 & 1 & 0 & 0 \\ 5 & 0 & 4 & 3 & 0 & 0 & 1 & 0 \\ 3 & 0 & 3 & 2 & 0 & 0 & 0 & 1 \end{bmatrix}$

$\underset{\substack{R2+(3)R1 \\ R3+(-5)R1 \\ R4+(-3)R1}}{\approx} \begin{bmatrix} 1 & 0 & 1 & 1 & -1 & 0 & 0 & 0 \\ 0 & -1 & 3 & 2 & -3 & 1 & 0 & 0 \\ 0 & 0 & -1 & -2 & 5 & 0 & 1 & 0 \\ 0 & 0 & 0 & -1 & 3 & 0 & 0 & 1 \end{bmatrix} \underset{\substack{(-1)R2 \\ (-1)R3 \\ (-1)R4}}{\approx} \begin{bmatrix} 1 & 0 & 1 & 1 & -1 & 0 & 0 & 0 \\ 0 & 1 & -3 & -2 & 3 & -1 & 0 & 0 \\ 0 & 0 & 1 & 2 & -5 & 0 & -1 & 0 \\ 0 & 0 & 0 & 1 & -3 & 0 & 0 & -1 \end{bmatrix}$

$\underset{\substack{R1+(-1)R3 \\ R2+(3)R3}}{\approx} \begin{bmatrix} 1 & 0 & 0 & -1 & 4 & 0 & 1 & 0 \\ 0 & 1 & 0 & 4 & -12 & -1 & -3 & 0 \\ 0 & 0 & 1 & 2 & -5 & 0 & -1 & 0 \\ 0 & 0 & 0 & 1 & -3 & 0 & 0 & -1 \end{bmatrix}$

$\underset{\substack{R1+R4 \\ R2+(-4)R4 \\ R3+(-2)R4}}{\approx} \begin{bmatrix} 1 & 0 & 0 & 0 & 1 & 0 & 1 & -1 \\ 0 & 1 & 0 & 0 & 0 & -1 & -3 & 4 \\ 0 & 0 & 1 & 0 & 1 & 0 & -1 & 2 \\ 0 & 0 & 0 & 1 & -3 & 0 & 0 & -1 \end{bmatrix}$, so the inverse is $\begin{bmatrix} 1 & 0 & 1 & -1 \\ 0 & -1 & -3 & 4 \\ 1 & 0 & -1 & 2 \\ -3 & 0 & 0 & -1 \end{bmatrix}$.

Section 2.4

7. (a) The inverse of the coefficient matrix is $\begin{bmatrix} -5 & 2 \\ 3 & -1 \end{bmatrix}$.

$$\begin{bmatrix} x_1 \\ x_2 \end{bmatrix} = \begin{bmatrix} -5 & 2 \\ 3 & -1 \end{bmatrix} \begin{bmatrix} 2 \\ 4 \end{bmatrix} = \begin{bmatrix} -2 \\ 2 \end{bmatrix}.$$

(b) $\begin{bmatrix} x_1 \\ x_2 \end{bmatrix} = \begin{bmatrix} -9 & 5 \\ 2 & -1 \end{bmatrix} \begin{bmatrix} -1 \\ 3 \end{bmatrix} = \begin{bmatrix} 24 \\ -5 \end{bmatrix}.$

(c) $\begin{bmatrix} x_1 \\ x_2 \end{bmatrix} = \begin{bmatrix} -1/5 & 3/5 \\ 2/5 & -1/5 \end{bmatrix} \begin{bmatrix} 5 \\ 10 \end{bmatrix} = \begin{bmatrix} 5 \\ 0 \end{bmatrix}.$

(d) $\begin{bmatrix} x_1 \\ x_2 \end{bmatrix} = \begin{bmatrix} 3/2 & -1/2 \\ -2 & 1 \end{bmatrix} \begin{bmatrix} 4 \\ 6 \end{bmatrix} = \begin{bmatrix} 3 \\ -2 \end{bmatrix}.$

(e) $\begin{bmatrix} x_1 \\ x_2 \end{bmatrix} = \begin{bmatrix} 2 & -1 \\ -3/4 & 1/2 \end{bmatrix} \begin{bmatrix} 6 \\ 1 \end{bmatrix} = \begin{bmatrix} 11 \\ -4 \end{bmatrix}.$

(f) $\begin{bmatrix} x_1 \\ x_2 \end{bmatrix} = \begin{bmatrix} 7/3 & -3 \\ -2/3 & 1 \end{bmatrix} \begin{bmatrix} 9 \\ 4 \end{bmatrix} = \begin{bmatrix} 9 \\ -2 \end{bmatrix}.$

8. (a) $\begin{bmatrix} x_1 \\ x_2 \\ x_3 \end{bmatrix} = \begin{bmatrix} 1/9 & 3/9 & 5/9 \\ 3/9 & 0 & -3/9 \\ -2/9 & 3/9 & -1/9 \end{bmatrix} \begin{bmatrix} 2 \\ 0 \\ 1 \end{bmatrix} = \begin{bmatrix} 7/9 \\ 3/9 \\ -5/9 \end{bmatrix}.$

(b) $\begin{bmatrix} x_1 \\ x_2 \\ x_3 \end{bmatrix} = \begin{bmatrix} 3/4 & -1/4 & 1/2 \\ -1/4 & -1/4 & 1/2 \\ -1/4 & 3/4 & -1/2 \end{bmatrix} \begin{bmatrix} 1 \\ 2 \\ 0 \end{bmatrix} = \begin{bmatrix} 1/4 \\ -3/4 \\ 5/4 \end{bmatrix}.$

(c) $\begin{bmatrix} x_1 \\ x_2 \\ x_3 \end{bmatrix} = \begin{bmatrix} -40 & 16 & 9 \\ 13 & -5 & -3 \\ 5 & -2 & -1 \end{bmatrix} \begin{bmatrix} 1 \\ 3 \\ 15 \end{bmatrix} = \begin{bmatrix} 143 \\ -47 \\ -16 \end{bmatrix}.$

(d) $\begin{bmatrix} x_1 \\ x_2 \\ x_3 \end{bmatrix} = \begin{bmatrix} -1 & -10 & -8 \\ -1 & -6 & -5 \\ 0 & -1 & -1 \end{bmatrix} \begin{bmatrix} 3 \\ 2 \\ -1 \end{bmatrix} = \begin{bmatrix} -15 \\ -10 \\ -1 \end{bmatrix}.$

(e) $\begin{bmatrix} x_1 \\ x_2 \\ x_3 \end{bmatrix} = \begin{bmatrix} 2 & 3 & 1 \\ 3 & 3 & 1 \\ 2 & 4 & 1 \end{bmatrix} \begin{bmatrix} 5 \\ -2 \\ 1 \end{bmatrix} = \begin{bmatrix} 5 \\ 10 \\ 3 \end{bmatrix}.$

9. $\begin{bmatrix} x_1 \\ x_2 \\ x_3 \\ x_4 \end{bmatrix} = \begin{bmatrix} -14/17 & 8/17 & 4/17 & 5/17 \\ 3/17 & -9/17 & 4/17 & 5/17 \\ 5/17 & 2/17 & 1/17 & -3/17 \\ 18/17 & -3/17 & -10/17 & -4/17 \end{bmatrix} \begin{bmatrix} 5 \\ 6 \\ 1 \\ 7 \end{bmatrix} = \begin{bmatrix} 1 \\ 0 \\ 1 \\ 2 \end{bmatrix}.$

10. We solve the three sets of equations simultaneously:

$$\begin{bmatrix} 1/9 & 3/9 & 5/9 \\ 3/9 & 0 & -3/9 \\ -2/9 & 3/9 & -1/9 \end{bmatrix} \begin{bmatrix} 1 & 0 & 5 \\ 2 & 1 & 2 \\ 3 & 4 & -3 \end{bmatrix} = \begin{bmatrix} 22/9 & 23/9 & -4/9 \\ -6/9 & -12/9 & 24/9 \\ 1/9 & -1/9 & -1/9 \end{bmatrix}.$$

Thus the solutions are, in turn,

$\begin{bmatrix} x_1 \\ x_2 \\ x_3 \end{bmatrix} = \begin{bmatrix} 22/9 \\ -6/9 \\ 1/9 \end{bmatrix}, \begin{bmatrix} x_1 \\ x_2 \\ x_3 \end{bmatrix} = \begin{bmatrix} 23/9 \\ -12/9 \\ -1/9 \end{bmatrix}, \begin{bmatrix} x_1 \\ x_2 \\ x_3 \end{bmatrix} = \begin{bmatrix} -4/9 \\ 24/9 \\ -1/9 \end{bmatrix}.$

11. $\begin{bmatrix} a & b \\ c & d \end{bmatrix} \dfrac{1}{ad-bc} \begin{bmatrix} d & -b \\ -c & a \end{bmatrix} = \dfrac{1}{ad-bc} \begin{bmatrix} a & b \\ c & d \end{bmatrix} \begin{bmatrix} d & -b \\ -c & a \end{bmatrix}$

$$= \frac{1}{ad-bc} \begin{bmatrix} ad-bc & 0 \\ 0 & ad-bc \end{bmatrix} = \begin{bmatrix} 1 & 0 \\ 0 & 1 \end{bmatrix}.$$

Likewise $\frac{1}{ad-bc} \begin{bmatrix} d & -b \\ -c & a \end{bmatrix} \begin{bmatrix} a & b \\ c & d \end{bmatrix} = \begin{bmatrix} 1 & 0 \\ 0 & 1 \end{bmatrix}.$ Thus $\frac{1}{ad-bc} \begin{bmatrix} d & -b \\ -c & a \end{bmatrix} = A^{-1}.$

The inverses of the given matrices are:

(a) $\begin{bmatrix} 3 & -8 \\ -1 & 3 \end{bmatrix},$

(b) $\frac{1}{2} \begin{bmatrix} 4 & -5 \\ -2 & 3 \end{bmatrix} = \begin{bmatrix} 2 & -5/2 \\ -1 & 3/2 \end{bmatrix},$

(c) $\frac{1}{2} \begin{bmatrix} 4 & -6 \\ -3 & 5 \end{bmatrix} = \begin{bmatrix} 2 & -3 \\ -3/2 & 5/2 \end{bmatrix},$

(d) $\frac{1}{4} \begin{bmatrix} -2 & 6 \\ -2 & 4 \end{bmatrix} = \begin{bmatrix} -1/2 & 3/2 \\ -1/2 & 1 \end{bmatrix}.$

12. (a) $I = (cA)(cA)^{-1}$. Multiply both sides by $\frac{1}{c} A^{-1}$.

$\frac{1}{c} A^{-1} = \frac{1}{c} A^{-1} (cA)(cA)^{-1} = \frac{1}{c} cA^{-1}A(cA)^{-1} = I (cA)^{-1} = (cA)^{-1}.$

(b) $I = A^n(A^n)^{-1}$. Multiply both sides by $(A^{-1})^n$.

$(A^{-1})^n = (A^{-1})^n A^n(A^n)^{-1} = (A^{-1}A^{-1} \ldots A^{-1}AA \ldots A)(A^n)^{-1} = I (A^n)^{-1} = (A^n)^{-1}.$

(c) $I = A^t(A^t)^{-1}$. Multiply both sides by $(A^{-1})^t$.

$(A^{-1})^t = (A^{-1})^t A^t(A^t)^{-1} = (AA^{-1})^t (A^t)^{-1} = I (A^t)^{-1} = (A^t)^{-1}.$

13. $A = \begin{bmatrix} 2 & 1 \\ 4 & 3 \end{bmatrix}^{-1} = \begin{bmatrix} 3/2 & -1/2 \\ -2 & 1 \end{bmatrix}.$

14. $A = 2 \begin{bmatrix} -3 & 2 \\ -10 & 6 \end{bmatrix}^{-1} = 2 \begin{bmatrix} 3 & -1 \\ 5 & -3/2 \end{bmatrix} = \begin{bmatrix} 6 & -2 \\ 10 & -3 \end{bmatrix}.$

15. (a) $(3A)^{-1} = \frac{1}{3} \begin{bmatrix} 2 & -1 \\ -5 & 3 \end{bmatrix} = \begin{bmatrix} 2/3 & -1/3 \\ -5/3 & 3/3 \end{bmatrix}.$

(b) $(A^2)^{-1} = (A^{-1})^2 = \begin{bmatrix} 9 & -5 \\ -25 & 14 \end{bmatrix}.$

(c) $A^{-2} = (A^{-1})^2 = \begin{bmatrix} 9 & -5 \\ -25 & 14 \end{bmatrix}.$ (d) $(A^t)^{-1} = (A^{-1})^t = \begin{bmatrix} 2 & -5 \\ -1 & 3 \end{bmatrix}.$

16. (a) $(2A^t)^{-1} = \frac{1}{2}(A^t)^{-1} = \frac{1}{2}(A^{-1})^t = \frac{1}{2}\begin{bmatrix} 2 & -9 \\ -1 & 5 \end{bmatrix}.$ (b) $A^{-3} = (A^{-1})^3 = \begin{bmatrix} 89 & -48 \\ -432 & 233 \end{bmatrix}.$

(c) $(AA^t)^{-1} = (A^t)^{-1}A^{-1} = (A^{-1})^t A^{-1} = \begin{bmatrix} 2 & -9 \\ -1 & 5 \end{bmatrix}\begin{bmatrix} 2 & -1 \\ -9 & 5 \end{bmatrix} = \begin{bmatrix} 85 & -47 \\ -47 & 26 \end{bmatrix}.$

17. $\begin{bmatrix} 2 & -7 \\ -1 & 4 \end{bmatrix}^{-1} = \begin{bmatrix} 4 & 7 \\ 1 & 2 \end{bmatrix},$ so $2x = 4$ and $x = 2.$

18. $\begin{bmatrix} 2x & x \\ 5 & 3 \end{bmatrix}^{-1} = \frac{1}{x}\begin{bmatrix} 3 & -x \\ -5 & 2x \end{bmatrix} = \frac{1}{2}\begin{bmatrix} 3 & -2 \\ -5 & 4 \end{bmatrix},$ so $x = 2.$

19. $4A^t = \begin{bmatrix} 2 & 3 \\ -4 & -4 \end{bmatrix}^{-1} = \frac{1}{4}\begin{bmatrix} -4 & -3 \\ 4 & 2 \end{bmatrix},$ so $A^t = \frac{1}{16}\begin{bmatrix} -4 & -3 \\ 4 & 2 \end{bmatrix} = \begin{bmatrix} -1/4 & -3/16 \\ 1/4 & 1/8 \end{bmatrix}$ and

$A = \begin{bmatrix} -1/4 & 1/4 \\ -3/16 & 1/8 \end{bmatrix}.$

20. $(ABC)^{-1} = C^{-1}(AB)^{-1} = C^{-1}B^{-1}A^{-1}.$

21. $(A^t B^t)^{-1} = (B^t)^{-1}(A^t)^{-1} = (B^{-1})^t(A^{-1})^t = (A^{-1}B^{-1})^t.$

22. Let $A = \begin{bmatrix} 3 & 1 \\ 5 & 2 \end{bmatrix}$ and $B = \begin{bmatrix} 2 & -1 \\ -9 & 5 \end{bmatrix}.$ Then $(A+B)^{-1} = \begin{bmatrix} 5 & 0 \\ -4 & 7 \end{bmatrix}^{-1} = \frac{1}{35}\begin{bmatrix} 7 & 0 \\ 4 & 5 \end{bmatrix}$

and $A^{-1} + B^{-1} = \begin{bmatrix} 2 & -1 \\ -5 & 3 \end{bmatrix} + \begin{bmatrix} 5 & 1 \\ 9 & 2 \end{bmatrix} = \begin{bmatrix} 7 & 0 \\ 4 & 5 \end{bmatrix}.$

23. Suppose A has no inverse but A^t has an inverse $(A^t)^{-1}$. Then $(A^t)^{-1}A^t = A^t(A^t)^{-1} = I$.

   Take the transpose. $A((A^t)^{-1})^t = ((A^t)^{-1})^t A = I^t = I$. i.e. $((A^t)^{-1})^t$ is the inverse of A. So if A has no inverse, then A^t has no inverse.

24. (a) $AB = AC$. Multiply by A^{-1} on the left. $A^{-1}AB = A^{-1}AC$. i.e. $B = C$.

   (b) $AB = O = AO$. Thus from part (a) $B = O$.

25. (a) If A is an nxn matrix with row i equal to row j, then A is not row equivalent to I_n and so cannot have an inverse.

   (b) The columns of A are the rows of A^t so from part (a) A^t has no inverse. From Exercise 23, that means that $(A^t)^t = A$ also has no inverse.

26. If a diagonal matrix has a zero diagonal element, then it has a row that is all zeros and is therefore not row equivalent to an identity matrix. Thus it cannot have an inverse.

   The inverse of an invertible diagonal martix is the diagonal matrix having for its diagonal elements the inverses of the diagonal elements in the original matrix.

27. (a) multiplications: $\dfrac{5^3}{3} + 5^2 - \dfrac{5}{3} = \dfrac{125+75-5}{3} = 65$;

   additions: $\dfrac{5^3}{3} + \dfrac{25}{2} - \dfrac{25}{6} = \dfrac{250+75-25}{6} = 50$

   (b) multiplications: $5^3 + 5^2 = 150$; additions: $5^3 - 5^2 = 100$

28. To find the inverse of A one finds the reduced echelon form of the matrix

Section 2.4

$$\begin{bmatrix} a_{11} & a_{12} & 1 & 0 \\ a_{21} & a_{22} & 0 & 1 \end{bmatrix}.$$ To find the first column of the inverse only, one would ignore the last column of the matrix. This is the same amount of computation as finding the reduced echelon form of the matrix $\begin{bmatrix} a_{11} & a_{12} & b_1 \\ a_{21} & a_{22} & b_2 \end{bmatrix}$ for any $B = \begin{bmatrix} b_1 \\ b_2 \end{bmatrix}$. That is, it is the same as finding the solution to the system of equations $AX = B$ using Gauss-Jordan elimination.

29.
R	E	T	R	E	A	T
18	5	20	18	5	1	20

The vectors are $\begin{bmatrix} 18 \\ 5 \end{bmatrix}, \begin{bmatrix} 20 \\ 18 \end{bmatrix}, \begin{bmatrix} 5 \\ 1 \end{bmatrix}$, and $\begin{bmatrix} 20 \\ 27 \end{bmatrix}$. The vectors obtained on multiplication by $\begin{bmatrix} 4 & -3 \\ 3 & -2 \end{bmatrix}$ are $\begin{bmatrix} 57 \\ 44 \end{bmatrix}, \begin{bmatrix} 26 \\ 24 \end{bmatrix}, \begin{bmatrix} 17 \\ 13 \end{bmatrix}$, and $\begin{bmatrix} -1 \\ 6 \end{bmatrix}$. The coded message is, therefore, 57, 44, 26, 24, 17, 13, -1, 6.

30.
T	H	E	*	B	R	I	T	I	S	H	*	A	R	E	*	C	O	M	I	N	G
20	8	5	27	2	18	9	20	9	19	8	27	1	18	5	27	3	15	13	9	14	7

The vectors are $\begin{bmatrix} 20 \\ 8 \\ 5 \end{bmatrix}, \begin{bmatrix} 27 \\ 2 \\ 18 \end{bmatrix}, \begin{bmatrix} 9 \\ 20 \\ 9 \end{bmatrix}, \begin{bmatrix} 19 \\ 8 \\ 27 \end{bmatrix}, \begin{bmatrix} 1 \\ 18 \\ 5 \end{bmatrix}, \begin{bmatrix} 27 \\ 3 \\ 15 \end{bmatrix}, \begin{bmatrix} 13 \\ 9 \\ 14 \end{bmatrix}$, and $\begin{bmatrix} 7 \\ 27 \\ 27 \end{bmatrix}$. The vectors obtained on multiplication by $\begin{bmatrix} 1 & 2 & 1 \\ 2 & 3 & 1 \\ -2 & 0 & 1 \end{bmatrix}$ are

$\begin{bmatrix} 41 \\ 69 \\ -35 \end{bmatrix}, \begin{bmatrix} 49 \\ 78 \\ -36 \end{bmatrix}, \begin{bmatrix} 58 \\ 87 \\ -9 \end{bmatrix}, \begin{bmatrix} 62 \\ 89 \\ -11 \end{bmatrix}, \begin{bmatrix} 42 \\ 61 \\ 3 \end{bmatrix}, \begin{bmatrix} 48 \\ 78 \\ -39 \end{bmatrix}, \begin{bmatrix} 45 \\ 67 \\ -12 \end{bmatrix}$, and $\begin{bmatrix} 88 \\ 122 \\ 13 \end{bmatrix}$. The coded message is, therefore, 41, 69, -35, 49, 78, -36, 58, 87, -9, 62, 89, -11, 42, 61, 3, 48, 78, -39, 45, 67, -12, 88, 122, 13.

Section 2.5

31. The decoding matrix is $\begin{bmatrix} 4 & -3 \\ 3 & -2 \end{bmatrix}^{-1} = \begin{bmatrix} -2 & 3 \\ -3 & 4 \end{bmatrix}$.

$\begin{bmatrix} -2 & 3 \\ -3 & 4 \end{bmatrix} \begin{bmatrix} 49 & -5 & -61 \\ 38 & -3 & -39 \end{bmatrix} = \begin{bmatrix} 16 & 1 & 5 \\ 5 & 3 & 27 \end{bmatrix}$, and the message is PEACE.

32. The decoding matrix is $\begin{bmatrix} 1 & 2 & 1 \\ 2 & 3 & 1 \\ -2 & 0 & 1 \end{bmatrix}^{-1} = \begin{bmatrix} 3 & -2 & -1 \\ -4 & 3 & 1 \\ 6 & -4 & -1 \end{bmatrix}$.

$\begin{bmatrix} 3 & -2 & -1 \\ -4 & 3 & 1 \\ 6 & -4 & -1 \end{bmatrix} \begin{bmatrix} 71 & 28 & 84 & 63 & 69 & 88 \\ 100 & 43 & 122 & 98 & 102 & 126 \\ -1 & -5 & -11 & -27 & -12 & -3 \end{bmatrix} = \begin{bmatrix} 14 & 3 & 19 & 20 & 15 & 15 \\ 15 & 12 & 19 & 15 & 18 & 23 \\ 27 & 1 & 27 & 13 & 18 & 27 \end{bmatrix}$,

and the message is NO CLASS TOMORROW.

Exercise Set 2.5

1. (a) $a_{32} = .25$ (b) $a_{21} = .40$ (c) electrical industry ($a_{43} = .30$)

   (d) steel industry ($a_{32} = .25$)   (e) steel industry ($a_{21} = .40$)

2. $(I - A)^{-1} = \begin{bmatrix} 15/8 & 5/4 \\ 5/6 & 5/3 \end{bmatrix}$. We do all the multiplications at once.

$\begin{bmatrix} 15/8 & 5/4 \\ 5/6 & 5/3 \end{bmatrix} \begin{bmatrix} 24 & 8 & 0 \\ 12 & 6 & 12 \end{bmatrix} = \begin{bmatrix} 60 & 45/2 & 15 \\ 40 & 50/3 & 20 \end{bmatrix}$, so the values of X, in turn, are

$\begin{bmatrix} 60 \\ 40 \end{bmatrix}, \begin{bmatrix} 45/2 \\ 50/3 \end{bmatrix}$, and $\begin{bmatrix} 15 \\ 20 \end{bmatrix}$.

3. $(I - A)^{-1} = \begin{bmatrix} 4/3 & 2/3 \\ 1/2 & 3/2 \end{bmatrix}$.

$$\begin{bmatrix} 4/3 & 2/3 \\ 1/2 & 3/2 \end{bmatrix} \begin{bmatrix} 6 & 18 & 24 \\ 12 & 6 & 12 \end{bmatrix} = \begin{bmatrix} 16 & 28 & 40 \\ 21 & 18 & 30 \end{bmatrix}$$, so the values of X, in turn, are

$$\begin{bmatrix} 16 \\ 21 \end{bmatrix}, \begin{bmatrix} 28 \\ 18 \end{bmatrix}, \text{ and } \begin{bmatrix} 40 \\ 30 \end{bmatrix}.$$

4. $(I - A)^{-1} = \begin{bmatrix} 15/7 & 10/7 \\ 5/6 & 5/3 \end{bmatrix}.$

$$\begin{bmatrix} 15/7 & 10/7 \\ 5/6 & 5/3 \end{bmatrix} \begin{bmatrix} 42 & 0 & 14 & 42 \\ 84 & 10 & 7 & 42 \end{bmatrix} = \begin{bmatrix} 210 & 100/7 & 40 & 150 \\ 175 & 50/3 & 70/3 & 105 \end{bmatrix}$$, so the values of X,

in turn, are $\begin{bmatrix} 210 \\ 175 \end{bmatrix}, \begin{bmatrix} 100/7 \\ 50/3 \end{bmatrix}, \begin{bmatrix} 40 \\ 70/3 \end{bmatrix},$ and $\begin{bmatrix} 150 \\ 105 \end{bmatrix}.$

5. $(I - A)^{-1} = \begin{bmatrix} 5/4 & 5/8 & 5/8 \\ 0 & 2 & 1 \\ 0 & 1 & 3 \end{bmatrix}.$

$$\begin{bmatrix} 5/4 & 5/8 & 5/8 \\ 0 & 2 & 1 \\ 0 & 1 & 3 \end{bmatrix} \begin{bmatrix} 4 & 0 & 8 \\ 8 & 8 & 24 \\ 8 & 16 & 8 \end{bmatrix} = \begin{bmatrix} 15 & 15 & 30 \\ 24 & 32 & 56 \\ 32 & 56 & 48 \end{bmatrix}$$, so the values of X, in turn, are

$$\begin{bmatrix} 15 \\ 24 \\ 32 \end{bmatrix}, \begin{bmatrix} 15 \\ 32 \\ 56 \end{bmatrix}, \text{ and } \begin{bmatrix} 30 \\ 56 \\ 48 \end{bmatrix}.$$

6. $(I - A)^{-1} = \begin{bmatrix} 25/12 & 5/6 & 5/6 \\ 10/3 & 10/3 & 10/3 \\ 35/18 & 10/9 & 25/9 \end{bmatrix}.$

$$\begin{bmatrix} 25/12 & 5/6 & 5/6 \\ 10/3 & 10/3 & 10/3 \\ 35/18 & 10/9 & 25/9 \end{bmatrix} \begin{bmatrix} 36 & 36 & 36 & 0 \\ 72 & 0 & 0 & 18 \\ 36 & 18 & 0 & 18 \end{bmatrix} = \begin{bmatrix} 165 & 90 & 75 & 30 \\ 480 & 180 & 120 & 120 \\ 250 & 120 & 70 & 70 \end{bmatrix}$$, so the values

Section 2.6

of X, in turn, are $\begin{bmatrix} 165 \\ 480 \\ 250 \end{bmatrix}$, $\begin{bmatrix} 90 \\ 180 \\ 120 \end{bmatrix}$, $\begin{bmatrix} 75 \\ 120 \\ 70 \end{bmatrix}$, and $\begin{bmatrix} 30 \\ 120 \\ 70 \end{bmatrix}$.

7. $(I - A)X = \begin{bmatrix} .8 & -.4 \\ -.5 & .9 \end{bmatrix} \begin{bmatrix} 8 \\ 10 \end{bmatrix} = \begin{bmatrix} 2.4 \\ 5 \end{bmatrix} = D.$

8. $(I - A)X = \begin{bmatrix} .9 & -.2 & -.3 \\ 0 & .9 & -.4 \\ -.5 & -.4 & .8 \end{bmatrix} \begin{bmatrix} 10 \\ 10 \\ 20 \end{bmatrix} = \begin{bmatrix} 1 \\ 1 \\ 7 \end{bmatrix} = D.$

9. $(I - A)X = \begin{bmatrix} .9 & -.1 & -.2 \\ -.2 & .9 & -.3 \\ -.4 & -.3 & .85 \end{bmatrix} \begin{bmatrix} 6 \\ 4 \\ 5 \end{bmatrix} = \begin{bmatrix} 4 \\ .9 \\ .65 \end{bmatrix} = D.$

10. We assume here that units are monetary values. a_{ij} is the amount of commodity i required to produce one unit of commodity j. Thus no a_{ij} can be negative. If the amount of commodity i required for one unit of commodity j were greater than 1, the value going into commodity j would be greater than the final value, an economically infeasible situation.

11. The sum of the elements of column j of the input-output matrix is the sum of the amounts of all the commodities required to produce one unit of commodity j. This sum will certainly be greater than or equal to zero since all elements are, and in an economically feasible situation, the sum of all that goes into producing one unit of commodity j should be worth less than one unit of commodity j.

Exercise Set 2.6

1. (a) stochastic

   (b) The elements in column 2 are not numbers between zero and 1, so the matrix is not stochastic.

   (c) The elements in column 2 do not add up to 1, so the matrix is not stochastic.

Section 2.6

(d) stochastic (e) stochastic

(f) There is a negative number in column 3, so the matrix is not stochastic.

2. $\begin{bmatrix} x & y \\ 1-x & 1-y \end{bmatrix}\begin{bmatrix} u & w \\ 1-u & 1-w \end{bmatrix} = \begin{bmatrix} xu+y(1-u) & xw+y(1-w) \\ (1-x)u+(1-y)(1-u) & (1-x)w+(1-y)(1-w) \end{bmatrix}.$

The elements of the product are all greater than or equal to zero and the sum of the terms in each column is 1, so the matrix is stochastic.

3. $\begin{bmatrix} 1 & 0 \\ 0 & 1 \end{bmatrix}$ and $\begin{bmatrix} 1/2 & 1/4 & 1/4 \\ 1/4 & 1/2 & 1/4 \\ 1/4 & 1/4 & 1/2 \end{bmatrix}$ are doubly stochastic matrices.

If A and B are doubly stochastic matrices, then AB is stochastic, and A^t and B^t are stochastic, so $B^t A^t = (AB)^t$ is stochastic. Thus AB is doubly stochastic.

4. (a) $p_{38} = .02.$ (b) $p_{26} = .05.$

(c) $p_{1\ 10} = .25.$ low density residential (d) $p_{55} = .70.$ auto commercial

5. (a) $P^2 = \begin{bmatrix} .96 & .01 \\ .04 & .99 \end{bmatrix}^2 = \begin{bmatrix} .922 & .0195 \\ .0780 & .9805 \end{bmatrix}$, so the probability of moving from city to suburb in two years is .0780.

(b) $P^3 = \begin{bmatrix} .96 & .01 \\ .04 & .99 \end{bmatrix}^3 = \begin{bmatrix} .8859 & .028525 \\ .1141 & .971475 \end{bmatrix}$, so the probability of moving from suburb to city in three years is .028525.

6. 2001: $\begin{bmatrix} .99 & .02 \\ .01 & .98 \end{bmatrix}\begin{bmatrix} 200 \\ 60 \end{bmatrix} = \begin{bmatrix} 199.2 \\ 60.8 \end{bmatrix}.$ Metro: 199.2 million
Nonmetro: 60.8 million

2002: $\begin{bmatrix} .99 & .02 \\ .01 & .98 \end{bmatrix}\begin{bmatrix} 199.2 \\ 60.8 \end{bmatrix} = \begin{bmatrix} 198.424 \\ 61.576 \end{bmatrix}.$ Metro: 198.424 million
Nonmetro: 61.576 million

Section 2.6

$$2003: \begin{bmatrix} .99 & .02 \\ .01 & .98 \end{bmatrix} \begin{bmatrix} 198.424 \\ 61.576 \end{bmatrix} = \begin{bmatrix} 197.6713 \\ 62.3287 \end{bmatrix}.$$

Metro: 197.6713, Nonmetro: 62.3287.

$$2004: \begin{bmatrix} .99 & .02 \\ .01 & .98 \end{bmatrix} \begin{bmatrix} 197.6713 \\ 62.3287 \end{bmatrix} = \begin{bmatrix} 196.9411 \\ 63.0589 \end{bmatrix}.$$

Metro: 196.9411, Nonmetro: 63.0589

$$2005: \begin{bmatrix} .99 & .02 \\ .01 & .98 \end{bmatrix} \begin{bmatrix} 196.9411 \\ 63.0589 \end{bmatrix} = \begin{bmatrix} 196.2329 \\ 63.7671 \end{bmatrix}.$$

Metro: 196.2329, Nonmetro: 63.7671

$\begin{bmatrix} .99 & .02 \\ .01 & .98 \end{bmatrix}^5 = \begin{bmatrix} .9529 & .0942 \\ .0471 & .9058 \end{bmatrix}$, so the probability that a person living in a metropolitan area in 2000 was still living in a metropolitan area in 2005 is .9529.

7. $2001: \begin{bmatrix} .96 & .01 & .015 \\ .03 & .98 & .005 \\ .01 & .01 & .98 \end{bmatrix} \begin{bmatrix} 58 \\ 142 \\ 60 \end{bmatrix} = \begin{bmatrix} 58 \\ 141.2 \\ 60.8 \end{bmatrix}.$

City: 58 million, Suburb: 141.2 million, Nonmetro: 60.8 million

$2002: \begin{bmatrix} .96 & .01 & .015 \\ .03 & .98 & .005 \\ .01 & .01 & .98 \end{bmatrix} \begin{bmatrix} 58 \\ 141.2 \\ 60.8 \end{bmatrix} = \begin{bmatrix} 58.0040 \\ 140.4200 \\ 61.5760 \end{bmatrix}.$

City: 58.0040 million, Suburb: 140.4200 million, Nonmetro: 61.5760 million

$\begin{bmatrix} .96 & .01 & .015 \\ .03 & .98 & .005 \\ .01 & .01 & .98 \end{bmatrix}^2 = \begin{bmatrix} .9221 & .0196 & .0292 \\ .0583 & .9608 & .0103 \\ .0197 & .0197 & .9606 \end{bmatrix}$, so the probability that a person living in the city in 2000 will be living in a nonmetropolitan area in 2002 is .0197.

8. $X_1 = 1.01 \begin{bmatrix} .96 & .01 \\ .04 & .99 \end{bmatrix} \begin{bmatrix} 58 \\ 142 \end{bmatrix} = \begin{bmatrix} .9696 & .0101 \\ .0404 & .9999 \end{bmatrix} \begin{bmatrix} 58 \\ 142 \end{bmatrix}.$ Likewise

Section 2.6

$$X_2 = \begin{bmatrix} .9696 & .0101 \\ .0404 & .9999 \end{bmatrix} X_1 = \begin{bmatrix} .9696 & .0101 \\ .0404 & .9999 \end{bmatrix}^2 \begin{bmatrix} 58 \\ 142 \end{bmatrix}$$

and in general $X_n = \begin{bmatrix} .9696 & .0101 \\ .0404 & .9999 \end{bmatrix}^n \begin{bmatrix} 58 \\ 142 \end{bmatrix}$.

$$X_5 = \begin{bmatrix} .9696 & .0101 \\ .0404 & .9999 \end{bmatrix}^5 \begin{bmatrix} 58 \\ 142 \end{bmatrix} = \begin{bmatrix} 56.6789 \\ 153.5231 \end{bmatrix}$$, so the predicted 2005 city

population is 56.6789 and the predicted 2005 suburban population is 153.5231.

9. $X_1 = \begin{bmatrix} 1.012 & 0 \\ 0 & 1.008 \end{bmatrix} \begin{bmatrix} .96 & .01 \\ .04 & .99 \end{bmatrix} \begin{bmatrix} 58 \\ 142 \end{bmatrix} = \begin{bmatrix} .9715 & .0101 \\ .0403 & .9980 \end{bmatrix} \begin{bmatrix} 58 \\ 142 \end{bmatrix}$.

Likewise $X_2 = \begin{bmatrix} .9715 & .0101 \\ .0403 & .9980 \end{bmatrix} X_1 = \begin{bmatrix} .9715 & .0101 \\ .0403 & .9980 \end{bmatrix}^2 \begin{bmatrix} 58 \\ 142 \end{bmatrix}$

and in general $X_n = \begin{bmatrix} .9715 & .0101 \\ .0403 & .9980 \end{bmatrix}^n \begin{bmatrix} 58 \\ 142 \end{bmatrix}$.

$$X_5 = \begin{bmatrix} .9715 & .0101 \\ .0403 & .9980 \end{bmatrix}^5 \begin{bmatrix} 58 \\ 142 \end{bmatrix} = \begin{bmatrix} 57.1669 \\ 152.1487 \end{bmatrix}$$, so the predicted 2005 city

population is 55.7812 and the predicted 2005 suburban population is 144.0534.

10. $X_{n-1} = A^{-1}X_n$. $A^{-1} = \begin{bmatrix} 1.0421 & -0.0105 \\ -0.0421 & 1.0105 \end{bmatrix}$. $X_{99} = \begin{bmatrix} 58.9474 \\ 141.0526 \end{bmatrix}$, $X_{98} = \begin{bmatrix} 59.9446 \\ 140.0554 \end{bmatrix}$,

$X_{97} = \begin{bmatrix} 60.9943 \\ 139.0057 \end{bmatrix}$, $X_{96} = \begin{bmatrix} 62.0993 \\ 137.9007 \end{bmatrix}$ City Suburb, $X_{95} = \begin{bmatrix} 63.2624 \\ 136.7376 \end{bmatrix}$ City Suburb

Not stochastic, A^{-1} has negative numbers, but sum of a column is 1.

11. (a) .2   (b) $\begin{bmatrix} 1 & .2 \\ 0 & .8 \end{bmatrix} \begin{bmatrix} 10000 \\ 20000 \end{bmatrix} = \begin{bmatrix} 14000 \\ 16000 \end{bmatrix}$ white collar manual

12. (a) .6   (b) next generation: $\begin{bmatrix} 1 & .4 \\ 0 & .6 \end{bmatrix} \begin{bmatrix} 10000 \\ 1000 \end{bmatrix} = \begin{bmatrix} 10400 \\ 600 \end{bmatrix}$ nonfarming farming

Section 2.7

$\begin{bmatrix} 1 & .4 \\ 0 & .6 \end{bmatrix}^4 \begin{bmatrix} 10000 \\ 1000 \end{bmatrix} = \begin{bmatrix} 10870.4 \\ 129.6 \end{bmatrix}$, so the distribution after four generations will be 10870 nonfarming and 130 farming.

(c) $\begin{bmatrix} 1 & .4 \\ 0 & .6 \end{bmatrix}^2 = \begin{bmatrix} 1 & .64 \\ 0 & .36 \end{bmatrix}$, so the probability the grandson will be a farmer is .36.

13. $\begin{bmatrix} .8 & .4 \\ .2 & .6 \end{bmatrix}^4 \begin{bmatrix} 40000 \\ 50000 \end{bmatrix} = \begin{bmatrix} 59488 \\ 30512 \end{bmatrix}$ small large

14. $\begin{bmatrix} .8 & .2 & .6 \\ .15 & .7 & .3 \\ .05 & .1 & .1 \end{bmatrix} \begin{bmatrix} 2.5 \\ 1.5 \\ .25 \end{bmatrix} = \begin{bmatrix} 2.45 \\ 1.5 \\ .3 \end{bmatrix}$ Democrats Republicans Independents

15.
    AA Aa aa
$\begin{bmatrix} 1/2 & 1/4 & 0 \\ 1/2 & 1/2 & 1/2 \\ 0 & 1/4 & 1/2 \end{bmatrix}$ AA Aa aa

after one generation
$\begin{bmatrix} 1/2 & 1/4 & 0 \\ 1/2 & 1/2 & 1/2 \\ 0 & 1/4 & 1/2 \end{bmatrix} \begin{bmatrix} 1/3 \\ 1/3 \\ 1/3 \end{bmatrix} = \begin{bmatrix} 1/4 \\ 1/2 \\ 1/4 \end{bmatrix}$ AA Aa aa

after two generations
$\begin{bmatrix} 1/2 & 1/4 & 0 \\ 1/2 & 1/2 & 1/2 \\ 0 & 1/4 & 1/2 \end{bmatrix} \begin{bmatrix} 1/4 \\ 1/2 \\ 1/4 \end{bmatrix} = \begin{bmatrix} 1/4 \\ 1/2 \\ 1/4 \end{bmatrix}$ AA Aa aa

after three generations
$\begin{bmatrix} 1/2 & 1/4 & 0 \\ 1/2 & 1/2 & 1/2 \\ 0 & 1/4 & 1/2 \end{bmatrix} \begin{bmatrix} 1/4 \\ 1/2 \\ 1/4 \end{bmatrix} = \begin{bmatrix} 1/4 \\ 1/2 \\ 1/4 \end{bmatrix}$ AA Aa aa

Exercise Set 2.7

1.   adjacency matrix   distance matrix

(a) $\begin{bmatrix} 0 & 1 & 0 \\ 1 & 0 & 1 \\ 1 & 0 & 0 \end{bmatrix}$ $\begin{bmatrix} 0 & 1 & 2 \\ 1 & 0 & 1 \\ 1 & 2 & 0 \end{bmatrix}$

Section 2.7

(b) $\begin{bmatrix} 0 & 1 & 0 & 0 & 0 \\ 1 & 0 & 1 & 0 & 0 \\ 0 & 0 & 0 & 1 & 1 \\ 0 & 1 & 0 & 0 & 0 \\ 0 & 0 & 0 & 0 & 0 \end{bmatrix}$ $\begin{bmatrix} 0 & 1 & 2 & 3 & 3 \\ 1 & 0 & 1 & 2 & 2 \\ 3 & 2 & 0 & 1 & 1 \\ 2 & 1 & 2 & 0 & 3 \\ x & x & x & x & 0 \end{bmatrix}$

(c) $\begin{bmatrix} 0 & 1 & 1 & 1 & 1 \\ 0 & 0 & 0 & 0 & 0 \\ 0 & 0 & 0 & 0 & 0 \\ 0 & 0 & 0 & 0 & 0 \\ 0 & 0 & 0 & 0 & 0 \end{bmatrix}$ $\begin{bmatrix} 0 & 1 & 1 & 1 & 1 \\ x & 0 & x & x & x \\ x & x & 0 & x & x \\ x & x & x & 0 & x \\ x & x & x & x & 0 \end{bmatrix}$

(d) $\begin{bmatrix} 0 & 0 & 1 & 1 & 0 \\ 1 & 0 & 1 & 0 & 1 \\ 0 & 1 & 0 & 1 & 0 \\ 0 & 0 & 0 & 0 & 1 \\ 1 & 0 & 0 & 0 & 0 \end{bmatrix}$ $\begin{bmatrix} 0 & 2 & 1 & 1 & 2 \\ 1 & 0 & 1 & 2 & 1 \\ 2 & 1 & 0 & 1 & 2 \\ 2 & 4 & 3 & 0 & 1 \\ 1 & 3 & 2 & 2 & 0 \end{bmatrix}$

2. (a) 2   (b) undefined   (c) undefined   (d) 4

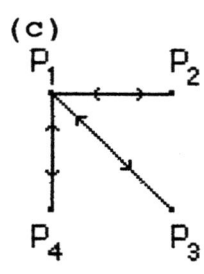

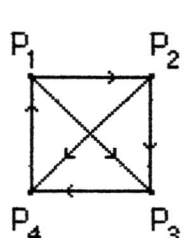

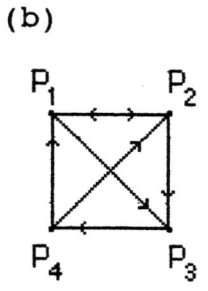

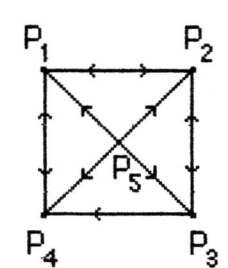

3. (a)   (b)
   (c)

   (d)   (e)

69

Section 2.7

4.   adjacency matrix

$$\begin{bmatrix} 0 & 1 & 0 & 0 & 0 & 0 \\ 0 & 0 & 1 & 0 & 1 & 0 \\ 0 & 0 & 0 & 1 & 0 & 0 \\ 0 & 0 & 0 & 0 & 1 & 0 \\ 0 & 1 & 0 & 0 & 0 & 1 \\ 1 & 0 & 0 & 0 & 0 & 0 \end{bmatrix}$$

distance matrix

$$\begin{bmatrix} 0 & 1 & 2 & 3 & 2 & 3 \\ 3 & 0 & 1 & 2 & 1 & 2 \\ 4 & 3 & 0 & 1 & 2 & 3 \\ 3 & 2 & 3 & 0 & 1 & 2 \\ 2 & 1 & 2 & 3 & 0 & 1 \\ 1 & 2 & 3 & 4 & 3 & 0 \end{bmatrix}$$

5.
$$\begin{array}{c} \\ \text{Raccoon} \\ \text{Bird} \\ \text{Deer} \\ \text{Grass} \\ \text{Insect} \end{array} \begin{array}{cccc} R & B & D & G & I \\ \begin{bmatrix} 0 & 0 & 0 & 0 & 1 \\ 0 & 0 & 0 & 1 & 1 \\ 0 & 0 & 0 & 1 & 0 \\ 0 & 0 & 0 & 0 & 0 \\ 0 & 0 & 0 & 1 & 0 \end{bmatrix} \end{array}$$

6.

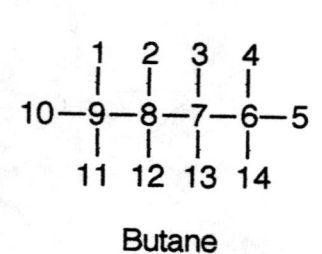

```
   1   2   3   4
   |   |   |   |
10—9—8—7—6—5
   |   |   |   |
  11  12  13  14
```
Butane

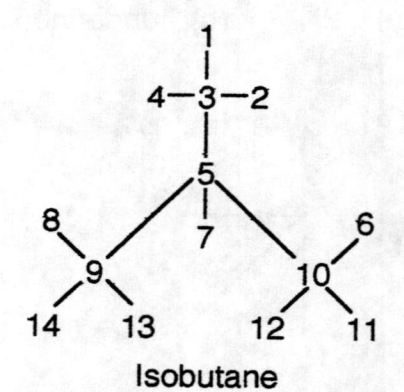

```
       1
       |
   4—3—2
       |
       5
  8   /|\   6
   \ / 7 \ /
    9     10
   / \   / \
  14 13 12 11
```
Isobutane

70

Section 2.7

$$\begin{bmatrix} 0&0&0&0&0&0&0&1&0&0&0&0&0&0 \\ 0&0&0&0&0&0&0&1&0&0&0&0&0&0 \\ 0&0&0&0&0&0&1&0&0&0&0&0&0&0 \\ 0&0&0&0&0&1&0&0&0&0&0&0&0&0 \\ 0&0&0&0&0&1&0&0&0&0&0&0&0&0 \\ 0&0&0&1&1&0&1&0&0&0&0&0&0&1 \\ 0&0&1&0&0&1&0&1&0&0&0&0&1&0 \\ 0&1&0&0&0&0&1&0&1&0&0&1&0&0 \\ 1&0&0&0&0&0&0&1&0&1&1&0&0&0 \\ 0&0&0&0&0&0&0&0&1&0&0&0&0&0 \\ 0&0&0&0&0&0&0&0&1&0&0&0&0&0 \\ 0&0&0&0&0&0&0&1&0&0&0&0&0&0 \\ 0&0&0&0&0&0&1&0&0&0&0&0&0&0 \\ 0&0&0&0&0&1&0&0&0&0&0&0&0&0 \end{bmatrix} \quad \begin{bmatrix} 0&0&1&0&0&0&0&0&0&0&0&0&0&0 \\ 0&0&1&0&0&0&0&0&0&0&0&0&0&0 \\ 1&1&0&1&1&0&0&0&0&0&0&0&0&0 \\ 0&0&1&0&0&0&0&0&0&0&0&0&0&0 \\ 0&0&1&0&0&0&1&0&1&1&0&0&0&0 \\ 0&0&0&0&0&0&0&0&0&1&0&0&0&0 \\ 0&0&0&0&1&0&0&0&0&0&0&0&0&0 \\ 0&0&0&0&0&0&0&0&1&0&0&0&0&0 \\ 0&0&0&0&1&0&0&1&0&0&0&0&1&1 \\ 0&0&0&0&1&1&0&0&0&0&1&1&0&0 \\ 0&0&0&0&0&0&0&0&0&1&0&0&0&0 \\ 0&0&0&0&0&0&0&0&0&1&0&0&0&0 \\ 0&0&0&0&0&0&0&0&1&0&0&0&0&0 \\ 0&0&0&0&0&0&0&0&1&0&0&0&0&0 \end{bmatrix}$$

7.

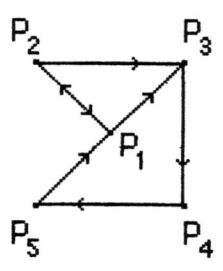

(a) $P_2 \to P_3 \to P_4 \to P_5$

length = 3

(b) $P_3 \to P_4 \to P_5 \to P_1 \to P_2$

length = 4

distance matrix
$$\begin{bmatrix} 0&1&1&2&3 \\ 1&0&1&2&3 \\ 3&4&0&1&2 \\ 2&3&3&0&1 \\ 1&2&2&3&0 \end{bmatrix}$$

8. (a) $(a_{22})^2 = 1$. There is one 2-path from P_2 to P_2.

$(a_{24})^2 = 0$. There is no 2-path from P_2 to P_4.

$(a_{31})^2 = 1$. There is one 2-path from P_3 to P_1.

$(a_{42})^2 = 1$. There is one 2-path from P_4 to P_2.

$(a_{12})^3 = 1$. There is one 3-path from P_1 to P_2.

$(a_{24})^3 = 0$. There is no 3-path from P_2 to P_4.

Section 2.7

$(a_{32})^3 = 1$. There is one 3-path from P_3 to P_2.
$(a_{41})^3 = 1$. There is one 3-path from P_4 to P_1.

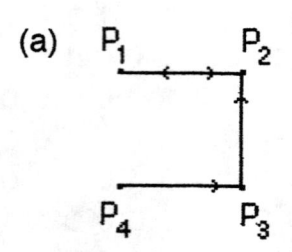

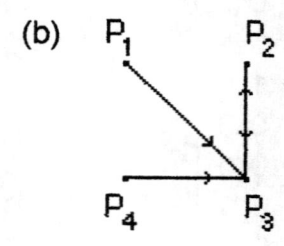

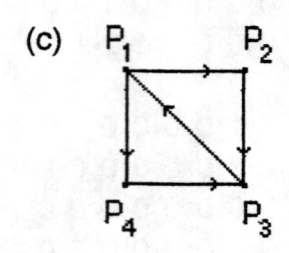

(a) $P_1 \quad P_2$
    $P_4 \quad P_3$

(b) $P_1 \quad P_2$
    $P_4 \quad P_3$

(c) $P_1 \quad P_2$
    $P_4 \quad P_3$

(b) $(a_{21})^2 = 0$. There is no 2-path from P_2 to P_1.
$(a_{33})^2 = 1$. There is one 2-path from P_3 to P_3.
$(a_{42})^2 = 1$. There is one 2-path from P_4 to P_2.
$(a_{13})^3 = 1$. There is one 3-path from P_1 to P_3.
$(a_{32})^3 = 1$. There is one 3-path from P_3 to P_2.
$(a_{34})^3 = 0$. There is no 3-path from P_3 to P_4.

(c) $(a_{13})^2 = 2$. There are two 2-paths from P_1 to P_3.
$(a_{21})^2 = 1$. There is one 2-path from P_2 to P_1.
$(a_{34})^2 = 1$. There is one 2-path from P_3 to P_4.
$(a_{11})^3 = 2$. There are two 3-paths from P_1 to P_1.
$(a_{24})^3 = 1$. There is one 3-path from P_2 to P_4.
$(a_{33})^3 = 2$. There are two 3-paths from P_3 to P_3.
$(a_{42})^3 = 1$. There is one 3-path from P_4 to P_2.

(d) $(a_{14})^2 = 2$. There are two 2-paths from P_1 to P_4.
$(a_{23})^2 = 1$. There is one 2-path from P_2 to P_3.
$(a_{53})^2 = 1$. There is one 2-path from P_5 to P_3.
$(a_{13})^3 = 2$. There are two 3-paths from P_1 to P_3.
$(a_{31})^3 = 0$. There is no 3-path from P_3 to P_1.

Section 2.7

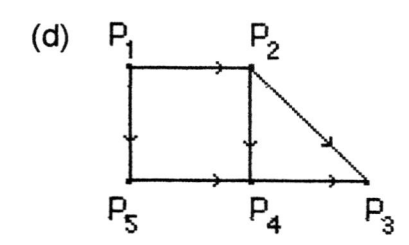

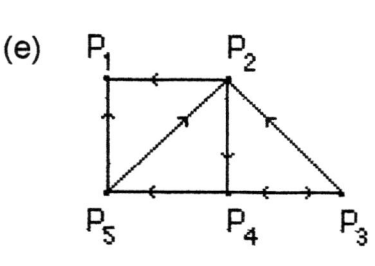

(d) [diagram with vertices P_1, P_2, P_3, P_4, P_5]

(e) [diagram with vertices P_1, P_2, P_3, P_4, P_5]

(e) $(a_{25})^2 = 1$. There is one 2-path from P_2 to P_5.
$(a_{31})^2 = 1$. There is one 2-path from P_3 to P_1.
$(a_{42})^2 = 2$. There are two 2-paths from P_4 to P_2.
$(a_{51})^2 = 1$. There is one 2-path from P_5 to P_1.
$(a_{53})^2 = 0$. There is no 2-path from P_5 to P_3.
$(a_{54})^2 = 1$. There is one 2-path from P_5 to P_4.

$(a_{21})^3 = 1$. There is one 3-path from P_2 to P_1.
$(a_{22})^3 = 2$. There are two 3-paths from P_2 to P_2.
$(a_{32})^3 = 2$. There are two 3-paths from P_3 to P_2.
$(a_{34})^3 = 1$. There is one 3-path from P_3 to P_4.
$(a_{35})^3 = 1$. There is one 3-path from P_3 to P_5.
$(a_{41})^3 = 2$. There are two 3-paths from P_4 to P_1.
$(a_{53})^3 = 1$. There is one 3-path from P_5 to P_3.

9. (a) There are no arcs from P_3 to any other vertex.
   (b) No arcs lead to P_4.
   (c) There are three arcs from P_5.
   (d) Two arcs lead to P_2.
   (e) There are no 3-paths from P_2.
   (f) No 4-paths lead to P_3.

10. (a) There are no arcs from P_2 to any other vertex.
    (b) No arcs lead to P_3.

Section 2.7

(c) There is an arc from P_4 to every other vertex.
(d) There is an arc from every other vertex to P_5.
(e) There are five arcs from P_3.
(f) Four arcs lead to P_2.
(g) There are seven arcs in the digraph.
(h) There are three 4-paths from P_2.
(i) Four 5-paths lead to P_3.
(j) There are no 2-paths from P_4.
(k) No 4-paths lead to P_3.

11. From A^2 it is known that there is one 2-path from P_1 to P_2, one 2-path from P_2 to P_3, one 2-path from P_3 to P_4, and one 2-path from P_4 to P_2. The possible 2-paths are

$P_1 \to P_3 \to P_2$ or $P_1 \to P_4 \to P_2$

$P_2 \to P_1 \to P_3$ or $P_2 \to P_4 \to P_3$

$P_3 \to P_1 \to P_4$ or $P_3 \to P_2 \to P_4$

$P_4 \to P_1 \to P_2$ or $P_4 \to P_3 \to P_2$

The only combination of these that does not produce additional 2-paths is
$P_1 \to P_3 \to P_2$, $P_2 \to P_4 \to P_3$, $P_3 \to P_2 \to P_4$, $P_4 \to P_3 \to P_2$.

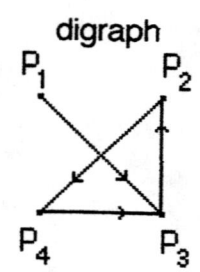

digraph

$$A^3 = \begin{bmatrix} 0 & 0 & 0 & 1 \\ 0 & 1 & 0 & 0 \\ 0 & 0 & 1 & 0 \\ 0 & 0 & 0 & 1 \end{bmatrix}$$

12. There is a 2-path from P_1 to P_3 and a 2-path from P_1 to P_4. The possible 2-paths are

$P_1 \to P_2 \to P_3$ or $P_1 \to P_4 \to P_3$

$P_1 \to P_2 \to P_4$ or $P_1 \to P_3 \to P_4$

Section 2.7

The only combination of these that does not produce additional 2-paths is
$P_1 \to P_2 \to P_3$ and $P_1 \to P_2 \to P_4$. The dashed lines in the digraph indicate arcs that
do not cause additional 2-paths and therefore may be in the digraph.

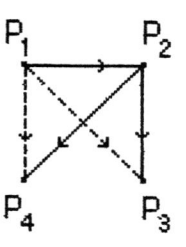

13. (a)

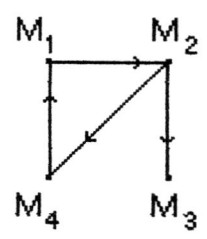

$$D = \begin{bmatrix} 0 & 1 & 2 & 2 \\ 2 & 0 & 1 & 1 \\ x & x & 0 & x \\ 1 & 2 & 3 & 0 \end{bmatrix} \begin{matrix} 5 \\ 4 \\ 3x \\ 6 \end{matrix}$$

most to least influential:
M_2, M_1, M_4, M_3

(b)

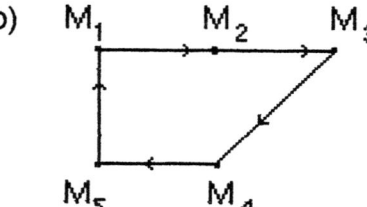

$$D = \begin{bmatrix} 0 & 1 & 2 & 3 & 4 \\ 4 & 0 & 1 & 2 & 3 \\ 3 & 4 & 0 & 1 & 2 \\ 2 & 3 & 4 & 0 & 1 \\ 1 & 2 & 3 & 4 & 0 \end{bmatrix}$$

All are equally influential.

(c)

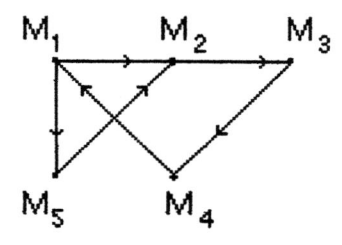

$$D = \begin{bmatrix} 0 & 1 & 2 & 3 & 1 \\ 3 & 0 & 1 & 2 & 4 \\ 2 & 3 & 0 & 1 & 3 \\ 1 & 2 & 3 & 0 & 2 \\ 4 & 1 & 2 & 3 & 0 \end{bmatrix} \begin{matrix} 7 \\ 10 \\ 9 \\ 8 \\ 10 \end{matrix}$$

most to least influential:

(d)

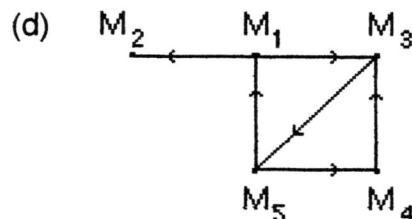

$$D = \begin{bmatrix} 0 & 1 & 1 & 3 & 2 \\ x & 0 & x & x & x \\ 2 & 3 & 0 & 2 & 1 \\ 3 & 4 & 1 & 0 & 2 \\ 1 & 2 & 2 & 1 & 0 \end{bmatrix} \begin{matrix} 7 \\ 4x \\ 8 \\ 9 \\ 6 \end{matrix}$$

most to least influential:

75

Section 2.7

M_1, M_4, M_3, with M_2 and M_5 equally uninfluential

M_5, M_1, M_3, M_4, M_2

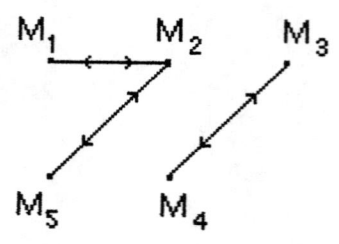

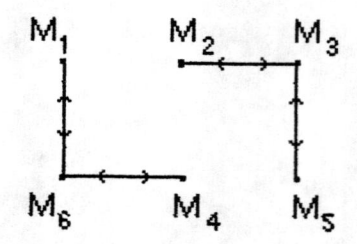

14. (a) [digraph with vertices M_1, M_2, M_3, M_5, M_4]   (b) [digraph with vertices $M_1, M_2, M_3, M_6, M_4, M_5$]

A symmetric matrix implies that all direct friendships are mutual.

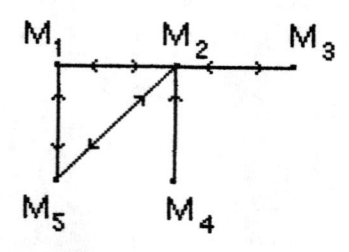

15. [digraph with vertices M_1, M_2, M_3, M_5, M_4]   In this digraph there is one clique: [triangle digraph with M_1, M_2, M_5]

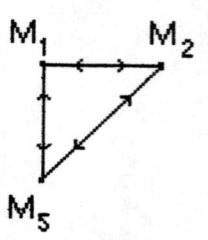

16. The number of rows and the number of columns in the adjacency matrix are both equal to the number of vertices in the digraph and are therefore the same.

17. If the digraph of A contains an arc from P_i to P_j, then the digraph of A^t contains an arc from P_j to P_i. If the digraph of A does not contain an arc from P_i to P_j, then the digraph of A^t does not contain an arc from P_j to P_i.

18. There is an arc from P_i to P_j if and only if there is an arc from P_j to P_i.

19. If a path from P_i to P_j contains P_k twice,

$$P_i \to \ldots \to P_k \to \ldots \to P_k \to \ldots \to P_j,$$

Section 2.7

then there is a shorter path from P_i to P_j obtained by omitting the path from P_k to P_k.

20. $n - 1$

21. $c_{ij} = a_{i1}a_{j1} + a_{i2}a_{j2} + \ldots + a_{in}a_{jn}$. $a_{ik}a_{jk} = 1$ if station k can receive messages directly from both station i and station j and $a_{ik}a_{jk} = 0$ otherwise. Thus c_{ij} is the number of stations that can receive messages directly from both station i and station j.

22. (a) $\begin{bmatrix} 1 & 1 & 0 & 0 \\ 1 & 1 & 0 & 0 \\ 1 & 1 & 1 & 0 \\ 1 & 1 & 1 & 1 \end{bmatrix}$ (b) $\begin{bmatrix} 1 & 1 & 1 & 0 \\ 0 & 1 & 1 & 0 \\ 0 & 1 & 1 & 0 \\ 0 & 1 & 1 & 1 \end{bmatrix}$ (c) $\begin{bmatrix} 1 & 1 & 1 & 1 \\ 1 & 1 & 1 & 1 \\ 1 & 1 & 1 & 1 \\ 1 & 1 & 1 & 1 \end{bmatrix}$

(d) $\begin{bmatrix} 1 & 1 & 1 & 1 & 1 \\ 0 & 1 & 1 & 1 & 0 \\ 0 & 0 & 1 & 0 & 0 \\ 0 & 0 & 1 & 1 & 0 \\ 0 & 0 & 1 & 1 & 1 \end{bmatrix}$ (e) $\begin{bmatrix} 1 & 0 & 0 & 0 & 0 \\ 1 & 1 & 1 & 1 & 1 \\ 1 & 1 & 1 & 1 & 1 \\ 1 & 1 & 1 & 1 & 1 \\ 1 & 1 & 1 & 1 & 1 \end{bmatrix}$

23. For each vertex, the adjacency matrix gives the vertices reachable by arcs or 1-paths, its square gives all vertices reachable by 2-paths, etc. Since there are n vertices, any vertex can be reached by a path of length at most n - 1 or else it cannot be reached at all. Thus all the information needed will be contained in the first n - 1 powers of the adjacency matrix.

24. (a) Yes. If P_j is reachable from P_i, then P_i is reachable from P_j by reversing the path.

(b) No. The digraph that has adjacency matrix $\begin{bmatrix} 0 & 1 & 0 \\ 0 & 0 & 1 \\ 1 & 1 & 0 \end{bmatrix}$ has reachability matrix $\begin{bmatrix} 1 & 1 & 1 \\ 1 & 1 & 1 \\ 1 & 1 & 1 \end{bmatrix}$.

25. r(i) is the number of stations that can receive a message from station i. c(j) is the number of stations that can send a message to station j.

77

26. $(r_{ij})^2 = r_{i1}r_{1j} + r_{i2}r_{2j} + \ldots + r_{in}r_{nj}$ is the number of vertices that can be reached from vertex i and that can reach vertex j; that is, the number of vertices that could be on paths from vertex i to vertex j.

27. The (i,j)th element of R^t is 1 if vertex i can be reached from vertex j or if i = j. It is zero if vertex i cannot be reached from vertex j.

28. (a) and (b) No. R has a 1 in every diagonal position and A has a zero in every diagonal position, and every 1 in A causes at least one 1 in R, so there cannot be as many ones in A as in R.

    (c) Yes. If $A = \begin{bmatrix} 0 & 1 & 0 \\ 0 & 0 & 1 \\ 0 & 0 & 0 \end{bmatrix}$, then $R = \begin{bmatrix} 1 & 1 & 1 \\ 0 & 1 & 1 \\ 0 & 0 & 1 \end{bmatrix}$. In fact R will always contain more 1s than A.

29. For processors in the same row or column, the maximum distance is 4, so the total maximum distance is 8.

Chapter 2 Review Exercises

1. (a) $2AB = \begin{bmatrix} 28 & 0 \\ 100 & -6 \end{bmatrix}$. (b) AB + C does not exist.

   (c) $BA + AB = \begin{bmatrix} 28 & 0 \\ 69 & -6 \end{bmatrix}$. (d) $AD - 3D = \begin{bmatrix} -6 \\ 26 \end{bmatrix}$.

   (e) $AC + BC = \begin{bmatrix} 54 & -9 & 27 \\ 46 & -6 & 14 \end{bmatrix}$. (f) 2DA + B does not exist.

2. (a) 2x2   (b) 2x3   (c) 2x2   (d) Does not exist.   (e) 3x2
   (f) Does not exist.   (g) 2x2

Chapter 2 Review Exercises

3. (a) $d_{12} = 2(1 \times 2 + (-3) \times 0) - 3(-4) = 16$.

   (b) $d_{23} = 2(0 \times (-3) + 4 \times (-1)) - 3 \times 0 = -8$.

4. $m_1 = (a_{12} - a_{22})(b_{21} + b_{22}) = (8)(-2) = -16$,
   $(2)(7) = 14$,

   $m_2 = (a_{11} + a_{22})(b_{11} + b_{22}) = (-3)(-3) = 9$,
   $5)(1) = -5$,

   $m_3 = (a_{11} - a_{21})(b_{11} + b_{12}) = (1)(4) = 4$,

   $m_4 = (a_{11} + a_{12})b_{22} = (5)(-4) = -20$,

   $m_5 = a_{11}(b_{12} - b_{22}) =$

   $m_6 = a_{22}(b_{21} - b_{11}) = (-5)(1) = -5$,

   $m_7 = (a_{21} + a_{22})b_{11} = (-4)(1) = -4$,

   so $AB = \begin{bmatrix} -16+9-(-20)+(-5) & -20+(14) \\ -5+(-4) & 9-(4)+(14)-(-4) \end{bmatrix} = \begin{bmatrix} 8 & -6 \\ -9 & 23 \end{bmatrix}$.

5. (a) $(A^t)^2 = (A^2)^t = \begin{bmatrix} 9 & 5 \\ 0 & 4 \end{bmatrix}^t = \begin{bmatrix} 9 & 0 \\ 5 & 4 \end{bmatrix}$.

   (b) $A^t - B^2 = \begin{bmatrix} 3 & 0 \\ 1 & 2 \end{bmatrix} - \begin{bmatrix} 7 & -1 \\ -3 & 4 \end{bmatrix} = \begin{bmatrix} -4 & 1 \\ 4 & -2 \end{bmatrix}$.

   (c) $AB^3 - 2C^2 = \begin{bmatrix} 3 & 1 \\ 0 & 2 \end{bmatrix}\begin{bmatrix} -17 & 6 \\ 18 & 1 \end{bmatrix} - 2\begin{bmatrix} 3 & 2 \\ 6 & 7 \end{bmatrix} = \begin{bmatrix} -39 & 15 \\ 24 & -12 \end{bmatrix}$.

   (d) $A^2 - 3A + 4I_2 = \begin{bmatrix} 9 & 5 \\ 0 & 4 \end{bmatrix} - \begin{bmatrix} 9 & 3 \\ 0 & 6 \end{bmatrix} + \begin{bmatrix} 4 & 0 \\ 0 & 4 \end{bmatrix} = \begin{bmatrix} 4 & 2 \\ 0 & 2 \end{bmatrix}$.

6. (a) $2 \times 2 \times 6 = 24$.   (b) $4 \times 2 \times 3 = 24$.   (c) $1 \times 7 \times 25 = 175$

   (d) $9 \times 5 \times 11 = 495$.

7. (a) $\begin{bmatrix} 1 & 4 & 1 & 0 \\ 2 & -1 & 0 & 1 \end{bmatrix} \underset{R2+(-2)R1}{\approx} \begin{bmatrix} 1 & 4 & 1 & 0 \\ 0 & -9 & -2 & 1 \end{bmatrix} \underset{(-1/9)R2}{\approx} \begin{bmatrix} 1 & 4 & 1 & 0 \\ 0 & 1 & 2/9 & -1/9 \end{bmatrix}$

   $\underset{R1+(-4)R2}{\approx} \begin{bmatrix} 1 & 0 & 1/9 & 4/9 \\ 0 & 1 & 2/9 & -1/9 \end{bmatrix}$, so the inverse is $\begin{bmatrix} 1/9 & 4/9 \\ 2/9 & -1/9 \end{bmatrix}$.

Chapter 2 Review Exercises

(b) $\begin{bmatrix} 0 & 3 & 3 & 1 & 0 & 0 \\ 1 & 2 & 3 & 0 & 1 & 0 \\ 1 & 4 & 6 & 0 & 0 & 1 \end{bmatrix} \underset{R1 \leftrightarrow R2}{\approx} \begin{bmatrix} 1 & 2 & 3 & 0 & 1 & 0 \\ 0 & 3 & 3 & 1 & 0 & 0 \\ 1 & 4 & 6 & 0 & 0 & 1 \end{bmatrix}$

$\underset{R3+(-1)R1}{\approx} \begin{bmatrix} 1 & 2 & 3 & 0 & 1 & 1 \\ 0 & 3 & 3 & 1 & 0 & 0 \\ 0 & 2 & 3 & 0 & -1 & 1 \end{bmatrix} \underset{(1/3)R2}{\approx} \begin{bmatrix} 1 & 2 & 3 & 0 & 1 & 0 \\ 0 & 1 & 1 & 1/3 & 0 & 0 \\ 0 & 2 & 3 & 0 & -1 & 1 \end{bmatrix}$

$\underset{\substack{R1+(-2)R2 \\ R3+(-2)R2}}{\approx} \begin{bmatrix} 1 & 0 & 1 & -2/3 & 1 & 0 \\ 0 & 1 & 1 & 1/3 & 0 & 0 \\ 0 & 0 & 1 & -2/3 & -1 & 1 \end{bmatrix} \underset{\substack{R1+(-1)R3 \\ R2+(-1)R3}}{\approx} \begin{bmatrix} 1 & 0 & 0 & 0 & 2 & -1 \\ 0 & 1 & 0 & 1 & 1 & -1 \\ 0 & 0 & 1 & -2/3 & -1 & 1 \end{bmatrix}$,

so the inverse is $\begin{bmatrix} 0 & 2 & -1 \\ 1 & 1 & -1 \\ -2/3 & -1 & 1 \end{bmatrix}$.

(c) $\begin{bmatrix} 1 & 2 & 3 & 1 & 0 & 0 \\ 2 & 5 & 3 & 0 & 1 & 0 \\ 1 & 0 & 8 & 0 & 0 & 1 \end{bmatrix} \underset{\substack{R2+(-2)R1 \\ R3+(-1)R1}}{\approx} \begin{bmatrix} 1 & 2 & 3 & 1 & 0 & 0 \\ 0 & 1 & -3 & -2 & 1 & 0 \\ 0 & -2 & 5 & -1 & 0 & 1 \end{bmatrix}$

$\underset{\substack{R1+(-2)R2 \\ R3+(2)R2}}{\approx} \begin{bmatrix} 1 & 0 & 9 & 5 & -2 & 0 \\ 0 & 1 & -3 & -2 & 1 & 0 \\ 0 & 0 & -1 & -5 & 2 & 1 \end{bmatrix} \underset{(-1)R3}{\approx} \begin{bmatrix} 1 & 0 & 9 & 5 & -2 & 0 \\ 0 & 1 & -3 & -2 & 1 & 0 \\ 0 & 0 & 1 & 5 & -2 & -1 \end{bmatrix}$

$\underset{\substack{R1+(-9)R3 \\ R2+(3)R3}}{\approx} \begin{bmatrix} 1 & 0 & 0 & -40 & 16 & 9 \\ 0 & 1 & 0 & 13 & -5 & -3 \\ 0 & 0 & 1 & 5 & -2 & -1 \end{bmatrix}$, so the inverse is $\begin{bmatrix} -40 & 16 & 9 \\ 13 & -5 & -3 \\ 5 & -2 & -1 \end{bmatrix}$.

8. The inverse of the coefficient matrix is $\begin{bmatrix} 14 & -8 & -1 \\ -17 & 10 & 1 \\ -19 & 11 & 1 \end{bmatrix}$.

$\begin{bmatrix} x_1 \\ x_2 \\ x_3 \end{bmatrix} = \begin{bmatrix} 14 & -8 & -1 \\ -17 & 10 & 1 \\ -19 & 11 & 1 \end{bmatrix} \begin{bmatrix} 1 \\ 5 \\ 7 \end{bmatrix} = \begin{bmatrix} -33 \\ 40 \\ 43 \end{bmatrix}$.

9. $A = 3 \begin{bmatrix} 5 & -6 \\ -2 & 3 \end{bmatrix}^{-1} = 3 \begin{bmatrix} 1 & 2 \\ 2/3 & 5/3 \end{bmatrix} = \begin{bmatrix} 3 & 6 \\ 2 & 5 \end{bmatrix}$.

Chapter 2 Review Exercises

10. The (i,j)th element of A(BC) is $A_i(BC_j)$, where A_i is the ith row of A and C_j is the jth column of C. Likewise, the (i,j)th element of (AB)C is $(A_iB)C_j$. We show that these elements are the same.

$$A_i(BC_j) = [a_{i1}\ a_{i2}\ \cdots\ a_{in}] \begin{bmatrix} b_{11}c_{1j}+b_{12}c_{2j}+\cdots+b_{1r}c_{rj} \\ b_{21}c_{1j}+b_{22}c_{2j}+\cdots+b_{2r}c_{rj} \\ \vdots \\ b_{n1}c_{1j}+b_{n2}c_{2j}+\cdots+b_{nr}c_{rj} \end{bmatrix}$$

$$= a_{i1}(b_{11}c_{1j}+b_{12}c_{2j}+\cdots+b_{1r}c_{rj}) + a_{i2}(b_{21}c_{1j}+b_{22}c_{2j}+\cdots+b_{2r}c_{rj})$$
$$+\ \cdot\quad\cdot\quad\cdot\quad + a_{in}(b_{n1}c_{1j}+b_{n2}c_{2j}+\cdots+b_{nr}c_{rj})$$

$$= (a_{i1}b_{11}+a_{i2}b_{21}+\cdots+a_{in}b_{n1})c_{1j} + (a_{i1}b_{12}+a_{i2}b_{22}+\cdots+a_{in}b_{n2})c_{2j}$$
$$+\ \cdot\quad\cdot\quad\cdot\quad + (a_{i1}b_{1r}+a_{i2}b_{2r}+\cdots+a_{in}b_{nr})c_{rj}$$

$$= (A_iB)C_j.$$

11. $A(AB + A^2) - B(A^2 + AB) + 3ABA - 4AB^2 = A^2B + A^3 - BA^2 - BAB + 3ABA - 4AB^2.$

12. $(cA)^n = (cA)(cA)\cdots(cA) = c^n A^n.$

13. The ith diagonal element of AA^t is
$$a_{i1}a_{i1} + a_{i2}a_{i2} + \cdots + a_{in}a_{in} = (a_{i1})^2 + (a_{i2})^2 + \cdots + (a_{in})^2.$$

Since each term is a square, the sum can equal zero only if each term is zero, i.e. if each $a_{ij} = 0$. Thus if $AA^t = O$, then $A = O$.

14. If A is a symmetric matrix then $A = A^t$, so $AA^t = A^2 = A^tA$ and A is normal.

Chapter 2 Review Exercises

15. If $A = A^2$ then $A^t = (A^2)^t = (A^t)^2$, so A^t is idempotent.

16. From Exercise 8 in Section 2.3, $(A^t)^n = (A^n)^t$. If $n < p$, $A^n \neq O$, so $(A^n)^t \neq O$, so $(A^t)^n \neq O$. $A^p = O$ so $(A^p)^t = (A^t)^p = O$. Thus A^t is nilpotent with degree of nilpotency $= p$.

17. $A = A^t$, so $A^{-1} = (A^t)^{-1} = (A^{-1})^t$, so A^{-1} is symmetric.

18. If row i of A is all zeros, then row i of AB is all zeros for any matrix B. The ith diagonal term of I_n is 1, so there is no matrix B for which $AB = I_n$. Likewise, if column j of A is all zeros, then column j of BA is all zeros for any B, so there is no matrix B with $BA = I_n$.

19. $A + B = \begin{bmatrix} 5+i & 5-5i \\ 6+10i & -3+i \end{bmatrix}$. $AB = \begin{bmatrix} 35+24i & -3+6i \\ 7+6i & 12-6i \end{bmatrix}$.

$\overline{A} = \begin{bmatrix} 2 & 4+3i \\ 4-3i & -1 \end{bmatrix}$, so $A^* = \begin{bmatrix} 2 & 4-3i \\ 4+3i & -1 \end{bmatrix} = A$; i.e., A is hermitian.

20. If A is a real symmetric matrix, then $\overline{A} = A = A^t$ so $A^* = A^t = A$. Thus A is hermitian.

21. $G = AA^t = \begin{bmatrix} 1 & 1 & 0 & 0 & 0 \\ 1 & 2 & 0 & 0 & 1 \\ 0 & 0 & 2 & 1 & 1 \\ 0 & 0 & 1 & 1 & 0 \\ 0 & 1 & 1 & 0 & 2 \end{bmatrix}$ $P = A^tA = \begin{bmatrix} 2 & 1 & 0 & 0 \\ 1 & 2 & 1 & 0 \\ 0 & 1 & 2 & 1 \\ 0 & 0 & 1 & 2 \end{bmatrix}$

$g_{12} = 1$, $g_{13} = 0$, $g_{14} = 0$, $g_{15} = 0$,
$g_{23} = 0$, $g_{24} = 0$, $g_{25} = 1$,
$g_{34} = 1$, $g_{35} = 1$, $g_{45} = 0$,
so $1 \to 2 \to 5 \to 3 \to 4$
or $4 \to 3 \to 5 \to 2 \to 1$.

$p_{12} = 1$, $p_{13} = 0$, $p_{14} = 0$,
$p_{23} = 1$, $p_{24} = 0$, $p_{34} = 1$,
so $1 \to 2 \to 3 \to 4$ or $4 \to 3 \to 2 \to 1$.

22. $P^2 = \begin{bmatrix} .9 & .25 \\ .1 & .75 \end{bmatrix}\begin{bmatrix} .9 & .25 \\ .1 & .75 \end{bmatrix} = \begin{bmatrix} .835 & .4125 \\ .165 & .5875 \end{bmatrix}.$

$\begin{bmatrix} .835 & .4125 \\ .165 & .5875 \end{bmatrix}\begin{bmatrix} 300000 \\ 750000 \end{bmatrix} = \begin{bmatrix} 559875 \\ 490125 \end{bmatrix}$, so in two generations the distribution will be 559,875 college-educated and 490,125 noncollege-educated.

The probability is .4125 that a couple with no college education will have at least one grandchild with a college education.

23. (a) No arcs lead to vertex 4.
    (b) There are two arcs from vertex 3.
    (c) There are four 3-paths from vertex 2.
    (d) No 2-paths lead to vertex 3.
    (e) There are two 3-paths from vertex 4 to vertex 4.
    (f) There are three pairs of vertices joined by 4-paths.

Chapter 3

Exercise Set 3.1

1. (a) $\begin{vmatrix} 2 & 1 \\ 3 & 5 \end{vmatrix} = (2 \times 5) - (1 \times 3) = 7.$
   (b) $\begin{vmatrix} 3 & -2 \\ 1 & 2 \end{vmatrix} = (3 \times 2) - (-2 \times 1) = 8.$

   (c) $\begin{vmatrix} 4 & 1 \\ -2 & 3 \end{vmatrix} = (4 \times 3) - (1 \times -2) = 14.$
   (d) $\begin{vmatrix} 5 & -2 \\ -3 & -4 \end{vmatrix} = (5 \times -4) - (-2 \times -3) = -26.$

2. (a) $\begin{vmatrix} 1 & -5 \\ 0 & 3 \end{vmatrix} = (1 \times 3) - (-5 \times 0) = 3.$
   (b) $\begin{vmatrix} 1 & 2 \\ 3 & 4 \end{vmatrix} = (1 \times 4) - (2 \times 3) = -2.$

   (c) $\begin{vmatrix} -3 & 1 \\ 2 & -5 \end{vmatrix} = (-3 \times -5) - (1 \times 2) = 13.$
   (d) $\begin{vmatrix} 3 & -2 \\ -1 & 0 \end{vmatrix} = (3 \times 0) - (-2 \times -1) = -2.$

3. (a) $M_{11} = \begin{vmatrix} 0 & 6 \\ 1 & -4 \end{vmatrix} = (0 \times -4) - (6 \times 1) = -6.$ $C_{11} = (-1)^{1+1}M_{11} = (-1)^2(-6) = -6.$

   (b) $M_{21} = \begin{vmatrix} 2 & -3 \\ 1 & -4 \end{vmatrix} = (2 \times -4) - (-3 \times 1) = -5.$ $C_{21} = (-1)^{2+1}M_{21} = (-1)^3(-5) = 5.$

   (c) $M_{23} = \begin{vmatrix} 1 & 2 \\ 7 & 1 \end{vmatrix} = (1 \times 1) - (2 \times 7) = -13.$ $C_{23} = (-1)^{2+3}M_{23} = (-1)^5(-13) = 13.$

   (d) $M_{33} = \begin{vmatrix} 1 & 2 \\ 5 & 0 \end{vmatrix} = (1 \times 0) - (2 \times 5) = -10.$ $C_{33} = (-1)^{3+3}M_{33} = (-1)^6(-10) = -10.$

4. (a) $M_{13} = \begin{vmatrix} -2 & 3 \\ 0 & -6 \end{vmatrix} = (-2 \times -6) - (3 \times 0) = 12.$ $C_{13} = (-1)^{1+3}M_{13} = (-1)^4(12) = 12.$

   (b) $M_{22} = \begin{vmatrix} 5 & 1 \\ 0 & 2 \end{vmatrix} = (5 \times 2) - (1 \times 0) = 10.$ $C_{22} = (-1)^{2+2}M_{22} = (-1)^4(10) = 10.$

   (c) $M_{31} = \begin{vmatrix} 0 & 1 \\ 3 & 7 \end{vmatrix} = (0 \times 7) - (1 \times 3) = -3.$ $C_{31} = (-1)^{3+1}M_{31} = (-1)^4(-3) = -3.$

Section 3.1

(d) $M_{33} = \begin{vmatrix} 5 & 0 \\ -2 & 3 \end{vmatrix} = (5 \times 3) - (0 \times -2) = 15.$ $\qquad C_{33} = (-1)^{3+3} M_{33} = (-1)^6 (15) = 15.$

5. (a) $M_{12} = \begin{vmatrix} 8 & 2 & 1 \\ 4 & -5 & 0 \\ 1 & 8 & 2 \end{vmatrix}$

$= (8 \times -5 \times 2) + (2 \times 0 \times 1) + (1 \times 4 \times 8) - (1 \times -5 \times 1) - (8 \times 0 \times 8) - (2 \times 4 \times 2)$

$= -80 + 0 + 32 - (-5) - 0 - 16 = -59.$

$C_{12} = (-1)^{1+2} M_{12} = (-1)^3 (-59) = 59.$

(b) $M_{24} = \begin{vmatrix} 2 & 0 & 1 \\ 4 & -3 & -5 \\ 1 & 4 & 8 \end{vmatrix}$

$= (2 \times -3 \times 8) + (0 \times -5 \times 1) + (1 \times 4 \times 4) - (1 \times -3 \times 1) - (2 \times -5 \times 4) - (0 \times 4 \times 8)$

$= -48 + 0 + 16 - (-3) - (-40) - 0 = 11.$

$C_{24} = (-1)^{2+4} M_{24} = (-1)^6 (11) = 11.$

(c) $M_{33} = \begin{vmatrix} 2 & 0 & -5 \\ 8 & -1 & 1 \\ 1 & 4 & 2 \end{vmatrix}$

$= (2 \times -1 \times 2) + (0 \times 1 \times 1) + (-5 \times 8 \times 4) - (-5 \times -1 \times 1) - (2 \times 1 \times 4) - (0 \times 8 \times 2)$

$= -4 + 0 + (-160) - 5 - 8 - 0 = -177.$

$C_{33} = (-1)^{3+3} M_{33} = (-1)^6 (-177) = -177.$

(d) $M_{43} = \begin{vmatrix} 2 & 0 & -5 \\ 8 & -1 & 1 \\ 4 & -3 & 0 \end{vmatrix}$

$= (2 \times -1 \times 0) + (0 \times 1 \times 4) + (-5 \times 8 \times -3) - (-5 \times -1 \times 4) - (2 \times 1 \times -3) - (0 \times 8 \times 0)$

$= 0 + 0 + 120 - 20 - (-6) - 0 = 106.$

Section 3.1

$$C_{43} = (-1)^{4+3}M_{43} = (-1)^7(106) = -106.$$

6. (a) $\begin{vmatrix} 1 & 2 & 4 \\ 4 & -1 & 5 \\ -2 & 2 & 1 \end{vmatrix} = \begin{vmatrix} -1 & 5 \\ 2 & 1 \end{vmatrix} - 2\begin{vmatrix} 4 & 5 \\ -2 & 1 \end{vmatrix} + 4\begin{vmatrix} 4 & -1 \\ -2 & 2 \end{vmatrix}$

$= [(-1 \times 1) - (5 \times 2)] - 2[(4 \times 1) - (5 \times -2)] + 4[(4 \times 2) - (-1 \times -2)]$

$= -11 - 2(14) + 4(6) = -15.$

"diagonals" method:

$(1 \times -1 \times 1) + (2 \times 5 \times -2) + (4 \times 4 \times 2) - (4 \times -1 \times -2) - (1 \times 5 \times 2) - (2 \times 4 \times 1)$

$= -1 + (-20) + 32 - 8 - 10 - 8 = -15.$

(b) $\begin{vmatrix} 3 & 1 & 4 \\ -7 & -2 & 1 \\ 9 & 1 & -1 \end{vmatrix} = 3\begin{vmatrix} -2 & 1 \\ 1 & -1 \end{vmatrix} - \begin{vmatrix} -7 & 1 \\ 9 & -1 \end{vmatrix} + 4\begin{vmatrix} -7 & -2 \\ 9 & 1 \end{vmatrix}$

$= 3[(-2 \times -1) - (1 \times 1)] - [(-7 \times -1) - (1 \times 9)] + 4[(-7 \times 1) - (-2 \times 9)]$

$= 3 - (-2) + 4(11) = 49.$

"diagonals" method:

$(3 \times -2 \times -1) + (1 \times 1 \times 9) + (4 \times -7 \times 1) - (4 \times -2 \times 9) - (3 \times 1 \times 1) - (1 \times -7 \times -1)$

$= 6 + 9 + (-28) - (-72) - 3 - 7 = 49.$

(c) $\begin{vmatrix} 4 & 1 & -2 \\ 5 & 3 & -1 \\ 2 & 4 & 1 \end{vmatrix} = 4\begin{vmatrix} 3 & -1 \\ 4 & 1 \end{vmatrix} - \begin{vmatrix} 5 & -1 \\ 2 & 1 \end{vmatrix} + (-2)\begin{vmatrix} 5 & 3 \\ 2 & 4 \end{vmatrix}$

$= 4[(3 \times 1) - (-1 \times 4)] - [(5 \times 1) - (-1 \times 2)] + (-2)[(5 \times 4) - (3 \times 2)]$

$= 4(7) - 7 + (-2)(14) = -7.$

"diagonals" method:

$(4 \times 3 \times 1) + (1 \times -1 \times 2) + (-2 \times 5 \times 4) - (-2 \times 3 \times 2) - (4 \times -1 \times 4) - (1 \times 5 \times 1)$

$= 12 + (-2) + (-40) - (-12) - (-16) - 5 = -7.$

7. (a) $\begin{vmatrix} 2 & 0 & 7 \\ 8 & -1 & -2 \\ 5 & 6 & 1 \end{vmatrix} = 2\begin{vmatrix} -1 & -2 \\ 6 & 1 \end{vmatrix} - 0\begin{vmatrix} 8 & -2 \\ 5 & 1 \end{vmatrix} + 7\begin{vmatrix} 8 & -1 \\ 5 & 6 \end{vmatrix}$

$= 2[(-1 \times 1) - (-2 \times 6)] - 0 + 7[(8 \times 6) - (-1 \times 5)] = 2(11) - 0 + 7(53) = 393.$

"diagonals" method:

$(2 \times -1 \times 1) + (0 \times -2 \times 5) + (7 \times 8 \times 6) - (7 \times -1 \times 5) - (2 \times -2 \times 6) - (0 \times 8 \times 1)$

$= -2 + 0 + (336) - (-35) - (-24) - 0 = 393.$

(b) $\begin{vmatrix} 2 & -1 & 3 \\ 4 & 0 & 2 \\ 1 & 1 & 1 \end{vmatrix} = 2\begin{vmatrix} 0 & 2 \\ 1 & 1 \end{vmatrix} - (-1)\begin{vmatrix} 4 & 2 \\ 1 & 1 \end{vmatrix} + 3\begin{vmatrix} 4 & 0 \\ 1 & 1 \end{vmatrix}$

$= 2[(0 \times 1) - (2 \times 1)] - (-1)[(4 \times 1) - (2 \times 1)] + 3[(4 \times 1) - (0 \times 1)]$

$= 2(-2) - (-1)(2) + 3(4) = 10.$

"diagonals" method:

$(2 \times 0 \times 1) + (-1 \times 2 \times 1) + (3 \times 4 \times 1) - (3 \times 0 \times 1) - (2 \times 2 \times 1) - (-1 \times 4 \times 1)$

$= 0 + (-2) + 12 - 0 - (4) - (-4) = 10.$

(c) $\begin{vmatrix} 0 & 0 & 5 \\ 1 & 1 & 1 \\ 2 & 2 & 2 \end{vmatrix} = 0\begin{vmatrix} 1 & 1 \\ 2 & 2 \end{vmatrix} - 0\begin{vmatrix} 1 & 1 \\ 2 & 2 \end{vmatrix} + 5\begin{vmatrix} 1 & 1 \\ 2 & 2 \end{vmatrix}$

$= 0 - 0 + 5[(1 \times 2) - (1 \times 2)] = 0$

"diagonals" method:

$(0 \times 1 \times 2) + (0 \times 1 \times 2) + (5 \times 1 \times 2) - (5 \times 1 \times 2) - (0 \times 1 \times 2) - (0 \times 1 \times 2) = 0.$

8. (a) $\begin{vmatrix} 0 & 3 & 2 \\ 1 & 5 & 7 \\ -2 & -6 & -1 \end{vmatrix}$

using row 2:

$$= -\begin{vmatrix} 3 & 2 \\ -6 & -1 \end{vmatrix} + 5 \begin{vmatrix} 0 & 2 \\ -2 & -1 \end{vmatrix} - 7 \begin{vmatrix} 0 & 3 \\ -2 & -6 \end{vmatrix} = -9 + 5 \times 4 - 7 \times 6 = -31.$$

using column 1:

$$= 0 \begin{vmatrix} 5 & 7 \\ -6 & -1 \end{vmatrix} - \begin{vmatrix} 3 & 2 \\ -6 & -1 \end{vmatrix} + (-2) \begin{vmatrix} 3 & 2 \\ 5 & 7 \end{vmatrix} = 0 - 9 + (-2)(11) = -31.$$

(b) $\begin{vmatrix} 4 & 2 & 1 \\ -6 & 3 & -2 \\ 7 & 1 & -1 \end{vmatrix}$

using row 3:

$$= 7 \begin{vmatrix} 2 & 1 \\ 3 & -2 \end{vmatrix} - \begin{vmatrix} 4 & 1 \\ -6 & -2 \end{vmatrix} + (-1) \begin{vmatrix} 4 & 2 \\ -6 & 3 \end{vmatrix} = -49 - (-2) + (-24) = -71.$$

using column 2:

$$= -2 \begin{vmatrix} -6 & -2 \\ 7 & -1 \end{vmatrix} + 3 \begin{vmatrix} 4 & 1 \\ 7 & -1 \end{vmatrix} - \begin{vmatrix} 4 & 1 \\ -6 & -2 \end{vmatrix} = -40 + (-33) - (-2) = -71.$$

(c) $\begin{vmatrix} 5 & -1 & 2 \\ 3 & 0 & 6 \\ -4 & 3 & 1 \end{vmatrix}$

using row 1:

$$= 5 \begin{vmatrix} 0 & 6 \\ 3 & 1 \end{vmatrix} - (-1) \begin{vmatrix} 3 & 6 \\ -4 & 1 \end{vmatrix} + 2 \begin{vmatrix} 3 & 0 \\ -4 & 3 \end{vmatrix} = -90 - (-27) + 18 = -45.$$

using row 3:

$$= (-4) \begin{vmatrix} -1 & 2 \\ 0 & 6 \end{vmatrix} - 3 \begin{vmatrix} 5 & 2 \\ 3 & 6 \end{vmatrix} + \begin{vmatrix} 5 & -1 \\ 3 & 0 \end{vmatrix} = 24 - (72) + (3) = -45.$$

(d) $\begin{vmatrix} 6 & 3 & 0 \\ -2 & -1 & 5 \\ 4 & 6 & -2 \end{vmatrix}$

using column 2:

$$= -3 \begin{vmatrix} -2 & 5 \\ 4 & -2 \end{vmatrix} + (-1) \begin{vmatrix} 6 & 0 \\ 4 & -2 \end{vmatrix} - 6 \begin{vmatrix} 6 & 0 \\ -2 & 5 \end{vmatrix} = 48 + 12 - 180 = -120.$$

using column 3:

$$= 0 \begin{vmatrix} -2 & -1 \\ 4 & 6 \end{vmatrix} - 5 \begin{vmatrix} 6 & 3 \\ 4 & 6 \end{vmatrix} + (-2) \begin{vmatrix} 6 & 3 \\ -2 & -1 \end{vmatrix} = 0 - 120 + 0 = -120.$$

9. (a) $\begin{vmatrix} 1 & 3 & -1 \\ 2 & 0 & 5 \\ 1 & 4 & 3 \end{vmatrix}$

using row 2:

$$= -2 \begin{vmatrix} 3 & -1 \\ 4 & 3 \end{vmatrix} + 0 \begin{vmatrix} 1 & -1 \\ 1 & 3 \end{vmatrix} - 5 \begin{vmatrix} 1 & 3 \\ 1 & 4 \end{vmatrix} = -26 + 0 - 5 = -31.$$

using column 1:

$$= \begin{vmatrix} 0 & 5 \\ 4 & 3 \end{vmatrix} - 2 \begin{vmatrix} 3 & -1 \\ 4 & 3 \end{vmatrix} + \begin{vmatrix} 3 & -1 \\ 0 & 5 \end{vmatrix} = -20 - 26 + 15 = -31.$$

(b) $\begin{vmatrix} -2 & -1 & 1 \\ 9 & 3 & 2 \\ 4 & 0 & 0 \end{vmatrix}$

using row 1:

$$= (-2) \begin{vmatrix} 3 & 2 \\ 0 & 0 \end{vmatrix} - (-1) \begin{vmatrix} 9 & 2 \\ 4 & 0 \end{vmatrix} + \begin{vmatrix} 9 & 3 \\ 4 & 0 \end{vmatrix} = 0 - 8 + (-12) = -20.$$

using column 3:

$$= \begin{vmatrix} 9 & 3 \\ 4 & 0 \end{vmatrix} - 2 \begin{vmatrix} -2 & -1 \\ 4 & 0 \end{vmatrix} + 0 \begin{vmatrix} -2 & -1 \\ 9 & 3 \end{vmatrix} = -12 - 8 + 0 = -20.$$

(c) $\begin{vmatrix} 1 & 0 & 2 \\ 3 & -2 & 1 \\ 4 & 0 & 2 \end{vmatrix}$

using column 1:

$$= \begin{vmatrix} -2 & 1 \\ 0 & 2 \end{vmatrix} - 3 \begin{vmatrix} 0 & 2 \\ 0 & 2 \end{vmatrix} + 4 \begin{vmatrix} 0 & 2 \\ -2 & 1 \end{vmatrix} = -4 - 0 + 16 = 12.$$

using column 2:

$$= -0 \begin{vmatrix} 3 & 1 \\ 4 & 2 \end{vmatrix} + (-2) \begin{vmatrix} 1 & 2 \\ 4 & 2 \end{vmatrix} - 0 \begin{vmatrix} 1 & 2 \\ 3 & 1 \end{vmatrix} = 0 + 12 - 0 = 12.$$

(d) $\begin{vmatrix} 1 & 2 & 3 \\ -1 & -4 & 0 \\ 0 & 0 & 4 \end{vmatrix}$

using row 3:

$$= 0 \begin{vmatrix} 2 & 3 \\ -4 & 0 \end{vmatrix} - 0 \begin{vmatrix} 1 & 3 \\ -1 & 0 \end{vmatrix} + 4 \begin{vmatrix} 1 & 2 \\ -1 & -4 \end{vmatrix} = 0 - 0 + (-8) = -8.$$

using column 1:

$$= \begin{vmatrix} -4 & 0 \\ 0 & 4 \end{vmatrix} - (-1) \begin{vmatrix} 2 & 3 \\ 0 & 4 \end{vmatrix} + 0 \begin{vmatrix} 2 & 3 \\ -4 & 0 \end{vmatrix} = -16 - (-8) + 0 = -8.$$

10. (a) Using column 3, $\begin{vmatrix} 1 & -2 & 3 \\ 1 & 4 & 0 \\ 2 & -1 & 0 \end{vmatrix} = 3 \begin{vmatrix} 1 & 4 \\ 2 & -1 \end{vmatrix} = -27.$

(b) Using row 2, $\begin{vmatrix} 3 & -1 & 2 \\ 0 & 4 & 0 \\ -5 & 1 & 9 \end{vmatrix} = 4 \begin{vmatrix} 3 & 2 \\ -5 & 9 \end{vmatrix} = 148.$

(c) Using row 3, $\begin{vmatrix} 9 & 2 & 1 \\ -3 & 2 & 6 \\ 0 & 0 & -3 \end{vmatrix} = (-3) \begin{vmatrix} 9 & 2 \\ -3 & 2 \end{vmatrix} = -72.$

Section 3.1

11. (a) Using column 4, $\begin{vmatrix} 1 & -2 & 3 & 0 \\ 4 & 0 & 5 & 0 \\ 7 & -3 & 8 & 4 \\ -3 & 0 & 4 & 0 \end{vmatrix} = -4 \begin{vmatrix} 1 & -2 & 3 \\ 4 & 0 & 5 \\ -3 & 0 & 4 \end{vmatrix}$

(using column 2 of the 3x3 matrix) $= -4(-(-2)) \begin{vmatrix} 4 & 5 \\ -3 & 4 \end{vmatrix} = -8 \times 31 = -248$.

(b) Using row 3, $\begin{vmatrix} 1 & 4 & 5 & 9 \\ 2 & 3 & -7 & 1 \\ 0 & 0 & 0 & -3 \\ 0 & 1 & 0 & 8 \end{vmatrix} = -(-3) \begin{vmatrix} 1 & 4 & 5 \\ 2 & 3 & -7 \\ 0 & 1 & 0 \end{vmatrix}$

(using row 3 of the 3x3 matrix) $= -(-3)(-1) \begin{vmatrix} 1 & 5 \\ 2 & -7 \end{vmatrix} = -3 \times (-17) = 51$.

(c) Using column 2, $\begin{vmatrix} 9 & 3 & 7 & -8 \\ 1 & 0 & 4 & 2 \\ 1 & 0 & 0 & -1 \\ -2 & 0 & -1 & 3 \end{vmatrix} = -3 \begin{vmatrix} 1 & 4 & 2 \\ 1 & 0 & -1 \\ -2 & -1 & 3 \end{vmatrix}$

(using row 2 of the 3x3 matrix) $= -3 \left(-\begin{vmatrix} 4 & 2 \\ -1 & 3 \end{vmatrix} - (-1) \begin{vmatrix} 1 & 4 \\ -2 & -1 \end{vmatrix} \right)$

$= -3(-14 - (-7)) = 21$.

12. $\begin{vmatrix} x+1 & x \\ 3 & x-2 \end{vmatrix} = (x+1)(x-2) - 3x = x^2 - x - 2 - 3x = 3$, so $x^2 - 4x - 5 = 0$.

$(x-5)(x+1) = 0$, so there are two solutions, $x = 5$ and $x = -1$.

13. $\begin{vmatrix} 2x & -3 \\ x-1 & x+2 \end{vmatrix} = (2x)(x+2) - (-3)(x-1) = 2x^2 + 4x - (-3x) - 3 = 1$, so $2x^2 + 7x - 4 = 0$.

$(2x-1)(x+4) = 0$, so there are two solutions, $x = 1/2$ and $x = -4$.

Section 3.1

14. $\begin{vmatrix} x-1 & -2 \\ x-2 & x-1 \end{vmatrix} = (x-1)(x-1) - (-2)(x-2) = x^2 - 2x + 1 - (-2x) - 4 = 0$,

    so $x^2 - 3 = 0$, and there are two solutions, $\sqrt{3}$ and $-\sqrt{3}$.

15. $\begin{vmatrix} x & 0 & 2 \\ 2x & x-1 & 4 \\ -x & x-1 & x+1 \end{vmatrix} = x(x-1)(x+1) + 0 + 2(2x)(x-1) - 2(x-1)(-x) - x(4)(x-1) - 0$

    $= x^3 - x + 4x^2 - 4x + 2x^2 - 2x - 4x^2 + 4x = x^3 + 2x^2 - 3x = x(x+3)(x-1) = 0$, so there are three solutions, $x = 0$, $x = -3$, and $x = 1$.

16. The cofactor expansion of each determinant using the third column gives $-3 \begin{vmatrix} 4 & -1 \\ 2 & 1 \end{vmatrix}$.

17. The cofactor expansion using the third row is independent of the numbers a and b.

18. (a) $4213 \to 4123 \to 1423 \to 1243 \to 1234$; even

    (b) $3142 \to 1342 \to 1324 \to 1234$; odd

    (c) $3214 \to 3124 \to 1324 \to 1234$; odd

    (d) $2413 \to 2143 \to 1243 \to 1234$; odd

    (e) $4321 \to 4312 \to 3412 \to 3142 \to 1342 \to 1324 \to 1234$; even

19. (a) $35241 \to 32541 \to 32451 \to 32415 \to 32145 \to 31245 \to 13245 \to 12345$; odd

    (b) $43152 \to 41352 \to 14352 \to 13452 \to 13425 \to 13245 \to 12345$; even

    (c) $54312 \to 45312 \to 43512 \to 43152 \to 43125 \to 34125 \to 31425 \to 31245$
    $\to 13245 \to 12345$; odd

    (d) $25143 \to 21543 \to 12543 \to 12453 \to 12435 \to 12345$; odd

    (e) $32514 \to 23514 \to 23154 \to 21354 \to 12354 \to 12345$; odd

Section 3.2

20.
1234	even	2134	odd
1243	odd	2143	even
1324	odd	2314	even
1342	even	2341	odd
1423	even	2413	odd
1432	odd	2431	even
3124	even	4123	odd
3142	odd	4132	even
3214	odd	4213	even
3241	even	4231	odd
3412	even	4312	odd
3421	odd	4321	even

21. Cofactor expansion gives $|A| = a_{11}C_{11} + a_{12}C_{12} + a_{13}C_{13}$

$= a_{11}(a_{22}a_{33} - a_{23}a_{32}) - a_{12}(a_{21}a_{33} - a_{23}a_{31}) + a_{13}(a_{21}a_{32} - a_{22}a_{31})$

$= a_{11}a_{22}a_{33} - a_{11}a_{23}a_{32} - a_{12}a_{21}a_{33} + a_{12}a_{23}a_{31} + a_{13}a_{21}a_{32} - a_{13}a_{22}a_{31}$.

The permutations of 1,2,3 are 123, 132, 213, 231, 312, and 321, so the second definition gives $|A| = a_{11}a_{22}a_{33} - a_{11}a_{23}a_{32} - a_{12}a_{21}a_{33} + a_{12}a_{23}a_{31} + a_{13}a_{21}a_{32} - a_{13}a_{22}a_{31}$.

The two expressions are the same.

Exercise Set 3.2

1. (a) $\begin{vmatrix} 1 & 2 & 3 \\ 2 & 4 & 1 \\ 1 & -1 & 1 \end{vmatrix} \underset{C2+(-2)C1}{=} \begin{vmatrix} 1 & 0 & 3 \\ 2 & 0 & 1 \\ 1 & -1 & 1 \end{vmatrix} = -(-1)\begin{vmatrix} 1 & 3 \\ 2 & 1 \end{vmatrix} = -5.$

(b) $\begin{vmatrix} 0 & 1 & 5 \\ 1 & 1 & 6 \\ 2 & 2 & 7 \end{vmatrix} \underset{C2+(-1)C1}{=} \begin{vmatrix} 0 & 1 & 5 \\ 1 & 0 & 6 \\ 2 & 0 & 7 \end{vmatrix} = -\begin{vmatrix} 1 & 6 \\ 2 & 7 \end{vmatrix} = 5.$

(c) $\begin{vmatrix} 2 & 1 & -1 \\ 3 & -1 & 1 \\ 1 & 4 & -4 \end{vmatrix} \underset{C2+C3}{=} \begin{vmatrix} 2 & 0 & -1 \\ 3 & 0 & 1 \\ 1 & 0 & -4 \end{vmatrix} = 0.$

(d) $\begin{vmatrix} 3 & -1 & 0 \\ 4 & 2 & 1 \\ 1 & 1 & 2 \end{vmatrix} \underset{R3+(-2)R2}{=} \begin{vmatrix} 3 & -1 & 0 \\ 4 & 2 & 1 \\ -7 & -3 & 0 \end{vmatrix} = -\begin{vmatrix} 3 & -1 \\ -7 & -3 \end{vmatrix} = 16.$

2. (a) $\begin{vmatrix} 2 & -1 & 2 \\ 1 & 2 & -4 \\ 3 & 1 & 2 \end{vmatrix} \underset{C3+(2)C2}{=} \begin{vmatrix} 2 & -1 & 0 \\ 1 & 2 & 0 \\ 3 & 1 & 4 \end{vmatrix} = -4\begin{vmatrix} 2 & -1 \\ 1 & 2 \end{vmatrix} = 20.$

(b) $\begin{vmatrix} 5 & 1 & 3 \\ 1 & 2 & 4 \\ -1 & 1 & -4 \end{vmatrix} \underset{R2+R3}{=} \begin{vmatrix} 5 & 1 & 3 \\ 0 & 3 & 0 \\ -1 & 1 & -4 \end{vmatrix} = 3\begin{vmatrix} 5 & 3 \\ -1 & -4 \end{vmatrix} = -51.$

(c) $\begin{vmatrix} 1 & -2 & 3 \\ -3 & 6 & -9 \\ 4 & 5 & 7 \end{vmatrix} \underset{R2+(3)R1}{=} \begin{vmatrix} 1 & -2 & 3 \\ 0 & 0 & 0 \\ 4 & 5 & 7 \end{vmatrix} = 0.$

(d) $\begin{vmatrix} -1 & 3 & 2 \\ 2 & 5 & -4 \\ 4 & 1 & -8 \end{vmatrix} \underset{C3+(2)C1}{=} \begin{vmatrix} -1 & 3 & 0 \\ 2 & 5 & 0 \\ 4 & 1 & 0 \end{vmatrix} = 0.$

3. (a) The given matrix can be obtained from A by multiplying the third row by 2, so its determinant is 2 |A| = −4.

   (b) The given matrix can be obtained from A by interchanging rows 1 and 2, so its determinant is −|A| = 2.

   (c) The given matrix can be obtained from A by adding twice row 1 to row 2, so its determinant is |A| = −2.

4. (a) The given matrix can be obtained from A by interchanging columns 2 and 3, so its determinant is −|A| = −5.

   (b) The given matrix can be obtained from A by adding −2 times column 3 to column 2, so its determinant is |A| = 5.

   (c) The given matrix is the transpose of A, and $|A^t|$ = |A| = 5.

Section 3.2

5. The second answer is correct.

6. (a) Row 3 is all zeros. (b) Columns 2 and 3 are equal.
   (c) Row 3 is -3 times row 1. (d) Row 3 is all zeros.

7. (a) Column 3 is all zeros. (b) Row 3 is all zeros.
   (c) Row 3 is 3 times row 1. (d) Column 3 is -3/2 times column 1.

8. (a) $|2A| = (2)^2|A| = 12$. (b) $|3A^t| = (3)^2|A^t| = 9|A| = 27$.
   (c) $|A^2| = |A||A| = 9$. (d) $|(A^t)^2| = |A^t||A^t| = |A||A| = 9$.
   (e) $|(A^2)^t| = |A^2| = 9$. (f) $|4A^{-1}| = (4)^2|A^{-1}| = 16/|A| = 16/3$.

9. (a) $|AB| = |A||B| = -6$. (b) $|AA^t| = |A||A^t| = |A||A| = 9$.
   (c) $|A^tB| = |A^t||B| = |A||B| = -6$. (d) $|3A^2B| = (3)^3|A||A||B| = 486$.
   (e) $|2AB^{-1}| = (2)^3|A||B^{-1}| = 8|A|/|B| = -12$. (f) $|(A^2B^{-1})^t| = |A^2B^{-1}| = |A||A|/|B| = 9/2$.

10. (a) The given matrix can be obtained from A by interchanging rows 1 and 2 and then interchanging rows 2 and 3. Thus its determinant is $(-1)(-1)|A| = 3$.

    (b) The given matrix can be obtained from A by interchanging rows 1 and 2 and then taking the transpose. Thus its determinant is $(-1)|A| = -3$.

    (c) The given matrix can be obtained from A by interchanging rows 1 and 2 and then interchanging columns 2 and 3 in the resulting matrix. Thus the determinant of the given matrix is $(-1)(-1)|A| = 3$.

    (d) The given matrix can be obtained from A by interchanging rows 1 and 2 of A and multiplying row 3 by 2. Thus the determinant is $(-1)(2)|A| = -6$.

11. $\begin{vmatrix} a+b & c+d & e+f \\ p & q & r \\ u & v & w \end{vmatrix} = (a+b)\begin{vmatrix} q & r \\ v & w \end{vmatrix} - (c+d)\begin{vmatrix} p & r \\ u & w \end{vmatrix} + (e+f)\begin{vmatrix} p & q \\ u & v \end{vmatrix}$

$= a\begin{vmatrix} q & r \\ v & w \end{vmatrix} - c\begin{vmatrix} p & r \\ u & w \end{vmatrix} + e\begin{vmatrix} p & q \\ u & v \end{vmatrix} + b\begin{vmatrix} q & r \\ v & w \end{vmatrix} - d\begin{vmatrix} p & r \\ u & w \end{vmatrix} + f\begin{vmatrix} p & q \\ u & v \end{vmatrix}$

$= \begin{vmatrix} a & c & e \\ p & q & r \\ u & v & w \end{vmatrix} + \begin{vmatrix} b & d & f \\ p & q & r \\ u & v & w \end{vmatrix}.$

12. Expand by row (or column) 1 at each stage.

$$\begin{vmatrix} a_{11} & 0 & 0 & \cdots & 0 \\ 0 & a_{22} & 0 & \cdots & 0 \\ 0 & 0 & a_{33} & \cdots & 0 \\ \vdots & \vdots & & & \vdots \\ 0 & 0 & 0 & \cdots & a_{nn} \end{vmatrix} = a_{11} \begin{vmatrix} a_{22} & 0 & \cdots & 0 \\ 0 & a_{33} & \cdots & 0 \\ \vdots & \vdots & & \vdots \\ 0 & 0 & \cdots & a_{nn} \end{vmatrix} = a_{11}a_{22} \begin{vmatrix} a_{33} & \cdots & 0 \\ \vdots & & \vdots \\ 0 & \cdots & a_{nn} \end{vmatrix}$$

$= \ldots = a_{11}a_{22}a_{33}\cdots a_{nn}.$

13. Suppose row i of B is row i of A plus k times row j of A, where k is any real number.

The cofactors of the ith rows of A and B are the same since only the elements in the ith rows are different.

$|B| = b_{i1}C_{i1} + b_{i2}C_{i2} + \ldots + b_{in}C_{in} = (a_{i1} + ka_{j1})C_{i1} + (a_{i2} + ka_{j2})C_{i2} + \ldots + (a_{in} + ka_{jn})C_{in}$

$= a_{i1}C_{i1} + a_{i2}C_{i2} + \ldots + a_{in}C_{in} + k(a_{j1}C_{i1} + a_{j2}C_{i2} + \ldots + a_{jn}C_{in}) = |A| + k|D|$, where D is

the matrix obtained from A by replacing the ith row of A by the jth row of A. Thus the ith and jth rows of D are equal, so $|D| = 0$ and $|B| = |A|$.

The proof for columns is similar.

Section 3.2

14. Suppose row j is k times row i. Let B be the matrix obtained from A by interchanging rows i and j. Call the cofactors of the ith row of A $C_{i1}, C_{i2}, \ldots, C_{in}$ and the cofactors of the ith row of B $D_{i1}, D_{i2}, \ldots, D_{in}$. Each cofactor of A is k times the corresponding cofactor of B, because row j in A is k times row j in B.

$$|A| = a_{i1}C_{i1} + a_{i2}C_{i2} + \ldots + a_{in}C_{in} = a_{i1}kD_{i1} + a_{i2}kD_{i2} + \ldots + a_{in}kD_{in}$$

and $|B| = a_{j1}D_{i1} + a_{j2}D_{i2} + \ldots + a_{jn}D_{in} = ka_{i1}D_{i1} + ka_{i2}D_{i2} + \ldots + ka_{in}D_{in}$, so $|A| = |B|$.

But $|B| = -|A|$ because B was obtained from A by interchanging two rows, so $|A| = -|A|$, and therefore $|A| = 0$.

The proof for proportional columns is similar.

15. Suppose the sum of the elements in each column is zero.

$$|A| = \begin{vmatrix} a_{11} & \cdots & a_{1n} \\ a_{21} & \cdots & a_{2n} \\ a_{31} & \cdots & a_{3n} \\ \vdots & & \vdots \\ a_{n1} & \cdots & a_{nn} \end{vmatrix} \underset{R1+R2}{=} \begin{vmatrix} a_{11}+a_{21} & \cdots & a_{1n}+a_{2n} \\ a_{21} & \cdots & a_{2n} \\ a_{31} & \cdots & a_{3n} \\ \vdots & & \vdots \\ a_{n1} & \cdots & a_{nn} \end{vmatrix}$$

$$\underset{R1+R3}{=} \begin{vmatrix} a_{11}+a_{21}+a_{31} & \cdots & a_{1n}+a_{2n}+a_{3n} \\ a_{21} & \cdots & a_{2n} \\ a_{31} & \cdots & a_{3n} \\ \vdots & & \vdots \\ a_{n1} & \cdots & a_{nn} \end{vmatrix} \underset{R1+R4}{=} \cdots$$

$$\underset{R1+Rn}{=} \begin{vmatrix} a_{11}+a_{21}+a_{31}+\ldots+a_{n1} & \cdots & a_{1n}+a_{2n}+a_{3n}+\ldots+a_{nn} \\ a_{21} & \cdots & a_{2n} \\ a_{31} & \cdots & a_{3n} \\ \vdots & & \vdots \\ a_{n1} & \cdots & a_{nn} \end{vmatrix} = 0, \text{ because the}$$

first row is all zeros.

Section 3.2

16. $|A^n| = |A||A|\ldots|A| = |A|^n$. If $A^n = O$, then $|A^n| = 0$, so $|A|^n = 0$, but that means $|A| = 0$.

17. $|AB| = |A||B| = |B||A| = |BA|$.

18. $|ABC| = |AB||C| = |A||B||C|$.

19. Let $A = \begin{bmatrix} 2 & 1 \\ 0 & 0 \end{bmatrix}$ and let $B = \begin{bmatrix} 0 & 0 \\ 4 & -1 \end{bmatrix}$. Then $A + B = \begin{bmatrix} 2 & 1 \\ 4 & -1 \end{bmatrix}$.

    $|A| = |B| = 0$ and $|A+B| = -6$, so $|A+B| \neq |A| + |B|$ for this example.

20. If B is obtained from A using an elementary row operation, then $|B| = |A|$ or $|B| = -|A|$ or $|B| = c|A|$. Since $c \neq 0$ by definition, $|B| \neq 0$ if and only if $|A| \neq 0$.

21. (a) Let E be the reduced echelon form of a 2x2 matrix A and suppose E has no zero rows. Then because E is a reduced echelon matrix, the first position in row 2 which can be nonzero is the (2,2) position, and it must be 1 in order for E to have no zero rows. This means the (1,2) position is zero and the (1,1) position is 1. So $E = I_2$.

    (b) Let E be the reduced echelon form of a 3x3 matrix A and suppose E has no zero rows. Then because E is a reduced echelon matrix, the first position in row 3 which can be nonzero is the (3,3) position, and it must be 1 in order for E to have no zero rows. This means the (1,3) and (2,3) positions are zero and the (2,2) position is 1, since it is now both the first and last position in row 2 which can be nonzero. Therefore the (1,2) position is zero and the (1,1) position is 1. So $E = I_3$.

    (c) Let E be the reduced echelon form of an nxn matrix A and suppose E has no zero rows. Then because E is a reduced echelon matrix, the first position in row n which can be nonzero is the (n,n) position, and it must be 1 in order for E to have no zero rows. This means the (1,n), (2,n),....,(n–1,n) positions are zero and the (n–1,n–1) position is 1, since it is now both the first and last position in row n–1 which can be nonzero. Therefore the (1,n–1), (2,n–1), . . . , (n–2,n–1) positions are zero and the (n–2,n–2) position is 1, ... Continuing in this way, we see that all diagonal elements of E must be 1 to avoid having a zero row, and therefore all other elements of E are zero. Thus $E = I_n$.

Section 3.3

(d) E is obtained from A by performing a sequence of elementary row operations.

|A| = $(-1)^k$c|E|, where k is the number of row interchanges required and c is 1 divided by the product of all constants multiplied by rows. Thus if |A| ≠ 0 then |E| ≠ 0, but if |E| ≠ 0, then E has no zero rows, so from part (c), E = I_n. On the other hand, if E = I_n, then |E| ≠ 0, so |A| ≠ 0.

Exercise Set 3.3

1. (a) 3x-1x4 = -12   (b) 3x1x5 = 15   (c) 9x3x1 = 27

2. (a) 2x3x5x-2 = -60   (b) 3x-2x5x1 = -30   (c) 7x6x0x8 = 0

3. (a) $\begin{vmatrix} 1 & 0 & -1 \\ 2 & 1 & 2 \\ -1 & 1 & 1 \end{vmatrix} \underset{\substack{R2+(-2)R1 \\ R3+R1}}{=} \begin{vmatrix} 1 & 0 & -1 \\ 0 & 1 & 4 \\ 0 & 1 & 0 \end{vmatrix} \underset{R3+(-1)R2}{=} \begin{vmatrix} 1 & 0 & -1 \\ 0 & 1 & 4 \\ 0 & 0 & -4 \end{vmatrix} = -4.$

(b) $\begin{vmatrix} 2 & 1 & 1 \\ 4 & 0 & 1 \\ -1 & 2 & 0 \end{vmatrix} \underset{R1 \leftrightarrow R3}{=} - \begin{vmatrix} -1 & 2 & 0 \\ 4 & 0 & 1 \\ 2 & 1 & 1 \end{vmatrix} \underset{\substack{R2+(4)R1 \\ R3+(2)R1}}{=} - \begin{vmatrix} -1 & 2 & 0 \\ 0 & 8 & 1 \\ 0 & 5 & 1 \end{vmatrix}$

$\underset{R3+(-5/8)R2}{=} - \begin{vmatrix} -1 & 2 & 0 \\ 0 & 8 & 1 \\ 0 & 0 & 3/8 \end{vmatrix} = 3.$

(c) $\begin{vmatrix} 1 & -2 & 3 \\ -1 & 2 & 1 \\ 2 & 1 & 3 \end{vmatrix} \underset{\substack{R2+R1 \\ R3+(-2)R1}}{=} \begin{vmatrix} 1 & -2 & 3 \\ 0 & 0 & 4 \\ 0 & 5 & -3 \end{vmatrix} \underset{R2 \leftrightarrow R3}{=} - \begin{vmatrix} 1 & -2 & 3 \\ 0 & 5 & -3 \\ 0 & 0 & 4 \end{vmatrix} = -20.$

4. (a) $\begin{vmatrix} 2 & 3 & 8 \\ -2 & -3 & 4 \\ 4 & 6 & -2 \end{vmatrix} \underset{\substack{R2+R1 \\ R3+(-2)R1}}{=} \begin{vmatrix} 2 & 3 & 8 \\ 0 & 0 & 12 \\ 0 & 0 & -18 \end{vmatrix} = 0.$

99

(b) $\begin{vmatrix} 1 & -1 & 2 \\ 3 & -1 & 7 \\ -2 & 2 & 0 \end{vmatrix} \underset{\substack{R2+(-3)R1 \\ R3+(2)R1}}{=} \begin{vmatrix} 1 & -1 & 2 \\ 0 & 2 & 1 \\ 0 & 0 & 4 \end{vmatrix} = 8.$

(c) $\begin{vmatrix} 2 & -1 & 4 \\ -2 & 1 & -3 \\ 0 & 3 & -2 \end{vmatrix} \underset{R2+R1}{=} \begin{vmatrix} 2 & -1 & 4 \\ 0 & 0 & 1 \\ 0 & 3 & -2 \end{vmatrix} \underset{R2 \leftrightarrow R3}{=} -\begin{vmatrix} 2 & -1 & 4 \\ 0 & 3 & -2 \\ 0 & 0 & 1 \end{vmatrix} = -6.$

5. (a) $\begin{vmatrix} 2 & 4 & 1 \\ 0 & 0 & 3 \\ 0 & 1 & 2 \end{vmatrix} \underset{R2 \leftrightarrow R3}{=} -\begin{vmatrix} 2 & 4 & 1 \\ 0 & 1 & 2 \\ 0 & 0 & 3 \end{vmatrix} = -6.$

(b) $\begin{vmatrix} 0 & 4 & 1 \\ 1 & 2 & 3 \\ 4 & 1 & 5 \end{vmatrix} \underset{R1 \leftrightarrow R2}{=} -\begin{vmatrix} 1 & 2 & 3 \\ 0 & 4 & 1 \\ 4 & 1 & 5 \end{vmatrix} \underset{R3+(-4)R1}{=} -\begin{vmatrix} 1 & 2 & 3 \\ 0 & 4 & 1 \\ 0 & -7 & -7 \end{vmatrix}$

$\underset{R3+(7/4)R2}{=} -\begin{vmatrix} 1 & 2 & 3 \\ 0 & 4 & 1 \\ 0 & 0 & -21/4 \end{vmatrix} = 21.$

(c) $\begin{vmatrix} 2 & -1 & 6 \\ 0 & 0 & 1 \\ 0 & 0 & 2 \end{vmatrix} = 0.$

6. (a) $\begin{vmatrix} -1 & 2 & 0 & 1 \\ 1 & 1 & -1 & 0 \\ 2 & 1 & 1 & 0 \\ -1 & -1 & 0 & 1 \end{vmatrix} \underset{\substack{R2+R1 \\ R3+(2)R1 \\ R4+(-1)R1}}{=} \begin{vmatrix} -1 & 2 & 0 & 1 \\ 0 & 3 & -1 & 1 \\ 0 & 5 & 1 & 2 \\ 0 & -3 & 0 & 0 \end{vmatrix}$

$\underset{\substack{R3+(-5/3)R2 \\ R4+R2}}{=} \begin{vmatrix} -1 & 2 & 0 & 1 \\ 0 & 3 & -1 & 1 \\ 0 & 0 & 8/3 & 1/3 \\ 0 & 0 & -1 & 1 \end{vmatrix} \underset{R4+(3/8)R3}{=} \begin{vmatrix} -1 & 2 & 0 & 1 \\ 0 & 3 & -1 & 1 \\ 0 & 0 & 8/3 & 1/3 \\ 0 & 0 & 0 & 9/8 \end{vmatrix} = -9.$

(b) $\begin{vmatrix} 1 & 0 & -2 & 1 \\ 2 & 1 & 0 & 2 \\ -1 & 1 & -2 & 1 \\ 3 & 1 & -1 & 0 \end{vmatrix} \underset{\substack{R2+(-2)R1 \\ R3+R1 \\ R4+(-3)R1}}{=} \begin{vmatrix} 1 & 0 & -2 & 1 \\ 0 & 1 & 4 & 0 \\ 0 & 1 & -4 & 2 \\ 0 & 1 & 5 & -3 \end{vmatrix}$

$\underset{\substack{R3+(-1)R2 \\ R4+(-1)R2}}{=} \begin{vmatrix} 1 & 0 & -2 & 1 \\ 0 & 1 & 4 & 0 \\ 0 & 0 & -8 & 2 \\ 0 & 0 & 1 & -3 \end{vmatrix} \underset{R4+(1/8)R3}{=} \begin{vmatrix} 1 & 0 & -2 & 1 \\ 0 & 1 & 4 & 0 \\ 0 & 0 & -8 & 2 \\ 0 & 0 & 0 & -11/4 \end{vmatrix} = 22.$

(c) $\begin{vmatrix} -1 & 1 & 2 & 1 \\ 1 & -1 & 3 & -1 \\ 2 & -2 & 3 & 1 \\ 1 & -1 & 0 & 1 \end{vmatrix} \underset{\substack{R2+R1 \\ R3+(2)R1 \\ R4+R1}}{=} \begin{vmatrix} -1 & 1 & 2 & 1 \\ 0 & 0 & 5 & 0 \\ 0 & 0 & 7 & 3 \\ 0 & 0 & 2 & 2 \end{vmatrix} = 0.$

7. (a) $\begin{vmatrix} 1 & -1 & 0 & 2 \\ -1 & 1 & 0 & 0 \\ 2 & -2 & 0 & 1 \\ 3 & 1 & 5 & -1 \end{vmatrix} \underset{\substack{R2+R1 \\ R3+(-2)R1 \\ R4+(-3)R1}}{=} \begin{vmatrix} 1 & -1 & 0 & 2 \\ 0 & 0 & 0 & 2 \\ 0 & 0 & 0 & -3 \\ 0 & 4 & 5 & -7 \end{vmatrix} \underset{R2 \leftrightarrow R4}{=} - \begin{vmatrix} 1 & -1 & 0 & 2 \\ 0 & 4 & 5 & -7 \\ 0 & 0 & 0 & -3 \\ 0 & 0 & 0 & 2 \end{vmatrix} = 0.$

(b) $\begin{vmatrix} 1 & 2 & -1 & 0 \\ 1 & 4 & 2 & 1 \\ -1 & 2 & 6 & 6 \\ 2 & 2 & -4 & -2 \end{vmatrix} \underset{\substack{R2+(-1)R1 \\ R3+R1 \\ R4+(-2)R1}}{=} \begin{vmatrix} 1 & 2 & -1 & 0 \\ 0 & 2 & 3 & 1 \\ 0 & 4 & 5 & 6 \\ 0 & -2 & -2 & -2 \end{vmatrix} \underset{\substack{R3+(-2)R2 \\ R4+R2}}{=} \begin{vmatrix} 1 & 2 & -1 & 0 \\ 0 & 2 & 3 & 1 \\ 0 & 0 & -1 & 4 \\ 0 & 0 & 1 & -1 \end{vmatrix}$

$\underset{R4+R3}{=} \begin{vmatrix} 1 & 2 & -1 & 0 \\ 0 & 2 & 3 & 1 \\ 0 & 0 & -1 & 4 \\ 0 & 0 & 0 & 3 \end{vmatrix} = -6.$

(c) $\begin{vmatrix} 2 & 1 & 3 & 1 \\ -2 & 3 & -1 & 2 \\ 2 & 1 & 2 & 3 \\ -4 & -2 & 0 & -1 \end{vmatrix} \underset{\substack{R2+R1 \\ R3+(-1)R1 \\ R4+(2)R1}}{=} \begin{vmatrix} 2 & 1 & 3 & 1 \\ 0 & 4 & 2 & 3 \\ 0 & 0 & -1 & 2 \\ 0 & 0 & 6 & 1 \end{vmatrix} \underset{R4+(6)R3}{=} \begin{vmatrix} 2 & 1 & 3 & 1 \\ 0 & 4 & 2 & 3 \\ 0 & 0 & -1 & 2 \\ 0 & 0 & 0 & 13 \end{vmatrix}$

$= -104.$

8. I_n is an upper triangular matrix, so its determinant is the product of its diagonal elements, all of which are 1.

9. $|A||A^{-1}| = |AA^{-1}| = |I_n| = 1$.

10. A diagonal matrix is an upper triangular matrix, so its determinant is the product of its diagonal elements.

11. $$\begin{vmatrix} a_{11} & 0 & 0 & \cdots & 0 \\ a_{21} & a_{22} & 0 & \cdots & 0 \\ a_{31} & a_{32} & a_{33} & \cdots & 0 \\ \vdots & \vdots & \vdots & & \vdots \\ a_{n1} & a_{n2} & a_{n3} & \cdots & a_{nn} \end{vmatrix} = a_{11} \begin{vmatrix} a_{22} & 0 & \cdots & 0 \\ a_{32} & a_{33} & \cdots & 0 \\ \vdots & \vdots & & \vdots \\ a_{n2} & a_{n3} & \cdots & a_{nn} \end{vmatrix} = a_{11}a_{22} \begin{vmatrix} a_{33} & \cdots & 0 \\ \vdots & & \vdots \\ a_{n3} & \cdots & a_{nn} \end{vmatrix}$$

$= \ldots = a_{11}a_{22}a_{33} \cdots a_{nn}$.

12. (a) $|A^{-1}| = 1/|A|$, but if $A = A^{-1}$ then $|A| = |A^{-1}|$, so $|A| = 1/|A|$ and $|A|$ must be ± 1.

    (b) $|A^t| = |A|$, but if $A^t = A^{-1}$ then $|A^t| = |A^{-1}| = 1/|A|$, so $|A| = 1/|A|$ and $|A|$ must be ± 1.

13. Since $|A||A^{-1}| = |AA^{-1}| = |I_n| = 1$, neither $|A|$ nor $|A^{-1}|$ can equal zero.

14. If $AB = I_n$ then $|A||B| = |AB| = |I_n| = 1$. If the product of two real numbers is not zero, then neither number is zero, so $|A| \neq 0$ and $|B| \neq 0$.

15. A square matrix is symmetric if $a_{ij} = a_{ji}$. The matrix is upper (lower) triangular if $a_{ij} = 0$ for $1 \leq j < i \leq n$ ($1 \leq i < j \leq n$). But since $a_{ji} = a_{ij}$, this means $a_{ji} = 0$, so that only the diagonal elements can be nonzero.

Exercise Set 3.4

1. (a) The determinant is 5. The matrix is invertible.

Section 3.4

(b) The determinant is zero. The matrix is singular. The inverse does not exist.

(c) The determinant is zero. The matrix is singular. The inverse does not exist.

(d) The determinant is 1. The matrix is invertible.

2. (a) The determinant is –6. The matrix is invertible.

   (b) The determinant is zero. The matrix is singular. The inverse does not exist.

   (c) The determinant is 7. The matrix is invertible.

   (d) The determinant is zero. The matrix is singular. The inverse does not exist.

3. (a) The determinant is 18. The matrix is invertible.

   (b) The determinant is zero. The matrix is singular. The inverse does not exist.

   (c) The determinant is –105. The matrix is invertible.

   (d) The determinant is zero. The matrix is singular. The inverse does not exist.

4. (a) The determinant is zero. The matrix is singular. The inverse does not exist.

   (b) The determinant is 42. The matrix is invertible.

   (c) The determinant is –27. The matrix is invertible.

   (d) The determinant is zero. The matrix is singular. The inverse does not exist.

5. (a) The determinant is –10.   (b) The determinant is 1.

$$\begin{bmatrix} 1 & 4 \\ 3 & 2 \end{bmatrix}^{-1} = \frac{-1}{10} \begin{bmatrix} 2 & -4 \\ -3 & 1 \end{bmatrix}.$$
$$\begin{bmatrix} -2 & -1 \\ 7 & 3 \end{bmatrix}^{-1} = \begin{bmatrix} 3 & 1 \\ -7 & -2 \end{bmatrix}.$$

   (c) The determinant is zero. The inverse does not exist.

   (d) The determinant is 2.

Section 3.4

$$\begin{bmatrix} 2 & 1 \\ 4 & 3 \end{bmatrix}^{-1} = \frac{1}{2}\begin{bmatrix} 3 & -1 \\ -4 & 2 \end{bmatrix}.$$

6. (a) The determinant is −3.

$$\begin{bmatrix} 1 & 2 & 3 \\ 0 & 1 & 2 \\ 4 & 5 & 3 \end{bmatrix}^{-1} = \frac{-1}{3}\begin{bmatrix} \begin{vmatrix} 1 & 2 \\ 5 & 3 \end{vmatrix} & -\begin{vmatrix} 2 & 3 \\ 5 & 3 \end{vmatrix} & \begin{vmatrix} 2 & 3 \\ 1 & 2 \end{vmatrix} \\ -\begin{vmatrix} 0 & 2 \\ 4 & 3 \end{vmatrix} & \begin{vmatrix} 1 & 3 \\ 4 & 3 \end{vmatrix} & -\begin{vmatrix} 1 & 3 \\ 0 & 2 \end{vmatrix} \\ \begin{vmatrix} 0 & 1 \\ 4 & 5 \end{vmatrix} & -\begin{vmatrix} 1 & 2 \\ 4 & 5 \end{vmatrix} & \begin{vmatrix} 1 & 2 \\ 0 & 1 \end{vmatrix} \end{bmatrix} = \frac{-1}{3}\begin{bmatrix} -7 & 9 & 1 \\ 8 & -9 & -2 \\ -4 & 3 & 1 \end{bmatrix}.$$

(b) The determinant is −3.

$$\begin{bmatrix} 0 & 3 & 3 \\ 1 & 2 & 3 \\ 1 & 4 & 6 \end{bmatrix}^{-1} = \frac{-1}{3}\begin{bmatrix} \begin{vmatrix} 2 & 3 \\ 4 & 6 \end{vmatrix} & -\begin{vmatrix} 3 & 3 \\ 4 & 6 \end{vmatrix} & \begin{vmatrix} 3 & 3 \\ 2 & 3 \end{vmatrix} \\ -\begin{vmatrix} 1 & 3 \\ 1 & 6 \end{vmatrix} & \begin{vmatrix} 0 & 3 \\ 1 & 6 \end{vmatrix} & -\begin{vmatrix} 0 & 3 \\ 1 & 3 \end{vmatrix} \\ \begin{vmatrix} 1 & 2 \\ 1 & 4 \end{vmatrix} & -\begin{vmatrix} 0 & 3 \\ 1 & 4 \end{vmatrix} & \begin{vmatrix} 0 & 3 \\ 1 & 2 \end{vmatrix} \end{bmatrix} = \frac{-1}{3}\begin{bmatrix} 0 & -6 & 3 \\ -3 & -3 & 3 \\ 2 & 3 & -3 \end{bmatrix}.$$

(c) The determinant is −4.

$$\begin{bmatrix} 1 & 2 & -1 \\ 2 & 4 & -3 \\ 1 & -2 & 0 \end{bmatrix}^{-1} = \frac{-1}{4}\begin{bmatrix} -6 & 2 & -2 \\ -3 & 1 & 1 \\ -8 & 4 & 0 \end{bmatrix}.$$

(d) The determinant is zero. The inverse does not exist.

7. (a) The determinant is −1.

$$\begin{bmatrix} 5 & 2 & 4 \\ 2 & 1 & 2 \\ 4 & 2 & 3 \end{bmatrix}^{-1} = -\begin{bmatrix} \begin{vmatrix} 1 & 2 \\ 2 & 3 \end{vmatrix} & -\begin{vmatrix} 2 & 4 \\ 2 & 3 \end{vmatrix} & \begin{vmatrix} 2 & 4 \\ 1 & 2 \end{vmatrix} \\ -\begin{vmatrix} 2 & 2 \\ 4 & 3 \end{vmatrix} & \begin{vmatrix} 5 & 4 \\ 4 & 3 \end{vmatrix} & -\begin{vmatrix} 5 & 4 \\ 2 & 2 \end{vmatrix} \\ \begin{vmatrix} 2 & 1 \\ 4 & 2 \end{vmatrix} & -\begin{vmatrix} 5 & 2 \\ 4 & 2 \end{vmatrix} & \begin{vmatrix} 5 & 2 \\ 2 & 1 \end{vmatrix} \end{bmatrix} = -\begin{bmatrix} -1 & 2 & 0 \\ 2 & -1 & -2 \\ 0 & -2 & 1 \end{bmatrix}.$$

(b) The determinant is 2.

$$\begin{bmatrix} -3 & -2 & -5 \\ 3 & 4 & 3 \\ 1 & 1 & 1 \end{bmatrix}^{-1} = \frac{1}{2} \begin{bmatrix} 1 & -3 & 14 \\ 0 & 2 & -6 \\ -1 & 1 & -6 \end{bmatrix}.$$

(c) The determinant is zero. The inverse does not exist.

(d) The determinant is –3.

$$\begin{bmatrix} 2 & 2 & 1 \\ 4 & -1 & 4 \\ 7 & 4 & 5 \end{bmatrix}^{-1} = \frac{-1}{3} \begin{bmatrix} -21 & -6 & 9 \\ 8 & 3 & -4 \\ 23 & 6 & -10 \end{bmatrix}.$$

8. (a) $x_1 = \dfrac{\begin{vmatrix} 8 & 2 \\ 19 & 5 \end{vmatrix}}{\begin{vmatrix} 1 & 2 \\ 2 & 5 \end{vmatrix}} = \dfrac{2}{1} = 2.$ $x_2 = \dfrac{\begin{vmatrix} 1 & 8 \\ 2 & 19 \end{vmatrix}}{\begin{vmatrix} 1 & 2 \\ 2 & 5 \end{vmatrix}} = \dfrac{3}{1} = 3.$

(b) $x_1 = \dfrac{\begin{vmatrix} 3 & 2 \\ -1 & 1 \end{vmatrix}}{\begin{vmatrix} 1 & 2 \\ 3 & 1 \end{vmatrix}} = \dfrac{5}{-5} = -1.$ $x_2 = \dfrac{\begin{vmatrix} 1 & 3 \\ 3 & -1 \end{vmatrix}}{\begin{vmatrix} 1 & 2 \\ 3 & 1 \end{vmatrix}} = \dfrac{-10}{-5} = 2.$

(c) $x_1 = \dfrac{\begin{vmatrix} 11 & 3 \\ -1 & 1 \end{vmatrix}}{\begin{vmatrix} 1 & 3 \\ -2 & 1 \end{vmatrix}} = \dfrac{14}{7} = 2.$ $x_2 = \dfrac{\begin{vmatrix} 1 & 11 \\ -2 & -1 \end{vmatrix}}{\begin{vmatrix} 1 & 3 \\ -2 & 1 \end{vmatrix}} = \dfrac{21}{7} = 3.$

9. (a) $x_1 = \dfrac{\begin{vmatrix} -1 & 1 \\ 3 & 1 \end{vmatrix}}{\begin{vmatrix} 3 & 1 \\ 1 & 1 \end{vmatrix}} = \dfrac{-4}{2} = -2.$ $x_2 = \dfrac{\begin{vmatrix} 3 & -1 \\ 1 & 3 \end{vmatrix}}{\begin{vmatrix} 3 & 1 \\ 1 & 1 \end{vmatrix}} = \dfrac{10}{2} = 5.$

(b) $x_1 = \dfrac{\begin{vmatrix} 11 & 2 \\ 14 & 3 \end{vmatrix}}{\begin{vmatrix} 3 & 2 \\ 2 & 3 \end{vmatrix}} = \dfrac{5}{5} = 1.$ $\qquad x_2 = \dfrac{\begin{vmatrix} 3 & 11 \\ 2 & 14 \end{vmatrix}}{\begin{vmatrix} 3 & 2 \\ 2 & 3 \end{vmatrix}} = \dfrac{20}{5} = 4.$

(c) $x_1 = \dfrac{\begin{vmatrix} -1 & 1 \\ 10 & 2 \end{vmatrix}}{\begin{vmatrix} 2 & 1 \\ -2 & 2 \end{vmatrix}} = \dfrac{-12}{6} = -2.$ $\qquad x_2 = \dfrac{\begin{vmatrix} 2 & -1 \\ -2 & 10 \end{vmatrix}}{\begin{vmatrix} 2 & 1 \\ -2 & 2 \end{vmatrix}} = \dfrac{18}{6} = 3.$

10. (a) $|A| = \begin{vmatrix} 1 & 3 & 4 \\ 2 & 6 & 9 \\ 3 & 1 & -2 \end{vmatrix} = 8$, $|A_1| = \begin{vmatrix} 3 & 3 & 4 \\ 5 & 6 & 9 \\ 7 & 1 & -2 \end{vmatrix} = 8$, $|A_2| = \begin{vmatrix} 1 & 3 & 4 \\ 2 & 5 & 9 \\ 3 & 7 & -2 \end{vmatrix} = 16$,

$|A_3| = \begin{vmatrix} 1 & 3 & 3 \\ 2 & 6 & 5 \\ 3 & 1 & 7 \end{vmatrix} = -8$, so $x_1 = \dfrac{8}{8} = 1$, $x_2 = \dfrac{16}{8} = 2$, $x_3 = \dfrac{-8}{8} = -1.$

(b) $|A| = \begin{vmatrix} 1 & 2 & 1 \\ 1 & 3 & -1 \\ 1 & 4 & -1 \end{vmatrix} = 2$, $|A_1| = \begin{vmatrix} 9 & 2 & 1 \\ 4 & 3 & -1 \\ 7 & 4 & -1 \end{vmatrix} = -2$, $|A_2| = \begin{vmatrix} 1 & 9 & 1 \\ 1 & 4 & -1 \\ 1 & 7 & -1 \end{vmatrix} = 6$,

$|A_3| = \begin{vmatrix} 1 & 2 & 9 \\ 1 & 3 & 4 \\ 1 & 4 & 7 \end{vmatrix} = 8$, so $x_1 = \dfrac{-2}{2} = -1$, $x_2 = \dfrac{6}{2} = 3$, $x_3 = \dfrac{8}{2} = 4.$

(c) $|A| = \begin{vmatrix} 2 & 1 & 3 \\ 3 & -2 & 4 \\ 1 & 4 & -2 \end{vmatrix} = 28$, $|A_1| = \begin{vmatrix} 2 & 1 & 3 \\ 2 & -2 & 4 \\ 1 & 4 & -2 \end{vmatrix} = 14$, $|A_2| = \begin{vmatrix} 2 & 2 & 3 \\ 3 & 2 & 4 \\ 1 & 1 & -2 \end{vmatrix} = 7$,

$|A_3| = \begin{vmatrix} 2 & 1 & 2 \\ 3 & -2 & 2 \\ 1 & 4 & 1 \end{vmatrix} = 7$, so $x_1 = \dfrac{14}{28} = \dfrac{1}{2}$, $x_2 = \dfrac{7}{28} = \dfrac{1}{4}$, $x_3 = \dfrac{7}{28} = \dfrac{1}{4}.$

Section 3.4

11. (a) $|A| = \begin{vmatrix} 1 & 4 & 2 \\ 1 & 4 & -1 \\ 2 & 6 & 1 \end{vmatrix} = -6$, $|A_1| = \begin{vmatrix} 5 & 4 & 2 \\ 2 & 4 & -1 \\ 7 & 6 & 1 \end{vmatrix} = -18$, $|A_2| = \begin{vmatrix} 1 & 5 & 2 \\ 1 & 2 & -1 \\ 2 & 7 & 1 \end{vmatrix} = 0$,

$|A_3| = \begin{vmatrix} 1 & 4 & 5 \\ 1 & 4 & 2 \\ 2 & 6 & 7 \end{vmatrix} = -6$, so $x_1 = \dfrac{-18}{-6} = 3$, $x_2 = \dfrac{0}{-6} = 0$, $x_3 = \dfrac{-6}{-6} = 1$.

(b) $|A| = \begin{vmatrix} 2 & -1 & 3 \\ 1 & 4 & 2 \\ 3 & 2 & 1 \end{vmatrix} = -35$, $|A_1| = \begin{vmatrix} 7 & -1 & 3 \\ 10 & 4 & 2 \\ 0 & 2 & 1 \end{vmatrix} = 70$, $|A_2| = \begin{vmatrix} 2 & 7 & 3 \\ 1 & 10 & 2 \\ 3 & 0 & 1 \end{vmatrix} = -35$,

$|A_3| = \begin{vmatrix} 2 & -1 & 7 \\ 1 & 4 & 10 \\ 3 & 2 & 0 \end{vmatrix} = -140$, so $x_1 = \dfrac{70}{-35} = -2$, $x_2 = \dfrac{-35}{-35} = 1$, $x_3 = \dfrac{-140}{-35} = 4$.

(c) $|A| = \begin{vmatrix} 8 & -2 & 1 \\ 2 & -1 & 6 \\ 6 & 1 & 4 \end{vmatrix} = -128$, $|A_1| = \begin{vmatrix} 1 & -2 & 1 \\ 3 & -1 & 6 \\ 3 & 1 & 4 \end{vmatrix} = -16$, $|A_2| = \begin{vmatrix} 8 & 1 & 1 \\ 2 & 3 & 6 \\ 6 & 3 & 4 \end{vmatrix} = -32$,

$|A_3| = \begin{vmatrix} 8 & -2 & 1 \\ 2 & -1 & 3 \\ 6 & 1 & 3 \end{vmatrix} = -64$, so $x_1 = \dfrac{-16}{-128} = \dfrac{1}{8}$, $x_2 = \dfrac{-32}{-128} = \dfrac{1}{4}$, $x_3 = \dfrac{-64}{-128} = \dfrac{1}{2}$.

12. (a) $|A| = 0$ so this system of equations cannot be solved using Cramer's rule.

(b) $|A| = \begin{vmatrix} 3 & 1 & -1 \\ 1 & 2 & 1 \\ 2 & 6 & 0 \end{vmatrix} = -18$, $|A_1| = \begin{vmatrix} 7 & 1 & -1 \\ 3 & 2 & 1 \\ -4 & 6 & 0 \end{vmatrix} = -72$, $|A_2| = \begin{vmatrix} 3 & 7 & -1 \\ 1 & 3 & 1 \\ 2 & -4 & 0 \end{vmatrix} = 36$,

$|A_3| = \begin{vmatrix} 3 & 1 & 7 \\ 1 & 2 & 3 \\ 2 & 6 & -4 \end{vmatrix} = -54$, so $x_1 = \dfrac{-72}{-18} = 4$, $x_2 = \dfrac{36}{-18} = -2$, $x_3 = \dfrac{-54}{-18} = 3$.

Section 3.4

(c) $|A| = \begin{vmatrix} 3 & 6 & -1 \\ 1 & -2 & 3 \\ 4 & -2 & 5 \end{vmatrix} = 24$, $|A_1| = \begin{vmatrix} 3 & 6 & -1 \\ 2 & -2 & 3 \\ 5 & -2 & 5 \end{vmatrix} = 12$, $|A_2| = \begin{vmatrix} 3 & 3 & -1 \\ 1 & 2 & 3 \\ 4 & 5 & 5 \end{vmatrix} = 9$,

$|A_3| = \begin{vmatrix} 3 & 6 & 3 \\ 1 & -2 & 2 \\ 4 & -2 & 5 \end{vmatrix} = 18$, so $x_1 = \frac{12}{24} = \frac{1}{2}$, $x_2 = \frac{9}{24} = \frac{3}{8}$, $x_3 = \frac{18}{24} = \frac{3}{4}$.

13. (a) The determinant of the coefficient matrix is zero, so there is not a unique solution.

    (b) The determinant of the coefficient matrix is −10, so there is a unique solution.

    (c) The determinant of the coefficient matrix is zero, so there is not a unique solution.

14. (a) The determinant of the coefficient matrix is 42, so there is a unique solution.

    (b) The determinant of the coefficient matrix is zero, so there is not a unique solution.

    (c) The determinant of the coefficient matrix is zero, so there is not a unique solution.

15. The system of equations will have nontrivial solutions if the determinant of the coefficient matrix is zero, i.e. if

$$\begin{vmatrix} 1-\lambda & 6 \\ 5 & 2-\lambda \end{vmatrix} = 0.$$

$\begin{vmatrix} 1-\lambda & 6 \\ 5 & 2-\lambda \end{vmatrix} = (1-\lambda)(2-\lambda) - 30 = 2 - 3\lambda + \lambda^2 - 30 = \lambda^2 - 3\lambda - 28 = 0$, so

$(\lambda-7)(\lambda+4) = 0$ and $\lambda = 7$ or $\lambda = -4$. Substituting $\lambda = 7$ in the given equations, one finds that the general solution is $x_1 = x_2 = r$. For $\lambda = -4$, the general solution is $x_1 = -6r/5$, $x_2 = r$.

16. The system of equations will have nontrivial solutions if the determinant of the coefficient matrix is zero, i.e. if

Section 3.4

$$\begin{vmatrix} \lambda+4 & \lambda-2 \\ 4 & \lambda-3 \end{vmatrix} = 0.$$

$\begin{vmatrix} \lambda+4 & \lambda-2 \\ 4 & \lambda-3 \end{vmatrix} = (\lambda+4)(\lambda-3) - (\lambda-2)4 = \lambda^2 + \lambda - 12 - 4\lambda + 8 = \lambda^2 - 3\lambda - 4 = 0,$

so $(\lambda-4)(\lambda+1) = 0$ and $\lambda = 4$ or $\lambda = -1$. For $\lambda = 4$, the general solution is

$x_1 = -r/4, x_2 = r$. For $\lambda = -1$, the general solution is $x_1 = x_2 = r$.

17. The system of equations will have nontrivial solutions if the determinant of the coefficient matrix is zero, i.e. if

$$\begin{vmatrix} 5-\lambda & 4 & 2 \\ 4 & 5-\lambda & 2 \\ 2 & 2 & 2-\lambda \end{vmatrix} = 0.$$

$\begin{vmatrix} 5-\lambda & 4 & 2 \\ 4 & 5-\lambda & 2 \\ 2 & 2 & 2-\lambda \end{vmatrix} = (5-\lambda)(5-\lambda)(2-\lambda) + 16 + 16 - 4(5-\lambda) - 4(5-\lambda) - 16(2-\lambda)$

$= 10 - 21\lambda + 12\lambda^2 - \lambda^3 = (1-\lambda)(1-\lambda)(10-\lambda) = 0$, so $\lambda = 1$ or $\lambda = 10$. For $\lambda = 1$,

the general solution is $x_1 = -s - r/2, x_2 = s, x_3 = r$. For $\lambda = 10$, the general solution is

$x_1 = x_2 = 2r, x_3 = r$.

18. $AX = \lambda X = \lambda I_n X$, so $AX - \lambda I_n X = 0$. Thus $(A - \lambda I_n)X = 0$, and this system of equations has a nontrivial solution if and only if $|A - \lambda I_n| = 0$.

19. If $A = A^t$, then the matrix obtained by deleting the ith row and jth column of A is the transpose of the matrix obtained by deleting the jth row and ith column. The determinant of the first of these matrices is C_{ij}, the (i,j)th element of $(\text{adj}(A))^t$, and the determinant of the second is C_{ji}, the (i,j)th element of $\text{adj}(A)$. Since these determinants are equal, $\text{adj}(A) = (\text{adj}(A))^t$, and so $\text{adj}(A)$ is symmetric.

109

Section 3.4

20. If A is the zero matrix then adj(A) is the zero matrix also. If A is not the zero matrix and if A is not invertible then $|A| = 0$ and $A\,\text{adj}(A) = |A|\,I_n = O$. If adj(A) is invertible then
$A = A\,I_n = A\,[\text{adj}(A)\,(\text{adj}(A))^{-1}] = [A\,\text{adj}(A)]\,(\text{adj}(A))^{-1} = O\,(\text{adj}(A))^{-1} = O$. Since A is not the zero matrix, this means that adj(A) has no inverse.

21. If A is invertible then $\frac{1}{|A|}\text{adj}(A) = A^{-1}$, so that $A\,\frac{1}{|A|}\text{adj}(A) = AA^{-1} = I_n$. Thus $\frac{1}{|A|}A = [\text{adj}(A)]^{-1}$.

22. $A = (A^{-1})^{-1} = \frac{1}{|A^{-1}|}\text{adj}(A^{-1})$, so $\text{adj}(A^{-1}) = |A^{-1}|A = \frac{1}{|A|}A = [\text{adj}(A)]^{-1}$ from Exercise 21.

23. If AB is invertible then $|AB| \neq 0$. $|A||B| = |AB|$, so $|A| \neq 0$ and $|B| \neq 0$ and both A and B are invertible. The converse is also true. If A and B are invertible then $|AB| = |A||B| \neq 0$, so AB is invertible.

24. Let $B = A^{-1}$. The (i,j)th element of the product $AB = I_n$ is $a_{i1}b_{1j} + a_{i2}b_{2j} + \ldots + a_{in}b_{nj}$.
For $i = n$, $a_{n1}b_{1j} + a_{n2}b_{2j} + \ldots + a_{nn}b_{nj} = a_{nn}b_{nj}$, because $a_{n1} = a_{n2} = \ldots = a_{n\,n-1} = 0$.
This means that $a_{nn}b_{nj} = 0$ for $j < n$. But $a_{nn} \neq 0$ so $b_{nj} = 0$ for $j < n$.
For $i = n-1$, $a_{n-1\,1}b_{1j} + a_{n-1\,2}b_{2j} + \ldots + a_{n-1\,n-1}b_{n-1\,j} + a_{n-1\,n}b_{nj} = a_{n-1\,n-1}b_{n-1\,j}$, for $j < n$, because $a_{n-1\,1} = a_{n-1\,2} = \ldots = a_{n-1\,n-2} = b_{nj} = 0$. This means that $a_{n-1\,n-1}b_{n-1\,j} = 0$ for $j < n-1$.
But $a_{n-1\,n-1} \neq 0$ so $b_{n-1\,j} = 0$ for $j < n-1$.
In the same manner each row of B is shown to have zero in all positions for which the column number is less than the row number. Thus B is upper triangular.

25. If $|A| = \pm 1$, then $A^{-1} = \frac{1}{|A|}\text{adj}(A) = \pm\text{adj}(A)$, and since all elements of A are integers, all elements of adj(A) are integers (because adding and multiplying integers gives integer results).

Chapter 3 Review Exercises

26. If $AX = 0$ has only the trivial solution, then $|A| \neq 0$ so that $|A^k| = |A|^k \neq 0$. Thus $A^k X = 0$ has only the trivial solution.

27. $AX = B_2$ has a unique solution if and only if $|A| \neq 0$ if and only if $AX = B_1$ has a unique solution.

Chapter 3 Review Exercises

1. (a) $3 \times 1 - 2 \times 5 = -7$. (b) $-3 \times 6 - 0 \times 1 = -18$. (c) $9 \times 4 - 7 \times 1 = 29$.

2. (a) $M_{12} = \begin{vmatrix} -3 & 1 \\ 7 & 2 \end{vmatrix} = -3 \times 2 - 1 \times 7 = -13.$ $C_{12} = (-1)^{1+2} M_{12} = 13.$

   (b) $M_{31} = \begin{vmatrix} 1 & 0 \\ 4 & 1 \end{vmatrix} = 1 \times 1 - 0 \times 4 = 1.$ $C_{31} = (-1)^{3+1} M_{31} = 1.$

   (c) $M_{22} = \begin{vmatrix} 2 & 0 \\ 7 & 2 \end{vmatrix} = 2 \times 2 - 0 \times 4 = 4.$ $C_{22} = (-1)^{2+2} M_{22} = 4.$

3. (a) $\begin{vmatrix} 1 & 2 & -3 \\ 0 & 2 & 5 \\ 4 & 1 & 2 \end{vmatrix}$ using row 1:
   $= \begin{vmatrix} 2 & 5 \\ 1 & 2 \end{vmatrix} - 2 \begin{vmatrix} 0 & 5 \\ 4 & 2 \end{vmatrix} + (-3) \begin{vmatrix} 0 & 2 \\ 4 & 1 \end{vmatrix} = -1 + 40 + 24 = 63.$

   using column 1:
   $= \begin{vmatrix} 2 & 5 \\ 1 & 2 \end{vmatrix} - 0 \begin{vmatrix} 2 & -3 \\ 1 & 2 \end{vmatrix} + 4 \begin{vmatrix} 2 & -3 \\ 2 & 5 \end{vmatrix} = -1 + 0 + 64 = 63.$

   (b) $\begin{vmatrix} 0 & 5 & 3 \\ 2 & -3 & 1 \\ 2 & 7 & 3 \end{vmatrix}$ using row 3:
   $= 2 \begin{vmatrix} 5 & 3 \\ -3 & 1 \end{vmatrix} - 7 \begin{vmatrix} 0 & 3 \\ 2 & 1 \end{vmatrix} + 3 \begin{vmatrix} 0 & 5 \\ 2 & -3 \end{vmatrix} = 28 + 42 - 30 = 40.$

   using column 2:

Chapter 3 Review Exercises

$$= -5 \begin{vmatrix} 2 & 1 \\ 2 & 3 \end{vmatrix} + (-3) \begin{vmatrix} 0 & 3 \\ 2 & 3 \end{vmatrix} - 7 \begin{vmatrix} 0 & 3 \\ 2 & 1 \end{vmatrix} = -20 + 18 + 42 = 40.$$

4. $\begin{vmatrix} x & x \\ 2 & x-3 \end{vmatrix} = x(x-3) - 2x = x^2 - 5x = -6$, $x^2 - 5x + 6 = 0$. Thus $(x-3)(x-2) = 0$, so $x = 3$ or $x = 2$.

5. (a) $\begin{vmatrix} 1 & 2 & -1 \\ 3 & 1 & 1 \\ 2 & 4 & 1 \end{vmatrix} \underset{R3+(-2)R1}{=} \begin{vmatrix} 1 & 2 & -1 \\ 3 & 1 & 1 \\ 0 & 0 & 3 \end{vmatrix} = 3 \begin{vmatrix} 1 & 2 \\ 3 & 1 \end{vmatrix} = -15.$

(b) $\begin{vmatrix} 5 & 3 & 4 \\ 4 & 6 & 1 \\ 2 & -3 & 7 \end{vmatrix} \underset{\substack{R2+(-2)R1 \\ R3+R1}}{=} \begin{vmatrix} 5 & 3 & 4 \\ -6 & 0 & -7 \\ 7 & 0 & 11 \end{vmatrix} = -3 \begin{vmatrix} -6 & -7 \\ 7 & 11 \end{vmatrix} = 51.$

(c) $\begin{vmatrix} 1 & 4 & -2 \\ 2 & 3 & 1 \\ -1 & 5 & 6 \end{vmatrix} \underset{\substack{R2+(-2)R1 \\ R3+R1}}{=} \begin{vmatrix} 1 & 4 & -2 \\ 0 & -5 & 5 \\ 0 & 9 & 4 \end{vmatrix} = \begin{vmatrix} -5 & 5 \\ 9 & 4 \end{vmatrix} = -65.$

6. (a) This matrix can be obtained from A by multiplying row 2 by 3, so its determinant is $3|A| = 6$.

(b) This matrix can be obtained from A by adding -2 times row 1 to row 2, so its determinant is $|A| = 2$.

(c) This matrix can be obtained from A by multiplying row 1 by 2, row 2 by -1, and row 3 by 3, so its determinant is $2 \times -1 \times 3 \times |A| = -12$.

7. (a) $\begin{vmatrix} 1 & 2 & 4 \\ -1 & 4 & 3 \\ 2 & 0 & 5 \end{vmatrix} \underset{R2+(-2)R1}{=} \begin{vmatrix} 1 & 2 & 4 \\ -3 & 0 & -5 \\ 2 & 0 & 5 \end{vmatrix} = -2 \begin{vmatrix} -3 & -5 \\ 2 & 5 \end{vmatrix} = 10.$

112

Chapter 3 Review Exercises

(b) $\begin{vmatrix} -1 & 3 & 2 \\ 0 & 5 & 2 \\ 1 & 7 & 6 \end{vmatrix} \underset{R3+R1}{=} \begin{vmatrix} -1 & 3 & 2 \\ 0 & 5 & 2 \\ 0 & 10 & 8 \end{vmatrix} = -1 \begin{vmatrix} 5 & 2 \\ 10 & 8 \end{vmatrix} = -20.$

(c) $\begin{vmatrix} 2 & -3 & 5 \\ 4 & 0 & 6 \\ 1 & 2 & 7 \end{vmatrix} \underset{R3+(2/3)R1}{=} \begin{vmatrix} 2 & -3 & 5 \\ 4 & 0 & 6 \\ 7/3 & 0 & 31/3 \end{vmatrix} = -(-3)\begin{vmatrix} 4 & 6 \\ 7/3 & 31/3 \end{vmatrix} = 82.$

8. (a) $|3A| = 3^3|A| = 27 \times -2 = -54.$ (b) $|2AA^t| = 2^3|A||A^t| = 8|A||A| = 32.$

   (c) $|A^3| = |A|^3 = (-2)^3 = -8.$

   (d) $|(A^tA)^2| = (|A^tA|)^2 = (|A^t||A|)^2 = (|A||A|)^2 = |A|^4 = 16.$

   (e) $|(A^t)^3| = (|A^t|)^3 = (|A|)^3 = (-2)^3 = -8.$

   (f) $|2A^t(A^{-1})^2| = (2)^3 |A^t||(A^{-1})^2| = 8|A^t||A^{-1}|^2 = 8|A|(1/|A|)^2 = 8/|A| = -4.$

9. If A is 2x2, then $cA = \begin{bmatrix} ca_{11} & ca_{12} \\ ca_{21} & ca_{22} \end{bmatrix}$, and $|cA| = c^2 a_{11}a_{22} - c^2 a_{12}a_{21} = c^2|A|.$

   If A is 3x3, then $cA = \begin{bmatrix} ca_{11} & ca_{12} & ca_{13} \\ ca_{21} & ca_{22} & ca_{23} \\ ca_{31} & ca_{32} & ca_{33} \end{bmatrix}$. The (i,j)th cofactor is 2x2 and so from above

   is $c^2 C_{ij}$, where C_{ij} is the (i,j)th cofactor of A. Thus

   $|cA| = ca_{11}c^2C_{11} + ca_{12}c^2C_{12} + ca_{13}c^2C_{13} = c^3(a_{11}C_{11} + a_{12}C_{12} + a_{13}C_{13}) = c^3|A|.$

   Repeat this argument for each size A, until A is nxn and the (i,j)th cofactor is $c^{n-1}C_{ij}$, where C_{ij} is the (i,j)th cofactor of A. Then

   $|cA| = ca_{11}c^{n-1}C_{11} + \ldots + ca_{1n}c^{n-1}C_{1n} = c^n(a_{11}C_{11} + \ldots + a_{1n}C_{1n}) = c^n|A|.$

10. $|B| \neq 0$ so B^{-1} exists, and A and B^{-1} can be multiplied. Let $C = AB^{-1}$. Then $CB = AB^{-1}B = A.$

113

Chapter 3 Review Exercises

11. $|C^{-1}AC| = |C^{-1}||AC| = |C^{-1}||A||C| = |C^{-1}||C||A| = |C^{-1}C||A| = |A|$

12. If A is upper triangular, then any element in A is zero if its row number is greater than its column number. If $i > j$, then for every k with $1 \le k \le n$, $a_{ik}a_{kj} = 0$ because either $i \ge k$ so that $a_{ik} = 0$, or $k \ge i > j$ so that $a_{kj} = 0$. Thus if $i > j$, the (i,j)th term $a_{i1}a_{1j} + a_{i2}a_{2j} + \ldots + a_{in}a_{nj}$ of A^2 is zero because each summand is zero. So A^2 is upper triangular. The proof is similar for lower triangular matrices.

13. If $A^2 = A$, then $|A||A| = |A|$ so $|A| = 1$ or zero. If A is also invertible, then $|A| \ne 0$ so $|A| = 1$.

14. (a) $\begin{vmatrix} 3 & 5 \\ 1 & 2 \end{vmatrix} = 1$, so $\begin{bmatrix} 3 & 5 \\ 1 & 2 \end{bmatrix}^{-1} = \begin{bmatrix} 2 & -5 \\ -1 & 3 \end{bmatrix}$.

(b) $\begin{vmatrix} 3 & 2 \\ -1 & 5 \end{vmatrix} = 17$, so $\begin{bmatrix} 3 & 2 \\ -1 & 5 \end{bmatrix}^{-1} = \frac{1}{17}\begin{bmatrix} 5 & -2 \\ 1 & 3 \end{bmatrix}$.

(c) $\begin{vmatrix} 1 & 4 & -1 \\ 0 & 2 & 0 \\ 1 & 6 & -1 \end{vmatrix} = 0$, so the inverse does not exist.

(d) $\begin{vmatrix} 2 & 1 & 3 \\ 0 & 2 & 9 \\ 4 & 2 & 11 \end{vmatrix} = 20$, so $\begin{bmatrix} 2 & 1 & 3 \\ 0 & 2 & 9 \\ 4 & 2 & 11 \end{bmatrix}^{-1}$

$= \frac{1}{20}\begin{bmatrix} \begin{vmatrix} 2 & 9 \\ 2 & 11 \end{vmatrix} & -\begin{vmatrix} 1 & 3 \\ 2 & 11 \end{vmatrix} & \begin{vmatrix} 1 & 3 \\ 2 & 9 \end{vmatrix} \\ -\begin{vmatrix} 0 & 9 \\ 4 & 11 \end{vmatrix} & \begin{vmatrix} 2 & 3 \\ 4 & 11 \end{vmatrix} & -\begin{vmatrix} 2 & 3 \\ 0 & 9 \end{vmatrix} \\ \begin{vmatrix} 0 & 2 \\ 4 & 2 \end{vmatrix} & -\begin{vmatrix} 2 & 1 \\ 4 & 2 \end{vmatrix} & \begin{vmatrix} 2 & 1 \\ 0 & 2 \end{vmatrix} \end{bmatrix} = \frac{1}{20}\begin{bmatrix} 4 & -5 & 3 \\ 36 & 10 & -18 \\ -8 & 0 & 4 \end{bmatrix}$.

Chapter 3 Review Exercises

15. (a) $x_1 = \dfrac{\begin{vmatrix} -1 & 1 \\ 18 & -5 \end{vmatrix}}{\begin{vmatrix} 2 & 1 \\ 3 & -5 \end{vmatrix}} = \dfrac{-13}{-13} = 1$, $\quad x_2 = \dfrac{\begin{vmatrix} 2 & -1 \\ 3 & 18 \end{vmatrix}}{\begin{vmatrix} 2 & 1 \\ 3 & -5 \end{vmatrix}} = \dfrac{39}{-13} = -3$.

(b) $|A| = \begin{vmatrix} 1 & 1 & 1 \\ 2 & -1 & 3 \\ 4 & 5 & 1 \end{vmatrix} = 8$, $|A_1| = \begin{vmatrix} 1 & 1 & 1 \\ 5 & -1 & 3 \\ 3 & 5 & 1 \end{vmatrix} = 16$, $|A_2| = \begin{vmatrix} 1 & 1 & 1 \\ 2 & 5 & 3 \\ 4 & 3 & 1 \end{vmatrix} = -8$,

$|A_3| = \begin{vmatrix} 1 & 1 & 1 \\ 2 & -1 & 5 \\ 4 & 5 & 3 \end{vmatrix} = 0$, so $x_1 = \dfrac{16}{8} = 2$, $x_2 = \dfrac{-8}{8} = -1$, and $x_3 = \dfrac{0}{8} = 0$.

16. If A is not invertible, $|A| = 0$. The (i,j)th term of A[adj(A)] is $a_{i1}C_{j1} + a_{i2}C_{j2} + \ldots + a_{in}C_{jn}$. If $i = j$, this term is $|A| = 0$ and if $i \neq j$ it is the determinant of the matrix obtained from A by replacing row j with row i; that is, it is the determinant of a matrix having two equal rows, and so it is zero. Thus A[adj(A)] is the zero matrix.

17. $|A|$ is the product of the diagonal elements, and since $|A| \neq 0$ all diagonal elements must be nonzero.

18. If $|A| = \pm 1$ then $A^{-1} = \pm\,\text{adj}(A)$, so $X = A^{-1}AX = A^{-1}B = \pm\,\text{adj}(A)B$. If all the elements of A and of B are integers, then all the elements of adj(A) are integers and all the elements of the product adj(A)B are integers, so X has all integer components.

Chapter 4

Exercise Set 4.1

1.

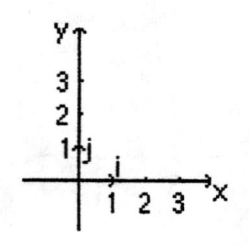

2.

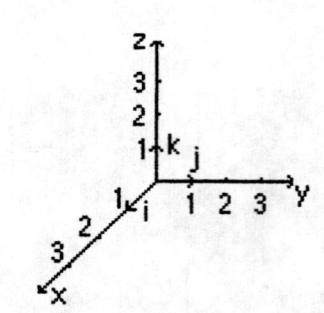

3. (a)

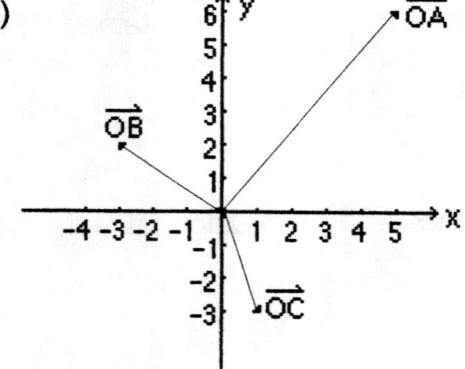

   (b)

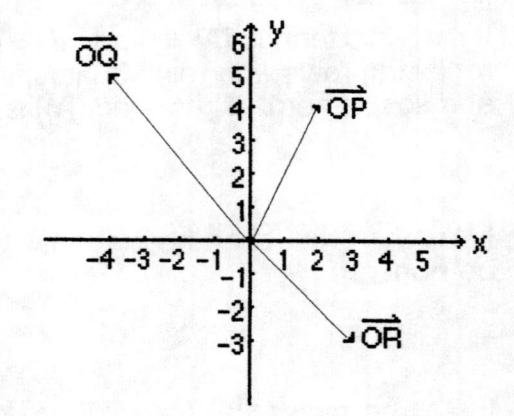

   (c)

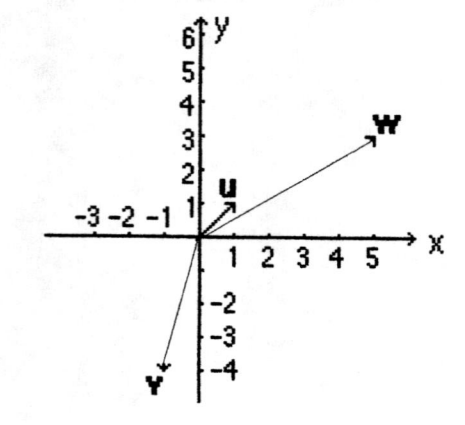

Section 4.1

4. (a)

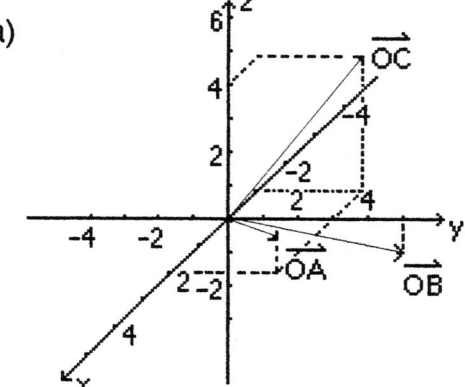

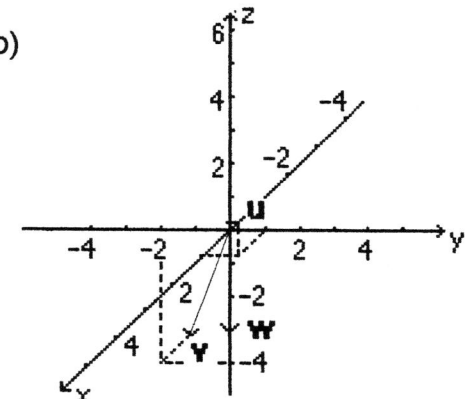

(b)

5. (a) $3(1,4) = (3,12)$.   (b) $-2(-1,3) = (2,-6)$.   (c) $(1/2)(2,6) = (1,3)$.

    (d) $(-1/2)(2,4,2) = (-1,-2,-1)$.   (e) $3(-1,2,3) = (-3,6,9)$.

    (f) $4(-1,2,3,-2) = (-4,8,12,-8)$.   (g) $-5(1,-4,3,-2,5) = (-5,20,-15,10,-25)$.

    (h) $3(3,0,4,2,-1) = (9,0,12,6,-3)$.

6. (a) $\mathbf{u} + \mathbf{w} = (1,2) + (-3,5) = (-2,7)$,   (b) $\mathbf{u} + 3\mathbf{v} = (1,2) + 3(4,-1) = (13,-1)$.

    (c) $\mathbf{v} + \mathbf{w} = (4,-1) + (-3,5) = (1,4)$.

    (d) $2\mathbf{u} + 3\mathbf{v} - \mathbf{w} = 2(1,2) + 3(4,-1) - (-3,5) = (17,-4)$.

    (e) $-3\mathbf{u} + 4\mathbf{v} - 2\mathbf{w} = -3(1,2) + 4(4,-1) - 2(-3,5) = (19,-20)$.

7. (a) $\mathbf{u} + \mathbf{w} = (2,1,3) + (2,4,-2) = (4,5,1)$.   (b) $2\mathbf{u} + \mathbf{v} = 2(2,1,3) + (-1,3,2) = (3,5,8)$.

    (c) $\mathbf{u} + 3\mathbf{w} = (2,1,3) + 3(2,4,-2) = (8,13,-3)$.

    (d) $5\mathbf{u} - 2\mathbf{v} + 6\mathbf{w} = 5(2,1,3) - 2(-1,3,2) + 6(2,4,-2) = (24,23,-1)$.

    (e) $2\mathbf{u} - 3\mathbf{v} - 4\mathbf{w} = 2(2,1,3) - 3(-1,3,2) - 4(2,4,-2) = (-1,-23,8)$.

8. (a) $\mathbf{u} + (\mathbf{v} + \mathbf{w}) = (u_1, u_2, \ldots, u_n) + ((v_1, v_2, \ldots, v_n) + (w_1, w_2, \ldots, w_n))$
       $= (u_1, u_2, \ldots, u_n) + (v_1+w_1, v_2+w_2, \ldots, v_n+w_n)$

117

Section 4.1

$$= (u_1+(v_1+w_1), u_2+(v_2+w_2), \ldots, u_n+(v_n+w_n))$$
$$= ((u_1+v_1)+w_1, (u_2+v_2)+w_2, \ldots, (u_n+v_n)+w_n)$$
$$= ((u_1+v_1), (u_2+v_2), \ldots, (u_n+v_n)) + (w_1, w_2, \ldots, w_n)$$
$$= ((u_1, u_2, \ldots, u_n) + (v_1, v_2, \ldots, v_n)) + (w_1, w_2, \ldots, w_n) = (\mathbf{u}+\mathbf{v})+\mathbf{w}.$$

(b) $\mathbf{u} + (-\mathbf{u}) = (u_1, u_2, \ldots, u_n) + (-1)(u_1, u_2, \ldots, u_n)$
$= (u_1, u_2, \ldots, u_n) + (-u_1, -u_2, \ldots, -u_n) = (u_1-u_1), (u_2-u_2), \ldots, (u_n-u_n)$
$= (0, 0, \ldots, 0) = \mathbf{0}.$

(c) $(c+d)\mathbf{u} = (c+d)(u_1, u_2, \ldots, u_n) = ((c+d)u_1, (c+d)u_2, \ldots, (c+d)u_n)$
$= (cu_1+du_1, cu_2+du_2, \ldots, cu_n+du_n)$
$= (cu_1, cu_2, \ldots, cu_n) + (du_1, du_2, \ldots, du_n)$
$= c(u_1, u_2, \ldots, u_n) + d(u_1, u_2, \ldots, u_n) = c\mathbf{u} + d\mathbf{u}.$

(d) $1\mathbf{u} = 1(u_1, u_2, \ldots, u_n) = (1 \times u_1, 1 \times u_2, \ldots, 1 \times u_n) = (u_1, u_2, \ldots, u_n) = \mathbf{u}.$

9. (a) $\mathbf{u} + \mathbf{v} = \begin{bmatrix} 2 \\ 3 \end{bmatrix} + \begin{bmatrix} -1 \\ -4 \end{bmatrix} = \begin{bmatrix} 1 \\ -1 \end{bmatrix}.$ (b) $2\mathbf{v} - 3\mathbf{w} = 2\begin{bmatrix} -1 \\ -4 \end{bmatrix} - 3\begin{bmatrix} 4 \\ -6 \end{bmatrix} = \begin{bmatrix} -14 \\ 10 \end{bmatrix}.$

(c) $2\mathbf{u} + 4\mathbf{v} - \mathbf{w} = 2\begin{bmatrix} 2 \\ 3 \end{bmatrix} + 4\begin{bmatrix} -1 \\ -4 \end{bmatrix} - \begin{bmatrix} 4 \\ -6 \end{bmatrix} = \begin{bmatrix} -4 \\ -4 \end{bmatrix}.$

(d) $-3\mathbf{u} - 2\mathbf{v} + 4\mathbf{w} = -3\begin{bmatrix} 2 \\ 3 \end{bmatrix} - 2\begin{bmatrix} -1 \\ -4 \end{bmatrix} + 4\begin{bmatrix} 4 \\ -6 \end{bmatrix} = \begin{bmatrix} 12 \\ -25 \end{bmatrix}.$

10. (a) $\mathbf{u} + 2\mathbf{v} = \begin{bmatrix} 1 \\ 2 \\ -1 \end{bmatrix} + 2\begin{bmatrix} 3 \\ 0 \\ 1 \end{bmatrix} = \begin{bmatrix} 7 \\ 2 \\ 1 \end{bmatrix}.$ (b) $-4\mathbf{v} + 3\mathbf{w} = -4\begin{bmatrix} 3 \\ 0 \\ 1 \end{bmatrix} + 3\begin{bmatrix} -1 \\ 0 \\ 5 \end{bmatrix} = \begin{bmatrix} -15 \\ 0 \\ 11 \end{bmatrix}.$

(c) $3\mathbf{u} - 2\mathbf{v} + 4\mathbf{w} = 3\begin{bmatrix} 1 \\ 2 \\ -1 \end{bmatrix} - 2\begin{bmatrix} 3 \\ 0 \\ 1 \end{bmatrix} + 4\begin{bmatrix} -1 \\ 0 \\ 5 \end{bmatrix} = \begin{bmatrix} -7 \\ 6 \\ 15 \end{bmatrix}.$

(d) $2\mathbf{u} + 3\mathbf{v} - 8\mathbf{w} = 2\begin{bmatrix} 1 \\ 2 \\ -1 \end{bmatrix} + 3\begin{bmatrix} 3 \\ 0 \\ 1 \end{bmatrix} - 8\begin{bmatrix} -1 \\ 0 \\ 5 \end{bmatrix} = \begin{bmatrix} 19 \\ 4 \\ -39 \end{bmatrix}.$

Section 4.2

11. The resultant forces are   (a)   (3,2) + (5,−5) = (8,−3),

     (b)   (−1,3) + (−2,3) = (−3,6),   (c)   (9,−1) + (−3,2) = (6,1).

12. The resultant forces are

     (a)   (1,4,6) + (2,5,1) = (3,9,7),   (b)   (3,−1,7) + (−2,5,−1) = (1,4,6).

13. The resultant force is (0,3) + (3,1) = (3,4).

    (6,8) = (3,1) + (3,7), so to double the magnitude, (0,3) must be replaced by (3,7).

14. The resultant force is (5,1) + (3,4) = (8,5).

    (9,7) = (5,1) + (4,6), so (3,4) must be replaced by (4,6).

15. The resultant force is (2,8) + (−1,3) = (1,11).

    (2,5) = (−1,3) + (3,2), so (2,8) must be replaced by (3,2).

16. (a) temperature - scalar   (b) acceleration - vector   (c) pressure - scalar
     (d) frequency - scalar   (e) gravity - vector   (f) position - vector
     (g) time - scalar   (h) sound - scalar   (i) cost - scalar

Exercise Set 4.2

1. (a) (2,1)·(3,4) = 2x3 + 1x4 = 6 + 4 = 10

    (b) (1,−4)·(3,0) = 1x3 + −4x0 = 3

    (c) (2,0)·(0,−1) = 2x0 + 0x−1 = 0

    (d) (5,−2)·(−3,−4) = 5x−3 + −2x−4 = −15 + 8 = −7

Section 4.2

2. (a) $(1,2,3) \cdot (4,1,0) = 1 \times 4 + 2 \times 1 + 3 \times 0 = 4 + 2 + 0 = 6$

   (b) $(3,4,-2) \cdot (5,1,-1) = 3 \times 5 + 4 \times 1 + -2 \times -1 = 15 + 4 + 2 = 21$

   (c) $(7,1,-2) \cdot (3,-5,8) = 7 \times 3 + 1 \times -5 + -2 \times 8 = 21 - 5 - 16 = 0$

3. (a) $(5,1) \cdot (2,-3) = 5 \times 2 + 1 \times -3 = 10 - 3 = 7$

   (b) $(-3,1,5) \cdot (2,0,4) = -3 \times 2 + 1 \times 0 + 5 \times 4 = -6 + 0 + 20 = 14$

   (c) $(7,1,2,-4) \cdot (3,0,-1,5) = 7 \times 3 + 1 \times 0 + 2 \times -1 + -4 \times 5 = 21 + 0 - 2 - 20 = -1$

   (d) $(2,3,-4,1,6) \cdot (-3,1,-4,5,-1) = 2 \times -3 + 3 \times 1 + -4 \times -4 + 1 \times 5 + 6 \times -1$
       $= -6 + 3 + 16 + 5 - 6 = 12$

   (e) $(1,2,3,0,0,0) \cdot (0,0,0,-2,-4,9) = 1 \times 0 + 2 \times 0 + 3 \times 0 + 0 \times -2 + 0 \times -4 + 0 \times 9 = 0$

4. (a) $\|(1,2)\| = \sqrt{(1,2) \cdot (1,2)} = \sqrt{1 \times 1 + 2 \times 2} = \sqrt{5}$

   (b) $\|(3,-4)\| = \sqrt{(3,-4) \cdot (3,-4)} = \sqrt{3 \times 3 + -4 \times -4} = \sqrt{25} = 5$

   (c) $\|(4,0)\| = \sqrt{(4,0) \cdot (4,0)} = \sqrt{4 \times 4 + 0 \times 0} = \sqrt{16} = 4$

   (d) $\|(-3,1)\| = \sqrt{(-3,1) \cdot (-3,1)} = \sqrt{-3 \times -3 + 1 \times 1} = \sqrt{10}$

   (e) $\|(0,27)\| = \sqrt{(0,27) \cdot (0,27)} = \sqrt{0 \times 0 + 27 \times 27} = 27$

5. (a) $\|(1,3,-1)\| = \sqrt{(1,3,-1) \cdot (1,3,-1)} = \sqrt{1 \times 1 + 3 \times 3 + -1 \times -1} = \sqrt{11}$

   (b) $\|(3,0,4)\| = \sqrt{(3,0,4) \cdot (3,0,4)} = \sqrt{3 \times 3 + 0 \times 0 + 4 \times 4} = \sqrt{25} = 5$

   (c) $\|(5,1,1)\| = \sqrt{(5,1,1) \cdot (5,1,1)} = \sqrt{5 \times 5 + 1 \times 1 + 1 \times 1} = \sqrt{27} = 3\sqrt{3}$

   (d) $\|(0,5,0)\| = \sqrt{(0,5,0) \cdot (0,5,0)} = \sqrt{0 \times 0 + 5 \times 5 + 0 \times 0} = \sqrt{25} = 5$

   (e) $\|(7,-2,-3)\| = \sqrt{(7,-2,-3) \cdot (7,-2,-3)} = \sqrt{7 \times 7 + -2 \times -2 + -3 \times -3} = \sqrt{62}$

6. (a) $\|(5,2)\| = \sqrt{(5,2) \cdot (5,2)} = \sqrt{5 \times 5 + 2 \times 2} = \sqrt{29}$

   (b) $\|(-4,2,3)\| = \sqrt{(-4,2,3) \cdot (-4,2,3)} = \sqrt{-4 \times -4 + 2 \times 2 + 3 \times 3} = \sqrt{29}$

Section 4.2

(c) $\|(1,2,3,4)\| = \sqrt{(1,2,3,4)\cdot(1,2,3,4)} = \sqrt{1\times1+2\times2+3\times3+4\times4} = \sqrt{30}$

(d) $\|(4,-2,1,3)\| = \sqrt{(4,-2,1,3)\cdot(4,-2,1,3)} = \sqrt{4\times4+-2\times-2+1\times1+3\times3} = \sqrt{30}$

(e) $\|(-3,0,1,4,2)\| = \sqrt{(-3,0,1,4,2)\cdot(-3,0,1,4,2)} = \sqrt{-3\times-3+0\times0+1\times1+4\times4+2\times2} = \sqrt{30}$

(f) $\|(0,0,0,7,0,0)\| = \sqrt{(0,0,0,7,0,0)\cdot(0,0,0,7,0,0)} = \sqrt{49} = 7$

7. (a) $\dfrac{(1,3)}{\|(1,3)\|} = \left(\dfrac{1}{\sqrt{10}}, \dfrac{3}{\sqrt{10}}\right)$

(b) $\dfrac{(2,-4)}{\|(2,-4)\|} = \left(\dfrac{2}{2\sqrt{5}}, \dfrac{-4}{2\sqrt{5}}\right) = \left(\dfrac{1}{\sqrt{5}}, \dfrac{-2}{\sqrt{5}}\right)$

(c) $\dfrac{(1,2,3)}{\|(1,2,3)\|} = \left(\dfrac{1}{\sqrt{14}}, \dfrac{2}{\sqrt{14}}, \dfrac{3}{\sqrt{14}}\right)$

(d) $\dfrac{(-2,4,0)}{\|(-2,4,0)\|} = \left(\dfrac{-2}{\sqrt{20}}, \dfrac{4}{\sqrt{20}}, 0\right) = \left(\dfrac{-1}{\sqrt{5}}, \dfrac{2}{\sqrt{5}}, 0\right)$

(e) $\dfrac{(0,5,0)}{\|(0,5,0)\|} = (0,1,0)$

8. (a) $\dfrac{(4,2)}{\|(4,2)\|} = \left(\dfrac{4}{2\sqrt{5}}, \dfrac{2}{2\sqrt{5}}\right) = \left(\dfrac{2}{\sqrt{5}}, \dfrac{1}{\sqrt{5}}\right)$

(b) $\dfrac{(4,1,1)}{\|(4,1,1)\|} = \left(\dfrac{4}{3\sqrt{2}}, \dfrac{1}{3\sqrt{2}}, \dfrac{1}{3\sqrt{2}}\right)$

(c) $\dfrac{(7,2,0,1)}{\|(7,2,0,1)\|} = \left(\dfrac{7}{3\sqrt{6}}, \dfrac{2}{3\sqrt{6}}, 0, \dfrac{1}{3\sqrt{6}}\right)$

(d) $\dfrac{(3,-1,1,2)}{\|(3,-1,1,2)\|} = \left(\dfrac{3}{\sqrt{15}}, \dfrac{-1}{\sqrt{15}}, \dfrac{1}{\sqrt{15}}, \dfrac{2}{\sqrt{15}}\right)$

(e) $\dfrac{(0,0,0,7,0,0)}{\|(0,0,0,7,0,0)\|} = (0,0,0,1,0,0)$

Section 4.2

9. (a) $\cos\theta = \dfrac{(-1,1)\cdot(0,1)}{\|(-1,1)\|\;\|(0,1)\|} = \dfrac{1}{\sqrt{2}}$, so $\theta = \dfrac{\pi}{4} = 45°$

   (b) $\cos\theta = \dfrac{(2,0)\cdot(1,\sqrt{3})}{\|(2,0)\|\;\|(1,\sqrt{3})\|} = \dfrac{2}{4} = \dfrac{1}{2}$, so $\theta = \dfrac{\pi}{3} = 60°$

   (c) $\cos\theta = \dfrac{(2,3)\cdot(3,-2)}{\|(2,3)\|\;\|(3,-2)\|} = 0$, so $\theta = \dfrac{\pi}{2} = 90°$

   (d) $\cos\theta = \dfrac{(5,2)\cdot(-5,-2)}{\|(5,2)\|\;\|(-5,-2)\|} = \dfrac{-29}{29} = -1$, so $\theta = \pi = 180°$

10. (a) $\cos\theta = \dfrac{(4,-1)\cdot(2,3)}{\|(4,-1)\|\;\|(2,3)\|} = \dfrac{5}{\sqrt{17}\sqrt{13}}$

    (b) $\cos\theta = \dfrac{(3,-1,2)\cdot(4,1,1)}{\|(3,-1,2)\|\;\|(4,1,1)\|} = \dfrac{13}{\sqrt{14}\sqrt{18}} = \dfrac{13}{6\sqrt{7}}$

    (c) $\cos\theta = \dfrac{(2,-1,0)\cdot(5,3,1)}{\|(2,-1,0)\|\;\|(5,3,1)\|} = \dfrac{7}{\sqrt{5}\sqrt{35}} = \dfrac{7}{5\sqrt{7}} = \dfrac{\sqrt{7}}{5}$

    (d) $\cos\theta = \dfrac{(7,1,0,0)\cdot(3,2,1,0)}{\|(7,1,0,0)\|\;\|(3,2,1,0)\|} = \dfrac{23}{\sqrt{50}\sqrt{14}} = \dfrac{23}{10\sqrt{7}}$

    (e) $\cos\theta = \dfrac{(1,2,-1,3,1)\cdot(2,0,1,0,4)}{\|(1,2,-1,3,1)\|\;\|(2,0,1,0,4)\|} = \dfrac{5}{4\sqrt{21}}$

11. (a) $(1,3)\cdot(3,-1) = 1\times 3 + 3\times -1 = 0$, so the vectors are orthogonal.

    (b) $(-2,4)\cdot(4,2) = -2\times 4 + 4\times 2 = 0$, so the vectors are orthogonal.

    (c) $(3,0)\cdot(0,-2) = 3\times 0 + 0\times -2 = 0$, so the vectors are orthogonal.

    (d) $(7,-1)\cdot(1,7) = 7\times 1 + -1\times 7 = 0$, so the vectors are orthogonal.

12. (a) $(3,-5)\cdot(5,3) = 3\times 5 + -5\times 3 = 0$, so the vectors are orthogonal.

    (b) $(1,2,-3)\cdot(4,1,2) = 1\times 4 + 2\times 1 + -3\times 2 = 0$, so the vectors are orthogonal.

    (c) $(7,1,0)\cdot(2,-14,3) = 7\times 2 + 1\times -14 + 0\times 3 = 0$, so the vectors are orthogonal.

    (d) $(5,1,0,2)\cdot(-3,7,9,4) = 5\times -3 + 1\times 7 + 0\times 9 + 2\times 4 = 0$, so the vectors are orthogonal.

Section 4.2

(e) $(1,-1,2,-5,9) \cdot (4,7,4,1,0) = 1 \times 4 + -1 \times 7 + 2 \times 4 + -5 \times 1 + 9 \times 0 = 0$, so the vectors are orthogonal.

13. (a) If (a,b) is orthogonal to $(1,3)$, then $(a,b) \cdot (1,3) = a + 3b = 0$, so $a = -3b$. Thus any vector of the form $(-3b, b)$ is orthogonal to $(1,3)$.

    (b) If (a,b) is orthogonal to $(7,-1)$, then $(a,b) \cdot (7,-1) = 7a - b = 0$, so $b = 7a$. Thus any vector of the form $(a, 7a)$ is orthogonal to $(7,-1)$.

    (c) If (a,b) is orthogonal to $(-4,-1)$, then $(a,b) \cdot (-4,-1) = -4a - b = 0$, so $b = -4a$. Thus any vector of the form $(a, -4a)$ is orthogonal to $(-4,-1)$.

    (d) If (a,b) is orthogonal to $(-3,0)$, then $(a,b) \cdot (-3,0) = -3a = 0$, so $a = 0$. Thus any vector of the form $(0, b)$ is orthogonal to $(-3,0)$.

14. (a) If (a,b) is orthogonal to $(5,-1)$, then $(a,b) \cdot (5,-1) = 5a - b = 0$, so $b = 5a$. Thus any vector of the form $(a, 5a)$ is orthogonal to $(5,-1)$.

    (b) If (a,b,c) is orthogonal to $(1,-2,3)$, then $(a,b,c) \cdot (1,-2,3) = a - 2b + 3c = 0$, so $a = 2b - 3c$. Thus any vector of the form $(2b-3c, b, c)$ is orthogonal to $(1,-2,3)$.

    (c) If (a,b,c) is orthogonal to $(5,1,-1)$, then $(a,b,c) \cdot (5,1,-1) = 5a + b - c = 0$, so $c = 5a + b$. Thus any vector of the form $(a, b, 5a+b)$ is orthogonal to $(5,1,-1)$.

    (d) If (a,b,c,d) is orthogonal to $(5,0,1,1)$, then $(a,b,c,d) \cdot (5,0,1,1) = 5a + c + d = 0$, so $d = -5a - c$. Thus any vector of the form $(a, b, c, -5a-c)$ is orthogonal to $(5,0,1,1)$.

    (e) If (a,b,c,d) is orthogonal to $(6,-1,2,3)$, then $(a,b,c,d) \cdot (6,-1,2,3) = 6a - b + 2c + 3d = 0$, so $b = 6a + 2c + 3d$. Thus any vector of the form $(a, 6a+2c+3d, c, d)$ is orthogonal to $(6,-1,2,3)$.

    (f) If (a,b,c,d,e) is orthogonal to $(0,-2,3,1,5)$, then $(a,b,c,d,e) \cdot (0,-2,3,1,5) = -2b + 3c + d + 5e = 0$, so $d = 2b - 3c - 5e$. Thus any vector of the form $(a, b, c, 2b-3c-5e, e)$ is orthogonal to $(0,-2,3,1,5)$.

15. If (a,b,c) is orthogonal to both $(1,2,-1)$ and $(3,1,0)$, then $(a,b,c) \cdot (1,2,-1) = a + 2b - c = 0$ and $(a,b,c) \cdot (3,1,0) = 3a + b = 0$. These equations yield the solution $b = -3a$ and $c = -5a$, so any vector of the form $(a, -3a, -5a)$ is orthogonal to both $(1,2,-1)$ and $(3,1,0)$.

16. (a) $d = \sqrt{(6-2)^2 + (5-2)^2} = 5$.   (b) $d = \sqrt{(3+4)^2 + (1-0)^2} = \sqrt{50} = 5\sqrt{2}$.

Section 4.2

(c) $d = \sqrt{(7-2)^2+(-3-2)^2} = \sqrt{50} = 5\sqrt{2}$. (d) $d = \sqrt{(1-5)^2+(-3-1)^2} = 4\sqrt{2}$.

17. (a) $d = \sqrt{(4-2)^2+(1+3)^2} = \sqrt{20} = 2\sqrt{5}$. (b) $d = \sqrt{(1-2)^2+(2-1)^2+(3-0)^2} = \sqrt{11}$.

(c) $d = \sqrt{(-3-4)^2+(1+1)^2+(2-1)^2} = \sqrt{54} = 3\sqrt{6}$.

(d) $d^2 = (5-2)^2+(1-0)^2+(0-1)^2+(0-3)^2 = 20$, so $d = \sqrt{20} = 2\sqrt{5}$.

(e) $d^2 = (-3-2)^2+(1-1)^2+(1-4)^2+(0-1)^2+(2+1)^2 = 44$, so $d = \sqrt{44} = 2\sqrt{11}$.

18. (a) $(\mathbf{u} + \mathbf{v})\cdot\mathbf{w} = (u_1 + v_1)w_1 + (u_2 + v_2)w_2 + \ldots + (u_n + v_n)w_n$
$= u_1 w_1 + v_1 w_1 + u_2 w_2 + v_2 w_2 + \ldots + u_n w_n + v_n w_n$
$= u_1 w_1 + u_2 w_2 + \ldots + u_n w_n + v_1 w_1 + v_2 w_2 + \ldots + v_n w_n = \mathbf{u}\cdot\mathbf{w} + \mathbf{v}\cdot\mathbf{w}$.

(b) $c\mathbf{u}\cdot\mathbf{v} = cu_1 v_1 + cu_2 v_2 + \ldots + cu_n v_n = c(u_1 v_1 + u_2 v_2 + \ldots + u_n v_n) = c(\mathbf{u}\cdot\mathbf{v})$, and
$cu_1 v_1 + cu_2 v_2 + \ldots + cu_n v_n = u_1 cv_1 + u_2 cv_2 + \ldots + u_n cv_n = \mathbf{u}\cdot c\mathbf{v}$.

19. $\mathbf{u}$ is a scalar multiple of $\mathbf{v}$ so it has the same direction as $\mathbf{v}$. The magnitude of $\mathbf{u}$ is

$\|\mathbf{u}\| = \dfrac{1}{\|\mathbf{v}\|}\sqrt{(v_1)^2+(v_2)^2+\ldots+(v_n)^2} = \dfrac{\|\mathbf{v}\|}{\|\mathbf{v}\|} = 1$, so $\mathbf{u}$ is a unit vector.

20. $\mathbf{u}$ and $\mathbf{v}$ are orthogonal if and only if the cosine of the angle θ between them is zero.

$\cos\theta = \dfrac{\mathbf{u}\cdot\mathbf{v}}{\|\mathbf{u}\|\ \|\mathbf{v}\|} = 0$ if and only if $\mathbf{u}\cdot\mathbf{v} = 0$.

21. If $\mathbf{u}\cdot\mathbf{v} = \mathbf{u}\cdot\mathbf{w}$ then $\mathbf{u}\cdot(\mathbf{v}-\mathbf{w}) = 0$ for all vectors $\mathbf{u}$ in U. Since $\mathbf{v}-\mathbf{w}$ is a vector in U this means that $(\mathbf{v}-\mathbf{w})\cdot(\mathbf{v}-\mathbf{w}) = 0$. Therefore $\mathbf{v}-\mathbf{w} = \mathbf{0}$, so $\mathbf{v} = \mathbf{w}$.

22. $\mathbf{u}\cdot(a_1\mathbf{v}_1 + a_2\mathbf{v}_2 + \ldots + a_n\mathbf{v}_n) = \mathbf{u}\cdot(a_1\mathbf{v}_1) + \mathbf{u}\cdot(a_2\mathbf{v}_2 + \ldots + a_n\mathbf{v}_n)$
$= \mathbf{u}\cdot(a_1\mathbf{v}_1) + \mathbf{u}\cdot(a_2\mathbf{v}_2) + \mathbf{u}\cdot(a_3\mathbf{v}_3 + \ldots + a_n\mathbf{v}_n) = \ldots = \mathbf{u}\cdot(a_1\mathbf{v}_1) + \mathbf{u}\cdot(a_2\mathbf{v}_2) + \ldots + \mathbf{u}\cdot(a_n\mathbf{v}_n)$

Section 4.2

$$= a_1(\mathbf{u}\cdot\mathbf{v}_1) + a_2(\mathbf{u}\cdot\mathbf{v}_2) + \ldots + a_n(\mathbf{u}\cdot\mathbf{v}_n) = a_1\mathbf{u}\cdot\mathbf{v}_1 + a_2\mathbf{u}\cdot\mathbf{v}_2 + \ldots + a_n\mathbf{u}\cdot\mathbf{v}_n.$$

23. (a) vector (b) makes no sense (c) makes no sense (d) scalar
    (e) makes no sense (f) scalar (g) makes no sense
    (h) makes no sense

24. $\|c(3,0,4)\| = \sqrt{3c\times 3c + 4c\times 4c} = |c|\sqrt{9+16} = 5|c| = 15$, so $|c| = 3$ and $c = \pm 3$.

25. $\|\mathbf{u}+\mathbf{v}\|^2 = (u_1+v_1)^2 + (u_2+v_2)^2 + \ldots + (u_n+v_n)^2$

    $= u_1^2 + 2u_1v_1 + v_1^2 + u_2^2 + 2u_2v_2 + v_2^2 + \ldots + u_n^2 + 2u_nv_n + v_n^2$

    $= u_1^2 + u_2^2 + \ldots + u_n^2 + 2u_1v_1 + 2u_2v_2 + \ldots + 2u_nv_n + v_1^2 + v_2^2 + \ldots + v_n^2$

    $= u_1^2 + u_2^2 + \ldots + u_n^2 + 2(u_1v_1 + u_2v_2 + \ldots + u_nv_n) + v_1^2 + v_2^2 + \ldots + v_n^2$

    $= \|\mathbf{u}\|^2 + 2(\mathbf{u}\cdot\mathbf{v}) + \|\mathbf{v}\|^2 = \|\mathbf{u}\|^2 + \|\mathbf{v}\|^2$ if and only if $\mathbf{u}\cdot\mathbf{v} = 0$, i.e., if and only if $\mathbf{u}$ and $\mathbf{v}$ are orthogonal.

26. $(a,b)\cdot(-b,a) = a\times -b + b\times a = 0$, so $(-b,a)$ is orthogonal to (a,b).

27. $(\mathbf{u}+\mathbf{v})\cdot(\mathbf{u}-\mathbf{v}) = (u_1+v_1)(u_1-v_1) + (u_2+v_2)(u_2-v_2) + \ldots + (u_n+v_n)(u_n-v_n)$

    $= u_1^2 - v_1^2 + u_2^2 - v_2^2 + \ldots + u_n^2 - v_n^2$

    $= u_1^2 + u_2^2 + \ldots + u_n^2 - v_1^2 - v_2^2 - \ldots - v_n^2 = \|\mathbf{u}\| - \|\mathbf{v}\|.$

    Thus $\|\mathbf{u}\| - \|\mathbf{v}\| = 0$ if and only if $(\mathbf{u}+\mathbf{v})\cdot(\mathbf{u}-\mathbf{v}) = 0$. That is $\|\mathbf{u}\| = \|\mathbf{v}\|$ if and only if $\mathbf{u}+\mathbf{v}$ and $\mathbf{u}-\mathbf{v}$ are orthogonal.

28. (a) $\|\mathbf{u}\|^2 = u_1^2 + u_2^2 + \ldots + u_n^2 \geq 0$, so $\|\mathbf{u}\| \geq 0$.

(b) $\|u\| = 0$ if and only if $u_1^2+u_2^2+\ldots+u_n^2 = 0$ if and only if $u_1 = u_2 = \ldots = u_n = 0$.

(c) $\|cu\|^2 = (cu_1)^2+(cu_2)^2+\ldots+(cu_n)^2 = c^2(u_1^2+u_2^2+\ldots+u_n^2) = c^2\|u\|^2$, so $\|cu\| = |c|\|u\|$.

29. (a) $\|u\| = |u_1|+|u_2|+\ldots+|u_n| \geq 0$ since each term is equal to or greater than zero.

$|u_1|+|u_2|+\ldots+|u_n| = 0$ if and only if each term is zero.

$\|cu\| = |cu_1|+|cu_2|+\ldots+|cu_n| = |c||u_1|+|c||u_2|+\ldots+|c||u_n|$

$= |c|(|u_1|+|u_2|+\ldots+|u_n|) = |c|\|u\|$.

$\|(1,2)\| = |1|+|2| = 3$, $\|(-3,4)\| = |-3|+|4| = 7$, $\|(1,2,-5)\| = |1|+|2|+|-5| = 8$,

and $\|(0,-2,7)\| = |0|+|-2|+|7| = 9$.

(b) $\|u\| = \max_{i=1,\ldots,n} |u_i| \geq 0$ since the absolute value of any number is equal to or greater than zero.

$\max_{i=1,\ldots,n} |u_i| = 0$ if and only if all $|u_i| = 0$.

$\|cu\| = \max_{i=1,\ldots,n} |cu_i| = |c| \max_{i=1,\ldots,n} |u_i| = |c|\|u\|$.

$\|(1,2)\| = |2| = 2$, $\|(-3,4)\| = |4| = 4$, $\|(1,2,-5)\| = |-5| = 5$, and $\|(0,-2,7)\| = |7| = 7$.

30. (a) $d(x,y) = \|x-y\| \geq 0$.

(b) $d(x,y) = \|x-y\| = 0$ if and only if $x-y = 0$ if and only if $x = y$.

(c) $d(x,z) = \|x-y+y-z\| \leq \|x-y\| + \|y-z\| = d(x,y) + d(y,z)$, from the triangle inequality.

Exercise Set 4.3

1. $T((x_1,y_1)+(x_2,y_2)) = T(x_1+x_2,y_1+y_2) = (2(x_1+x_2), x_1+x_2-y_1-y_2) = (2x_1+2x_2, x_1-y_1+x_2-y_2)$

$= (2x_1,x_1-y_1) + (2x_2,x_2-y_2) = T(x_1,y_1) + T(x_2,y_2)$ and $T(c(x,y)) = T(cx,cy) = (2cx,cx-cy)$

$= c(2x,x-y) = cT(x,y)$, so T is linear. $T(1,2) = (2,-1)$ and $T(-1,4) = (-2,-5)$.

2. $T((x_1,y_1)+(x_2,y_2)) = T(x_1+x_2,y_1+y_2) = (3(x_1+x_2)+y_1+y_2, 2(y_1+y_2), x_1+x_2-y_1-y_2)$

$= (3x_1+3x_2+y_1+y_2, 2y_1+2y_2, x_1+x_2-y_1-y_2) = (3x_1+y_1, 2y_1, x_1-y_1) + (3x_2+y_2, 2y_2, x_2-y_2)$

$= T(x_1,y_1) + T(x_2,y_2)$ and $T(c(x,y)) = T(cx,cy) = (3cx+cy, 2cy, cx-cy)$

$= c(3x+y, 2y, x-y) = cT(x,y)$, so T is linear. $T(1,2) = (5,4,-1)$ and $T(2,-5) = (1,-10,7)$.

3. $T((x_1,y_1,z_1)+(x_2,y_2,z_2)) = T(x_1+x_2, y_1+y_2, z_1+z_2) = (0,y_1+y_2,0) = (0,y_1,0) + (0,y_2,0)$

$= T((x_1,y_1,z_1)+T(x_2,y_2,z_2)$ and $T(c(x,y,z) = T(cx,cy,cz) = (0,cy,0) = c(0,y,0) = cT(x,y,z)$, so

T is linear. The image of (x,y,z) under T is the projection of (x,y,z) on the y axis.

4. (a) $T(c(x,y,z)) = T(cx,cy,cz) = (3cx,(cy)^2) = (3cx,c^2y^2) \neq c(3x,y^2) = cT(x,y,z)$,

so T is not linear.

(b) $T(c(x,y,z)) = T(cx,cy,cz) = (cx+2,4cy) \neq c(x+2,4y) = cT(x,y,z)$, so T is not linear.

5. $T(c(x,y)) = T(cx,cy) = cx + a \neq cT(x,y)$, so T is not linear.

6. $T((x_1,y_1,z_1)+(x_2,y_2,z_2)) = T(x_1+x_2,y_1+y_2,z_1+z_2) = (2(x_1+x_2),y_1+y_2) = (2x_1+2x_2,y_1+y_2)$

$= (2x_1,y_1) = (2x_2,y_2) = T((x_1,y_1,z_1)+T(x_2,y_2,z_2)$ and $T(c(x,y,z) = T(cx,cy,cz) = (2cx,cy)$

$= c(2x,y) = cT(x,y,z)$, so T is linear.

Section 4.3

7. $T((x_1,y_1)+(x_2,y_2)) = T(x_1+x_2, y_1+y_2) = x_1+x_2-y_1-y_2 = x_1-y_1+x_2-y_2 = T(x_1,y_1) + T(x_2,y_2)$ and

   $T(c(x,y)) = T(cx,cy) = cx-cy = c(x-y) = cT(x,y)$, so T is linear.

8. (a) $T(x_1+x_2, y_1+y_2) = (x_1+x_2, y_1+y_2, 0) = (x_1,y_1,0) + (x_2,y_2,0) = T(x_1,y_1) + T(x_2,y_2)$, and
   $T(c(x,y)) = T(cx,cy) = (cx,cy,0) = c(x,y,0) = cT(x,y)$, so T is linear.

   (b) $T(x_1+x_2, y_1+y_2) = (x_1+x_2, y_1+y_2, 1) \neq (x_1,y_1,1) + (x_2,y_2,1) = T(x_1,y_1) + T(x_2,y_2)$, so T is not linear.

9. $T(x_1+x_2) = (x_1+x_2, 2(x_1+x_2), 3(x_1+x_2)) = (x_1+x_2, 2x_1+2x_2, 3x_1+3x_2) = (x_1, 2x_1, 3x_1)$

   $+ (x_2, 2x_2, 3x_2) = T(x_1)+T(x_2)$ and $T(cx) = (cx, 2cx, 3cx) = c(x, 2x, 3x) = cT(x)$, so T is linear.

10. If $c \neq 0$ or 1, $T(c(x,y)) = T(cx,cy) = ((cx)^2, cy) = (c^2x^2, cy) = c(cx^2, y) \neq cT(x,y)$, so T is not linear.

11. $T((x_1,y_1,z_1)+(x_2,y_2,z_2)) = T(x_1+x_2, y_1+y_2, z_1+z_2)$

    $= (x_1+x_2+2(y_1+y_2), x_1+x_2+y_1+y_2+z_1+z_2, 3(z_1+z_2))$

    $= (x_1+x_2+2y_1+2y_2, x_1+x_2+y_1+y_2+z_1+z_2, 3z_1+3z_2)$

    $= (x_1+2y_1, x_1+y_1+z_1, 3z_1) + (x_2+2y_2, x_2+y_2+z_2, 3z_2) = T((x_1,y_1,z_1))+T(x_2,y_2,z_2)$ and $T(c(x,y,z))$

    $= T(cx,cy,cz) = (cx+2cy, cx+cy+cz, 3cz) = c(x+2y, x+y+z, 3z) = cT(x,y,z)$, so T is linear.

12. $A\mathbf{x} = \begin{bmatrix} 1 \\ 4 \\ 1 \end{bmatrix}$, $A\mathbf{y} = \begin{bmatrix} 8 \\ 7 \\ 8 \end{bmatrix}$, and $A\mathbf{z} = \begin{bmatrix} 9 \\ 1 \\ 9 \end{bmatrix}$.

Section 4.3

13. $A\mathbf{x} = \begin{bmatrix} 8 \\ 1 \end{bmatrix}$, $A\mathbf{y} = \begin{bmatrix} 0 \\ 4 \end{bmatrix}$, and $A\mathbf{z} = \begin{bmatrix} -8 \\ 7 \end{bmatrix}$.

14. $T(\mathbf{x}_1+\mathbf{x}_2) = A(\mathbf{x}_1+\mathbf{x}_2) + \mathbf{c} = A\mathbf{x}_1+A\mathbf{x}_2 + \mathbf{c} \neq T(\mathbf{x}_1)+T(\mathbf{x}_2)$ if $\mathbf{c} \neq 0$, so T is not linear.

15. (a) For any scalar c and linear transformation T, $T(c\mathbf{v}) = cT(\mathbf{v})$, so with $c = -1$,
    $T(-\mathbf{v}) = T((-1)\mathbf{v}) = (-1)T(\mathbf{v}) = -T(\mathbf{v})$.

    (b) $T(\mathbf{v}-\mathbf{w}) = T(\mathbf{v}+(-1)\mathbf{w}) = T(\mathbf{v})+T((-1)\mathbf{w}) = T(\mathbf{v}) +(-1)T(\mathbf{w}) = T(\mathbf{v}) -T(\mathbf{w})$.

16. (a) $T(\mathbf{x}) = A_2 A_1 \mathbf{x} = \begin{bmatrix} -1 & 0 \\ 1 & 5 \end{bmatrix} \begin{bmatrix} 1 & 2 \\ 3 & 0 \end{bmatrix} \mathbf{x} = \begin{bmatrix} -1 & -2 \\ 16 & 2 \end{bmatrix} \mathbf{x}$, so

    $T\left(\begin{bmatrix} 5 \\ 2 \end{bmatrix}\right) = \begin{bmatrix} -1 & -2 \\ 16 & 2 \end{bmatrix} \begin{bmatrix} 5 \\ 2 \end{bmatrix} = \begin{bmatrix} -9 \\ 84 \end{bmatrix}$.

    (b) $T(\mathbf{x}) = A_2 A_1 \mathbf{x} = \begin{bmatrix} 2 & 2 \\ 1 & -1 \end{bmatrix} \begin{bmatrix} 0 & 1 & 2 \\ 3 & 4 & -1 \end{bmatrix} \mathbf{x} = \begin{bmatrix} 6 & 10 & 2 \\ -3 & -3 & 3 \end{bmatrix} \mathbf{x}$, so

    $T\left(\begin{bmatrix} 0 \\ 1 \\ 3 \end{bmatrix}\right) = \begin{bmatrix} 6 & 10 & 2 \\ -3 & -3 & 3 \end{bmatrix} \begin{bmatrix} 0 \\ 1 \\ 3 \end{bmatrix} = \begin{bmatrix} 16 \\ 6 \end{bmatrix}$.

    (c) $T(\mathbf{x}) = A_2 A_1 \mathbf{x} = \begin{bmatrix} 2 & 2 \\ 1 & -1 \\ 0 & 4 \end{bmatrix} \begin{bmatrix} 3 & -2 \\ 0 & 1 \end{bmatrix} \mathbf{x} = \begin{bmatrix} 6 & -2 \\ 3 & -3 \\ 0 & 4 \end{bmatrix} \mathbf{x}$, so

    $T\left(\begin{bmatrix} -3 \\ 2 \end{bmatrix}\right) = \begin{bmatrix} 6 & -2 \\ 3 & -3 \\ 0 & 4 \end{bmatrix} \begin{bmatrix} -3 \\ 2 \end{bmatrix} = \begin{bmatrix} -22 \\ -15 \\ 8 \end{bmatrix}$.

17. $T(x,y) = T_2 \circ T_1(x,y) = T_2(2x,-y) = (0,2x-y)$, so $T(2,-3) = (0,7)$.

18. $T(x,y) = T_2 \circ T_1(x,y) = T_2(3x+y, 4y) = (8y, 3x-3y)$, so $T(1,2) = (16,-3)$.

19. $T(x,y) = T_2 \circ T_1(x,y) = T_2(x+y, 2x, 3y) = (x+y, 3x+y, 2x-y)$, so $T(-2,5) = (3,-1,-9)$.

20. In general $T_2 \circ T_1 \neq T_1 \circ T_2$. In fact, one of $T_2 \circ T_1$ and $T_1 \circ T_2$ may not exist as in Exercise 16(b) above, and if both do exist they will generally not be equal because composition of functions is not commutative.

Exercise Set 4.4

1. (a) $A = \begin{bmatrix} 0 & -1 \\ 1 & 0 \end{bmatrix}$ $\qquad A\begin{bmatrix} 2 \\ 1 \end{bmatrix} = \begin{bmatrix} -1 \\ 2 \end{bmatrix}$

   (b) $A = \begin{bmatrix} 0 & 1 \\ -1 & 0 \end{bmatrix}$ $\qquad A\begin{bmatrix} 2 \\ 1 \end{bmatrix} = \begin{bmatrix} 1 \\ -2 \end{bmatrix}$

   (c) $A = \begin{bmatrix} \frac{1}{\sqrt{2}} & \frac{-1}{\sqrt{2}} \\ \frac{1}{\sqrt{2}} & \frac{1}{\sqrt{2}} \end{bmatrix}$ $\qquad A\begin{bmatrix} 2 \\ 1 \end{bmatrix} = \begin{bmatrix} \frac{1}{\sqrt{2}} \\ \frac{3}{\sqrt{2}} \end{bmatrix}$

   (d) $A = \begin{bmatrix} -1 & 0 \\ 0 & -1 \end{bmatrix}$ $\qquad A\begin{bmatrix} 2 \\ 1 \end{bmatrix} = \begin{bmatrix} -2 \\ -1 \end{bmatrix}$

   (e) $A = \begin{bmatrix} 0 & -1 \\ 1 & 0 \end{bmatrix}$ $\qquad A\begin{bmatrix} 2 \\ 1 \end{bmatrix} = \begin{bmatrix} -1 \\ 2 \end{bmatrix}$

   (f) $A = \begin{bmatrix} \frac{\sqrt{3}}{2} & \frac{-1}{2} \\ \frac{1}{2} & \frac{\sqrt{3}}{2} \end{bmatrix}$ $\qquad A\begin{bmatrix} 2 \\ 1 \end{bmatrix} = \begin{bmatrix} \sqrt{3} - \frac{1}{2} \\ 1 + \frac{\sqrt{3}}{2} \end{bmatrix}$

(g) $A = \begin{bmatrix} \frac{1}{2} & \frac{\sqrt{3}}{2} \\ -\frac{\sqrt{3}}{2} & \frac{1}{2} \end{bmatrix}$ $A\begin{bmatrix} 2 \\ 1 \end{bmatrix} = \begin{bmatrix} 1 + \frac{\sqrt{3}}{2} \\ -\sqrt{3} + \frac{1}{2} \end{bmatrix}$

2. $T\left(\begin{bmatrix} 1 \\ 0 \end{bmatrix}\right) = \begin{bmatrix} \cos\theta \\ \sin\theta \end{bmatrix}$ and $T\left(\begin{bmatrix} 0 \\ 1 \end{bmatrix}\right) = \begin{bmatrix} -\sin\theta \\ \cos\theta \end{bmatrix}$, so $A = \begin{bmatrix} \cos\theta & -\sin\theta \\ \sin\theta & \cos\theta \end{bmatrix}$.

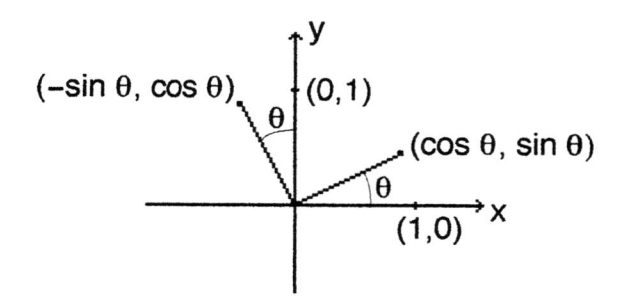

3. $\begin{bmatrix} 0 & -1 \\ 1 & 0 \end{bmatrix}\begin{bmatrix} x \\ y \end{bmatrix} = \begin{bmatrix} -y \\ x \end{bmatrix} = \begin{bmatrix} x' \\ y' \end{bmatrix}$. $\frac{x^2}{4} + \frac{y^2}{9} = 1$, so that $\frac{y'^2}{4} + \frac{x'^2}{9} = 1$. Thus the images of the points on the ellipse $\frac{x^2}{4} + \frac{y^2}{9} = 1$ are the points on the ellipse $\frac{x^2}{9} + \frac{y^2}{4} = 1$.

4. $\begin{bmatrix} 2 & 0 \\ 0 & 2 \end{bmatrix}\begin{bmatrix} 0 & -1 \\ 1 & 0 \end{bmatrix} = \begin{bmatrix} 0 & -2 \\ 2 & 0 \end{bmatrix}$
   dilation rotation

5. $\begin{bmatrix} r & 0 \\ 0 & r \end{bmatrix}\begin{bmatrix} \cos\theta & -\sin\theta \\ \sin\theta & \cos\theta \end{bmatrix} = \begin{bmatrix} r\cos\theta & -r\sin\theta \\ r\sin\theta & r\cos\theta \end{bmatrix}$

Section 4.4

6. $\begin{bmatrix} 3 & 0 \\ 0 & 3 \end{bmatrix} \begin{bmatrix} x \\ y \end{bmatrix} = \begin{bmatrix} 3x \\ 3y \end{bmatrix} = \begin{bmatrix} x' \\ y' \end{bmatrix}$, so $\frac{x'^2}{9} + \frac{y'^2}{9} = 1$. Thus the images of the points on the circle $x^2 + y^2 = 1$ are the points on the circle $x^2 + y^2 = 9$.

7.

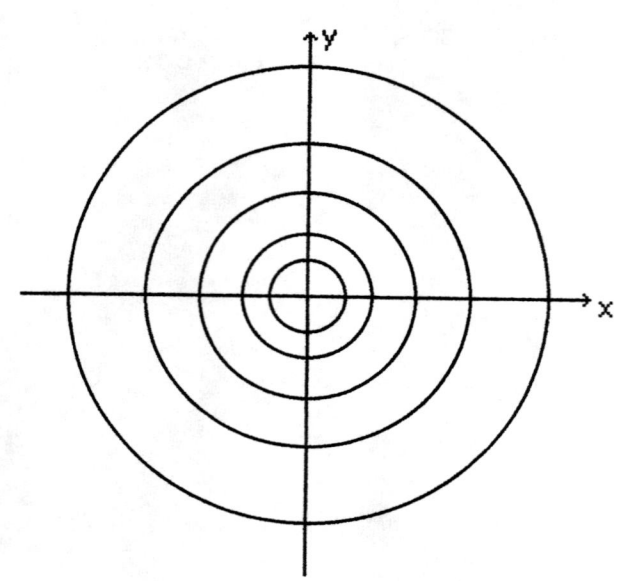

Radii are 1, 1.5, 2.25, 3.375 and 5.0625.

8. $\begin{bmatrix} 1 \\ 0 \end{bmatrix} \mapsto \begin{bmatrix} -1 \\ 0 \end{bmatrix}$ and $\begin{bmatrix} 0 \\ 1 \end{bmatrix} \mapsto \begin{bmatrix} 0 \\ 1 \end{bmatrix}$, so $A = \begin{bmatrix} -1 & 0 \\ 0 & 1 \end{bmatrix}$. $A \begin{bmatrix} 2 \\ 1 \end{bmatrix} = \begin{bmatrix} -2 \\ 1 \end{bmatrix}$.

9. $\begin{bmatrix} 1 \\ 0 \end{bmatrix} \mapsto \begin{bmatrix} 0 \\ 1 \end{bmatrix}$ and $\begin{bmatrix} 0 \\ 1 \end{bmatrix} \mapsto \begin{bmatrix} 1 \\ 0 \end{bmatrix}$, so $A = \begin{bmatrix} 0 & 1 \\ 1 & 0 \end{bmatrix}$

10. Vertices of the image of the unit square ((1,0), (1,1), (0,1),(0,0)) are

    (a) (0,1), (−1,1), (−1,0), (0,0)
    (b) (2,0), (2,2), (0,2), (0,0)
    (c) (3,1), (3,5), (0,4), (0,0)
    (d) (4,1), (3,6), (−1,5), (0,0)
    (e) (−2,0), (−5,4), (−3,4), (0,0)
    (f) (−2,−4), (−6,−5), (−4,−1), (0,0)

132

(g) (0,2), (−2,2), (−2,0), (0,0)   (h) (0,−3), (3,−3), (3,0), (0,0)

11. (a) $\begin{bmatrix} 1 \\ 0 \end{bmatrix} \mapsto \begin{bmatrix} 2 \\ 1 \end{bmatrix}$ and $\begin{bmatrix} 0 \\ 1 \end{bmatrix} \mapsto \begin{bmatrix} 0 \\ -1 \end{bmatrix}$, so $A = \begin{bmatrix} 2 & 0 \\ 1 & -1 \end{bmatrix}$.

(b) $\begin{bmatrix} 1 \\ 0 \end{bmatrix} \mapsto \begin{bmatrix} 1 \\ 1 \end{bmatrix}$ and $\begin{bmatrix} 0 \\ 1 \end{bmatrix} \mapsto \begin{bmatrix} -1 \\ 1 \end{bmatrix}$, so $A = \begin{bmatrix} 1 & -1 \\ 1 & 1 \end{bmatrix}$.

(c) $\begin{bmatrix} 1 \\ 0 \end{bmatrix} \mapsto \begin{bmatrix} 2 \\ 0 \end{bmatrix}$ and $\begin{bmatrix} 0 \\ 1 \end{bmatrix} \mapsto \begin{bmatrix} -5 \\ 3 \end{bmatrix}$, so $A = \begin{bmatrix} 2 & -5 \\ 0 & 3 \end{bmatrix}$.

(d) $\begin{bmatrix} 1 \\ 0 \end{bmatrix} \mapsto \begin{bmatrix} 0 \\ -3 \end{bmatrix}$ and $\begin{bmatrix} 0 \\ 1 \end{bmatrix} \mapsto \begin{bmatrix} 2 \\ 0 \end{bmatrix}$, so $A = \begin{bmatrix} 0 & 2 \\ -3 & 0 \end{bmatrix}$.

12. $\begin{bmatrix} 1 \\ 0 \end{bmatrix} \mapsto \begin{bmatrix} 1 \\ 0 \end{bmatrix}$ and $\begin{bmatrix} 0 \\ 1 \end{bmatrix} \mapsto \begin{bmatrix} 0 \\ 0 \end{bmatrix}$, so $A = \begin{bmatrix} 1 & 0 \\ 0 & 0 \end{bmatrix}$.

13. $T\left(\begin{bmatrix} x \\ y \end{bmatrix}\right) = \begin{bmatrix} 0 \\ y \end{bmatrix}$, so $\begin{bmatrix} 1 \\ 0 \end{bmatrix} \mapsto \begin{bmatrix} 0 \\ 0 \end{bmatrix}$ and $\begin{bmatrix} 0 \\ 1 \end{bmatrix} \mapsto \begin{bmatrix} 0 \\ 1 \end{bmatrix}$, and $A = \begin{bmatrix} 0 & 0 \\ 0 & 1 \end{bmatrix}$.

14. $T\left(\begin{bmatrix} x \\ y \end{bmatrix}\right) = \begin{bmatrix} \frac{x+y}{2} \\ \frac{x+y}{2} \end{bmatrix}$, so $\begin{bmatrix} 1 \\ 0 \end{bmatrix} \mapsto \begin{bmatrix} \frac{1}{2} \\ \frac{1}{2} \end{bmatrix}$ and $\begin{bmatrix} 0 \\ 1 \end{bmatrix} \mapsto \begin{bmatrix} \frac{1}{2} \\ \frac{1}{2} \end{bmatrix}$, and $A = \begin{bmatrix} \frac{1}{2} & \frac{1}{2} \\ \frac{1}{2} & \frac{1}{2} \end{bmatrix}$.

$\begin{bmatrix} 4 \\ 2 \end{bmatrix} \mapsto \begin{bmatrix} 3 \\ 3 \end{bmatrix}$.

15. $\begin{bmatrix} 1 \\ 0 \end{bmatrix} \mapsto \begin{bmatrix} a \\ 0 \end{bmatrix}$ and $\begin{bmatrix} 0 \\ 1 \end{bmatrix} \mapsto \begin{bmatrix} 0 \\ b \end{bmatrix}$, so $A = \begin{bmatrix} a & 0 \\ 0 & b \end{bmatrix}$.

If a = 3 and b = 2, the unit square becomes a 3x2 rectangle.

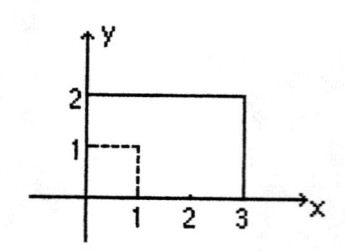

16. $\begin{bmatrix} 2 & 0 \\ 0 & 3 \end{bmatrix}\begin{bmatrix} x \\ y \end{bmatrix} = \begin{bmatrix} 2x \\ 3y \end{bmatrix} = \begin{bmatrix} x' \\ y' \end{bmatrix}$. $y = 2x$, so $\frac{y'}{3} = 2\frac{x'}{2}$, and $y' = 3x'$. Thus the images of the points on the line $y = 2x$ are the points on the line $y = 3x$.

17. $\begin{bmatrix} 4 & 0 \\ 0 & 3 \end{bmatrix}\begin{bmatrix} x \\ y \end{bmatrix} = \begin{bmatrix} 4x \\ 3y \end{bmatrix} = \begin{bmatrix} x' \\ y' \end{bmatrix}$. $x^2 + y^2 = 1$, so $\left(\frac{x'}{4}\right)^2 + \left(\frac{y'}{3}\right)^2 = 1$. Thus the images of the points on the circle $x^2 + y^2 = 1$ are the points on the ellipse $\frac{x^2}{16} + \frac{y^2}{9} = 1$.

18. $\begin{bmatrix} 1 \\ 0 \end{bmatrix} \mapsto \begin{bmatrix} 1 \\ 0 \end{bmatrix}$ and $\begin{bmatrix} 0 \\ 1 \end{bmatrix} \mapsto \begin{bmatrix} c \\ 1 \end{bmatrix}$, so $A = \begin{bmatrix} 1 & c \\ 0 & 1 \end{bmatrix}$. If $c = 2$, then

$\begin{bmatrix} 1 \\ 0 \end{bmatrix} \mapsto \begin{bmatrix} 1 \\ 0 \end{bmatrix}$, $\begin{bmatrix} 1 \\ 1 \end{bmatrix} \mapsto \begin{bmatrix} 3 \\ 1 \end{bmatrix}$, and $\begin{bmatrix} 0 \\ 1 \end{bmatrix} \mapsto \begin{bmatrix} 2 \\ 1 \end{bmatrix}$.

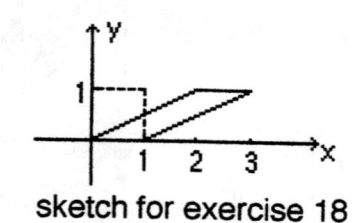

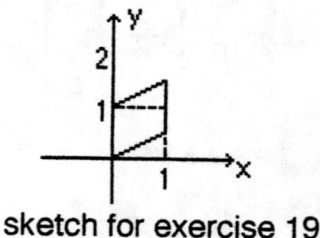

sketch for exercise 18   sketch for exercise 19

19. $T\left(\begin{bmatrix} x \\ y \end{bmatrix}\right) = \begin{bmatrix} x \\ y+cx \end{bmatrix}$, so $\begin{bmatrix} 1 \\ 0 \end{bmatrix} \mapsto \begin{bmatrix} 1 \\ c \end{bmatrix}$ and $\begin{bmatrix} 0 \\ 1 \end{bmatrix} \mapsto \begin{bmatrix} 0 \\ 1 \end{bmatrix}$, and $A = \begin{bmatrix} 1 & 0 \\ c & 1 \end{bmatrix}$.

If $c = .5$, then $\begin{bmatrix} 1 \\ 0 \end{bmatrix} \mapsto \begin{bmatrix} 1 \\ .5 \end{bmatrix}$, $\begin{bmatrix} 1 \\ 1 \end{bmatrix} \mapsto \begin{bmatrix} 1 \\ 1.5 \end{bmatrix}$, and $\begin{bmatrix} 0 \\ 1 \end{bmatrix} \mapsto \begin{bmatrix} 0 \\ 1 \end{bmatrix}$.

Section 4.4

20. $\begin{bmatrix} 1 & 5 \\ 0 & 1 \end{bmatrix} \begin{bmatrix} x \\ y \end{bmatrix} = \begin{bmatrix} x+5y \\ y \end{bmatrix} = \begin{bmatrix} x' \\ y' \end{bmatrix}$. $y = 3x$, so $y' = 3(x' - 5y')$ and $16y' = 3x'$. Thus the images of the points on the line $y = 3x$ are the points on the line $16y = 3x$.

21. $\begin{bmatrix} a & b \\ c & d \end{bmatrix} \begin{bmatrix} x \\ y \end{bmatrix} = \begin{bmatrix} ax+by \\ cx+dy \end{bmatrix} = \begin{bmatrix} x' \\ y' \end{bmatrix}$. If A is nonsingular, then $y = \dfrac{cx' - ay'}{cb - ad}$ and $x = \dfrac{by' - dx'}{cb - ad}$, so that if $rx + sy = t$ then $r(by' - dx') + s(cx' - ay') = t(cb - ad)$ and $(sc - rd)x' + (rb - sa)y' = t(cb - ad)$. Thus the images of the points on the line $rx + sy = t$ are the points on the line $(sc - rd)x + (rb - sa)y = t(cb - ad)$.

22. Let $T: \mathbf{R}^n \to \mathbf{R}^n$ be defined by $T(\mathbf{u}) = A\mathbf{u} + \mathbf{v}$, where $\mathbf{v}$ is fixed, $\mathbf{v} \neq \mathbf{0}$, and A is an nxn matrix. T is a translation if $A = I_n$ (so that $A\mathbf{u} = \mathbf{u}$), and T is an affine transformation if $A \neq I_n$. $T(\mathbf{u}_1) + T(\mathbf{u}_2) = A\mathbf{u}_1 + \mathbf{v} + A\mathbf{u}_2 + \mathbf{v} = A(\mathbf{u}_1 + \mathbf{u}_2) + 2\mathbf{v}$ and $T(\mathbf{u}_1 + \mathbf{u}_2) = A(\mathbf{u}_1 + \mathbf{u}_2) + \mathbf{v}$, so $T(\mathbf{u}_1) + T(\mathbf{u}_2) \neq T(\mathbf{u}_1 + \mathbf{u}_2)$ and T is not linear.

23. (a) $T\left(\begin{bmatrix} x \\ y \end{bmatrix}\right) = \begin{bmatrix} x \\ y \end{bmatrix} + \begin{bmatrix} 2 \\ 5 \end{bmatrix} = \begin{bmatrix} x+2 \\ y+5 \end{bmatrix} = \begin{bmatrix} x' \\ y' \end{bmatrix}$. $y = 3x + 1$, so $y' - 5 = 3(x' - 2) + 1$ and thus $y' = 3x'$, so the image of the line $y = 3x + 1$ is the line $y = 3x$.

(b) $T\left(\begin{bmatrix} x \\ y \end{bmatrix}\right) = \begin{bmatrix} x \\ y \end{bmatrix} + \begin{bmatrix} -1 \\ 1 \end{bmatrix} = \begin{bmatrix} x-1 \\ y+1 \end{bmatrix} = \begin{bmatrix} x' \\ y' \end{bmatrix}$. $y = 3x + 1$, so $y' - 1 = 3(x' + 1) + 1$ and thus $y' = 3x' + 5$, so the image of the line $y = 3x + 1$ is the line $y = 3x + 5$.

24. (a) $\begin{bmatrix} 1 \\ 0 \end{bmatrix} \mapsto \begin{bmatrix} 6 \\ 4 \end{bmatrix}, \begin{bmatrix} 1 \\ 1 \end{bmatrix} \mapsto \begin{bmatrix} 6 \\ 6 \end{bmatrix}, \begin{bmatrix} 0 \\ 1 \end{bmatrix} \mapsto \begin{bmatrix} 4 \\ 6 \end{bmatrix}$, and $\begin{bmatrix} 0 \\ 0 \end{bmatrix} \mapsto \begin{bmatrix} 4 \\ 4 \end{bmatrix}$.

$\begin{bmatrix} x \\ y \end{bmatrix} \mapsto \begin{bmatrix} 2x+4 \\ 2y+4 \end{bmatrix} = \begin{bmatrix} x' \\ y' \end{bmatrix}$, so the image of $x^2 + y^2 = 1$ is $(x-4)^2 + (y-4)^2 = 4$.

135

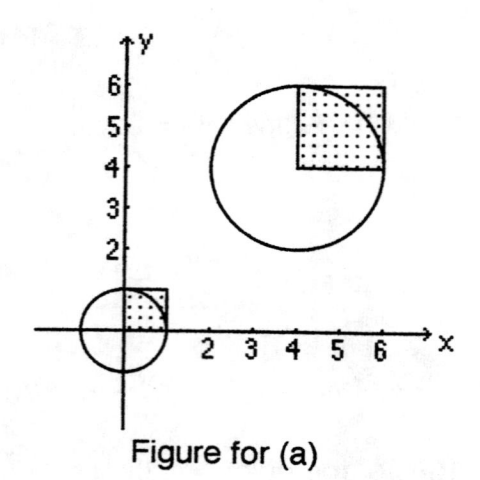

Figure for (a)

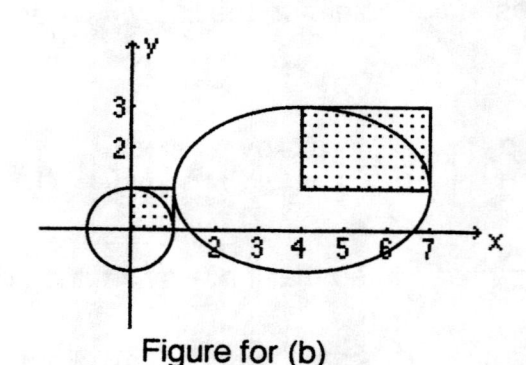

Figure for (b)

(b) $\begin{bmatrix} 1 \\ 0 \end{bmatrix} \mapsto \begin{bmatrix} 7 \\ 1 \end{bmatrix}, \begin{bmatrix} 1 \\ 1 \end{bmatrix} \mapsto \begin{bmatrix} 7 \\ 3 \end{bmatrix}, \begin{bmatrix} 0 \\ 1 \end{bmatrix} \mapsto \begin{bmatrix} 4 \\ 3 \end{bmatrix}$, and $\begin{bmatrix} 0 \\ 0 \end{bmatrix} \mapsto \begin{bmatrix} 4 \\ 1 \end{bmatrix}$.

$\begin{bmatrix} x \\ y \end{bmatrix} \mapsto \begin{bmatrix} 3x+4 \\ 2y+1 \end{bmatrix} = \begin{bmatrix} x' \\ y' \end{bmatrix}$, so the image of $x^2 + y^2 = 1$ is $\dfrac{(x-4)^2}{9} + \dfrac{(y-1)^2}{4} = 1$.

(c) $\begin{bmatrix} 1 \\ 0 \end{bmatrix} \mapsto \begin{bmatrix} \frac{1}{\sqrt{2}} + 3 \\ \frac{1}{\sqrt{2}} + 1 \end{bmatrix}, \begin{bmatrix} 1 \\ 1 \end{bmatrix} \mapsto \begin{bmatrix} 3 \\ \frac{2}{\sqrt{2}} + 1 \end{bmatrix}, \begin{bmatrix} 0 \\ 1 \end{bmatrix} \mapsto \begin{bmatrix} \frac{-1}{\sqrt{2}} + 3 \\ \frac{1}{\sqrt{2}} + 1 \end{bmatrix}$, and

$\begin{bmatrix} 0 \\ 0 \end{bmatrix} \mapsto \begin{bmatrix} 3 \\ 1 \end{bmatrix}$. $\begin{bmatrix} x \\ y \end{bmatrix} \mapsto \begin{bmatrix} \frac{x}{\sqrt{2}} - \frac{y}{\sqrt{2}} + 3 \\ \frac{x}{\sqrt{2}} + \frac{y}{\sqrt{2}} + 1 \end{bmatrix} = \begin{bmatrix} x' \\ y' \end{bmatrix}$, so the image of the circle

$x^2 + y^2 = 1$ is $\dfrac{(x+y-4)^2}{2} + \dfrac{(y-x+2)^2}{2} = 1$, which is the circle $(x-3)^2 + (y-1)^2 = 1$.

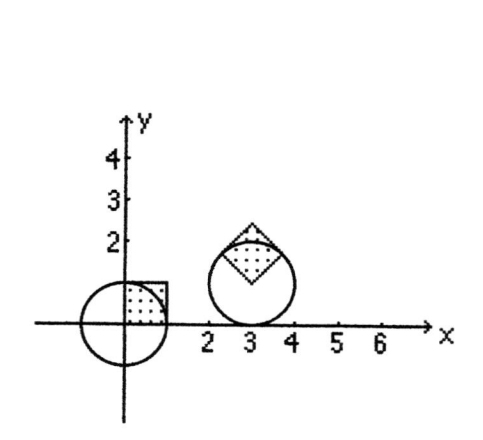

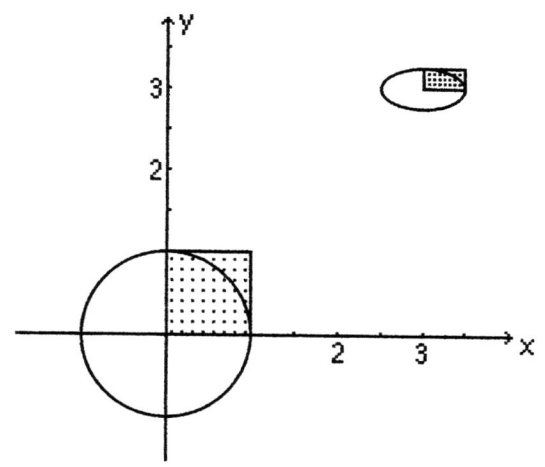

Figure for (c)

Figure for (d)

(d) $\begin{bmatrix} 1 \\ 0 \end{bmatrix} \mapsto \begin{bmatrix} 3.5 \\ 3 \end{bmatrix}, \begin{bmatrix} 1 \\ 1 \end{bmatrix} \mapsto \begin{bmatrix} 3.5 \\ 3.25 \end{bmatrix}, \begin{bmatrix} 0 \\ 1 \end{bmatrix} \mapsto \begin{bmatrix} 3 \\ 3.25 \end{bmatrix},$ and $\begin{bmatrix} 0 \\ 0 \end{bmatrix} \mapsto \begin{bmatrix} 3 \\ 3 \end{bmatrix}.$

$\begin{bmatrix} x \\ y \end{bmatrix} \mapsto \begin{bmatrix} .5x+3 \\ .25y+3 \end{bmatrix} = \begin{bmatrix} x' \\ y' \end{bmatrix}$, so the image of $x^2 + y^2 = 1$ is $4(x-3)^2 + 16(y-3)^2 = 1$.

25. (a) $\begin{bmatrix} \cos\theta & -\sin\theta \\ \sin\theta & \cos\theta \end{bmatrix} \begin{bmatrix} \cos(-\theta) & -\sin(-\theta) \\ \sin(-\theta) & \cos(-\theta) \end{bmatrix}$
rotation through θ rotation through $(-\theta)$

$= \begin{bmatrix} \cos\theta & -\sin\theta \\ \sin\theta & \cos\theta \end{bmatrix} \begin{bmatrix} \cos\theta & \sin\theta \\ -\sin\theta & \cos\theta \end{bmatrix} = \begin{bmatrix} 1 & 0 \\ 0 & 1 \end{bmatrix}.$

(b) $\begin{bmatrix} 1 & 0 \\ 0 & -1 \end{bmatrix} \begin{bmatrix} 1 & 0 \\ 0 & -1 \end{bmatrix} = \begin{bmatrix} 1 & 0 \\ 0 & 1 \end{bmatrix}.$

26. To reverse the operations on **u**, one must first subtract **v**, and then multiply by A^{-1}, so $T^{-1}(\mathbf{u}) = A^{-1}(\mathbf{u}-\mathbf{v})$. Checking, $T^{-1} \circ T(\mathbf{u}) = T^{-1}(A\mathbf{u}+\mathbf{v}) = A^{-1}(A\mathbf{u}+\mathbf{v}-\mathbf{v}) = A^{-1}A\mathbf{u} = \mathbf{u}$ and $T \circ T^{-1}(\mathbf{u}) = T(A^{-1}(\mathbf{u}-\mathbf{v})) = AA^{-1}(\mathbf{u}-\mathbf{v}) + \mathbf{v} = \mathbf{u} - \mathbf{v} + \mathbf{v} = \mathbf{u}$. T^{-1} is also an affine transformation since $T^{-1}(\mathbf{u}) = A^{-1}\mathbf{u} + (-A^{-1}\mathbf{v}) = A^{-1}\mathbf{u} + \mathbf{w}$, where $\mathbf{w} = -A^{-1}\mathbf{v}$.

Section 4.4

27. (a) $\begin{bmatrix} 2 & 0 \\ 0 & 2 \end{bmatrix} \begin{bmatrix} 0 & -1 \\ 1 & 0 \end{bmatrix} = \begin{bmatrix} 0 & -2 \\ 2 & 0 \end{bmatrix}$ and $\begin{bmatrix} 2 \\ 1 \end{bmatrix} \mapsto \begin{bmatrix} -2 \\ 4 \end{bmatrix}$.
   dilation   rotation

(b) $\begin{bmatrix} 1 & 0 \\ 0 & -1 \end{bmatrix} \begin{bmatrix} 4 & 0 \\ 0 & 4 \end{bmatrix} = \begin{bmatrix} 4 & 0 \\ 0 & -4 \end{bmatrix}$ and $\begin{bmatrix} 2 \\ 1 \end{bmatrix} \mapsto \begin{bmatrix} 8 \\ -4 \end{bmatrix}$.
   reflection   dilation

(c) $\begin{bmatrix} -1 & 0 \\ 0 & -1 \end{bmatrix} \begin{bmatrix} 0 & 1 \\ 1 & 0 \end{bmatrix} = \begin{bmatrix} 0 & -1 \\ -1 & 0 \end{bmatrix}$ and $\begin{bmatrix} 2 \\ 1 \end{bmatrix} \mapsto \begin{bmatrix} -1 \\ -2 \end{bmatrix}$.
   rotation   reflection

28. (a) $\begin{bmatrix} 1 & 2 \\ 0 & 1 \end{bmatrix} \begin{bmatrix} 3 & 0 \\ 0 & 3 \end{bmatrix} = \begin{bmatrix} 3 & 6 \\ 0 & 3 \end{bmatrix}$ and $\begin{bmatrix} 3 \\ 2 \end{bmatrix} \mapsto \begin{bmatrix} 21 \\ 6 \end{bmatrix}$.
    shear   dilation

(b) $\begin{bmatrix} 0 & 1 \\ 1 & 0 \end{bmatrix} \begin{bmatrix} 3 & 0 \\ 0 & 2 \end{bmatrix} = \begin{bmatrix} 0 & 2 \\ 3 & 0 \end{bmatrix}$ and $\begin{bmatrix} 3 \\ 2 \end{bmatrix} \mapsto \begin{bmatrix} 4 \\ 9 \end{bmatrix}$.
    reflection scaling

(c) $\begin{bmatrix} 0 & -1 \\ 1 & 0 \end{bmatrix} \begin{bmatrix} 1 & 3 \\ 0 & 1 \end{bmatrix} \begin{bmatrix} 2 & 0 \\ 0 & 2 \end{bmatrix} = \begin{bmatrix} 0 & -2 \\ 2 & 6 \end{bmatrix}$ and $\begin{bmatrix} 3 \\ 2 \end{bmatrix} \mapsto \begin{bmatrix} -4 \\ 18 \end{bmatrix}$.
    rotation   shear   dilation

29. $A \begin{bmatrix} 1 \\ 0 \\ \vdots \\ 0 \end{bmatrix} = A_1$, $A \begin{bmatrix} 0 \\ 1 \\ \vdots \\ 0 \end{bmatrix} = A_2$, ..., $A \begin{bmatrix} 0 \\ 0 \\ \vdots \\ 1 \end{bmatrix} = A_n$, where $A_1, A_2, \ldots, A_n$ are the columns of A. A' is the matrix that has the images of the standard basis vectors for its columns, so A' = A.

30. $A = \begin{bmatrix} \frac{1}{\sqrt{2}} & \frac{-1}{\sqrt{2}} \\ \frac{1}{\sqrt{2}} & \frac{1}{\sqrt{2}} \end{bmatrix}$, $A^2 = \begin{bmatrix} 0 & -1 \\ 1 & 0 \end{bmatrix}$, $A^4 = \begin{bmatrix} -1 & 0 \\ 0 & -1 \end{bmatrix}$, and $A^8 = \begin{bmatrix} 1 & 0 \\ 0 & 1 \end{bmatrix}$.

Eight successive rotations of $\frac{\pi}{4}$ gives a total rotation of 2π, taking every point back to where it started.

Section 4.4

31. $A^2 BA^2 = \begin{bmatrix} 0 & -1 \\ 1 & 0 \end{bmatrix} \begin{bmatrix} -1 & 0 \\ 0 & -1 \end{bmatrix} \begin{bmatrix} 0 & -1 \\ 1 & 0 \end{bmatrix} = \begin{bmatrix} 1 & 0 \\ 0 & 1 \end{bmatrix}.$

These three matrices A^2, B, and A^2 represent successive rotations of $\frac{\pi}{2}$, π, and $\frac{\pi}{2}$, for a total rotation of 2π, taking every point back to where it started.

32.

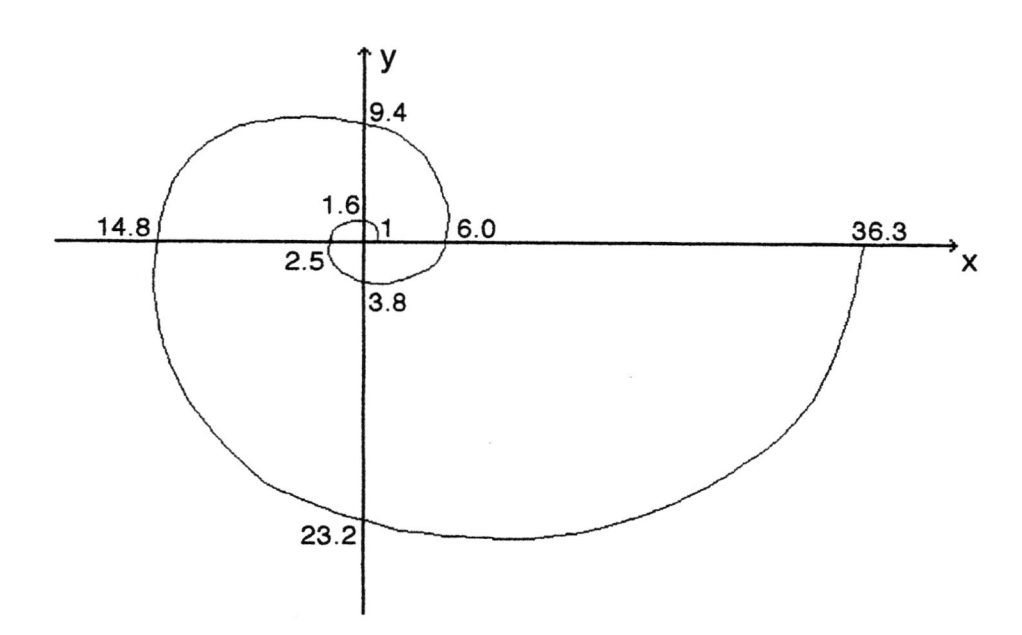

33. (a) $A = \begin{bmatrix} .5 & 0 \\ 0 & .5 \end{bmatrix}$ $\qquad A^2 = \begin{bmatrix} .25 & 0 \\ 0 & .25 \end{bmatrix}$ $\qquad A^3 = \begin{bmatrix} .0625 & 0 \\ 0 & .0625 \end{bmatrix}$

$\begin{bmatrix} 2.5 \\ 2 \end{bmatrix} = A \begin{bmatrix} 1 \\ 0 \end{bmatrix} + \begin{bmatrix} 2 \\ 2 \end{bmatrix}$, $\begin{bmatrix} 2.5 \\ 2.5 \end{bmatrix} = A \begin{bmatrix} 1 \\ 1 \end{bmatrix} + \begin{bmatrix} 2 \\ 2 \end{bmatrix}$, $\begin{bmatrix} 2 \\ 2.5 \end{bmatrix} = A \begin{bmatrix} 0 \\ 1 \end{bmatrix} + \begin{bmatrix} 2 \\ 2 \end{bmatrix}$,

$\begin{bmatrix} 2 \\ 2 \end{bmatrix} = A \begin{bmatrix} 0 \\ 0 \end{bmatrix} + \begin{bmatrix} 2 \\ 2 \end{bmatrix}$. Thus the image of the unit square is the square with vertices (2.5,2), (2.5,2.5), (2,2.5), (2,2), when T = Au + v. Likewise the images of the unit square are the squares with vertices (2.25,2), (2.25,2.25), (2,2.25), (2,2), when T = A^2u + v and (2.125,2), (2.125,2.125), (2,2.125), (2,2), when T = A^3u + v.

(b) $A = \begin{bmatrix} \frac{1}{\sqrt{2}} & \frac{-1}{\sqrt{2}} \\ \frac{1}{\sqrt{2}} & \frac{1}{\sqrt{2}} \end{bmatrix}$ $\qquad A^2 = \begin{bmatrix} 0 & -1 \\ 1 & 0 \end{bmatrix}$ $\qquad A^3 = \begin{bmatrix} \frac{-1}{\sqrt{2}} & \frac{-1}{\sqrt{2}} \\ \frac{1}{\sqrt{2}} & \frac{-1}{\sqrt{2}} \end{bmatrix}$

139

Section 4.4

$(1,0), (1,1), (0,1), (1,0) \to \left(\frac{1}{\sqrt{2}}+3, \frac{1}{\sqrt{2}}+2\right), \left(3, \frac{2}{\sqrt{2}}+2\right), \left(\frac{-1}{\sqrt{2}}+3, \frac{1}{\sqrt{2}}+2\right), (3,2) \to$

$(3,3), (2,3), (2,2), (3,2) \to \left(\frac{-1}{\sqrt{2}}+3, \frac{1}{\sqrt{2}}+2\right), \left(\frac{-2}{\sqrt{2}}+3, 2\right), \left(\frac{-1}{\sqrt{2}}+3, \frac{-1}{\sqrt{2}}+2\right), (3,2).$

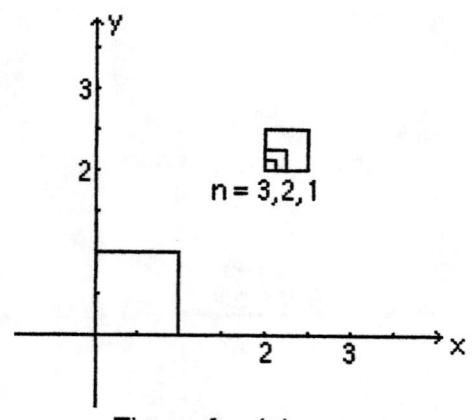

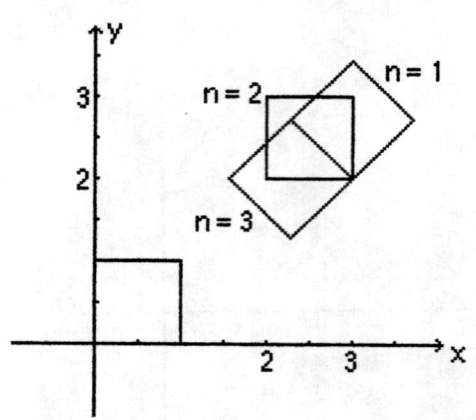

Figure for (a)

Figure for (b)

34. $T(\mathbf{u}) = \mathbf{u} + \mathbf{v}$, so $\mathbf{v} = T(\mathbf{u}) - \mathbf{u} = T(1,2) - (1,2) = (2,-3) - (1,2) = (1,-5)$.

$T(3,4) = (3,4) + (1,-5) = (4,-1)$ and $T(4,6) = (4,6) + (1,-5) = (5,1)$, so image of triangle with vertices $(1,2)$, $(3,4)$, and $(4,6)$ is triangle with vertices $(2,-3)$, $(4,-1)$, and $(5,1)$.

35. The pairs which commute are D and R, D and F, D and S, and D and H.

36. If $\begin{bmatrix} x \\ y \\ z \end{bmatrix} \mapsto \begin{bmatrix} -y \\ x \\ z \end{bmatrix}$ then $\begin{bmatrix} 1 \\ 0 \\ 0 \end{bmatrix} \mapsto \begin{bmatrix} 0 \\ 1 \\ 0 \end{bmatrix}, \begin{bmatrix} 0 \\ 1 \\ 0 \end{bmatrix} \mapsto \begin{bmatrix} -1 \\ 0 \\ 0 \end{bmatrix}$, and $\begin{bmatrix} 0 \\ 0 \\ 1 \end{bmatrix} \mapsto \begin{bmatrix} 0 \\ 0 \\ 1 \end{bmatrix}$, so that

$A = \begin{bmatrix} 0 & -1 & 0 \\ 1 & 0 & 0 \\ 0 & 0 & 1 \end{bmatrix}$. If $\begin{bmatrix} x \\ y \\ z \end{bmatrix} \mapsto \begin{bmatrix} y \\ -x \\ z \end{bmatrix}$ then $A = \begin{bmatrix} 0 & 1 & 0 \\ -1 & 0 & 0 \\ 0 & 0 & 1 \end{bmatrix}$. 37. $\begin{bmatrix} 3 & 0 & 0 \\ 0 & 3 & 0 \\ 0 & 0 & 3 \end{bmatrix}$

38. Let $\theta = \frac{\pi}{2}$, $h = 5$ and $k = 1$ in the matrix $\begin{bmatrix} \cos\theta & -\sin\theta & -h\cos\theta+k\sin\theta+h \\ \sin\theta & \cos\theta & -h\sin\theta-k\cos\theta+k \\ 0 & 0 & 1 \end{bmatrix}$.

140

Section 4.4

The resulting matrix is $\begin{bmatrix} 0 & -1 & 6 \\ 1 & 0 & -4 \\ 0 & 0 & 1 \end{bmatrix}$, and $\begin{bmatrix} 1 \\ 0 \\ 1 \end{bmatrix} \mapsto \begin{bmatrix} 6 \\ -3 \\ 1 \end{bmatrix}$, $\begin{bmatrix} 1 \\ 1 \\ 1 \end{bmatrix} \mapsto \begin{bmatrix} 5 \\ -3 \\ 1 \end{bmatrix}$,

$\begin{bmatrix} 0 \\ 1 \\ 1 \end{bmatrix} \mapsto \begin{bmatrix} 5 \\ -4 \\ 1 \end{bmatrix}$, and $\begin{bmatrix} 0 \\ 0 \\ 1 \end{bmatrix} \mapsto \begin{bmatrix} 6 \\ -4 \\ 1 \end{bmatrix}$, so that the image of the unit square is the square

with vertices (6,−3), (5,−3), (5,−4), and (6,−4).

39. $T^{-1} = \begin{bmatrix} 1 & 0 & -h \\ 0 & 1 & -k \\ 0 & 0 & 1 \end{bmatrix}$.

40. $S^{-1} = \begin{bmatrix} 1/c & 0 & 0 \\ 0 & 1/d & 0 \\ 0 & 0 & 1 \end{bmatrix}$.

41. $SRT = \begin{bmatrix} 3 & 0 & 0 \\ 0 & 5 & 0 \\ 0 & 0 & 1 \end{bmatrix} \begin{bmatrix} 0 & 1 & 0 \\ -1 & 0 & 0 \\ 0 & 0 & 1 \end{bmatrix} \begin{bmatrix} 1 & 0 & 4 \\ 0 & 1 & -3 \\ 0 & 0 & 1 \end{bmatrix} = \begin{bmatrix} 0 & 3 & -9 \\ -5 & 0 & -20 \\ 0 & 0 & 1 \end{bmatrix}$.

$\begin{bmatrix} 1 \\ 6 \\ 1 \end{bmatrix} \mapsto \begin{bmatrix} 9 \\ -25 \\ 1 \end{bmatrix}$, $\begin{bmatrix} 3 \\ 0 \\ 1 \end{bmatrix} \mapsto \begin{bmatrix} -9 \\ -35 \\ 1 \end{bmatrix}$, and $\begin{bmatrix} 4 \\ 6 \\ 1 \end{bmatrix} \mapsto \begin{bmatrix} 9 \\ -40 \\ 1 \end{bmatrix}$, so that the image of the

given triangle is the triangle with vertices (9,−25), (−9,−35), and (9,−40).

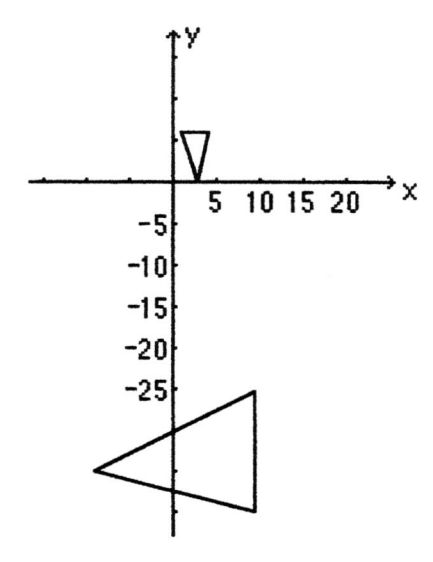

Chapter 4 Review Exercises

1.

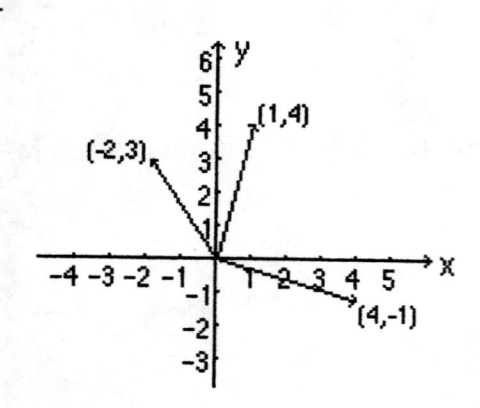

2.

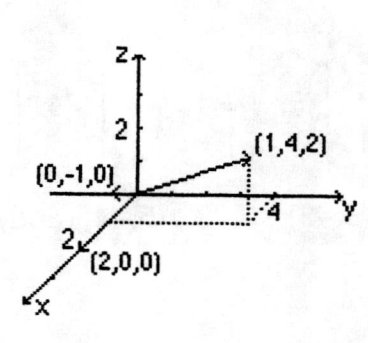

3. (a) **u** + **w** = (3,−1,5) + (0,1,−3) = (3,0,2).

    (b) 3**u** + **v** = 3(3,−1,5) + (2,3,7) = (11,0,22).

    (c) **u** − 2**w** = (3,−1,5) − 2(0,1,−3) = (3,−3,11).

    (d) 4**u** − 2**v** + 3**w** = 4(3,−1,5) − 2(2,3,7) + 3(0,1,−3) = (8,−7,−3).

    (e) 2**u** − 5**v** − **w** = 2(3,−1,5) − 5(2,3,7) − (0,1,−3) = (−4,−18,−22).

4. (a) $\mathbf{u} + \mathbf{v} = \begin{bmatrix} -2 \\ 4 \end{bmatrix} + \begin{bmatrix} 1 \\ -5 \end{bmatrix} = \begin{bmatrix} -1 \\ -1 \end{bmatrix}$

    (b) $\mathbf{v} - 2\mathbf{w} = \begin{bmatrix} 1 \\ -5 \end{bmatrix} - 2\begin{bmatrix} -7 \\ 1 \end{bmatrix} = \begin{bmatrix} 15 \\ -7 \end{bmatrix}$

    (c) $2\mathbf{u} - \mathbf{v} + 3\mathbf{w} = 2\begin{bmatrix} -2 \\ 4 \end{bmatrix} - \begin{bmatrix} 1 \\ -5 \end{bmatrix} + 3\begin{bmatrix} -7 \\ 1 \end{bmatrix} = \begin{bmatrix} -26 \\ 16 \end{bmatrix}$

    (d) $4\mathbf{u} - 2\mathbf{v} + 3\mathbf{w} = 4\begin{bmatrix} -2 \\ 4 \end{bmatrix} - 2\begin{bmatrix} 1 \\ -5 \end{bmatrix} + 3\begin{bmatrix} -26 \\ 1 \end{bmatrix} = \begin{bmatrix} -88 \\ 29 \end{bmatrix}$

5. (a) (1,2)·(3,−4) = 1×3 + 2×−4 = 3 − 8 = −5.

    (b) (1,−2,3)·(4,2,−7) = 1×4 + −2×2 + 3×−7 = 4 − 4 − 21 = −21.

    (c) (2,2,−5)·(3,2,−1) = 2×3 + 2×2 + −5×−1 = 6 + 4 + 5 = 15.

Chapter 4 Review Exercises

6. (a) $\|(1,-4)\| = \sqrt{(1,-4)\cdot(1,-4)} = \sqrt{1\times1 + -4\times-4} = \sqrt{17}$.

    (b) $\|(-2,1,3)\| = \sqrt{(-2,1,3)\cdot(-2,1,3)} = \sqrt{-2\times-2+1\times1+3\times3} = \sqrt{14}$.

    (c) $\|(1,-2,3,4)\| = \sqrt{(1,-2,3,4)\cdot(1,-2,3,4)} = \sqrt{1\times1+-2\times-2+3\times3+4\times4} = \sqrt{30}$.

7. (a) $\cos\theta = \dfrac{(-1,1)\cdot(2,3)}{\|(-1,1)\|\,\|(2,3)\|} = \dfrac{1}{\sqrt{2}\sqrt{13}} = \dfrac{1}{\sqrt{26}}$.

    (b) $\cos\theta = \dfrac{(1,2,-3)\cdot(4,1,2)}{\|(1,2,-3)\|\,\|(4,1,2)\|} = 0$.

8. The vector (1,2,0) is orthogonal to (-2,1,5).

9. (a) $d = \sqrt{(1-5)^2+(-2-3)^2} = \sqrt{16+25} = \sqrt{41}$.

    (b) $d = \sqrt{(3-7)^2+(2-1)^2+(1-2)^2} = \sqrt{18} = 3\sqrt{2}$.

    (c) $d^2 = (3-4)^2+(1-1)^2+(-1-6)^2+(2-2)^2 = 50$, so $d = \sqrt{50} = 5\sqrt{2}$.

10. $\|c(1,2,3)\| = |c|\,\|(1,2,3)\| = |c|\sqrt{(1,2,3)\cdot(1,2,3)} = |c|\sqrt{1\times1+2\times2+3\times3} = |c|\sqrt{14}$, so if $\|c(1,2,3)\| = 196$ then $c = \pm\dfrac{196}{\sqrt{14}} = \pm 14\sqrt{14}$.

11. (a) $T((x_1,y_1)+(x_2,y_2)) = T(x_1+x_2, y_1+y_2) = (2(x_1+x_2), y_1+y_2, y_1+y_2-x_1-x_2)$

    $= (2x_1+2x_2, y_1+y_2, y_1-x_1+y_2-x_2) = (2x_1,y_1,y_1-x_1) + (2x_2,y_2,y_2-x_2) = T(x_1,y_1) + T(x_2,y_2)$

    and $T(c(x,y)) = T(cx,cy) = (2cx, cy, cy-cx) = c(2x,y,y-x) = cT(x,y)$, so T is linear.

    (b) $T(c(x,y)) = T(cx,cy) = (cx+cy, 2cy+3)$ and $cT(x,y) = c(x+y, 2y+3) = (cx+cy, 2cy+3c)$, so T is not linear.

143

Chapter 4 Review Exercises

12. $\begin{bmatrix} a & b \\ c & d \end{bmatrix}\begin{bmatrix} 1 \\ 2 \end{bmatrix} = \begin{bmatrix} 5 \\ 1 \end{bmatrix}$, so $a + 2b = 5$ and $c + 2d = 1$, and $\begin{bmatrix} a & b \\ c & d \end{bmatrix}\begin{bmatrix} 3 \\ -2 \end{bmatrix} = \begin{bmatrix} -1 \\ 11 \end{bmatrix}$, so

    $3a - 2b = -1$ and $3c - 2d = 11$. Thus $a = 1$, $b = 2$, $c = 3$, and $d = -1$, and the matrix is

    $\begin{bmatrix} a & b \\ c & d \end{bmatrix} = \begin{bmatrix} 1 & 2 \\ 3 & -1 \end{bmatrix}$.

13. If T is linear, then $T(a\mathbf{u} + b\mathbf{v}) = T(a\mathbf{u}) + T(b\mathbf{v}) = aT(\mathbf{u}) + bT(\mathbf{v})$.

    If $T(a\mathbf{u} + b\mathbf{v}) = aT(\mathbf{u}) + bT(\mathbf{v})$, then $T(\mathbf{u} + \mathbf{v}) = T(\mathbf{u}) + T(\mathbf{v})$ and $T(c\mathbf{u}) = cT(\mathbf{u})$, so T is linear.

14. $\begin{bmatrix} 3 & 0 \\ 0 & 3 \end{bmatrix}\begin{bmatrix} \frac{\sqrt{3}}{2} & \frac{-1}{2} \\ \frac{1}{2} & \frac{\sqrt{3}}{2} \end{bmatrix} = \begin{bmatrix} \frac{3\sqrt{3}}{2} & \frac{-3}{2} \\ \frac{3}{2} & \frac{3\sqrt{3}}{2} \end{bmatrix}$.

15. $\begin{bmatrix} 1 \\ 0 \end{bmatrix} \mapsto \begin{bmatrix} 0 \\ -1 \end{bmatrix}$ and $\begin{bmatrix} 0 \\ 1 \end{bmatrix} \mapsto \begin{bmatrix} -1 \\ 0 \end{bmatrix}$, so $A = \begin{bmatrix} 0 & -1 \\ -1 & 0 \end{bmatrix}$.

16. $\begin{bmatrix} 1 \\ 0 \end{bmatrix} \mapsto \begin{bmatrix} \frac{1}{2} \\ \frac{-1}{2} \end{bmatrix}$ and $\begin{bmatrix} 0 \\ 1 \end{bmatrix} \mapsto \begin{bmatrix} \frac{-1}{2} \\ \frac{1}{2} \end{bmatrix}$, so $A = \begin{bmatrix} \frac{1}{2} & \frac{-1}{2} \\ \frac{-1}{2} & \frac{1}{2} \end{bmatrix}$.

17. $\begin{bmatrix} 5 & 0 \\ 0 & 2 \end{bmatrix}\begin{bmatrix} x \\ y \end{bmatrix} = \begin{bmatrix} 5x \\ 2y \end{bmatrix} = \begin{bmatrix} x' \\ y' \end{bmatrix}$. $y = -5x + 1$, so $\frac{y'}{2} = -5\frac{x'}{5} + 1$. Thus the images of the

    points on the line $y = -5x + 1$ are the points on the line $\frac{y}{2} = -x + 1$.

18. $\begin{bmatrix} 1 & 0 \\ 3 & 1 \end{bmatrix}\begin{bmatrix} x \\ y \end{bmatrix} = \begin{bmatrix} x \\ 3x+y \end{bmatrix} = \begin{bmatrix} x' \\ y' \end{bmatrix}$. $y = 2x + 3$, so $y' - 3x' = 2x' + 3$ and $y' = 5x' + 3$. Thus the

    images of the points on the line $y = 2x + 3$ are the points on the line $y = 5x + 3$.

19. $\begin{bmatrix} -1 & 0 \\ 0 & 1 \end{bmatrix}\begin{bmatrix} 2 & 0 \\ 0 & 1 \end{bmatrix}\begin{bmatrix} 1 & 0 \\ 3 & 1 \end{bmatrix} = \begin{bmatrix} -2 & 0 \\ 3 & 1 \end{bmatrix}$.

Chapter 5

Exercise Set 5.1

1. $\mathbf{u} = \begin{bmatrix} a & b \\ c & d \end{bmatrix}$ and $\mathbf{v} = \begin{bmatrix} e & f \\ g & h \end{bmatrix}$ in M_{22}; k and l are scalars.

   axiom 2 $k\mathbf{u} = k\begin{bmatrix} a & b \\ c & d \end{bmatrix} = \begin{bmatrix} ka & kb \\ kc & kd \end{bmatrix}$ is in M_{22}.

   axiom 7 $k(\mathbf{u}+\mathbf{v}) = k\begin{bmatrix} a+e & b+f \\ c+g & d+h \end{bmatrix} = \begin{bmatrix} k(a+e) & k(b+f) \\ k(c+g) & k(d+h) \end{bmatrix} = \begin{bmatrix} ka+ke & kb+kf \\ kc+kg & kd+kh \end{bmatrix}$

   $= \begin{bmatrix} ka & kb \\ kc & kd \end{bmatrix} + \begin{bmatrix} ke & kf \\ kg & kh \end{bmatrix} = k\mathbf{u} + k\mathbf{v}$.

   axiom 8 $(k+l)\mathbf{u} = (k+l)\begin{bmatrix} a & b \\ c & d \end{bmatrix} = \begin{bmatrix} (k+l)a & (k+l)b \\ (k+l)c & (k+l)d \end{bmatrix} = \begin{bmatrix} ka+la & kb+lb \\ kc+lc & kd+ld \end{bmatrix}$

   $= \begin{bmatrix} ka & kb \\ kc & kd \end{bmatrix} + \begin{bmatrix} la & lb \\ lc & ld \end{bmatrix} = k\mathbf{u} + l\mathbf{u}$.

   axiom 9 $k(l\mathbf{u}) = k\begin{bmatrix} la & lb \\ lc & ld \end{bmatrix} = \begin{bmatrix} kla & klb \\ klc & kld \end{bmatrix} = \begin{bmatrix} (kl)a & (kl)b \\ (kl)c & (kl)d \end{bmatrix} = (kl)\begin{bmatrix} a & b \\ c & d \end{bmatrix}$

   $= (kl)\mathbf{u}$.

   axiom 10 $1\mathbf{u} = 1\begin{bmatrix} a & b \\ c & d \end{bmatrix} = \begin{bmatrix} a & b \\ c & d \end{bmatrix} = \mathbf{u}$.

2. (a) $(f+g)(x) = f(x) + g(x) = x + 2 + x^2 - 1 = x^2 + x + 1$,

   $(2f)(x) = 2(f(x)) = 2(x+2) = 2x + 4$, and $(3g)(x) = 3(g(x)) = 3(x^2 - 1) = 3x^2 - 3$.

   (b) $(f+g)(x) = f(x) + g(x) = 2x + 4 - 2x = 4$, $(3f)(x) = 3(f(x)) = 3(2x) = 6x$, and

   $-g(x) = -(4 - 2x) = -4 + 2x$.

Section 5.1

3. f, g, and h are functions and c and d are scalars.

   axiom 3 $(f + g)(x) = f(x) + g(x) = g(x) + f(x) = (g + f)(x)$, so $f + g = g + f$.

   axiom 4 $((f + g) + h)(x) = (f + g)(x) + h(x) = (f(x) + g(x)) + h(x) = f(x) + (g(x) + h(x))$
   $= f(x) + (g + h)(x) = (f + (g + h))(x)$, so $(f + g) + h = f + (g + h)$.

   axiom 7 $(c(f + g))(x) = c(f(x) + g(x)) = c(f(x)) + c(g(x)) = (cf)(x) + (cg)(x) = (cf + cg)(x)$,
   so $c(f + g) = cf + cg$.

   axiom 8 $((c + d)f)(x) = (c + d)(f(x)) = c(f(x)) + d(f(x)) = (cf)(x) + (df)(x) = (cf + df)(x)$, so
   $(c + d)f = cf + df$.

   axiom 9 $c((df)(x)) = c(d(f(x))) = (cd)f(x) = ((cd)f)(x)$, so $(cd)f = c(df)$.

   axiom 10 $(1f)(x) = 1(f(x)) = f(x)$, so $1f = f$.

4. $\mathbf{u} = (u_1, \ldots, u_n)$, $\mathbf{v} = (v_1, \ldots, v_n)$ and $\mathbf{w} = (w_1, \ldots, w_n)$ are vectors in $\mathbf{C}^n$ and c and d are (complex) scalars.

   axiom 1 $(u_1, \ldots, u_n) + (v_1, \ldots, v_n) = (u_1 + v_1, \ldots, u_n + v_n)$ is in $\mathbf{C}^n$, so the set is closed under vector addition.

   axiom 2 $c(u_1, \ldots, u_n) = (cu_1, \ldots, cu_n)$ is in $\mathbf{C}^n$, so the set is closed under scalar multiplication.

   axiom 3 $\mathbf{u} + \mathbf{v} = (u_1, \ldots, u_n) + (v_1, \ldots, v_n) = (u_1 + v_1, \ldots, u_n + v_n) = (v_1 + u_1, \ldots, v_n + u_n)$
   $= (v_1, \ldots, v_n) + (u_1, \ldots, u_n) = \mathbf{v} + \mathbf{u}$.

   axiom 4 $(\mathbf{u} + \mathbf{v}) + \mathbf{w} = ((u_1, \ldots, u_n) + (v_1, \ldots, v_n)) + (w_1, \ldots, w_n)$
   $= (u_1 + v_1, \ldots, u_n + v_n) + (w_1, \ldots, w_n) = ((u_1 + v_1) + w_1, \ldots, (u_n + v_n) + w_n)$
   $= (u_1 + (v_1 + w_1), \ldots, u_n + (v_n + w_n)) = (u_1, \ldots, u_n) + (v_1 + w_1, \ldots, v_n + w_n)$
   $= (u_1, \ldots, u_n) + ((v_1, \ldots, v_n) + (w_1, \ldots, w_n)) = \mathbf{u} + (\mathbf{v} + \mathbf{w})$.

   axiom 5 $\mathbf{u} + \mathbf{0} = (u_1, \ldots, u_n) + (0, \ldots, 0) = (u_1, \ldots, u_n) = \mathbf{u}$.

   axiom 6 $\mathbf{u} + (-\mathbf{u}) = (u_1, \ldots, u_n) + (-u_1, \ldots, -u_n) = (u_1 - u_1, \ldots, u_n - u_n) = (0, \ldots, 0) = \mathbf{0}$.

Section 5.1

axiom 7 $c(\mathbf{u} + \mathbf{v}) = c(u_1 + v_1, \ldots, u_n + v_n) = (c(u_1 + v_1), \ldots, c(u_n + v_n))$

$= (cu_1 + cv_1, \ldots, cu_n + cv_n) = (cu_1, \ldots, cu_n) + (cv_1, \ldots, cv_n) = c\mathbf{u} + c\mathbf{v}$.

axiom 8 $(c + d)\mathbf{u} = (c + d)(u_1, \ldots, u_n) = ((c+d)u_1, \ldots, (c+d)u_n)$

$= (cu_1 + du_1, \ldots, cu_n + du_n) = (cu_1, \ldots, cu_n) + (du_1, \ldots, du_n) = c\mathbf{u} + d\mathbf{u}$.

axiom 9 $c(d\mathbf{u}) = c(du_1, \ldots, du_n) = (cdu_1, \ldots, cdu_n) = ((cd)u_1, \ldots, (cd)u_n) =$

$= (cd)(u_1, \ldots, u_n) = (cd)\mathbf{u}$.

axiom 10 $1\mathbf{u} = 1(u_1, \ldots, u_n) = (1u_1, \ldots, 1u_n) = (u_1, \ldots, u_n) = \mathbf{u}$.

(a) $\mathbf{u} + \mathbf{v} = (2 - i, 3 + 4i) + (5, 1 + 3i) = (7 - i, 4 + 7i)$ and

$c\mathbf{u} = (3 - 2i)(2 - i, 3 + 4i) = (6 - 2 - 3i - 4i, 9 + 8 + 12i - 6i) = (4 - 7i, 17 + 6i)$.

(b) $\mathbf{u} + \mathbf{v} = (1 + 5i, -2 - 3i) + (2i, 3 - 2i) = (1 + 7i, 1 - 5i)$ and

$c\mathbf{u} = (4 + i)(1 + 5i, -2 - 3i) = (4 - 5 + 20i + i, -8 + 3 - 12i - 2i) = (-1 + 21i, -5 - 14i)$.

5.   (a)   The set of all continuous functions on [0,1] is closed under addition and scalar multiplication. It is a subset of the set of all functions on [0,1], which is a vector space by the same reasoning as that for the set of all functions with domain the real numbers. Thus the set of all continuous functions on [0,1] is a vector space.

      (b)   This set is not closed under addition. For example, if $f(x) = 2$ for $0 \le x \le 1/2$ and $f(x) = 3$ for $1/2 < x \le 1$, and $g(x) = 3$ for $0 \le x \le 1/2$ and $g(x) = 2$ for $1/2 < x \le 1$ then both f and g are discontinuous functions on [0,1], but $f + g$ is the constant function $(f + g)(x) = 5$, which is continuous.

6. $f(x) = x^2 + 2x + 3$ and $g(x) = -x^2 + x + 1$ are polynomials of degree 2. Neither $f(x) + g(x) = 3x + 4$ nor $0f(x) = 0$ is a polynomial of degree 2, so the set is not closed under addition or under scalar multiplication.

7.   (a) $c\mathbf{0} \underset{\text{axiom 5}}{=} c\mathbf{0} + \mathbf{0} \underset{\text{axiom 6}}{=} c\mathbf{0} + (c\mathbf{0} + -(c\mathbf{0})) \underset{\text{axiom 4}}{=} (c\mathbf{0} + c\mathbf{0}) + -(c\mathbf{0})$

147

Section 5.2

$$\underset{\text{axiom7}}{=} c(\mathbf{0} + \mathbf{0}) + -(c\mathbf{0}) \underset{\text{axiom5}}{=} c\mathbf{0} + -(c\mathbf{0}) \underset{\text{axiom6}}{=} \mathbf{0}.$$

(b) $c\mathbf{v} = \mathbf{0}$. Suppose $c \neq 0$. Then $\mathbf{v} \underset{\text{axiom10}}{=} 1\mathbf{v} = (\frac{1}{c}c)\mathbf{v} \underset{\text{axiom9}}{=} \frac{1}{c}(c\mathbf{v}) = \frac{1}{c}(\mathbf{0}) \underset{\text{(a)}}{=} \mathbf{0}$.

Now suppose $\mathbf{v} \neq \mathbf{0}$. It was just shown that if $c \neq 0$ then $\mathbf{v} = \mathbf{0}$. Since $\mathbf{v} \neq \mathbf{0}$, it must be that $c = 0$.

(c) By axiom 6 there is a vector $-(-\mathbf{v})$ such that $(-\mathbf{v}) + (-(-\mathbf{v})) = \mathbf{0}$. However,

$$(-\mathbf{v}) + \mathbf{v} \underset{\text{axiom3}}{=} \mathbf{v} + (-\mathbf{v}) \underset{\text{axiom6}}{=} \mathbf{0}. \text{ Thus the vector } -(-\mathbf{v}) \text{ is } \mathbf{v}.$$

(d) $\mathbf{u} \underset{\text{axiom5}}{=} \mathbf{u} + \mathbf{0} \underset{\text{axiom6}}{=} \mathbf{u} + (\mathbf{w} + (-\mathbf{w})) \underset{\text{axiom4}}{=} (\mathbf{u} + \mathbf{w}) + (-\mathbf{w}) \underset{\text{given}}{=} (\mathbf{v} + \mathbf{w}) + (-\mathbf{w})$

$\underset{\text{axiom4}}{=} \mathbf{v} + (\mathbf{w} + (-\mathbf{w})) \underset{\text{axiom6}}{=} \mathbf{v} + \mathbf{0} \underset{\text{axiom5}}{=} \mathbf{v}.$

(e) $\mathbf{0} \underset{\text{axiom6}}{=} a\mathbf{u} + (-a\mathbf{u}) \underset{\text{given}}{=} b\mathbf{u} + (-a\mathbf{u}) \underset{\text{axiom8}}{=} (b + -a)\mathbf{u}$, so from part (b), $b + -a = 0$

and therefore $b = a$.

Exercise Set 5.2

1. (a) $(a,0) + (b,0) = (a+b,0)$ and $c(a,0) = (ca,0)$; thus the sum and scalar product of vectors in the set are also in the set, and so the set is a subspace of $\mathbf{R}^2$. The set is a line, the x axis.

   (b) $(a,a) + (b,b) = (a+b,a+b)$ and $c(a,a) = (ca,ca)$; thus the sum and scalar product of vectors in the set are also in the set, and so the set is a subspace of $\mathbf{R}^2$. The set is the line given by the equation $x = y$.

   (c) $(a,2a) + (b,2b) = (a+b,2(a+b))$ and $c(a,2a) = (ca,2ca)$; thus the sum and scalar product of vectors in the set are also in the set, and so the set is a subspace of $\mathbf{R}^2$. The set is the line given by the equation $y = 2x$.

   (d) $(a,a+3b) + (d,d+3e) = (a+d,a+d+3(b+e))$ and $c(a,a+3b) = (ca,ca+3cb)$; thus the sum and scalar product of vectors in the set are also in the set. This set is all of $\mathbf{R}^2$.

Section 5.2

2. (a) $(a,b,0) + (d,e,0) = (a+d,b+e,0)$ and $c(a,b,0) = (ca,cb,0)$; thus the sum and scalar product of vectors in the set are also in the set, and so the set is a subspace of $\mathbf{R}^3$. The set is the xy plane.

   (b) $(0,a,0) + (0,b,0) = (0,a+b,0)$ and $c(0,a,0) = (0,ca,0)$; thus the sum and scalar product of vectors in the set are also in the set, and so the set is a subspace of $\mathbf{R}^3$. The set is the y axis.

   (c) $(a,2a,b) + (d,2d,e) = (a+d,2(a+d),b+e)$ and $c(a,2a,b) = (ca,2ca,cb)$; thus the sum and scalar product of vectors in the set are also in the set, and so the set is a subspace of $\mathbf{R}^3$. The set is the plane given by the equation $y = 2x$.

   (d) $(a,b,a+b) + (d,e,d+e) = (a+d,b+e,a+d+b+e)$ and $c(a,b,a+b) = (ca,cb,ca+cb)$; thus the sum and scalar product of vectors in the set are also in the set, and so the set is a subspace of $\mathbf{R}^3$. The set is the plane given by the equation $z = x+y$.

3. (a) $(a,a,a) + (b,b,b) = (a+b,a+b,a+b)$ and $c(a,a,a) = (ca,ca,ca)$; thus the sum and scalar product of vectors in the set are also in the set, and so the set is a subspace of $\mathbf{R}^3$. The set is the line given by $x = y = z$.

   (b) $(0,a,2a) + (0,b,2b) = (0,a+b,2(a+b))$ and $c(0,a,2a) = (0,ca,2ca)$; thus the sum and scalar product of vectors in the set are also in the set, and so the set is a subspace of $\mathbf{R}^3$. The set is the line in the yz plane ($x = 0$) given by the equation $z = 2y$.

   (c) $(a,a+b,3a) + (d,d+e,3d) = (a+d,a+d+b+e,3(a+d))$ and $c(a,a+b,3a) = (ca,ca+cb,3ca)$; thus the sum and scalar product of vectors in the set are also in the set, and so the set is a subspace of $\mathbf{R}^3$. The set is the plane given by the equation $z = 3x$.

   (d) $(a,2a,3a+5b) + (d,2d,3d+5e) = (a+d,2(a+d),3(a+d)+5(b+e))$ and $c(a,2a,3a+5b) = (ca,2ca,3ca+5cb)$; thus the sum and scalar product of vectors in the set are also in the set, and so the set is a subspace of $\mathbf{R}^3$. The set is the plane given by the equation $y = 2x$.

4. (a) $(a,2a,b,3b) + (d,2d,e,3e) = (a+d,2(a+d),b+e,3(b+e))$ and $c(a,2a,b,3b) = (ca,2ca,cb,3cb)$; thus the sum and scalar product of vectors in the set are also in the set, and so the set is a subspace of $\mathbf{R}^4$.

   (b) $(a,2a,3a,4a) + (b,2b,3b,4b) = (a+b,2(a+b),3(a+b),4(a+b))$ and $c(a,2a,3a,4a) = (ca,2ca,3ca,4ca)$; thus the sum and scalar product of vectors in the set are also in the set, and so the set is a subspace of $\mathbf{R}^4$.

Section 5.2

(c) $(0,a,b,a+2b) + (0,d,e,d+2e) = (0,a+d,b+e,a+d+2(b+e))$ and $c(0,a,b,a+2b) = (0,ca,cb,ca+2cb)$; thus the sum and scalar product of vectors in the set are also in the set, and so the set is a subspace of $\mathbf{R}^4$.

(d) $(a,b,c,a+2b+3c) + (d,e,f,d+2e+3f) = (a+d,b+e,c+f,a+d+2(b+e)+3(c+f))$ and $h(a,b,c,a+2b+3c) = (ha,hb,hc,ha+2hb+3hc)$; thus the sum and scalar product of vectors in the set are also in the set, and so the set is a subspace of $\mathbf{R}^4$.

5. A is a subspace of $\mathbf{R}^2$. (See Exercise 1(c).)

B is not a subspace of $\mathbf{R}^2$, because $c(a,a^2) = (ca,ca^2)$ is not of the form (a,a^2).

C is not a subspace of $\mathbf{R}^2$, because $c(a,a^2+3) = (ca,ca^2+3c)$ is not of the form (a,a^2+3).

6. (a) $(a,b,a+3)+(c,d,c+3) = (a+c,b+d,a+c+6)$, so the sum is not in the set. Thus the set is not a subspace of $\mathbf{R}^3$.

(b) $(a,4a,-3a)+(b,4b,-3b) = (a+b,4(a+b),-3(a+b))$ and $c(a,4a,-3a) = (ca,4ca,-3ca)$ so the sum and scalar product of the vectors are in the set. Thus the set is a subspace of $\mathbf{R}^3$.

(c) $(a,b,2)+(c,d,2) = (a+c,b+d,4)$, so the sum is not in the set. Thus the set is not a subspace of $\mathbf{R}^3$.

(d) $(a,b,4a-1)+(c,d,4c-1) = (a+c,b+d,4(a+c)-2)$, so the sum is not in the set. Thus the set is not a subspace of $\mathbf{R}^3$.

7. (a) $(a,b,a-4b)+(c,d,c-4d) = (a+c,b+d,a+c-4(b+d))$ and $h(a,b,a-4b) = (ha,ha,ha-4hb)$, so the sum and the scalar product of the vectors are in the set. Thus the set is a subspace of $\mathbf{R}^3$.

(b) $(a,a^2,5a)+(b,b^2,5b) = (a+b,a^2+b^2,5(a+b))$. a^2+b^2 is not the square of $a+b$ so the sum is not in the set and therefore the set is not a subspace of $\mathbf{R}^3$.

(c) $(a,1,1)+(b,1,1) = (a+b,2,2)$, so the sum is not in the set. Thus the set is not a subspace of $\mathbf{R}^3$.

(d) $(a,b,2a+3b+6)+(c,d,2c+3d+6) = (a+c,b+d,2(a+c)+3(b+d)+12)$, so the sum is not in the set. Thus the set is not a subspace of $\mathbf{R}^3$.

Section 5.2

8. (a) Yes, this set is a subspace of $\mathbf{R}^3$. $(a,b,c)+(d,e,f) = (a+d,b+e,c+f)$ and $(a+d)+(b+e)+(c+f) = (a+b+c)+(d+e+f) = 0$. Also $k(a,b,c) = (ka,kb,kc)$ and $ka+kb+kc = k(a+b+c) = 0$.

(b) No, this set is not a subspace of $\mathbf{R}^3$. Neither the sum nor the scalar product of these vectors is in the set. $(a+b+c)+(d+e+f) = 1+1 = 2$ and $k(a+b+c) = k$.

(c) No, this set is not a subspace of $\mathbf{R}^3$. The sum of two such vectors is not necessarily such a vector. $(0,1,1)+(1,0,1) = (1,1,2)$, which is not in the set.

(d) No, this set is not a subspace of $\mathbf{R}^3$. Neither the sum nor the scalar product of these vectors is necessarily in the set. $(5,1,1)+(5,1,1) = 2(5,1,1) = (10,2,2)$, which is not in the set.

(e) No, this set is not a subspace of $\mathbf{R}^3$. The sum of two such vectors is not necessarily in the set. $(0,1,2)+(2,3,3) = (2,4,5)$, which is not in the set.

(f) Yes, this set is a subspace of $\mathbf{R}^3$. $(a,b,c)+(d,e,f) = (a+d,b+e,c+f)$ and $k(a,b,c) = (ka,kb,kc)$. If $a = b+c$ and $d = e+f$ then $a+d = (b+c)+(e+f) = (b+e)+(c+f)$ and $ka = kb+kc$.

9. (a) No, this set is not a subspace of $\mathbf{R}^3$. If $k = 1/2$, then $k(1,b,c) = (1/2,kb,kc)$ and the first component is not an integer.

(b) No, this set is not a subspace of $\mathbf{R}^3$. If k is negative, then $k(1,b,c) = (k,kb,kc)$ and the first component is negative.

(c) No, this set is not a subspace of $\mathbf{R}^3$. If k is irrational, then $k(1,b,c) = (k,kb,kc)$ and the first component is irrational.

10. (a) No, this set is not a subspace of $\mathbf{R}^2$. If $b \neq 0$ and k is negative, then $k(a,b^2) = (ka,kb^2)$, and kb^2 is negative and so cannot be the square of a real number.

(b) Yes, this set is a subspace of $\mathbf{R}^2$. $(a,b^3)+(d,e^3) = (a+d,b^3+e^3)$ and $c(a,b^3) = (ca,cb^3)$. b^3+e^3 is the cube of some real number and so is cb^3.

(c) No, this set is not a subspace of $\mathbf{R}^2$. If k is negative, then $k(a,b) = (ka,kb)$ and ka is negative.

(d) No, this set is not a subspace of $\mathbf{R}^2$. If $k = 0$, then $k(a,b) = (0,0)$, which is not in the set.

(e) No, this set is not a subspace of R^2. If k is negative, then k(a,b) = (ka,kb) and ka is positive if a is negative and kb is negative if b is positive.

11. (a) The set of vectors (a,0,0), where a is nonnegative, is closed under addition but not under scalar multiplication. If k is negative, then k(a,0,0) = (ka,0,0) and ka is negative if a is positive.

   (b) The set of vectors (a,b,0), where a ≠ b unless a = b = 0, is closed under scalar multiplication but not under addition. (1,2,0)+(2,1,0) = (3,3,0).

12. (a) If (a,a+1,b) = (0,0,0), then a = a+1 = b = 0. a = a+1 is impossible. Therefore (0,0,0) is not in the set.

   (b) If (a,3,2a) = (0,0,0), then 3 = 0, so (0,0,0) is not in the set.

   (c) If (a,b,a+b−4) = (0,0,0), then a = b = a+b−4 = 0, but if a = b = 0 then a+b−4 = −4, not zero, so (0,0,0) is not in the set.

   (d) If a > 0, then a ≠ 0 so the first component of a vector in this set cannot be zero. Thus (0,0,0) is not in the set.

13. (a) The set of vectors of the form (a,a^2) contains the zero vector but is not a subspace of R^2. (See Exercise 5.)

   (b) The set of vectors of the form $(a,a^2,5a)$ contains the zero vector but is not a subspace of R^3. (See Exercise 7.)

14. (a,2a,b) = (p,2p,q) if and only if a = p and b = q, so U and V consist of the same vectors. (a,2a,b)+(d,2d,e) = (a+d,2(a+d),b+e) and c(a,2a,b) = (ca,2ca,cb), so the set is a subspace of R^3.

15. Let (x,y,z) be any vector in R^3. If a = x, b = y, c = z, then (x,y,z) = (a,b,c) and is in U. If a = x, b = y − x, c = z, then (x,y,z) = (a,a+b,c) and is in V. Thus U = V = R^3.

16. Let a = p and b+2 = r^3. (Every real number is the cube of some real number.) U = V. (a,3a,b+2)+(d,3d,e+2) = (a+d,3(a+d),b+2+e+2). Thus the second term is three times the first and the third term is two more than some number, so the sum is in the set. c(a,3a,b+2) = (ca,3ca,cb+2c). Thus again the second term is three times the first and the third term can be written h+2, where h = cb+2c−2, so U = V is a subspace of R^3.

Section 5.2

17. Let W be the set of all vectors (a,y,2a) in $\mathbf{R}^3$. If b = y then (a,y,2a) = (a,b,2a) and if b = y−a then (a,y,2a) = (a,a+b,2a), so every vector in W is in both U and V. If (a,b,2a) is in U and y = b then (a,b,2a) = (a,y,2a) and if (a,a+b,2a) is in V and y = a+b then (a,a+b,2a) = (a,y,2a), so all vectors in U and all vectors in V are in W. Thus U = W and V = W, and so U = V. (a,b,2a)+(d,e,2d) = (a+d,b+e,2(a+d)) and c(a,b,2a) = (ca,cb,2ca), so the sum and scalar product of vectors in U are in U and U = V is a subspace of $\mathbf{R}^3$.

18. If U is a subspace of $\mathbf{R}^3$, then the sum of every pair of vectors in U is in U and cu is in U for every scalar c and every vector u in U. If u_1 and u_2 are vectors in U and a and b are scalars, then au_1 is in U and bu_2 is in U and their sum $au_1 + bu_2$ is in U.

Let U be a subset of $\mathbf{R}^3$ such that for every pair of vectors u_1 and u_2 in U and every pair of scalars a and b, the vector $au_1 + bu_2$ is in U. Let a = b = 1. Then $au_1 + bu_2 = u_1 + u_2$ is in U for every pair of vectors u_1 and u_2 in U. Next let a = c, any scalar, and b = 0. Then $au_1 + bu_2 = cu_1$ is in U for any scalar c and any vector u_1 in U. Thus U is a subspace of $\mathbf{R}^3$.

19. (a) This set is a subspace. $\begin{bmatrix} 0 & a \\ b & 0 \end{bmatrix} + \begin{bmatrix} 0 & c \\ d & 0 \end{bmatrix} = \begin{bmatrix} 0 & a+c \\ b+d & 0 \end{bmatrix}$ and

$k \begin{bmatrix} 0 & a \\ b & 0 \end{bmatrix} = \begin{bmatrix} 0 & ka \\ kb & 0 \end{bmatrix}$.

(b) $2 \begin{bmatrix} 2 & -1 \\ 0 & 5 \end{bmatrix} = \begin{bmatrix} 4 & -2 \\ 0 & 10 \end{bmatrix}$, which is not in the set, so the set is not a subspace.

(c) For k ≠ 0 or 1, $k \begin{bmatrix} a & a^2 \\ b & b^2 \end{bmatrix} = \begin{bmatrix} ka & ka^2 \\ kb & kb^2 \end{bmatrix} \neq \begin{bmatrix} ka & (ka)^2 \\ kb & (kb)^2 \end{bmatrix}$, so the set is not a subspace.

(d) $2 \begin{bmatrix} a & a+2 \\ b & c \end{bmatrix} = \begin{bmatrix} 2a & 2a+4 \\ 2b & 2c \end{bmatrix}$, and 2a + 4 ≠ 2a + 2, so the set is not a subspace.

(e) This set is a subspace. $\begin{bmatrix} a & c \\ c & b \end{bmatrix} + \begin{bmatrix} e & g \\ g & f \end{bmatrix} = \begin{bmatrix} a+e & c+g \\ c+g & b+f \end{bmatrix}$ and

$$k\begin{bmatrix} a & c \\ c & b \end{bmatrix} = \begin{bmatrix} ka & kc \\ kc & kb \end{bmatrix}.$$

(f) The matrices $\begin{bmatrix} 1 & 1 \\ 0 & 1 \end{bmatrix}$ and $\begin{bmatrix} -1 & 0 \\ 0 & 2 \end{bmatrix}$ are invertible but their sum $\begin{bmatrix} 0 & 1 \\ 0 & 3 \end{bmatrix}$ is not, so the set is not a subspace.

20. (a) $\begin{bmatrix} a & b & 0 \\ c & d & 0 \end{bmatrix} + \begin{bmatrix} e & f & 0 \\ g & h & 0 \end{bmatrix} = \begin{bmatrix} a+e & b+f & 0 \\ c+g & d+h & 0 \end{bmatrix}$ and $k\begin{bmatrix} a & b & 0 \\ c & d & 0 \end{bmatrix} = \begin{bmatrix} ka & kb & 0 \\ kc & kd & 0 \end{bmatrix}$, so the set is closed under addition and scalar multiplication and is therefore a subspace.

(b) $\begin{bmatrix} a & 2a & 3a \\ b & 2b & 3b \end{bmatrix} + \begin{bmatrix} c & 2c & 3c \\ d & 2d & 3d \end{bmatrix} = \begin{bmatrix} a+c & 2a+2c & 3a+3c \\ b+d & 2b+2d & 3b+3d \end{bmatrix} = \begin{bmatrix} a+c & 2(a+c) & 3(a+c) \\ b+d & 2(b+d) & 3(b+d) \end{bmatrix}$

and $k\begin{bmatrix} a & 2a & 3a \\ b & 2b & 3b \end{bmatrix} = \begin{bmatrix} ka & 2ka & 3ka \\ kb & 2kb & 3kb \end{bmatrix}$, so the set is closed under addition and scalar multiplication and is therefore a subspace.

21. Every element of P_2 is an element of P_3. Both P_2 and P_3 are vector spaces with the same operations and the same set of scalars, so P_2 is a subspace of P_3.

22. If $f(x) = ax^2 + bx + 3$ then $2f(x) = 2ax^2 + 2bx + 6$, which is not in S, so S is not closed under scalar multiplication and therefore not a subspace of P_2.

23. $\mathbf{R}^n$ is a subset of $\mathbf{C}^n$, but the scalars for $\mathbf{C}^n$ are the complex numbers. If a nonzero element of $\mathbf{R}^n$ is multiplied by a complex number such as $1 + 2i$, the result is an element of $\mathbf{C}^n$ that is not an element of $\mathbf{R}^n$. Thus $\mathbf{R}^n$ is not closed under multiplication by the scalars in $\mathbf{C}^n$.

Section 5.2

24. (a) $\begin{bmatrix} a & 1 \\ b & c \end{bmatrix} \neq \begin{bmatrix} 0 & 0 \\ 0 & 0 \end{bmatrix}$ for any a,b,c, so the zero vector $\begin{bmatrix} 0 & 0 \\ 0 & 0 \end{bmatrix}$ is not in the subset.

   (b) ax + 2 ≠ 0 for any a and all x, so the zero vector is not in the subset.

25. (a) (f+g)(x) = f(x) + g(x) so (f+g)(0) = 0 + 0 = 0, and (cf)(x) = c(f(x)) so (cf)(0) = c0 = 0. Thus this subset is a subspace.

   (b) (f+g)(0) = f(0) + g(0) = 3 + 3 = 6, so the subset is not closed under addition, and the subset is not a subspace.

   (c) If f(x) = a and g(x) = b then (f+g)(x) = a + b, a constant, and (cf)(x) = ca, a constant, so the subset is closed under addition and scalar multiplication and is therefore a subspace.

26. If **u** and **v** are both in V but not in U, then it is possible for **u** + **v** to be in U. For example, let U be the subspace of V = R^2 with basis (1,1). Then (1,0) and (0,1) are in R^2 but not in U. However (1,0) + (0,1) = (1,1) is in U.

   If **u** is in U then –**u** is in U and (**u** + **v**) + –**u** = **v**, so if **u** + **v** and **u** are in U then so is **v**. Therefore if **u** is in U and **v** is not, then their sum is not in U.

   If c**u** is in U then $\frac{1}{c}$ c**u** = **u** is in U, so if **u** is not in U, then c**u** is also not in U for all c ≠ 0.

27. If a**u** + b**v** is in U for all vectors **u** and **v** in U and all scalars a and b, then in particular if a = b = 1, **u** + **v** is in U and if a = c and b = 0, c**u** is in U, so U is a subspace of V.

   If U is a subspace of V, then **u** + **v** and c**u** are in U for all vectors **u** and **v** in U and all scalars c. Therefore if **u** is in U and a is a scalar, then a**u** is in U. Likewise b**v** is in U, so the sum a**u** + b**v** is in U.

28. (a) Let U be the subspace of R^2 with basis (1,0) and let V be the subspace of R^2 with basis (0,1). (1,0) and (0,1) are in the union of U and V but their sum, (1,1), is not.

   (b) If **u** and **v** are two vectors in the intersection of two subspaces U and V, then **u** + **v** is in U and in V, so **u** + **v** is in the intersection. Likewise if c is any scalar, then c**u** is in U and in V, so c**u** is in the intersection of U and V. Thus the intersection of U and V is a subspace.

Section 5.3

Exercise Set 5.3

1. (a) To find a and b such that $(-1,7) = a(1,-1) + b(2,4)$ we solve the equations $-1 = a + 2b$ and $7 = -a + 4b$ and find that $a = -3$ and $b = 1$, so that $(-1,7) = -3(1,-1) + (2,4)$. Thus $(-1,7)$ is a linear combination of $(1,-1)$ and $(2,4)$.

    (b) $(8,13) = 2(1,2) + 3(2,3)$.

    (c) If $(-1,15) = a(-1,4) + b(2,-8)$ then $-1 = -a + 2b$ and $15 = 4a - 8b$. This system of equations has no solution, so $(-1,15)$ is not a linear combination of $(-1,4)$ and $(2,-8)$.

    (d) $(13,6) = (1,3) + 3(4,1)$.

2. (a) $(7,5) = -3(1,-1) + 2(5,1)$.   (b) $(6,22) = 4(2,3) + 2(-1,5)$.

    (c) $(2,1) = \frac{-13}{3}(3,-1) + \frac{5}{3}(9,-2)$.   (d) $(4,0) = -(-1,2) + (3,2) + 0(6,4)$.

3. (a) $(-3,3,7) = 2(1,-1,2) - (2,1,0) + 3(-1,2,1)$.

    (b) $(-2,11,7) = 2(1,-1,0) + (2,1,4) + 3(-2,4,1)$.

    (c) If $(2,7,13) = a(1,2,3) + b(-1,2,4) + c(1,6,10)$ then $2 = a - b + c$, $7 = 2a + 2b + 6c$, and $13 = 3a + 4b + 10c$. This system of equations has no solution, so $(2,7,13)$ is not a linear combination of $(1,2,3)$, $(-1,2,4)$, and $(1,6,10)$.

    (d) The system of equations $0 = -a + b + c$, $10 = 2a + 3b + 8c$, $8 = 3a + b + 5c$ has many solutions. The general solution is $a = 2 - c$, $b = 2 - 2c$, $c =$ any real number. Thus $(0,10,8) = (2-c)(-1,2,3) + (2-2c)(1,3,1) + c(1,8,5)$, where c is any real number.

4. (a) $(4,-4,6) = (1,2,-3) + \frac{c+3}{2}(2,-4,6) + c(-1,2,-3)$.

    (b) The system of equations $4 = -a + 2b$, $3 = b + c$, $8 = a + 3b + 5c$ has no solution. Therefore $(4,3,8)$ is not a linear combination of $(-1,0,1)$, $(2,1,3)$, and $(0,1,5)$.

    (c) $(-1,4,-9) = 2(-1,3,1) + (1,1,1) - 3(0,1,4)$.

    (d) The system of equations $1 = a - b - c$, $5 = 2a + 3b + 8c$, $3 = a + 2b + 5c$ has no solution. Therefore $(1,5,3)$ is not a linear combination of $(1,2,1)$, $(-1,3,2)$, and $(-1,8,5)$.

156

Section 5.3

5. (a) To show that the vectors (1,1) and (1,−1) span $\mathbf{R}^2$ one must show that any vector (x_1, x_2) in $\mathbf{R}^2$ can be written as a linear combination of (1,1) and (1,−1). $(x_1, x_2) = a(1,1) + b(1,−1)$ if and only if $x_1 = a + b$ and $x_2 = a − b$. The determinant of the coefficient matrix for this system of equations is $\begin{vmatrix} 1 & 1 \\ 1 & -1 \end{vmatrix} = -2$, so the system has a unique solution and we have shown that the given vectors span $\mathbf{R}^2$. The solution of this system of equations is $a = \dfrac{x_1 + x_2}{2}$, $b = \dfrac{x_1 - x_2}{2}$, so that

$(x_1, x_2) = \dfrac{x_1 + x_2}{2}(1,1) + \dfrac{x_1 - x_2}{2}(1,-1)$ and $(3,5) = 4(1,1) + (-1)(1,-1)$.

(b) $(x_1, x_2) = a(1,4) + b(-2,0)$ if and only if $x_1 = a - 2b$ and $x_2 = 4a$. The unique solution of this system of equations is $a = \dfrac{x_2}{4}$, $b = \dfrac{-4x_1 + x_2}{8}$, so the given vectors span $\mathbf{R}^2$ and $(x_1, x_2) = \dfrac{x_2}{4}(1,4) + \dfrac{-4x_1 + x_2}{8}(-2,0)$. $(3,5) = \dfrac{5}{4}(1,4) + \dfrac{-7}{8}(-2,0)$.

(c) $(x_1, x_2) = a(1,3) + b(3,10)$ if and only if $x_1 = a + 3b$ and $x_2 = 3a + 10b$. The unique solution of this system of equations is $a = 10x_1 - 3x_2$, $b = -3x_1 + x_2$, so the given vectors span $\mathbf{R}^2$ and $(x_1, x_2) = (10x_1 - 3x_2)(1,3) + (-3x_1 + x_2)(3,10)$. $(3,5) = 15(1,3) + (-4)(3,10)$.

(d) $(x_1, x_2) = x_1(1,0) + x_2(0,1)$ and $(3,5) = 3(1,0) + 5(0,1)$.

6. (a) $(x_1, x_2, x_3) = a(1,2,3) + b(-1,-1,0) + c(2,5,4)$ if and only if $x_1 = a - b + 2c$, $x_2 = 2a - b + 5c$, and $x_3 = 3a + 4c$. The determinant of the coefficient matrix for this system of equations is $\begin{vmatrix} 1 & -1 & 2 \\ 2 & -1 & 5 \\ 3 & 0 & 4 \end{vmatrix} = -5$, so the system of equations has a unique solution and the given vectors span $\mathbf{R}^3$. Using Cramer's rule, the solution is

$$a = \frac{1}{-5} \begin{vmatrix} x_1 & -1 & 2 \\ x_2 & -1 & 5 \\ x_3 & 0 & 4 \end{vmatrix} = \frac{1}{-5}(-4x_1 + 4x_2 - 3x_3),$$

$$b = \frac{1}{-5} \begin{vmatrix} 1 & x_1 & 2 \\ 2 & x_2 & 5 \\ 3 & x_3 & 4 \end{vmatrix} = \frac{1}{-5}(7x_1 - 2x_2 - x_3),$$

$$c = \frac{1}{-5} \begin{vmatrix} 1 & -1 & x_1 \\ 2 & -1 & x_2 \\ 3 & 0 & x_3 \end{vmatrix} = \frac{1}{-5}(3x_1 - 3x_2 + x_3).$$

$(1,3,-2) = \frac{14}{-5}(1,2,3) + \frac{3}{-5}(-1,-1,0) + \frac{8}{5}(2,5,4).$

(b) $(x_1,x_2,x_3) = a(1,3,1) + b(-1,1,0) + c(4,1,1)$ if and only if $x_1 = a - b + 4c$, $x_2 = 3a + b + c$, $x_3 = a + c$. The determinant of the coefficient matrix for this system of equations is $\begin{vmatrix} 1 & -1 & 4 \\ 3 & 1 & 1 \\ 1 & 0 & 1 \end{vmatrix} = -1$, so the system of equations has a unique solution and the given vectors span $\mathbf{R}^3$. The solution of the system is $a = -x_1 - x_2 + 5x_3$, $b = 2x_1 + 3x_2 - 11x_3$, $c = x_1 + x_2 - 4x_3$, and $(1,3,-2) = -14(1,3,1) + 33(-1,1,0) + 12(4,1,1)$.

(c) $(x_1,x_2,x_3) = a(5,1,3) + b(2,0,1) + c(-2,-3,-1)$ if and only if $x_1 = 5a + 2b - 2c$, $x_2 = a - 3c$, $x_3 = 3a + b - c$. The determinant of the coefficient matrix for this system of equations is $\begin{vmatrix} 5 & 2 & -2 \\ 1 & 0 & -3 \\ 3 & 1 & -1 \end{vmatrix} = -3$, so the system of equations has a unique solution and the given vectors span $\mathbf{R}^3$. The solution of the system is $a = \frac{1}{-3}(3x_1 - 6x_3)$, $b = \frac{1}{-3}(-8x_1 + x_2 + 13x_3)$, $c = \frac{1}{-3}(x_1 + x_2 - 2x_3)$, and $(1,3,-2) = -5(5,1,3) + \frac{31}{3}(2,0,1) + \frac{8}{-3}(-2,-3,-1).$

(d) $(x_1,x_2,x_3) = x_1(1,0,0) + x_2(0,1,0) + x_3(0,0,1)$ and $(1,3,-2) = (1,0,0) + 3(0,1,0) - 2(0,0,1).$

Section 5.3

7. (a) $(x_1, x_2) = a(1,-3) + b(2,-5)$ if and only if $x_1 = a + 2b$ and $x_2 = -3a - 5b$. The determinant of the coefficient matrix for this system of equations is $\begin{vmatrix} 1 & 2 \\ -3 & -5 \end{vmatrix} \neq 0$, so the system has a unique solution and the given vectors span $\mathbf{R}^2$.

(b) $(x_1, x_2) = a(1,1) + b(-2,1)$ if and only if $x_1 = a - 2b$ and $x_2 = a + b$. The determinant of the coefficient matrix for this system of equations is $\begin{vmatrix} 1 & -2 \\ 1 & 1 \end{vmatrix} \neq 0$, so the system has a unique solution and the given vectors span $\mathbf{R}^2$.

(c) $(3,-1) = -(-3,1)$, so all the vectors spanned by these two vectors are just the multiples of $(-3,1)$. These vectors do not span all of $\mathbf{R}^2$.

(d) Any one of the three given vectors is a linear combination of the other two, so we'll be clever and make one of the coefficients zero. $(x_1, x_2) = a(3,2) + 0(1,1) + c(1,0)$ if and only if $x_1 = 3a + c$ and $x_2 = 2a$. This system of equations has the unique solution $a = x_2/2$ and $c = x_1 - (3/2)x_2$, so the given vectors span $\mathbf{R}^2$. (We could find a more general solution by not making one of the coefficients zero, but we don't need it.)

(e) As in (d), we make one of the coefficients zero. $(x_1, x_2) = 0(4,-1) + b(2,3) + c(6,5)$ if and only if $x_1 = 2b + 6c$ and $x_2 = 3b + 5c$. The determinant of the coefficient matrix for this system of equations is $\begin{vmatrix} 2 & 6 \\ 3 & 5 \end{vmatrix} \neq 0$, so the system has a unique solution and the given vectors span $\mathbf{R}^2$.

8. (a) $(x_1, x_2, x_3) = a(2,1,0) + b(-1,3,1) + c(4,5,0)$ if and only if $x_1 = 2a - b + 4c$, $x_2 = a + 3b + 5c$, $x_3 = b$. The determinant of the coefficient matrix for this system of equations is $\begin{vmatrix} 2 & -1 & 4 \\ 1 & 3 & 5 \\ 0 & 1 & 0 \end{vmatrix} \neq 0$, so the system of equations has a unique solution and the given vectors span $\mathbf{R}^3$.

(b) $(x_1, x_2, x_3) = a(4,0,1) + b(0,1,0) + c(0,0,1)$ if and only if

159

$x_1 = 4a$, $x_2 = b$, $x_3 = a + c$. This system of equations has the unique solution $a = x_1/4$, $b = x_2$, $c = -x_1 + x_3$, so the given vectors span $\mathbf{R}^3$.

(c) $(x_1, x_2, x_3) = a(1,2,1) + b(-1,3,0) + c(0,5,1)$ if and only if $x_1 = a - b$, $x_2 = 2a + 3b + 5c$, $x_3 = a + c$. The determinant of the coefficient matrix for this system of equations is $\begin{vmatrix} 1 & -1 & 0 \\ 2 & 3 & 5 \\ 1 & 0 & 1 \end{vmatrix} = 0$, so the system of equations does not have a unique solution. It has either many solutions or no solutions depending on the values of x_1, x_2, and x_3. If the system is solved using Gaussian elimination, the last row of the coefficient matrix becomes all zeros. Thus if $(x_1, x_2, x_3) = (0,0,1)$, the last row gives $0 = 1$, and there is no solution. $(0,0,1)$ cannot be expressed as a linear combination of $(1,2,1)$, $(-1,3,0)$ and $(0,5,1)$, so these vectors do not span $\mathbf{R}^3$.

(d) $(x_1, x_2, x_3) = a(1,-1,-1) + b(0,1,2) + c(1,2,1)$ if and only if $x_1 = a + c$, $x_2 = -a + b + 2c$, $x_3 = -a + 2b + c$. The determinant of the coefficient matrix for this system of equations is $\begin{vmatrix} 1 & 0 & 1 \\ -1 & 1 & 2 \\ -1 & 2 & 1 \end{vmatrix} \neq 0$, so the system of equations has a unique solution and the given vectors span $\mathbf{R}^3$.

9. $(1,2,3) + (1,2,0) = (2,4,3)$, $(1,2,3) - (1,2,0) = (0,0,3)$, $2(1,2,3) = (2,4,6)$.

10. $(1,2,1) + (2,1,4) = (3,3,5)$, $(1,2,1) - (2,1,4) = (-1,1,-3)$, $2(1,2,1) = (2,4,2)$.

11. $-(1,2,3) = (-1,-2,-3)$, $2(1,2,3) = (2,4,6)$, $(1/2)(1,2,3) = (1/2,1,3/2)$.

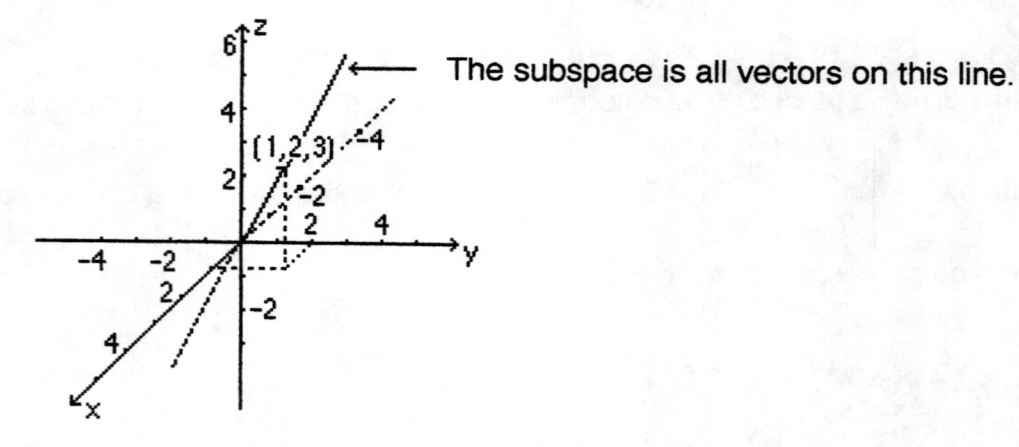

The subspace is all vectors on this line.

Section 5.3

12. −(4,−1,3) = (−4,1,−3), 2(4,−1,3) = (8,−2,6), (1/3)(4,−1,3) = (4/3,−1/3,1).

13. −(1,2) = (−1,−2), 2(1,2) = (2,4), 5(1,2) = (5,10).

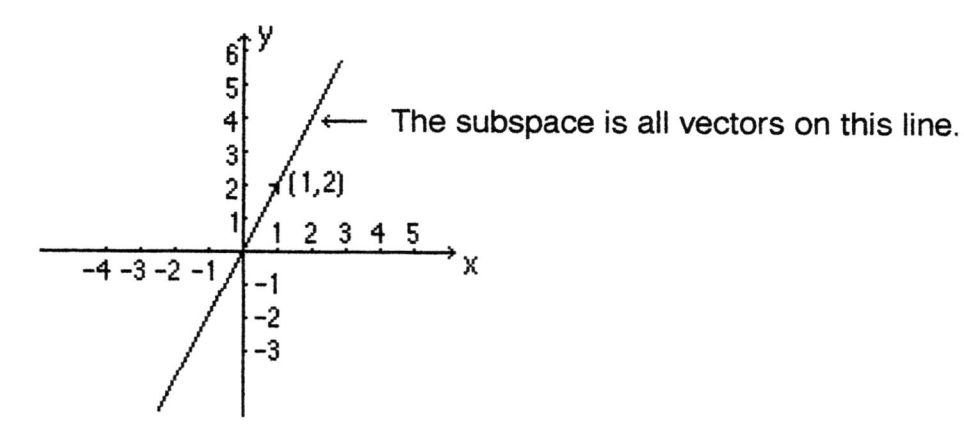

← The subspace is all vectors on this line.

14. −(1,2,−1,3) = (−1,−2,1,−3), 2(1,2,−1,3) = (2,4,−2,6), (.1)(1,2,−1,3) = (.1,.2,−.1,.3).

15. −(2,1,−3,4) = (−2,−1,3,−4), (2,1,−3,4) + (−3,0,1,5) = (−1,1,−2,9), 5(4,1,2,0) = (20,5,10,0).

16. U is the subspace of all vectors $a(-1,3)$ and V is the subspace of all vectors $b(-2,6)$. $a(-1,3) = b(-2,6)$ if and only if $2a = b$. Thus every vector in U is also in V and every vector in V is also in U, so U = V.

17. The subspace U consists of all vectors $a(1,2,3) + b(-1,2,5) = (a - b, 2a+2b, 3a+5b)$, and the subspace V consists of all vectors $c(1,6,11) + d(2,0,-2) = (c+2d, 6c, 11c-2d)$. Therefore U = V if and only if $a-b = c+2d$, $2a+2b = 6c$, and $3a+5b = 11c-2d$. This system of equations can be solved for a and b in terms of c and d or for c and d in terms of a and b: $c = (a+b)/3$ and $d = (a-2b)/3$. Thus U = V.

18. The subspace U consists of all vectors $a(-2,1,1) + b(0,1,3) = (-2a, a+b, a+3b)$, and the subspace V consists of all vectors $c(-2,2,4) + d(-2,3,7) = (-2c-2d, 2c+3d, 4c+7d)$. Therefore U = V if and only if $-2a = -2c-2d$, $a+b = 2c+3d$, and $a+3b = 4c+7d$. This system of equations can be solved for c and d in terms of a and b or for a and b in terms of c and d: $a = c+d$ and $b = c+2d$. Thus U = V.

Section 5.3

19. The subspace U consists of all vectors $a(3,-1,2) + b(1,0,4) = (3a+b,-a,2a+4b)$, and the subspace V consists of all vectors $c(4,-1,6) + d(1,-1,-6) = (4c+d,-c-d,6c-6d)$. Therefore $U = V$ if and only if $3a+b = 4c+d$, $-a = -c-d$, and $2a+4b = 6c-6d$. This system of equations can be solved for c and d in terms of a and b or for a and b in terms of c and d: $a = c+d$ and $b = c-2d$. Thus $U = V$.

20. By definition, the subspace generated by **u** is the set of all vectors a**u**, where a is a real number. This set is the set of all vectors on the line through the origin determined by **u**.

21. (a) $\begin{bmatrix} 5 & 7 \\ 5 & -10 \end{bmatrix} = 2\begin{bmatrix} 1 & 2 \\ 3 & -4 \end{bmatrix} - \begin{bmatrix} 0 & 3 \\ 1 & 2 \end{bmatrix} + 3\begin{bmatrix} 1 & 2 \\ 0 & 0 \end{bmatrix}.$

    (b) $\begin{bmatrix} 7 & 6 \\ -5 & -3 \end{bmatrix} = \begin{bmatrix} 3 & 0 \\ 1 & 1 \end{bmatrix} - 2\begin{bmatrix} 0 & 1 \\ 3 & 4 \end{bmatrix} + 4\begin{bmatrix} 1 & 2 \\ 0 & 1 \end{bmatrix}.$

    (c) If $\begin{bmatrix} 4 & 1 \\ 7 & 10 \end{bmatrix} = a\begin{bmatrix} 1 & 1 \\ 1 & 1 \end{bmatrix} + b\begin{bmatrix} 3 & 1 \\ 0 & 0 \end{bmatrix} + c\begin{bmatrix} -1 & -1 \\ 2 & 3 \end{bmatrix}$, then $4 = a + 3b - c$, $1 = a + b - c$, $7 = a + 2c$, and $10 = a + 3c$. This system of equations has no solution, so the first matrix is not a linear combination of the others.

22. (a) $3x^2 + 2x + 9 = 3(x^2 + 1) + 2(x + 3).$

    (b) If $2x^2 + x - 3 = a(x^2 - x + 1) + b(x^2 + 2x - 2)$, then $2 = a + b$, $1 = -a + 2b$, and $-3 = a - 2b$. This system of equations has no solution, so the first function is not a linear combination of the others.

    (c) $x^2 + 4x + 5 = -2(x^2 + x - 1) + 3(x^2 + 2x + 1).$

23. (a) Yes; $f(x) = 2h(x) - g(x).$

    (b) If $f(x) = ag(x) + bh(x)$ then $3x^2 + 5x + 1 = (2a+b)x^2 + 3bx + 3a - b$, so that $3 = 2a + b$, $5 = 3b$, and $1 = 3a - b$. This system of equations has no solution, so f(x) is not in the subspace generated by g(x) and h(x).

    (c) $g(x) + h(x) = (2x^2 + 3) + (x^2 + 3x - 1) = 3x^2 + 3x + 2$,
    $g(x) - h(x) = (2x^2 + 3) - (x^2 + 3x - 1) = x^2 - 3x + 4$, and
    $2g(x) = 2(2x^2 + 3) = 4x^2 + 6$ are functions in the space generated by g(x) and h(x).

24. If $v = av_1 + bv_2$ then $v = \dfrac{a}{c_1} c_1 v_1 + \dfrac{b}{c_2} c_2 v_2$.

25. Suppose v is not a linear combination of v_1 and v_2 but $v = a(c_1 v_1) + b(c_2 v_2)$. Then $v = (ac_1)v_1 + (bc_2)v_2$; i.e., v is a linear combination of v_1 and v_2. Thus v must not be a linear combination of $c_1 v_1$ and $c_2 v_2$.

26. If v_1 and v_2 do not span V, then there is a vector v in V which is not a linear combination of v_1 and v_2. If $c_1 v_1$ and $c_2 v_2$ span V, then $v = a(c_1 v_1) + b(c_2 v_2) = (ac_1)v_1 + (bc_2)v_2$, so that v is a linear combination of v_1 and v_2. Thus v must not be a linear combination of $c_1 v_1$ and $c_2 v_2$ and so they do not span V.

27. If v_1 and v_2 span V, then any vector v in V can be written as a linear combination of v_1 and v_2: $v = a_1 v_1 + a_2 v_2$. In particular $v_3 = b_1 v_1 + b_2 v_2$. Thus $v = a_1 v_1 + a_2 v_2 = (a_1 - b_1)v_1 + (a_2 - b_2)v_2 + v_3$, so that v_1, v_2, and v_3 span V. (It is also trivially true that $v = a_1 v_1 + a_2 v_2 + 0 v_3$.)

Exercise Set 5.4

1. For each pair of vectors we need to find constants a and b, not both zero, such that a times the first vector plus b times the second vector is the zero vector.

    (a) $2(-1,2) + (2,-4) = \mathbf{0}$.
    (b) $-3(3,1) + (9,3) = \mathbf{0}$.
    (c) $3(-2,3) + (6,-9) = \mathbf{0}$.
    (d) $0(1,5) + 2(0,0) = \mathbf{0}$.

2. (a) If $(0,0) = a(1,0) + b(0,1) = (a,b)$, then $a = b = 0$, so the vectors are linearly independent.

    (b) We must show that if $(0,0) = a(1,2) + b(3,2) = (a+3b, 2a+2b)$, then $a = b = 0$. The system of homogeneous equations $a+3b = 0$, $2a+2b = 0$ has coefficient matrix $\begin{bmatrix} 1 & 3 \\ 2 & 2 \end{bmatrix}$. The determinant of this matrix is not zero, so the system of equations has the unique solution $a = b = 0$. Thus the vectors are linearly independent.

Section 5.5

(c) We must show that if $(0,0) = a(-1,3) + b(2,5) = (-a+2b, 3a+5b)$, then $a = b = 0$. The system of homogeneous equations $-a+2b = 0$, $3a+5b = 0$ has coefficient matrix $\begin{bmatrix} -1 & 2 \\ 3 & 5 \end{bmatrix}$. The determinant of this matrix is not zero, so the system of equations has the unique solution $a = b = 0$. Thus the vectors are linearly independent.

(d) We must show that if $(0,0) = a(2,-4) + b(5,3) = (2a+5b, -4a+3b)$, then $a = b = 0$. The system of homogeneous equations $2a+5b = 0$, $-4a+3b = 0$ has coefficient matrix $\begin{bmatrix} 2 & 5 \\ -4 & 3 \end{bmatrix}$. The determinant of this matrix is not zero, so the system of equations has the unique solution $a = b = 0$. Thus the vectors are linearly independent.

3. For each set of vectors **u,v,w** we need to find constants a, b, and c, not all zero, such that $a\mathbf{u} + b\mathbf{v} + c\mathbf{w} = \mathbf{0}$.

   (a) $-2(1,-2,3) - 3(-2,4,1) + (-4,8,9) = \mathbf{0}$, so the vectors are linearly dependent.

   $(-4,8,9) = 2(1,-2,3) + 3(-2,4,1)$.

   (b) $3(1,0,2) - 2(2,6,4) + (1,12,2) = \mathbf{0}$, so the vectors are linearly dependent.

   $(1,12,2) = -3(1,0,2) + 2(2,6,4)$.

   (c) $(3,4,1) + 3(2,1,0) - (9,7,1) = \mathbf{0}$, so the vectors are linearly dependent.

   $(3,4,1) = -3(2,1,0) + (9,7,1)$.

   (d) $0(1,2,-3) + 0(2,1,-1) + 5(0,0,0) = \mathbf{0}$, so the vectors are linearly dependent.

   $(0,0,0) = 0(1,2,-3) + 0(2,1,-1)$.

4. (a) We must show that if $(0,0,0) = a(1,2,5) + b(1,-2,1) + c(2,1,4)$
   $= (a+b+2c, 2a-2b+c, 5a+b+4c)$, then $a = b = c = 0$. The system of homogeneous equations $a+b+2c = 0$, $2a-2b+c = 0$, $5a+b+4c = 0$ has coefficient matrix $\begin{bmatrix} 1 & 1 & 2 \\ 2 & -2 & 1 \\ 5 & 1 & 4 \end{bmatrix}$. The determinant of this matrix is not zero, so the system of equations has the unique solution $a = b = c = 0$. Thus the vectors are linearly independent.

(b) We must show that if $(0,0,0) = a(1,1,1) + b(-4,3,2) + c(4,1,2)$
$= (a-4b+4c, a+3b+c, a+2b+2c)$, then $a = b = c = 0$. The system of homogeneous equations $a-4b+4c = 0$, $a+3b+c = 0$, $a+2b+2c = 0$ has coefficient matrix
$$\begin{bmatrix} 1 & -4 & 4 \\ 1 & 3 & 1 \\ 1 & 2 & 2 \end{bmatrix}.$$ The determinant of this matrix is not zero, so the system of equations has the unique solution $a = b = c = 0$. Thus the vectors are linearly independent.

(c) We must show that if $(0,0,0) = a(1,3,-4) + b(3,-1,4) + c(1,0,-2)$
$= (a+3b+c, 3a-b, -4a+4b-2c)$, then $a = b = c = 0$. The system of homogeneous equations $a+3b+c = 0$, $3a-b = 0$, $-4a+4b-2c = 0$ has coefficient matrix
$$\begin{bmatrix} 1 & 3 & 1 \\ 3 & -1 & 0 \\ -4 & 4 & -2 \end{bmatrix}.$$ The determinant of this matrix is not zero, so the system of equations has the unique solution $a = b = c = 0$. Thus the vectors are linearly independent.

(d) We must show that if $(0,0,0) = a(3,4,7) + b(2,-1,1) + c(4,1,3)$
$= (3a+2b+4c, 4a-b+c, 7a+b+3c)$, then $a = b = c = 0$. The system of homogeneous equations $3a+2b+4c = 0$, $4a-b+c = 0$, $7a+b+3c = 0$ has coefficient matrix
$$\begin{bmatrix} 3 & 2 & 4 \\ 4 & -1 & 1 \\ 7 & 1 & 3 \end{bmatrix}.$$ The determinant of this matrix is not zero, so the system of equations has the unique solution $a = b = c = 0$. Thus the vectors are linearly independent.

(e) If $(0,0,0) = a(1,0,0) + b(0,1,0) + c(0,0,1) = (a,b,c)$, then $a = b = c = 0$, so the vectors are linearly independent.

5. (a) $2(2,-1,3) + (-4,2,-6) = \mathbf{0}$, so the vectors $(2,-1,3)$ and $(-4,2,-6)$ are linearly dependent and any set of vectors containing these vectors is linearly dependent.

(b) $-3(1,-2,3) + (3,-6,9) = \mathbf{0}$, so the vectors $(1,-2,3)$ and $(3,-6,9)$ are linearly dependent and any set of vectors containing these vectors is linearly dependent.

(c) $(3,0,4) + (-3,0,-4) = \mathbf{0}$, so the vectors $(3,0,4)$ and $(-3,0,-4)$ are linearly dependent and any set of vectors containing these vectors is linearly dependent.

(d) $-2(1,1,1) + (2,2,2) = \mathbf{0}$, so the vectors $(1,1,1)$ and $(2,2,2)$ are linearly dependent and any set of vectors containing these vectors is linearly dependent.

6. (a) $2(1,2) + 3(-1,4) - (-1,16) = \mathbf{0}$, so the vectors are linearly dependent.

   (b) If $(0,0) = a(1,3) + b(-2,1) = (a-2b, 3a+b)$, then $a-2b = 0$ and $3a+b = 0$. This system of homogeneous equations has coefficient matrix $\begin{bmatrix} 1 & -2 \\ 3 & 1 \end{bmatrix}$. The determinant of this matrix is not zero, so the system of equations has the unique solution $a = b = 0$. Thus the vectors are linearly independent.

   (c) $(1,-1,3) + 2(0,2,3) - 3(1,-1,2) - (-2,6,3) = \mathbf{0}$, so the vectors are linearly dependent.

   (d) If $(0,0,0) = a(1,2,8) + b(1,-1,-1) + c(1,0,3) = (a+b+c, 2a-b, 8a-b+3c)$, then $a+b+c = 0$, $2a-b = 0$, and $8a-b+3c = 0$. This system of homogeneous equations has coefficient matrix $\begin{bmatrix} 1 & 1 & 1 \\ 2 & -1 & 0 \\ 8 & -1 & 3 \end{bmatrix}$. The determinant of this matrix is not zero, so the system of equations has the unique solution $a = b = c = 0$. Thus the vectors are linearly independent.

   (e) If $(0,0,0,0) = a(1,0,0,0) + b(0,1,0,0) + c(0,0,1,0) + d(0,0,0,1) = (a,b,c,d)$, then $a = b = c = d = 0$, and so the vectors are linearly independent.

7. (a) $-4 = -2 \times 2$, so $t = -2x-1 = 2$.   (b) $6 = 2 \times 3$, so $t-1 = 2xt$ and therefore $t = -1$.

   (c) The vectors are linearly dependent if $(2t+6, 4t) = c(2,-t)$, i.e., if $2c = 2t+6$ and $-tc = 4t$. From the second equation, $c = -4$, and therefore, from the first equation, $t = -7$.

8. (a) If $(0,0) = a(1,1) + b(0,2) = (a, a+2b)$, then $a = 0$ and $a+2b = 0+2b = 0$, so $b = 0$. Thus the vectors are linearly independent.

   (b) If $(0,0,0) = a(1,1,2) + b(0,-1,3) + c(0,0,5) = (a, a-b, 2a+3b+5c)$, then $a = 0$, $a-b = 0-b = 0$, so $b = 0$ and $2a+3b+5c = 0+0+5c = 0$, so $c = 0$. Thus the vectors are linearly independent.

   (c) If $(0,0,0,0) = a(3,-2,4,5) + b(0,2,3,-4) + c(0,0,2,7) + d(0,0,0,4)$

Section 5.5

= (3a, −2a+2b, 4a+3b+2c, 5a−4b+7c+4d), then 3a = 0 so a = 0, −2a+2b = 0 so b = 0, 4a+3b+2c = 0 so c = 0, and 5a−4b+7c+4d = 0 so d = 0. Thus the vectors are linearly independent.

(d) The first vector has all nonzero components, the second vector has its first component zero and all other components nonzero, the third vector has its first and second components zero and all other components nonzero. In general, the kth vector has its first k−1 components zero and all others nonzero. The set {(1,2,4,−1,3), (0,1,−1,3,3), (0,0,2,4,10), (0,0,0,−1,−1), (0,0,0,0,1)} is linearly independent in $\mathbf{R}^5$.

9. If (0,0,0,0) = a(1,0,0,7) + b(0,1,0,4) + c(0,0,1,3) = (a, b, c, 7a+4b+3c), then equating the first three components of (a, b, c, 7a+4b+3c) and (0,0,0,0) we see that a = 0, b = 0, and c = 0. Thus the vectors are linearly independent.

The set of nonzero row vectors of any matrix in reduced echelon form is linearly independent using the same reasoning as above.

10. (a) $1(2x^2 + 1) + (-1)(x^2 + 4x) + (-1)(x^2 - 4x + 1) = 0$, so the set is linearly dependent.

(b) $-2(x^2 + 3) + 3(x + 1) + (2x^2 - 3x + 3) = 0$, so the set is linearly dependent.

(c) If $a(x^2 + 3x - 1) + b(x + 3) + c(2x^2 - x + 1) = 0$, then a + 2c = 0, 3a + b − c = 0, and −a + 3b + c = 0. This system of homogeneous equations has the unique solution a = b = c = 0, so the functions are linearly independent.

(d) If $a \begin{bmatrix} 1 & 0 \\ 0 & 0 \end{bmatrix} + b \begin{bmatrix} 0 & 2 \\ 0 & 0 \end{bmatrix} + c \begin{bmatrix} 0 & 0 \\ 3 & 0 \end{bmatrix} + d \begin{bmatrix} 0 & 0 \\ 0 & 4 \end{bmatrix} = \begin{bmatrix} 0 & 0 \\ 0 & 0 \end{bmatrix}$,

then $\begin{bmatrix} a & 2b \\ 3c & 4d \end{bmatrix} = \begin{bmatrix} 0 & 0 \\ 0 & 0 \end{bmatrix}$. a=b=c=d=0. Matrices are linearly independent.

(e) If $a \begin{bmatrix} 1 & 2 \\ 3 & 1 \end{bmatrix} + b \begin{bmatrix} 1 & 1 \\ 1 & 1 \end{bmatrix} + c \begin{bmatrix} 2 & 1 \\ 4 & 2 \end{bmatrix} = \begin{bmatrix} 0 & 0 \\ 0 & 0 \end{bmatrix}$, then a + b + 2c = 0,

2a + b + c = 0, 3a + b + 4c = 0, and a + b + 2c = 0. This system of equations has the unique solution a = b = c = 0, so the set is linearly independent.

(f) $2 \begin{bmatrix} 1 & 2 \\ -1 & 0 \end{bmatrix} + (-3) \begin{bmatrix} 1 & 2 \\ 1 & 1 \end{bmatrix} + \begin{bmatrix} 1 & 2 \\ 5 & 3 \end{bmatrix} = \begin{bmatrix} 0 & 0 \\ 0 & 0 \end{bmatrix}$.

11. Let $c_1 \mathbf{v}_1 + c_2 \mathbf{v}_2 = \mathbf{0}$, where c_1 and c_2 are not both zero. Any linear combination of

$v_1 + v_2$ and $v_1 - v_2$ is a linear combination of v_1 and v_2: $a_1(v_1 + v_2) + a_2(v_1 - v_2)$
$= (a_1 + a_2)v_1 + (a_1 - a_2)v_2$. Thus if $a_1 + a_2 = c_1$ and $a_1 - a_2 = c_2$, i.e., $a_1 = \frac{c_1+c_2}{2}$ and $a_2 = \frac{c_1-c_2}{2}$, then at least one of a_1 and a_2 is not zero and $a_1(v_1 + v_2) + a_2(v_1 - v_2) = 0$.

12. (a) Let $c_1v_1 + c_2v_2 + c_3v_3 = 0$, where $c_1, c_2,$ and c_3 are not all zero.
$a_1v_1 + a_2(v_1 + v_2) + a_3v_3 = (a_1 + a_2)v_1 + a_2v_2 + a_3v_3 = 0$ if $a_1 + a_2 = c_1$, $a_2 = c_2$, and $a_3 = c_3$; i.e., $a_1 = c_1 - c_2$, $a_2 = c_2$, and $a_3 = c_3$. At least one of $a_1, a_2,$ and a_3 is not zero, so the set $\{v_1, v_1 + v_2, v_3\}$ is linearly dependent.

(b) Let $c_1v_1 + c_2v_2 + c_3v_3 = 0$, where $c_1, c_2,$ and c_3 are not all zero.
$a_1v_1 + a_2cv_2 + a_3v_3 = 0$, if $a_1 = c_1$, $a_2 = c_2/c$, and $a_3 = c_3$, at least one of which is not zero. Thus the set $\{v_1, cv_2, v_3\}$ is linearly dependent.

(c) Let $c_1v_1 + c_2v_2 + c_3v_3 = 0$, where $c_1, c_2,$ and c_3 are not all zero.
$a_1v_1 + a_2(v_1 + cv_2) + a_3v_3 = (a_1 + a_2)v_1 + a_2cv_2 + a_3v_3 = 0$ if $(a_1 + a_2) = c_1$, $a_2 = c_2/c$, and $a_3 = c_3$, i.e., $a_1 = c_1 - c_2/c$, $a_2 = c_2/c$, and $a_3 = c_3$, at least one of which is not zero. Thus the set $\{v_1, v_1 + cv_2, v_3\}$ is linearly dependent.

13. (a) $a_1v_1 + a_2(v_1 + v_2) + a_3v_3 = (a_1 + a_2)v_1 + a_2v_2 + a_3v_3 = 0$ only if $a_1 + a_2 = 0$, $a_2 = 0$, and $a_3 = 0$ (because $v_1, v_2,$ and v_3 are linearly independent), i.e. only if $a_1 = a_2 = a_3 = 0$. Thus the set $\{v_1, v_1 + v_2, v_3\}$ is linearly independent.

(b) $a_1v_1 + a_2cv_2 + a_3v_3 = 0$, only if $a_1 = 0$, $a_2c = 0$, and $a_3 = 0$ (because $v_1, v_2,$ and v_3 are linearly independent), i.e., only if $a_1 = a_2 = a_3 = 0$. Thus the set $\{v_1, cv_2, v_3\}$ is linearly independent.

(c) $a_1v_1 + a_2(v_1 + cv_2) + a_3v_3 = (a_1 + a_2)v_1 + a_2cv_2 + a_3v_3 = 0$ only if $(a_1 + a_2) = 0$, $a_2c = 0$, and $a_3 = 0$, i.e., only if $a_1 = a_2 = a_3 = 0$. Thus the set $\{v_1, v_1 + cv_2, v_3\}$ is linearly independent.

14. Let $V = \{v_1, v_2, \ldots, v_k\}$ be a subset of S and let $a_1, a_2, \ldots, a_k$ be scalars such that $a_1v_1 + a_2v_2 + \ldots + a_kv_k = 0$. Then $a_1v_1 + a_2v_2 + \ldots + a_kv_k + 0v_{k+1} + \ldots + 0v_n = 0$, where $v_{k+1}, \ldots, v_n$ are the vectors in S but not in V. A linear combination of all the vectors in S can equal zero only if all the coefficients are zero, so $a_1, a_2, \ldots, a_k$ must all

Section 5.5

are not already in the same plane, but is more likely to move one of the vectors out of the plane of the other two if they are in fact in the same plane.

Exercise Set 5.5

1. For two vectors to be linearly dependent one must be a multiple of the other. In each case below, neither vector is a multiple of the other, so the vectors are linearly independent. We show that each set spans R^2.

   (a) $(x_1, x_2) = \dfrac{3x_2 - x_1}{5}(1,2) + \dfrac{2x_1 - x_2}{5}(3,1)$.

   (b) $(x_1, x_2) = \dfrac{-5x_1 + 2x_2}{13}(-1,4) + \dfrac{4x_1 + x_2}{13}(2,5)$.

   (c) $(x_1, x_2) = \dfrac{x_1 + x_2}{2}(1,1) + \dfrac{-x_1 + x_2}{2}(-1,1)$.

   (d) $(x_1, x_2) = (x_1 - x_2)(1,0) + x_2(1,1)$.

2. It is necessary to show either that the set of vectors is linearly independent or that the set spans R^2. For two vectors to be linearly dependent, one must be a multiple of the other. In each case neither vector is a multiple of the other, so the vectors are linearly independent and therefore a basis for R^2.

3. (a) These two vectors are linearly independent since neither is a multiple of the other. Thus they are a basis for R^2.

   (b) $(-2,6) = -2(1,-3)$, so these vectors are not linearly independent and therefore not a basis for R^2.

   (c) These two vectors are linearly independent since neither is a multiple of the other. Thus they are a basis for R^2.

   (d) $(3,-6) = -3(-1,2)$, so these vectors are not linearly independent and therefore not a basis for R^2.

4. (a) $a(1,1,1) + b(0,1,2) + c(3,0,1) = (0,0,0)$ if and only if $a+3c = 0$, $a+b = 0$, and

Section 5.5

be zero, and therefore V is a linearly independent set.

If P is a linearly dependent set, it may contain linearly independent subsets. For example, the set $\{(1,-2,3), (-1,2,-3), (1,0,0)\}$ is linearly dependent since $(1,-2,3) + (-1,2,-3) + 0(1,0,0) = \mathbf{0}$, but the subset $\{(-1,2,-3), (1,0,0)\}$ is linearly independent.

15. Suppose $c_1\mathbf{v}_1 + c_2\mathbf{v}_2 + c_3\mathbf{v}_3 = \mathbf{0}$. If $c_3 \neq 0$, then $\mathbf{v}_3 = -\frac{c_1}{c_3}\mathbf{v}_1 - \frac{c_2}{c_3}\mathbf{v}_2$, but if $c_3 = 0$, then $c_1\mathbf{v}_1 + c_2\mathbf{v}_2 = \mathbf{0}$, and therefore $c_1 = c_2 = 0$. Thus if $c_1\mathbf{v}_1 + c_2\mathbf{v}_2 + c_3\mathbf{v}_3 = \mathbf{0}$ and if $\mathbf{v}_3$ is not a linear combination of $\mathbf{v}_1$ and $\mathbf{v}_2$, then $c_1 = c_2 = c_3 = 0$; that is, the three vectors are linearly independent.

16. Suppose the set $V = \{\mathbf{v}_1, \mathbf{v}_2, \ldots, \mathbf{v}_n\}$ is linearly independent. If one vector is a linear combination of the others, call that vector $\mathbf{v}_1$. If $\mathbf{v}_1 = c_2\mathbf{v}_2 + \ldots + c_n\mathbf{v}_n$, then $-\mathbf{v}_1 + c_2\mathbf{v}_2 + \ldots + c_n\mathbf{v}_n = \mathbf{0}$. But the set is linearly independent, so all the coefficients, including the -1, must be zero. Since $-1 \neq 0$, there must not be a vector that is a linear combination of the others.

    Suppose no vector in the set V is a linear combination of the others. If $c_1\mathbf{v}_1 + c_2\mathbf{v}_2 + \ldots + c_n\mathbf{v}_n = \mathbf{0}$ and some coefficient $c_i \neq 0$, then $-c_i\mathbf{v}_i = c_1\mathbf{v}_1 + c_2\mathbf{v}_2 + \ldots + c_n\mathbf{v}_n$, where of course there is no term involving $\mathbf{v}_i$ on the right-hand side of the equation. This gives $\mathbf{v}_i = a_1\mathbf{v}_1 + a_2\mathbf{v}_2 + \ldots + a_n\mathbf{v}_n$, where $a_k = -\frac{c_k}{c_i}$; that is, $\mathbf{v}_i$ is a linear combination of the other vectors in the set. But no vector in the set is a linear combination of the others, so all the coefficients must be zero. Thus the set is linearly independent.

17. It is more likely that the vectors will be independent. For them to be dependent, one has to be a multiple of the other.

18. It is more likely that the vectors will be independent. For them to be dependent, they must all lie in the same plane.

19. The computer is more likely to state that the vectors are linearly independent when they are linearly dependent, because round-off error causes zero to look like a nonzero number close to zero to the computer. Thinking geometrically, a small move by one or more of the vectors (round-off error) is not likely to place them in the same plane if they

Section 5.5

$a+2b+c = 0$. The coefficient matrix for this homogeneous system of equations is
$$\begin{bmatrix} 1 & 0 & 3 \\ 1 & 1 & 0 \\ 1 & 2 & 1 \end{bmatrix}.$$ Its determinant is nonzero, so the system has the unique solution $a = b = c = 0$, and the three vectors are linearly independent. Thus they are a basis for $\mathbf{R}^3$.

(b) $a(1,2,3) + b(2,4,1) + c(3,0,0) = (0,0,0)$ if and only if $a+2b+3c = 0$, $2a+4b = 0$, and $3a+b = 0$. The coefficient matrix for this homogeneous system of equations is
$$\begin{bmatrix} 1 & 2 & 3 \\ 2 & 4 & 0 \\ 3 & 1 & 0 \end{bmatrix}.$$ Its determinant is nonzero, so the system has the unique solution $a = b = c = 0$, and the three vectors are linearly independent. Thus they are a basis for $\mathbf{R}^3$.

(c) $a(0,0,1) + b(2,3,1) + c(4,1,2) = (0,0,0)$ if and only if $2b+4c = 0$, $3b+c = 0$, and $a+b+2c = 0$. The coefficient matrix for this homogeneous system of equations is
$$\begin{bmatrix} 0 & 2 & 4 \\ 0 & 3 & 1 \\ 1 & 1 & 2 \end{bmatrix}.$$ Its determinant is nonzero, so the system has the unique solution $a = b = c = 0$, and the three vectors are linearly independent. Thus they are a basis for $\mathbf{R}^3$.

(d) $a(1,1,4) + b(2,1,3) + c(0,1,6) = (0,0,0)$ if and only if $a+2b = 0$, $a+b+c = 0$, and $4a+3b+6c = 0$. The coefficient matrix for this homogeneous system of equations is
$$\begin{bmatrix} 1 & 2 & 0 \\ 1 & 1 & 1 \\ 4 & 3 & 6 \end{bmatrix}.$$ Its determinant is nonzero, so the system has the unique solution $a = b = c = 0$, and the three vectors are linearly independent. Thus they are a basis for $\mathbf{R}^3$.

5. (a) $a(1,-1,2) + b(2,0,1) + c(3,0,0) = (0,0,0)$ if and only if $a+2b+3c = 0$, $-a = 0$, and $2a+b = 0$. The coefficient matrix for this homogeneous system of equations is
$$\begin{bmatrix} 1 & 2 & 3 \\ -1 & 0 & 0 \\ 2 & 1 & 0 \end{bmatrix}.$$ Its determinant is nonzero, so the system has the unique solution $a = b = c = 0$, and the three vectors are linearly independent. Thus they are a basis for $\mathbf{R}^3$.

(b) $2(2,1,0) + (-1,1,1) = (3,3,1)$, so the vectors are linearly dependent and therefore not a basis for $\mathbf{R}^3$.

Section 5.5

(c) $2(3,1,-1) + 2(-1,-1,0) = (4,0,-2)$, so the vectors are linearly dependent and therefore not a basis for $\mathbf{R}^3$.

(d) $a(1,2,2) + b(-1,0,1) + c(-3,1,-1) = (0,0,0)$ if and only if $a-b-3c = 0$, $2a+c = 0$, and $2a+b-c = 0$. The coefficient matrix for this homogeneous system of equations is $\begin{bmatrix} 1 & -1 & -3 \\ 2 & 0 & 1 \\ 2 & 1 & -1 \end{bmatrix}$. Its determinant is nonzero, so the system has the unique solution $a = b = c = 0$, and the three vectors are linearly independent. Thus they are a basis for $\mathbf{R}^3$.

6. (a) These vectors are linearly dependent.

   (b) A basis for $\mathbf{R}^2$ can contain only two vectors.

   (c) These vectors are linearly dependent.

   (d) The third vector is a multiple of the second, so the set is not linearly independent.

   (e) A basis for $\mathbf{R}^3$ can contain only three vectors.

   (f) $(1,4)$ is not in $\mathbf{R}^3$.

   (g) None of these vectors is in $\mathbf{R}^4$.

7. $(1,4,3) = 3(-1,2,1) + 2(2,-1,0)$, and $(-1,2,1)$ and $(2,-1,0)$ are linearly independent, so the dimension is 2 and $(-1,2,1)$ and $(2,-1,0)$ are a basis for the subspace.

8. $(1,2,-1) = 2(1,3,1) - (1,4,3)$.

9. $(2,1,4) = (1,0,2) + (1,1,2)$.

10. $(-3,3,-6) = -\frac{3}{2}(2,-2,4)$.

11. No, the three vectors $(1,2,-1)$, $(1,-1,0)$ and $(3,-1,2)$ are linearly independent.

12. The set $\{(1,2),(0,1)\}$ is a basis for $\mathbf{R}^2$.

13. The set $\{(1,1,1),(1,0,-2),(1,0,0)\}$ is a basis for $\mathbf{R}^3$.

Section 5.5

14. The set $\{(-1,0,2),(0,1,1),(1,0,0)\}$ is a basis for $\mathbf{R}^3$.

15. (a) $(a,a,b) = a(1,1,0) + b(0,0,1)$, so the linearly independent set $\{(1,1,0),(0,0,1)\}$ spans the subspace of vectors of the form (a,a,b) and is therefore a basis. The dimension of the space is 2 since there are 2 vectors in the basis.

    (b) $(a,a,2a) = a(1,1,2)$, so the single vector $(1,1,2)$ is a basis for the subspace of vectors of the form $(a,a,2a)$. The dimension of the space is 1 since there is 1 vector in the basis.

    (c) $(a, b, a+b) = a(1,0,1) + b(0,1,1)$, so the linearly independent set $\{(1,0,1),(0,1,1)\}$ spans the subspace of vectors of the form $(a,b,a+b)$ and is therefore a basis. The dimension of the space is 2 since there are 2 vectors in the basis.

    (d) $(a, 2b, a+3b) = a(1,0,1) + b(0,2,3)$, so the linearly independent set $\{(1,0,1),(0,2,3)\}$ spans the subspace of vectors of the form $(a,2b,a+3b)$ and is therefore a basis. The dimension of the space is 2 since there are 2 vectors in the basis.

    (e) $(a,b,c) = (a, b, -a-b) = a(1,0,-1) + b(0,1,-1)$, so the linearly independent set $\{(1,0,-1),(0,1,-1)\}$ spans the subspace of vectors of the form (a,b,c) where $a+b+c = 0$. Thus the set $\{(1,0,-1),(0,1,-1)\}$ is a basis for the subspace. The dimension of the space is 2 since there are 2 vectors in the basis.

16. (a) $(a, b, a+b, a-b) = a(1,0,1,1) + b(0,1,1,-1)$, so the linearly independent set $\{(1,0,1,1),(0,1,1,-1)\}$ spans the subspace of vectors of the form $(a,b,a+b,a-b)$ and is therefore a basis. The dimension of the space is 2 since there are 2 vectors in the basis.

    (b) $(a,2a,b,0) = a(1,2,0,0) + b(0,0,1,0)$, so the linearly independent set $\{(1,2,0,0),(0,0,1,0)\}$ spans the subspace of vectors of the form $(a,2a,b,0)$ and is therefore a basis. The dimension of the space is 2 since there are 2 vectors in the basis.

    (c) $(2a, b, a+3b, c) = a(2,0,1,0) + b(0,1,3,0)+c(0,0,0,1)$, so the linearly independent set $\{(2,0,1,0), (0,1,3,0),(0,0,0,1)\}$ spans the subspace of vectors of the form $(2a, b, a+3b, 0)$ and is therefore a basis. The dimension of the space is 3 since there are 3 vectors in the basis.

    (d) $(a,a,a,a) = a(1,1,1,1)$, so the single vector $(1,1,1,1)$ is a basis for the subspace of vectors of the form (a,a,a,a). The dimension of the space is 1 since there is 1 vector in the basis.

17. (a) The set $\{x^3, x^2, x, 1\}$ is a basis. The dimension is 4.

(b) $\begin{bmatrix} 1 & 0 & 0 \\ 0 & 0 & 0 \\ 0 & 0 & 0 \end{bmatrix}, \begin{bmatrix} 0 & 1 & 0 \\ 0 & 0 & 0 \\ 0 & 0 & 0 \end{bmatrix}, \begin{bmatrix} 0 & 0 & 1 \\ 0 & 0 & 0 \\ 0 & 0 & 0 \end{bmatrix}, \begin{bmatrix} 0 & 0 & 0 \\ 1 & 0 & 0 \\ 0 & 0 & 0 \end{bmatrix}, \begin{bmatrix} 0 & 0 & 0 \\ 0 & 1 & 0 \\ 0 & 0 & 0 \end{bmatrix}, \begin{bmatrix} 0 & 0 & 0 \\ 0 & 0 & 1 \\ 0 & 0 & 0 \end{bmatrix},$

$\begin{bmatrix} 0 & 0 & 0 \\ 0 & 0 & 0 \\ 1 & 0 & 0 \end{bmatrix}, \begin{bmatrix} 0 & 0 & 0 \\ 0 & 0 & 0 \\ 0 & 1 & 0 \end{bmatrix},$ and $\begin{bmatrix} 0 & 0 & 0 \\ 0 & 0 & 0 \\ 0 & 0 & 1 \end{bmatrix}$ are a basis. The dimension is 9.

(c) $\begin{bmatrix} 1 & 0 & 0 \\ 0 & 0 & 0 \end{bmatrix}, \begin{bmatrix} 0 & 1 & 0 \\ 0 & 0 & 0 \end{bmatrix}, \begin{bmatrix} 0 & 0 & 1 \\ 0 & 0 & 0 \end{bmatrix}, \begin{bmatrix} 0 & 0 & 0 \\ 1 & 0 & 0 \end{bmatrix}, \begin{bmatrix} 0 & 0 & 0 \\ 0 & 1 & 0 \end{bmatrix},$ and $\begin{bmatrix} 0 & 0 & 0 \\ 0 & 0 & 1 \end{bmatrix}$

are a basis. The dimension is 6.

(d) $\begin{bmatrix} 1 & 0 \\ 0 & 0 \end{bmatrix}$ and $\begin{bmatrix} 0 & 0 \\ 0 & 1 \end{bmatrix}$ are a basis. The dimension is 2.

(e) $\begin{bmatrix} 1 & 0 \\ 0 & 0 \end{bmatrix}, \begin{bmatrix} 0 & 0 \\ 0 & 1 \end{bmatrix},$ and $\begin{bmatrix} 0 & 1 \\ 1 & 0 \end{bmatrix}$ are a basis. The dimension is 3.

18. (a) Yes; $f(x) = 2h(x) - g(x)$.

(b) If $f(x) = ag(x) + bh(x)$, then $3x^2 + 5x + 1 = (2a+b)x^2 + 3bx + 3a - b$, so that $3 = 2a + b$, $5 = 3b$, and $1 = 3a - b$. This system of equations has no solution, so $f(x)$ is not in the subspace spanned by $g(x)$ and $h(x)$.

(c) $g(x) + h(x) = (2x^2 + 3) + (x^2 + 3x - 1) = 3x^2 + 3x + 2$,
$g(x) - h(x) = (2x^2 + 3) - (x^2 + 3x - 1) = x^2 - 3x + 4$, and
$2g(x) = 2(2x^2 + 3) = 4x^2 + 6$ are functions in the space spanned by $g(x)$ and $h(x)$.

(d) $f(x)$ and $g(x)$ are linearly independent and $h(x) = g(x) - f(x)$, so the set $\{f(x), g(x)\}$ is a basis.

19. (a) A basis for P_2 must consist of three linearly independent elements of P_2.
$f(x) + g(x) - h(x) = 0$ so the three given functions are not linearly independent and therefore not a basis for P_2.

(b) These three functions are linearly independent, so they are a basis for P_2.

Section 5.5

(c) $\begin{bmatrix} 1 & 2 \\ 0 & 1 \end{bmatrix} - \begin{bmatrix} 3 & 4 \\ 1 & 1 \end{bmatrix} + 2\begin{bmatrix} 1 & 2 \\ 1 & 1 \end{bmatrix} - \begin{bmatrix} 0 & 2 \\ 1 & 2 \end{bmatrix} = \begin{bmatrix} 0 & 0 \\ 0 & 0 \end{bmatrix}$, so these matrices are linearly dependent and therefore not a basis for M_{22}.

(d) These four matrices are linearly independent, so they are a basis for M_{22}.

(e) The set $\{(1,0), (0,1)\}$ is a basis for $\mathbf{C}^2$, so the dimension of $\mathbf{C}^2$ is 2, and any set of two linearly independent vectors in $\mathbf{C}^2$ is a basis. The given vectors are linearly independent since neither is a multiple of the other, so they are a basis.

(f) $2(1+2i, 3-i, 1) - (4+i, 3i, 1+i) = (-2+3i, 6-5i, 1-i)$, so the vectors are linearly dependent and therefore not a basis.

20. Since dim(V) = 2, any basis for V must contain 2 vectors. In Example 4 in Section 5.4, it was shown that the vectors $\mathbf{u}_1 = \mathbf{v}_1 + \mathbf{v}_2$ and $\mathbf{u}_2 = \mathbf{v}_1 - \mathbf{v}_2$ are linearly independent. Thus from Theorem 5.11 the set $\{\mathbf{u}_1, \mathbf{u}_2\}$ is a basis for V.

21. Since V is three-dimensional, any basis for V must contain three vectors. We need only show that the vectors $\mathbf{u}_1 = \mathbf{v}_1$, $\mathbf{u}_2 = \mathbf{v}_1 + \mathbf{v}_2$, and $\mathbf{u}_3 = \mathbf{v}_1 + \mathbf{v}_2 + \mathbf{v}_3$ are linearly independent. If $a\mathbf{v}_1 + b(\mathbf{v}_1 + \mathbf{v}_2) + c(\mathbf{v}_1 + \mathbf{v}_2 + \mathbf{v}_3) = (a+b+c)\mathbf{v}_1 + (b+c)\mathbf{v}_2 + c\mathbf{v}_3 = \mathbf{0}$, then a+b+c = b+c = c = 0 so that a = b = c = 0. Thus $\mathbf{u}_1$, $\mathbf{u}_2$, and $\mathbf{u}_3$ are linearly independent and a basis for V.

22. Since V is n-dimensional any basis for V must contain n vectors. We need only show that the vectors $c\mathbf{v}_1, c\mathbf{v}_2, \ldots, c\mathbf{v}_n$ are linearly independent.
If $a_1 c\mathbf{v}_1 + a_2 c\mathbf{v}_2 + \ldots + a_n c\mathbf{v}_n = \mathbf{0}$, then $a_1 c = a_2 c = \ldots = a_n c = 0$ and since $c \neq 0$, $a_1 = a_2 = \ldots = a_n = 0$. Thus the vectors $c\mathbf{v}_1, c\mathbf{v}_2, \ldots, c\mathbf{v}_n$ are linearly independent and a basis for V.

23. (a) This statement is false. Any set of vectors containing the zero vector is linearly dependent.

   (b) true   (c) true   (d) true

Section 5.5

24. If a set of n−1 vectors spans V, then either that set is linearly independent and is a basis for V, so that dim(V) = n−1, or there is a linearly independent subset of those vectors that is a basis for V and dim(v) < n−1. Both possibilities contradict dim(V) = n.

25. If m > n, then any basis of W is linearly dependent by Theorem 5.7. But a basis is linearly independent, so m ≤ n.

26. (a) False. The vectors are linearly dependent.

    (b) False. Any set which spans $\mathbf{R}^3$ must contain at least three vectors.

    (c) True. (1,2,3) + (0,1,4) = (1,3,7), and (1,2,3) and (0,1,4) are linearly independent, so the dimension is 2.

    (d) True. In fact any set of two linearly independent vectors in $\mathbf{R}^2$ spans $\mathbf{R}^2$ and any set of two vectors that spans $\mathbf{R}^2$ is linearly independent.

27. (a) False. $(200,567,0) = (100 - \frac{1701}{8})(2,0,0) + \frac{567}{4}(3,4,0)$.

    (b) False. If the two vectors are linearly dependent you can't get a basis no matter what vector you add to them.

    (c) False. In fact any set of more than two vectors in $\mathbf{R}^2$ is a linearly dependent set.

    (d) True. The number of vectors in any set of linearly independent vectors is less than or equal to the dimension of the vector space.

28. Suppose (a,b) is orthogonal to (u_1,u_2). Then $(a,b)\cdot(u_1,u_2) = au_1 + bu_2 = 0$, so $au_1 = -bu_2$. Thus if $u_1 \neq 0$ (a,b) is orthogonal to (u_1,u_2) if and only if (a,b) is of the form $\left(\frac{-bu_2}{u_1}, b\right) = \frac{b}{u_1}(-u_2, u_1)$; that is, (a,b) is orthogonal to (u_1,u_2) if and only if (a,b) is in the vector space with basis $(-u_2, u_1)$. If $u_1 = 0$, then b = 0 and (a,b) is orthogonal to (u_1,u_2) if and only if (a,b) is of the form a(1,0); that is, (a,b) is in the vector space with basis (1,0).

29. Suppose (a,b,c) is orthogonal to (u_1,u_2,u_3). Then $(a,b,c)\cdot(u_1,u_2,u_3) = au_1 + bu_2 + cu_3 = 0$, so $au_1 = -bu_2 - cu_3$. Thus if $u_1 \neq 0$, (a,b,c) is orthogonal to (u_1,u_2,u_3) if and only if

Section 5.6

(a,b,c) is of the form $\left(\dfrac{-bu_2 - cu_3}{u_1}, b, c\right) = \dfrac{b}{u_1}(-u_2, u_1, 0) + \dfrac{c}{u_1}(-u_3, 0, u_1)$; that is, (a,b,c) is orthogonal to (u_1, u_2, u_3) if and only if (a,b,c) is in the vector space with basis vectors $(-u_2, u_1, 0)$ and $(-u_3, 0, u_1)$. If $u_1 = 0$ then $-bu_2 = cu_3$ and (a,b,c) is orthogonal to (u_1, u_2, u_3) if and only if (a,b,c) is of the form $\left(a, b, \dfrac{-bu_2}{u_3}\right) = a(1,0,0) - \dfrac{b}{u_3}(0, -u_3, u_2)$ or

$\left(a, \dfrac{-cu_3}{u_2}, c\right) = a(1,0,0) + \dfrac{c}{u_2}(0, -u_3, u_2)$; that is, (a,b,c) is in the vector space with basis vectors $(1,0,0)$ and $(0, -u_3, u_2)$.

30. Suppose (a,b,c) is orthogonal to $\mathbf{u} = (u_1, u_2, u_3)$ and $\mathbf{v} = (v_1, v_2, v_3)$. Then $(a,b,c)\cdot(u_1, u_2, u_3) = au_1 + bu_2 + cu_3 = 0$ and $(a,b,c)\cdot(v_1, v_2, v_3) = av_1 + bv_2 + cv_3 = 0$. At least one of $u_2v_1 - u_1v_2$, $u_3v_1 - u_1v_3$, and $u_3v_2 - u_2v_3$ is nonzero. (Otherwise $\mathbf{v} = k\mathbf{u}$.) We assume $u_2v_1 - u_1v_2$ is nonzero and solve for a and b in terms of c. $a = \dfrac{c(u_3v_2 - u_2v_3)}{u_2v_1 - u_1v_2}$ and $b = \dfrac{c(u_1v_3 - u_3v_1)}{u_2v_1 - u_1v_2}$, so $(a,b,c) = c\left(\dfrac{u_3v_2 - u_2v_3}{u_2v_1 - u_1v_2}, \dfrac{u_1v_3 - u_3v_1}{u_2v_1 - u_1v_2}, 1\right)$.

So (a,b,c) is orthogonal to both $\mathbf{u}$ and $\mathbf{v}$ if and only if (a,b,c) is in the one-dimensional vector space with basis vector $\left(\dfrac{u_3v_2 - u_2v_3}{u_2v_1 - u_1v_2}, \dfrac{u_1v_3 - u_3v_1}{u_2v_1 - u_1v_2}, 1\right)$.

31. The two-dimensional subspace with basis $(1,0,0)$ and $(0,1,0)$ is orthogonal to the one-dimensional subspace with basis $(0,0,1)$.

Exercise Set 5.6

1. (a) The row vectors are linearly independent, so the rank of the matrix is 2.

   (b) The row vectors are multiples of one another, so the rank of the matrix is 1.

   (c) The row vectors are linearly independent, so the rank of the matrix is 2.

   (d) The row vectors are multiples of one another, so the rank of the matrix is 1.

2. (a) The first row is in the space spanned by the other two linearly independent rows. The rank of the matrix is 2.

(b) The rows are the standard basis for $\mathbf{R}^3$, so the rank of the matrix is 3.

(c) Rows 2 and 3 are multiples of row 1, so the rank of the matrix is 1.

(d) The rows are linearly independent, so the rank of the matrix is 3.

(e) The first row is in the space spanned by the other two linearly independent rows. The rank of the matrix is 2.

(f) The first row is in the space spanned by the other two linearly independent rows. The rank of the matrix is 2.

3. (a) $\begin{bmatrix} 1 & 2 & -1 \\ 2 & 5 & 2 \\ 0 & 2 & 9 \end{bmatrix} \approx \begin{bmatrix} 1 & 2 & -1 \\ 0 & 1 & 4 \\ 0 & 2 & 9 \end{bmatrix} \approx \begin{bmatrix} 1 & 0 & -9 \\ 0 & 1 & 4 \\ 0 & 0 & 1 \end{bmatrix} \approx \begin{bmatrix} 1 & 0 & 0 \\ 0 & 1 & 0 \\ 0 & 0 & 1 \end{bmatrix}$, so the vectors (1,0,0), (0,1,0), and (0,0,1) are a basis for the row space and the rank of the matrix is 3.

(b) $\begin{bmatrix} 1 & 1 & 8 \\ 0 & 1 & 3 \\ -1 & 1 & -2 \end{bmatrix} \approx \begin{bmatrix} 1 & 1 & 8 \\ 0 & 1 & 3 \\ 0 & 2 & 6 \end{bmatrix} \approx \begin{bmatrix} 1 & 0 & 5 \\ 0 & 1 & 3 \\ 0 & 0 & 0 \end{bmatrix}$, so the vectors (1,0,5) and (0,1,3) are a basis for the row space and the rank of the matrix is 2.

(c) $\begin{bmatrix} 1 & -3 & 2 \\ -2 & 6 & -4 \\ -1 & 3 & -2 \end{bmatrix} \approx \begin{bmatrix} 1 & -3 & 2 \\ 0 & 0 & 0 \\ 0 & 0 & 0 \end{bmatrix}$, so the vector (1,-3,2) is a basis for the row space and the rank of the matrix is 1.

4. (a) $\begin{bmatrix} 1 & 4 & 0 \\ -1 & -3 & 3 \\ 2 & 9 & 5 \end{bmatrix} \approx \begin{bmatrix} 1 & 4 & 0 \\ 0 & 1 & 3 \\ 0 & 1 & 5 \end{bmatrix} \approx \begin{bmatrix} 1 & 0 & -12 \\ 0 & 1 & 3 \\ 0 & 0 & 2 \end{bmatrix} \approx \begin{bmatrix} 1 & 0 & 0 \\ 0 & 1 & 0 \\ 0 & 0 & 1 \end{bmatrix}$, so the vectors (1,0,0), (0,1,0), and (0,0,1) are a basis for the row space and the rank of the matrix is 3.

(b) $\begin{bmatrix} 1 & 2 & 0 \\ 0 & 1 & 1 \\ -1 & 2 & 3 \end{bmatrix} \approx \begin{bmatrix} 1 & 2 & 0 \\ 0 & 1 & 1 \\ 0 & 4 & 3 \end{bmatrix} \approx \begin{bmatrix} 1 & 0 & -2 \\ 0 & 1 & 1 \\ 0 & 0 & -1 \end{bmatrix} \approx \begin{bmatrix} 1 & 0 & 0 \\ 0 & 1 & 0 \\ 0 & 0 & 1 \end{bmatrix}$, so the vectors (1,0,0),

(0,1,0), and (0,0,1) are a basis for the row space and the rank of the matrix is 3.

(c) $\begin{bmatrix} 1 & 2 & 3 \\ 0 & -1 & -1 \\ 3 & 4 & 7 \end{bmatrix} \approx \begin{bmatrix} 1 & 2 & 3 \\ 0 & 1 & 1 \\ 0 & -2 & -2 \end{bmatrix} \approx \begin{bmatrix} 1 & 0 & 1 \\ 0 & 1 & 1 \\ 0 & 0 & 0 \end{bmatrix}$, so the vectors (1,0,1) and (0,1,1) are a basis for the row space and the rank of the matrix is 2.

5. (a) $\begin{bmatrix} 1 & 2 & 3 & 4 \\ -1 & 2 & 0 & 1 \\ 0 & 1 & 0 & 2 \end{bmatrix} \approx \begin{bmatrix} 1 & 2 & 3 & 4 \\ 0 & 4 & 3 & 5 \\ 0 & 1 & 0 & 2 \end{bmatrix} \approx \begin{bmatrix} 1 & 2 & 3 & 4 \\ 0 & 1 & .75 & 1.25 \\ 0 & 1 & 0 & 2 \end{bmatrix} \approx \begin{bmatrix} 1 & 0 & 1.5 & 1.5 \\ 0 & 1 & .75 & 1.25 \\ 0 & 0 & -.75 & .75 \end{bmatrix} \approx$

$\begin{bmatrix} 1 & 0 & 1.5 & 1.5 \\ 0 & 1 & .75 & 1.25 \\ 0 & 0 & 1 & -1 \end{bmatrix} \approx \begin{bmatrix} 1 & 0 & 0 & 3 \\ 0 & 1 & 0 & 2 \\ 0 & 0 & 1 & -1 \end{bmatrix}$, so the vectors (1,0,0,3), (0,1,0,2), and

(0,0,1,–1) are a basis for the row space and the rank of the matrix is 3.

(b) $\begin{bmatrix} 1 & 2 & -1 & 4 \\ 0 & 1 & -2 & 3 \\ -1 & 0 & -3 & 2 \end{bmatrix} \approx \begin{bmatrix} 1 & 2 & -1 & 4 \\ 0 & 1 & -2 & 3 \\ 0 & 2 & -4 & 6 \end{bmatrix} \approx \begin{bmatrix} 1 & 0 & 3 & -2 \\ 0 & 1 & -2 & 3 \\ 0 & 0 & 0 & 0 \end{bmatrix}$, so the vectors (1,0,3,-2) and (0,1,-2,3) are a basis for the row space and the rank of the matrix is 2.

(c) $\begin{bmatrix} 1 & 1 & 0 & -1 \\ 2 & 1 & 0 & 0 \\ 3 & 2 & 0 & -1 \\ -1 & 0 & 1 & 1 \end{bmatrix} \approx \begin{bmatrix} 1 & 1 & 0 & -1 \\ 0 & -1 & 0 & 2 \\ 0 & -1 & 0 & 2 \\ 0 & 1 & 1 & 0 \end{bmatrix} \approx \begin{bmatrix} 1 & 1 & 0 & -1 \\ 0 & 1 & 0 & -2 \\ 0 & -1 & 0 & 2 \\ 0 & 1 & 1 & 0 \end{bmatrix} \approx \begin{bmatrix} 1 & 0 & 0 & 1 \\ 0 & 1 & 0 & -2 \\ 0 & 0 & 0 & 0 \\ 0 & 0 & 1 & 2 \end{bmatrix} \approx$

$\begin{bmatrix} 1 & 0 & 0 & 1 \\ 0 & 1 & 0 & -2 \\ 0 & 0 & 1 & 2 \\ 0 & 0 & 0 & 0 \end{bmatrix}$, so the vectors (1,0,0,1), (0,1,0,–2), and (0,0,1,2) are a basis for the

row space and the rank of the matrix is 3.

6. (a) The given vectors are linearly independent so they are a basis for the space they

Section 5.6

span, which is $\mathbf{R}^3$. Also $\begin{bmatrix} 1 & 3 & 2 \\ 0 & 1 & 4 \\ 1 & 4 & 9 \end{bmatrix} \approx \begin{bmatrix} 1 & 3 & 2 \\ 0 & 1 & 4 \\ 0 & 1 & 7 \end{bmatrix} \approx \begin{bmatrix} 1 & 0 & -10 \\ 0 & 1 & 4 \\ 0 & 0 & 3 \end{bmatrix} \approx \begin{bmatrix} 1 & 0 & 0 \\ 0 & 1 & 0 \\ 0 & 0 & 1 \end{bmatrix}$, so the row vectors of any of these matrices are a basis for the same vector space.

(b) The given vectors are linearly independent so they are a basis for the space they span, which is $\mathbf{R}^3$. Also $\begin{bmatrix} 1 & -2 & 5 \\ 1 & -1 & 4 \\ 2 & -5 & 14 \end{bmatrix} \approx \begin{bmatrix} 1 & -2 & 5 \\ 0 & 1 & -1 \\ 0 & -1 & 4 \end{bmatrix} \approx \begin{bmatrix} 1 & 0 & 3 \\ 0 & 1 & -1 \\ 0 & 0 & 1 \end{bmatrix} \approx \begin{bmatrix} 1 & 0 & 0 \\ 0 & 1 & 0 \\ 0 & 0 & 1 \end{bmatrix}$,

so the row vectors of any of these matrices are a basis for the same vector space.

(c) $\begin{bmatrix} 1 & -1 & 3 \\ 1 & 0 & 1 \\ -2 & 1 & -4 \end{bmatrix} \approx \begin{bmatrix} 1 & -1 & 3 \\ 0 & 1 & -2 \\ 0 & -1 & 2 \end{bmatrix} \approx \begin{bmatrix} 1 & 0 & 1 \\ 0 & 1 & -2 \\ 0 & 0 & 0 \end{bmatrix}$, so the vectors (1,0,1) and (0,1,-2) are a basis for the space spanned by the given vectors.

(d) All three of the given vectors are multiples of the vector (1,-3,2), so it is a basis for the vector space.

7. (a) $\begin{bmatrix} 1 & 3 & -1 & 4 \\ 1 & 3 & 0 & 6 \\ -1 & -3 & 0 & -8 \end{bmatrix} \approx \begin{bmatrix} 1 & 3 & -1 & 4 \\ 0 & 0 & 1 & 2 \\ 0 & 0 & -1 & -4 \end{bmatrix} \approx \begin{bmatrix} 1 & 3 & 0 & 6 \\ 0 & 0 & 1 & 2 \\ 0 & 0 & 0 & 0 \end{bmatrix} \approx \begin{bmatrix} 1 & 3 & 0 & 0 \\ 0 & 0 & 1 & 0 \\ 0 & 0 & 0 & 1 \end{bmatrix}$, so the

vectors (1,3,0,0), (0,0,1,0), and (0,0,0,1) are a basis for the subspace. The given vectors are linearly independent so they are a basis also.

(b) $\begin{bmatrix} 1 & 2 & 0 & 1 \\ -1 & -1 & 3 & 1 \\ 2 & 3 & -2 & 4 \end{bmatrix} \approx \begin{bmatrix} 1 & 2 & 0 & 1 \\ 0 & 1 & 3 & 2 \\ 0 & -1 & -2 & 2 \end{bmatrix} \approx \begin{bmatrix} 1 & 0 & -6 & -3 \\ 0 & 1 & 3 & 2 \\ 0 & 0 & 1 & 4 \end{bmatrix} \approx \begin{bmatrix} 1 & 0 & 0 & 21 \\ 0 & 1 & 0 & -10 \\ 0 & 0 & 1 & 4 \end{bmatrix}$, so the

vectors (1,0,0,21), (0,1,0,-10), and (0,0,1,4) are a basis for the subspace. The given vectors are linearly independent so they are a basis also.

(c) $\begin{bmatrix} 1 & 2 & 3 & 4 \\ 0 & -1 & 2 & 3 \\ 2 & 3 & 8 & 11 \\ 2 & 3 & 6 & 8 \end{bmatrix} \approx \begin{bmatrix} 1 & 2 & 3 & 4 \\ 0 & 1 & -2 & -3 \\ 0 & -1 & 2 & 3 \\ 0 & -1 & 0 & 0 \end{bmatrix} \approx \begin{bmatrix} 1 & 0 & 7 & 10 \\ 0 & 1 & -2 & -3 \\ 0 & 0 & 0 & 0 \\ 0 & 0 & -2 & -3 \end{bmatrix} \approx \begin{bmatrix} 1 & 0 & 7 & 10 \\ 0 & 1 & -2 & -3 \\ 0 & 0 & 1 & 1.5 \\ 0 & 0 & 0 & 0 \end{bmatrix}$

$\approx \begin{bmatrix} 1 & 0 & 0 & -.5 \\ 0 & 1 & 0 & 0 \\ 0 & 0 & 1 & 1.5 \\ 0 & 0 & 0 & 0 \end{bmatrix}$, so the vectors (1,0,0,-.5), (0,1,0,0), and (0,0,1,1.5) are a basis for the subspace.

(d) $\begin{bmatrix} 1 & -3 & -1 & 2 \\ 0 & 1 & -4 & 1 \\ 1 & -4 & 5 & 1 \\ 2 & -5 & -6 & 5 \end{bmatrix} \approx \begin{bmatrix} 1 & -3 & -1 & 2 \\ 0 & 1 & -4 & 1 \\ 0 & -1 & 6 & -1 \\ 0 & 1 & -4 & 1 \end{bmatrix} \approx \begin{bmatrix} 1 & 0 & -13 & 5 \\ 0 & 1 & -4 & 1 \\ 0 & 0 & 2 & 0 \\ 0 & 0 & 0 & 0 \end{bmatrix} \approx \begin{bmatrix} 1 & 0 & 0 & 5 \\ 0 & 1 & 0 & 1 \\ 0 & 0 & 1 & 0 \\ 0 & 0 & 0 & 0 \end{bmatrix}$, so the

vectors (1,0,0,5), (0,1,0,1), and (0,0,1,0) are a basis for the subspace.

8. $A = \begin{bmatrix} 1 & 2 & -1 \\ 0 & 1 & 3 \\ 1 & 4 & 6 \end{bmatrix} \approx \begin{bmatrix} 1 & 2 & -1 \\ 0 & 1 & 3 \\ 0 & 2 & 7 \end{bmatrix} \approx \begin{bmatrix} 1 & 0 & -7 \\ 0 & 1 & 3 \\ 0 & 0 & 1 \end{bmatrix} \approx \begin{bmatrix} 1 & 0 & 0 \\ 0 & 1 & 0 \\ 0 & 0 & 1 \end{bmatrix}$, so the vectors (1,0,0),

(0,1,0), and (0,0,1) are a basis for the row space of A.

$A^t = \begin{bmatrix} 1 & 0 & 1 \\ 2 & 1 & 4 \\ -1 & 3 & 6 \end{bmatrix} \approx \begin{bmatrix} 1 & 0 & 1 \\ 0 & 1 & 2 \\ 0 & 3 & 7 \end{bmatrix} \approx \begin{bmatrix} 1 & 0 & 1 \\ 0 & 1 & 2 \\ 0 & 0 & 1 \end{bmatrix} \approx \begin{bmatrix} 1 & 0 & 0 \\ 0 & 1 & 0 \\ 0 & 0 & 1 \end{bmatrix}$, so the vectors (1,0,0), (0,1,0),

and (0,0,1) are a basis for the row space of A^t. Therefore the column vectors

$\begin{bmatrix} 1 \\ 0 \\ 0 \end{bmatrix}, \begin{bmatrix} 0 \\ 1 \\ 0 \end{bmatrix}$, and $\begin{bmatrix} 0 \\ 0 \\ 1 \end{bmatrix}$ are a basis for the column space of A. Both the row space and

the column space of A have dimension 3.

9. $A = \begin{bmatrix} 1 & 3 & 2 \\ 1 & 4 & 1 \\ 2 & 5 & 5 \end{bmatrix} \approx \begin{bmatrix} 1 & 3 & 2 \\ 0 & 1 & -1 \\ 0 & -1 & 1 \end{bmatrix} \approx \begin{bmatrix} 1 & 0 & 5 \\ 0 & 1 & -1 \\ 0 & 0 & 0 \end{bmatrix}$, so the vectors (1,0,5) and (0,1,-1)

181

are a basis for the row space of A.

$$A^t = \begin{bmatrix} 1 & 1 & 2 \\ 3 & 4 & 5 \\ 2 & 1 & 5 \end{bmatrix} \approx \begin{bmatrix} 1 & 1 & 2 \\ 0 & 1 & -1 \\ 0 & -1 & 1 \end{bmatrix} \approx \begin{bmatrix} 1 & 0 & 3 \\ 0 & 1 & -1 \\ 0 & 0 & 0 \end{bmatrix}$$, so the vectors (1,0,3) and (0,1,−1)

are a basis for the row space of A^t. Therefore the column vectors

$$\begin{bmatrix} 1 \\ 0 \\ 3 \end{bmatrix} \text{ and } \begin{bmatrix} 0 \\ 1 \\ -1 \end{bmatrix}$$ are a basis for the column space of A. Both the row space and the

column space of A have dimension 2.

10. (a) There are only two rows, so the row space must have dimension less than or equal to 2; i.e., rank(A) ≤ 2.

(b) The largest possible rank is the smaller of the two numbers m and n.

11. dim(column space of A) = dim(row space of A) ≤ number of rows of A = m < n

12. (a) The rank of A cannot be greater than 3, so a basis for the column space of A can contain no more than three vectors. The four column vectors in A must therefore be linearly dependent by Theorem 5.7.

(b) The rank of A cannot be greater than 3, so a basis for the row space of A can contain no more than three vectors. The seven row vectors in A must therefore be linearly dependent by Theorem 5.7.

(c) The rank of A cannot be greater than m, so a basis for the column space of A can contain no more than m vectors. The n column vectors in A must therefore be linearly dependent by Theorem 5.7.

13. (a) If |A| ≠ 0, then A is row equivalent to I_n. So rank(A) = dim(row space of A) = n. The n row vectors of A span the row space of A, so by Theorem 5.11 they are a basis and are linearly independent. Thus the n row vectors of A are n linearly independent vectors in $\mathbf{R}^n$. They are therefore a basis for $\mathbf{R}^n$.

(b) A is invertible if and only if |A| ≠ 0, so the proof is the same as in part (a).

Section 5.7

14. If the n columns of A are linearly independent, they are a basis for the column space of A and dim(column space of A) is n, so rank(A) = n.

    If rank(A) = n, then the n column vectors of A are basis for the column space of A because they span the column space of A (Theorem 5.11), but this means they are linearly independent.

15. If the rows of A are linearly independent, then dim(row space of A) = rank(A) = n. Thus the n columns of A are a basis for the column space of A. But the columns are linearly independent vectors in $\mathbf{R}^n$, which has dimension n. Thus by Theorem 4.11 they are a basis for $\mathbf{R}^n$.

    If the columns of A span $\mathbf{R}^n$, then they are linearly independent and rank(A) = n = dim(row space of A) so the n rows of A are linearly independent.

16. Each row in A + B is a sum of a row in A and a row in B. Each row in A is a linear combination of rank(A) basis vectors and each row in B is a linear combination of rank(B) basis vectors. Thus the row space of A + B is spanned by rank(A) + rank(B) vectors. A basis for the row space of A + B therefore consists of no more than rank(A) + rank(B) vectors, so rank(A + B) ≤ rank(A) + rank(B).

Exercise Set 5.7

1. (a) (1,2)·(2,−1) = 0, so the vectors are orthogonal.

    (b) (3,−1)·(0,5) = −5, so the vectors are not orthogonal.

    (c) (0,−2)·(3,0) = 0, so the vectors are orthogonal.

    (d) (4,1)·(2,−3) = 5, so the vectors are not orthogonal.

    (e) (−3,2)·(2,3) = 0, so the vectors are orthogonal.

2. (a) (1,2,1)·(4,−2,0) = 0, (1,2,1)·(2,4,−10) = 0, and (4,−2,0)·(2,4,−10) = 0, so the set of vectors is orthogonal.

Section 5.7

(b) $(3,-1,1) \cdot (2,0,1) = 7$, so the set of vectors is not orthogonal.

(c) $(-2,-1,3) \cdot (6,-1,-1) = -14$, so the set of vectors is not orthogonal.

(d) $(1,2,-1,1) \cdot (3,1,4,-1) = 0$, $(1,2,-1,1) \cdot (0,1,-1,-3) = 0$, and $(3,1,4,-1) \cdot (0,1,-1,-3) = 0$, so the set of vectors is orthogonal.

3. (a) $\left(\frac{1}{3},\frac{2}{3},\frac{2}{3}\right) \cdot \left(\frac{2}{3},-\frac{2}{3},\frac{1}{3}\right) = 0$, $\left(\frac{1}{3},\frac{2}{3},\frac{2}{3}\right) \cdot \left(\frac{2}{3},\frac{1}{3},-\frac{2}{3}\right) = 0$, and

$\left(\frac{2}{3},-\frac{2}{3},\frac{1}{3}\right) \cdot \left(\frac{2}{3},\frac{1}{3},-\frac{2}{3}\right) = 0$, so the set of vectors is orthogonal.

$\left(\frac{1}{3},\frac{2}{3},\frac{2}{3}\right) \cdot \left(\frac{1}{3},\frac{2}{3},\frac{2}{3}\right) = 1$, $\left(\frac{2}{3},-\frac{2}{3},\frac{1}{3}\right) \cdot \left(\frac{2}{3},-\frac{2}{3},\frac{1}{3}\right) = 1$, and

$\left(\frac{2}{3},\frac{1}{3},-\frac{2}{3}\right) \cdot \left(\frac{2}{3},\frac{1}{3},-\frac{2}{3}\right) = 1$, so all the vectors are unit vectors. Thus the vectors are an orthonormal set.

(b) $\left(\frac{1}{\sqrt{10}},\frac{3}{\sqrt{10}}\right) \cdot \left(\frac{-3}{\sqrt{10}},\frac{1}{\sqrt{10}}\right) = 0$, so the vectors are orthogonal.

$\left(\frac{1}{\sqrt{10}},\frac{3}{\sqrt{10}}\right) \cdot \left(\frac{1}{\sqrt{10}},\frac{3}{\sqrt{10}}\right) = 1$ and $\left(\frac{-3}{\sqrt{10}},\frac{1}{\sqrt{10}}\right) \cdot \left(\frac{-3}{\sqrt{10}},\frac{1}{\sqrt{10}}\right) = 1$,

so the vectors are unit vectors. Thus the vectors are an orthonormal set.

(c) $\left(\frac{1}{\sqrt{2}},0,\frac{1}{\sqrt{2}}\right) \cdot \left(\frac{1}{\sqrt{2}},0,\frac{-1}{\sqrt{2}}\right) = 0$, $\left(\frac{1}{\sqrt{2}},0,\frac{1}{\sqrt{2}}\right) \cdot (0,1,0) = 0$, and

$\left(\frac{1}{\sqrt{2}},0,\frac{-1}{\sqrt{2}}\right) \cdot (0,1,0) = 0$, so the set of vectors is orthogonal.

$\left(\frac{1}{\sqrt{2}},0,\frac{1}{\sqrt{2}}\right) \cdot \left(\frac{1}{\sqrt{2}},0,\frac{1}{\sqrt{2}}\right) = 1$, $\left(\frac{1}{\sqrt{2}},0,\frac{-1}{\sqrt{2}}\right) \cdot \left(\frac{1}{\sqrt{2}},0,\frac{-1}{\sqrt{2}}\right) = 1$, and

$(0,1,0) \cdot (0,1,0) = 1$, so the vectors are unit vectors. Thus the vectors are an orthonormal set.

(d) $\left(\frac{1}{\sqrt{6}},\frac{-1}{\sqrt{6}},\frac{2}{\sqrt{6}}\right) \cdot \left(0,\frac{2}{\sqrt{5}},\frac{1}{\sqrt{5}}\right) = 0$, $\left(\frac{1}{\sqrt{6}},\frac{-1}{\sqrt{6}},\frac{2}{\sqrt{6}}\right) \cdot \left(\frac{5}{\sqrt{30}},\frac{1}{\sqrt{30}},\frac{-2}{\sqrt{30}}\right) = 0$, and

Section 5.7

$\left(0, \frac{2}{\sqrt{5}}, \frac{1}{\sqrt{5}}\right) \cdot \left(\frac{5}{\sqrt{30}}, \frac{1}{\sqrt{30}}, \frac{-2}{\sqrt{30}}\right) = 0$, so the set of vectors is orthogonal.

$\left(\frac{1}{\sqrt{6}}, \frac{-1}{\sqrt{6}}, \frac{2}{\sqrt{6}}\right) \cdot \left(\frac{1}{\sqrt{6}}, \frac{-1}{\sqrt{6}}, \frac{2}{\sqrt{6}}\right) = 1$, $\left(0, \frac{2}{\sqrt{5}}, \frac{1}{\sqrt{5}}\right) \cdot \left(0, \frac{2}{\sqrt{5}}, \frac{1}{\sqrt{5}}\right) = 1$, and

$\left(\frac{5}{\sqrt{30}}, \frac{1}{\sqrt{30}}, \frac{-2}{\sqrt{30}}\right) \cdot \left(\frac{5}{\sqrt{30}}, \frac{1}{\sqrt{30}}, \frac{-2}{\sqrt{30}}\right) = 1$, so the vectors are unit vectors. Thus the set of vectors is an orthonormal set.

(e) $\left(\frac{1}{\sqrt{32}}, \frac{-2}{\sqrt{32}}, \frac{5}{\sqrt{32}}\right) \cdot \left(\frac{1}{\sqrt{32}}, \frac{-2}{\sqrt{32}}, \frac{5}{\sqrt{32}}\right) = \frac{30}{32} \ne 1$, so $\left(\frac{1}{\sqrt{32}}, \frac{-2}{\sqrt{32}}, \frac{5}{\sqrt{32}}\right)$ is not a unit vector. Therefore the set is not orthonormal.

4. $\mathbf{v} = (\mathbf{v} \cdot \mathbf{u}_1)\mathbf{u}_1 + (\mathbf{v} \cdot \mathbf{u}_2)\mathbf{u}_2 + (\mathbf{v} \cdot \mathbf{u}_3)\mathbf{u}_3$.

$\mathbf{v} \cdot \mathbf{u}_1 = (2,-3,1) \cdot (0,-1,0) = 3$, $\mathbf{v} \cdot \mathbf{u}_2 = (2,-3,1) \cdot \left(\frac{3}{5}, 0, \frac{-4}{5}\right) = \frac{2}{5}$, and

$\mathbf{v} \cdot \mathbf{u}_3 = (2,-3,1) \cdot \left(\frac{4}{5}, 0, \frac{3}{5}\right) = \frac{11}{5}$, so $\mathbf{v} = 3\mathbf{u}_1 + \frac{2}{5}\mathbf{u}_2 + \frac{11}{5}\mathbf{u}_3$.

5. $\mathbf{v} = (\mathbf{v} \cdot \mathbf{u}_1)\mathbf{u}_1 + (\mathbf{v} \cdot \mathbf{u}_2)\mathbf{u}_2 + (\mathbf{v} \cdot \mathbf{u}_3)\mathbf{u}_3$.

$\mathbf{v} \cdot \mathbf{u}_1 = (7,5,-1) \cdot (1,0,0) = 7$, $\mathbf{v} \cdot \mathbf{u}_2 = (7,5,-1) \cdot \left(0, \frac{1}{\sqrt{2}}, \frac{1}{\sqrt{2}}\right) = \frac{4}{\sqrt{2}} = 2\sqrt{2}$, and

$\mathbf{v} \cdot \mathbf{u}_3 = (7,5,-1) \cdot \left(0, \frac{1}{\sqrt{2}}, \frac{-1}{\sqrt{2}}\right) = \frac{6}{\sqrt{2}} = 3\sqrt{2}$, so $\mathbf{v} = 7\mathbf{u}_1 + 2\sqrt{2}\,\mathbf{u}_2 + 3\sqrt{2}\,\mathbf{u}_3$.

6. $\text{proj}_\mathbf{u}\mathbf{v} = \frac{\mathbf{v} \cdot \mathbf{u}}{\mathbf{u} \cdot \mathbf{u}}\,\mathbf{u}$.

(a) $\text{proj}_\mathbf{u}\mathbf{v} = \frac{(7,4) \cdot (1,2)}{(1,2) \cdot (1,2)}(1,2) = 3(1,2) = (3,6)$.

(b) $\text{proj}_\mathbf{u}\mathbf{v} = \frac{(-1,5) \cdot (3,-2)}{(3,-2) \cdot (3,-2)}(3,-2) = -(3,-2) = (-3,2)$.

(c) $\text{proj}_\mathbf{u}\mathbf{v} = \frac{(4,6,4) \cdot (1,2,3)}{(1,2,3) \cdot (1,2,3)}(1,2,3) = 2(1,2,3) = (2,4,6)$.

(d) $\text{proj}_u v = \dfrac{(6,-8,7)\cdot(-1,3,0)}{(-1,3,0)\cdot(-1,3,0)}(-1,3,0) = -3(-1,3,0) = (3,-9,0).$

(e) $\text{proj}_u v = \dfrac{(1,2,3,0)\cdot(1,-1,2,3)}{(1,-1,2,3)\cdot(1,-1,2,3)}(1,-1,2,3) = \dfrac{1}{3}(1,-1,2,3) = \left(\dfrac{1}{3},\dfrac{-1}{3},\dfrac{2}{3},1\right).$

7. (a) $\text{proj}_u v = \dfrac{(1,2)\cdot(2,5)}{(2,5)\cdot(2,5)}(2,5) = \dfrac{12}{29}(2,5) = \left(\dfrac{24}{29},\dfrac{60}{29}\right).$

(b) $\text{proj}_u v = \dfrac{(-1,3)\cdot(2,4)}{(2,4)\cdot(2,4)}(2,4) = \dfrac{1}{2}(2,4) = (1,2).$

(c) $\text{proj}_u v = \dfrac{(1,2,3)\cdot(1,2,0)}{(1,2,0)\cdot(1,2,0)}(1,2,0) = (1,2,0).$

(d) $\text{proj}_u v = \dfrac{(2,1,4)\cdot(-1,-3,2)}{(-1,-3,2)\cdot(-1,-3,2)}(-1,-3,2) = \dfrac{3}{14}(-1,-3,2) = \left(\dfrac{-3}{14},\dfrac{-9}{14},\dfrac{6}{14}\right).$

(e) $\text{proj}_u v = \dfrac{(2,-1,3,1)\cdot(-1,2,1,3)}{(-1,2,1,3)\cdot(-1,2,1,3)}(-1,2,1,3) = \dfrac{2}{15}(-1,2,1,3) = \left(\dfrac{-2}{15},\dfrac{4}{15},\dfrac{2}{15},\dfrac{2}{5}\right).$

8. (a) $u_1 = (1,2)$, $u_2 = (-1,3) - \dfrac{(-1,3)\cdot(1,2)}{(1,2)\cdot(1,2)}(1,2) = (-1,3) - (1,2) = (-2,1)$ is an orthogonal basis for $\mathbf{R}^2$. $\|u_1\| = \sqrt{5}$ and $\|u_2\| = \sqrt{5}$, so the set $\left\{\left(\dfrac{1}{\sqrt{5}},\dfrac{2}{\sqrt{5}}\right),\left(\dfrac{-2}{\sqrt{5}},\dfrac{1}{\sqrt{5}}\right)\right\}$ is an orthonormal basis for $\mathbf{R}^2$.

(b) $u_1 = (1,1)$, $u_2 = (6,2) - \dfrac{(6,2)\cdot(1,1)}{(1,1)\cdot(1,1)}(1,1) = (6,2) - \dfrac{8}{2}(1,1) = (2,-2)$ is an orthogonal basis for $\mathbf{R}^2$. $\|u_1\| = \sqrt{2}$ and $\|u_2\| = \sqrt{8} = 2\sqrt{2}$, so the set $\left\{\left(\dfrac{1}{\sqrt{2}},\dfrac{1}{\sqrt{2}}\right),\left(\dfrac{1}{\sqrt{2}},\dfrac{-1}{\sqrt{2}}\right)\right\}$ is an orthonormal basis for $\mathbf{R}^2$.

(c) $u_1 = (1,-1)$, $u_2 = (4,-2) - \dfrac{(4,-2)\cdot(1,-1)}{(1,-1)\cdot(1,-1)}(1,-1) = (1,1)$ is an orthogonal

Section 5.7

basis for $\mathbf{R}^2$. $\|\mathbf{u}_1\| = \sqrt{2}$ and $\|\mathbf{u}_2\| = \sqrt{2}$, so the set $\left\{ \left(\frac{1}{\sqrt{2}}, \frac{-1}{\sqrt{2}} \right), \left(\frac{1}{\sqrt{2}}, \frac{1}{\sqrt{2}} \right) \right\}$ is an orthonormal basis for $\mathbf{R}^2$.

9. (a) $\mathbf{u}_1 = (1,1,1)$, $\mathbf{u}_2 = (2,0,1) - \frac{(2,0,1) \cdot (1,1,1)}{(1,1,1) \cdot (1,1,1)} (1,1,1) = (1,-1,0)$,

$\mathbf{u}_3 = (2,4,5) - \frac{(2,4,5) \cdot (1,1,1)}{(1,1,1) \cdot (1,1,1)} (1,1,1) - \frac{(2,4,5) \cdot (1,-1,0)}{(1,-1,0) \cdot (1,-1,0)} (1,-1,0)$

$= (2,4,5) - \frac{11}{3}(1,1,1) - (-1)(1,-1,0) = \left(\frac{-2}{3}, \frac{-2}{3}, \frac{4}{3} \right)$ is an orthogonal basis for

$\mathbf{R}^3$. $\|\mathbf{u}_1\| = \sqrt{3}$, $\|\mathbf{u}_2\| = \sqrt{2}$, and $\|\mathbf{u}_3\| = \frac{\sqrt{24}}{3} = \frac{2\sqrt{6}}{3}$, so the set

$\left\{ \left(\frac{1}{\sqrt{3}}, \frac{1}{\sqrt{3}}, \frac{1}{\sqrt{3}} \right), \left(\frac{1}{\sqrt{2}}, \frac{-1}{\sqrt{2}}, 0 \right), \left(\frac{-1}{\sqrt{6}}, \frac{-1}{\sqrt{6}}, \frac{2}{\sqrt{6}} \right) \right\}$ is an orthonormal basis for $\mathbf{R}^3$.

(b) $\mathbf{u}_1 = (3,2,0)$, $\mathbf{u}_2 = (1,5,-1) - \frac{(1,5,-1) \cdot (3,2,0)}{(3,2,0) \cdot (3,2,0)} (3,2,0) = (-2,3,-1)$,

$\mathbf{u}_3 = (5,-1,2) - \frac{(5,-1,2) \cdot (3,2,0)}{(3,2,0) \cdot (3,2,0)} (3,2,0) - \frac{(5,-1,2) \cdot (-2,3,-1)}{(-2,3,-1) \cdot (-2,3,-1)} (-2,3,-1)$

$= (5,-1,2) - (3,2,0) - \left(\frac{-15}{14} \right) (-2,3,-1) = \left(\frac{-2}{14}, \frac{3}{14}, \frac{13}{14} \right)$ is an orthogonal basis

for $\mathbf{R}^3$. $\|\mathbf{u}_1\| = \sqrt{13}$, $\|\mathbf{u}_2\| = \sqrt{14}$, and $\|\mathbf{u}_3\| = \frac{\sqrt{182}}{14}$, so the set

$\left\{ \left(\frac{3}{\sqrt{13}}, \frac{2}{\sqrt{13}}, 0 \right), \left(\frac{-2}{\sqrt{14}}, \frac{3}{\sqrt{14}}, \frac{-1}{\sqrt{14}} \right), \left(\frac{-2}{\sqrt{182}}, \frac{3}{\sqrt{182}}, \frac{13}{\sqrt{182}} \right) \right\}$

is an orthonormal basis for $\mathbf{R}^3$.

10. (a) $\mathbf{u}_1 = (1,0,2)$, $\mathbf{u}_2 = (-1,0,1) - \frac{(-1,0,1) \cdot (1,0,2)}{(1,0,2) \cdot (1,0,2)} (1,0,2) = (-1,0,1) - \frac{1}{5}(1,0,2)$

$= \left(\frac{-6}{5}, 0, \frac{3}{5} \right)$ is an orthogonal basis for the subspace of $\mathbf{R}^3$. $\|\mathbf{u}_1\| = \sqrt{5}$ and

187

Section 5.7

$\|u_2\| = \frac{3}{5}\sqrt{5}$, so the set $\left\{\left(\frac{1}{\sqrt{5}}, 0, \frac{2}{\sqrt{5}}\right), \left(\frac{-2}{\sqrt{5}}, 0, \frac{1}{\sqrt{5}}\right)\right\}$ is an orthonormal basis for the subspace.

(b) $u_1 = (1,-1,1)$, $u_2 = (1,2,-1) - \frac{(1,2,-1)\cdot(1,-1,1)}{(1,-1,1)\cdot(1,-1,1)}(1,-1,1) = (1,2,-1) - \frac{-2}{3}(1,-1,1)$

$= \left(\frac{5}{3}, \frac{4}{3}, \frac{-1}{3}\right)$ is an orthogonal basis for the subspace of $\mathbf{R}^3$. $\|u_1\| = \sqrt{3}$ and

$\|u_2\| = \frac{1}{3}\sqrt{42}$, so the set $\left\{\left(\frac{1}{\sqrt{3}}, \frac{-1}{\sqrt{3}}, \frac{1}{\sqrt{3}}\right), \left(\frac{5}{\sqrt{42}}, \frac{4}{\sqrt{42}}, \frac{-1}{\sqrt{42}}\right)\right\}$ is an

orthonormal basis for the subspace.

11. (a) $u_1 = (1,2,3,4)$, $u_2 = (-1,1,0,1) - \frac{(-1,1,0,1)\cdot(1,2,3,4)}{(1,2,3,4)\cdot(1,2,3,4)}(1,2,3,4)$

$= (-1,1,0,1) - \frac{1}{6}(1,2,3,4) = \left(\frac{-7}{6}, \frac{4}{6}, \frac{-3}{6}, \frac{2}{6}\right)$ is an orthogonal basis for the

subspace of $\mathbf{R}^4$. $\|u_1\| = \sqrt{30}$ and $\|u_2\| = \frac{1}{6}\sqrt{78}$, so the set

$\left\{\left(\frac{1}{\sqrt{30}}, \frac{2}{\sqrt{30}}, \frac{3}{\sqrt{30}}, \frac{4}{\sqrt{30}}\right), \left(\frac{-7}{\sqrt{78}}, \frac{4}{\sqrt{78}}, \frac{-3}{\sqrt{78}}, \frac{2}{\sqrt{78}}\right)\right\}$ is an orthonormal

basis for the subspace.

(b) $u_1 = (3,0,0,0)$, $u_2 = (0,1,2,1)$ (because $u_1 \cdot u_2 = 0$),

$u_3 = (0,-1,3,2) - 0(3,0,0,0) - \frac{(0,-1,3,2)\cdot(0,1,2,1)}{(0,1,2,1)\cdot(0,1,2,1)}(0,1,2,1) = \left(0, \frac{-13}{6}, \frac{4}{6}, \frac{5}{6}\right)$ is an

orthogonal basis for the subspace of $\mathbf{R}^4$. $\|u_1\| = 3$, $\|u_2\| = \sqrt{6}$, and $\|u_3\| = \frac{1}{6}\sqrt{210}$,

so the set $\left\{(1,0,0,0), \left(0, \frac{1}{\sqrt{6}}, \frac{2}{\sqrt{6}}, \frac{1}{\sqrt{6}}\right), \left(0, \frac{-13}{\sqrt{210}}, \frac{4}{\sqrt{210}}, \frac{5}{\sqrt{210}}\right)\right\}$

is an orthonormal basis for the subspace.

12. To actually construct such a vector, start with any vector that is not a multiple of the given vector (1,2,−1,−1) and use the Gram-Schmidt process to find a vector orthogonal to (1,2,−1,−1). The vector (−2,1,0,0) is orthogonal to (1,2,−1,−1).

13. To actually construct such a vector, start with any vector that is not a multiple of the given vector (2,0,1,1) and use the Gram-Schmidt process to find a vector orthogonal to (2,0,1,1). The vector (1,1,−2,0) is orthogonal to (2,0,1,1).

14. First find an orthonormal basis for W. $\mathbf{u}_1 = (0,-1,3)$,

$$\mathbf{u}_2 = (1,1,2) - \frac{(1,1,2)\cdot(0,-1,3)}{(0,-1,3)\cdot(0,-1,3)}(0,-1,3) = (1,1,2) - \frac{1}{2}(0,-1,3) = \left(1, \frac{3}{2}, \frac{1}{2}\right)$$

is an orthogonal basis for the subspace of $\mathbf{R}^3$. $\|\mathbf{u}_1\| = \sqrt{10}$ and $\|\mathbf{u}_2\| = \frac{1}{2}\sqrt{14}$, so the set

$$\left\{\left(0, \frac{-1}{\sqrt{10}}, \frac{3}{\sqrt{10}}\right), \left(\frac{2}{\sqrt{14}}, \frac{3}{\sqrt{14}}, \frac{1}{\sqrt{14}}\right)\right\}$$

is an orthonormal basis for W.

(a) $\text{proj}_W(3,-1,2) = \left((3,-1,2)\cdot\left(0, \frac{-1}{\sqrt{10}}, \frac{3}{\sqrt{10}}\right)\right)\left(0, \frac{-1}{\sqrt{10}}, \frac{3}{\sqrt{10}}\right)$

$+ \left((3,-1,2)\cdot\left(\frac{2}{\sqrt{14}}, \frac{3}{\sqrt{14}}, \frac{1}{\sqrt{14}}\right)\right)\left(\frac{2}{\sqrt{14}}, \frac{3}{\sqrt{14}}, \frac{1}{\sqrt{14}}\right) = \frac{7}{10}(0,-1,3) + \frac{5}{14}(2,3,1)$

$= \left(\frac{5}{7}, \frac{13}{35}, \frac{86}{35}\right).$

(b) $\text{proj}_W(1,1,1) = \left((1,1,1)\cdot\left(0, \frac{-1}{\sqrt{10}}, \frac{3}{\sqrt{10}}\right)\right)\left(0, \frac{-1}{\sqrt{10}}, \frac{3}{\sqrt{10}}\right)$

$+ \left((1,1,1)\cdot\left(\frac{2}{\sqrt{14}}, \frac{3}{\sqrt{14}}, \frac{1}{\sqrt{14}}\right)\right)\left(\frac{2}{\sqrt{14}}, \frac{3}{\sqrt{14}}, \frac{1}{\sqrt{14}}\right) = \frac{1}{5}(0,-1,3) + \frac{3}{7}(2,3,1)$

$= \left(\frac{6}{7}, \frac{38}{35}, \frac{36}{35}\right).$

(c) $\text{proj}_W(4,2,1) = \left((4,2,1)\cdot\left(0, \frac{-1}{\sqrt{10}}, \frac{3}{\sqrt{10}}\right)\right)\left(0, \frac{-1}{\sqrt{10}}, \frac{3}{\sqrt{10}}\right)$

$$+ \left((4,2,1) \cdot \left(\frac{2}{\sqrt{14}}, \frac{3}{\sqrt{14}}, \frac{1}{\sqrt{14}} \right) \right) \left(\frac{2}{\sqrt{14}}, \frac{3}{\sqrt{14}}, \frac{1}{\sqrt{14}} \right) = \frac{1}{10}(0,-1,3) + \frac{15}{14}(2,3,1)$$

$$= \left(\frac{15}{7}, \frac{109}{35}, \frac{48}{35} \right).$$

15. First find an orthonormal basis for V. $\mathbf{u}_1 = (-1,0,2,1)$,

$$\mathbf{u}_2 = (1,-1,0,3) - \frac{(1,-1,0,3)\cdot(-1,0,2,1)}{(-1,0,2,1)\cdot(-1,0,2,1)}(-1,0,2,1) = (1,-1,0,3) - \frac{1}{3}(-1,0,2,1)$$

$= \left(\frac{4}{3}, -1, \frac{-2}{3}, \frac{8}{3} \right)$ is an orthogonal basis for the subspace of $\mathbf{R}^3$. $\|\mathbf{u}_1\| = \sqrt{6}$ and

$\|\mathbf{u}_2\| = \frac{1}{3}\sqrt{93}$, so the set $\left\{ \left(\frac{-1}{\sqrt{6}}, 0, \frac{2}{\sqrt{6}}, \frac{1}{\sqrt{6}} \right), \left(\frac{4}{\sqrt{93}}, \frac{-3}{\sqrt{93}}, \frac{-2}{\sqrt{93}}, \frac{8}{\sqrt{93}} \right) \right\}$

is an orthonormal basis for V.

(a) $\text{proj}_V(1,-1,1,-1) = \left((1,-1,1,-1) \cdot \left(\frac{-1}{\sqrt{6}}, 0, \frac{2}{\sqrt{6}}, \frac{1}{\sqrt{6}} \right) \right) \left(\frac{-1}{\sqrt{6}}, 0, \frac{2}{\sqrt{6}}, \frac{1}{\sqrt{6}} \right)$

$+ \left((1,-1,1,-1) \cdot \left(\frac{4}{\sqrt{93}}, \frac{-3}{\sqrt{93}}, \frac{-2}{\sqrt{93}}, \frac{8}{\sqrt{93}} \right) \right) \left(\frac{4}{\sqrt{93}}, \frac{-3}{\sqrt{93}}, \frac{-2}{\sqrt{93}}, \frac{8}{\sqrt{93}} \right)$

$= 0(-1,0,2,1) + \frac{-1}{31}(4,-3,-2,8) = \left(\frac{-4}{31}, \frac{3}{31}, \frac{2}{31}, \frac{-8}{31} \right).$

(b) $\text{proj}_V(2,0,1,-1) = \left((2,0,1,-1) \cdot \left(\frac{-1}{\sqrt{6}}, 0, \frac{2}{\sqrt{6}}, \frac{1}{\sqrt{6}} \right) \right) \left(\frac{-1}{\sqrt{6}}, 0, \frac{2}{\sqrt{6}}, \frac{1}{\sqrt{6}} \right)$

$+ \left((2,0,1,-1) \cdot \left(\frac{4}{\sqrt{93}}, \frac{-3}{\sqrt{93}}, \frac{-2}{\sqrt{93}}, \frac{8}{\sqrt{93}} \right) \right) \left(\frac{4}{\sqrt{93}}, \frac{-3}{\sqrt{93}}, \frac{-2}{\sqrt{93}}, \frac{8}{\sqrt{93}} \right)$

$= \frac{-1}{6}(-1,0,2,1) + \frac{-2}{93}(4,-3,-2,8) = \left(\frac{5}{62}, \frac{2}{31}, \frac{-9}{31}, \frac{-21}{62} \right).$

(c) $\text{proj}_V(3,2,1,0) = \left((3,2,1,0) \cdot \left(\frac{-1}{\sqrt{6}}, 0, \frac{2}{\sqrt{6}}, \frac{1}{\sqrt{6}} \right) \right) \left(\frac{-1}{\sqrt{6}}, 0, \frac{2}{\sqrt{6}}, \frac{1}{\sqrt{6}} \right)$

$+ \left((3,2,1,0) \cdot \left(\frac{4}{\sqrt{93}}, \frac{-3}{\sqrt{93}}, \frac{-2}{\sqrt{93}}, \frac{8}{\sqrt{93}} \right) \right) \left(\frac{4}{\sqrt{93}}, \frac{-3}{\sqrt{93}}, \frac{-2}{\sqrt{93}}, \frac{8}{\sqrt{93}} \right)$

Section 5.7

$$= \frac{-1}{6}(-1,0,2,1) + \frac{4}{93}(4,-3,-2,8) = \left(\frac{21}{62}, \frac{-4}{31}, \frac{-13}{31}, \frac{11}{62}\right).$$

16. $V = \{a(1,1,0) + b(0,0,1)\}$. The set $\{(1,1,0), (0,0,1)\}$ is an orthogonal basis. The vectors $\left(\frac{1}{\sqrt{2}}, \frac{1}{\sqrt{2}}, 0\right)$ and $(0,0,1)$ are therefore an orthonormal basis. $\mathbf{v} = \mathbf{w} + \mathbf{w}_\perp$, where

$$\mathbf{w} = \text{proj}_V \mathbf{v} = \left((1,2,-1) \cdot \left(\frac{1}{\sqrt{2}}, \frac{1}{\sqrt{2}}, 0\right)\right)\left(\frac{1}{\sqrt{2}}, \frac{1}{\sqrt{2}}, 0\right) + ((1,2,-1) \cdot (0,0,1))(0,0,1)$$

$$= \frac{3}{2}(1,1,0) + (-1)(0,0,1) = \left(\frac{3}{2}, \frac{3}{2}, -1\right) \text{ and}$$

$$\mathbf{w}_\perp = \mathbf{v} - \text{proj}_V \mathbf{v} = (1,2,-1) - \left(\frac{3}{2}, \frac{3}{2}, -1\right) = \left(\frac{-1}{2}, \frac{1}{2}, 0\right).$$

17. $V = \{a(1,2,0) + b(0,0,1)\}$. The set $\{(1,2,0), (0,0,1)\}$ is an orthogonal basis. The vectors $\left(\frac{1}{\sqrt{5}}, \frac{2}{\sqrt{5}}, 0\right)$ and $(0,0,1)$ are therefore an orthonormal basis. $\mathbf{v} = \mathbf{w} + \mathbf{w}_\perp$, where

$$\mathbf{w} = \text{proj}_V \mathbf{v} = \left((4,1,-2) \cdot \left(\frac{1}{\sqrt{5}}, \frac{2}{\sqrt{5}}, 0\right)\right)\left(\frac{1}{\sqrt{5}}, \frac{2}{\sqrt{5}}, 0\right) + ((4,1,-2) \cdot (0,0,1))(0,0,1)$$

$$= \frac{6}{5}(1,2,0) + (-2)(0,0,1) = \left(\frac{6}{5}, \frac{12}{5}, -2\right) \text{ and}$$

$$\mathbf{w}_\perp = \mathbf{v} - \text{proj}_V \mathbf{v} = (4,1,-2) - \left(\frac{6}{5}, \frac{12}{5}, -2\right) = \left(\frac{14}{5}, \frac{-7}{5}, 0\right).$$

18. $W = \{a(1,0,1) + b(0,1,1)\}$. We must find an orthogonal basis. $\mathbf{u}_1 = (1,0,1)$ and

$$\mathbf{u}_2 = (0,1,1) - \frac{(0,1,1) \cdot (1,0,1)}{(1,0,1) \cdot (1,0,1)}(1,0,1) = (0,1,1) - \frac{1}{2}(1,0,1) = \left(\frac{-1}{2}, 1, \frac{1}{2}\right) \text{ are an}$$

orthogonal basis. The vectors $\left(\frac{1}{\sqrt{2}}, 0, \frac{1}{\sqrt{2}}\right)$ and $\left(\frac{-1}{\sqrt{6}}, \frac{2}{\sqrt{6}}, \frac{1}{\sqrt{6}}\right)$ are therefore an orthonormal basis. $\mathbf{v} = \mathbf{w} + \mathbf{w}_\perp$, where

$$\mathbf{w} = \text{proj}_W \mathbf{v} = \left((3,2,1) \cdot \left(\frac{1}{\sqrt{2}}, 0, \frac{1}{\sqrt{2}}\right)\right)\left(\frac{1}{\sqrt{2}}, 0, \frac{1}{\sqrt{2}}\right)$$

Section 5.7

$$+ \left((3,2,1) \cdot \left(\frac{-1}{\sqrt{6}}, \frac{2}{\sqrt{6}}, \frac{1}{\sqrt{6}} \right) \right) \left(\frac{-1}{\sqrt{6}}, \frac{2}{\sqrt{6}}, \frac{1}{\sqrt{6}} \right) = 2(1,0,1) + \frac{1}{3}(-1,2,1) = \left(\frac{5}{3}, \frac{2}{3}, \frac{7}{3} \right)$$

and $\mathbf{w}_\perp = \mathbf{v} - \text{proj}_W \mathbf{v} = (3,2,1) - \left(\frac{5}{3}, \frac{2}{3}, \frac{7}{3} \right) = \left(\frac{4}{3}, \frac{4}{3}, \frac{-4}{3} \right)$.

19. $W = \{a(1,1,0) + b(0,0,1)\}$. The set $\{(1,1,0), (0,0,1)\}$ is an orthogonal basis. The vectors $\left(\frac{1}{\sqrt{2}}, \frac{1}{\sqrt{2}}, 0 \right)$ and $(0,0,1)$ are therefore an orthonormal basis.

$$\text{proj}_W \mathbf{x} = \left((1,3,-2) \cdot \left(\frac{1}{\sqrt{2}}, \frac{1}{\sqrt{2}}, 0 \right) \right) \left(\frac{1}{\sqrt{2}}, \frac{1}{\sqrt{2}}, 0 \right) + ((1,3,-2) \cdot (0,0,1))(0,0,1)$$

$$= 2(1,1,0) + (-2)(0,0,1) = (2,2,-2), \text{ so}$$

$$d(\mathbf{x}, W) = \|\mathbf{x} - \text{proj}_W \mathbf{x}\| = \|(1,3,-2) - (2,2,-2)\| = \|(-1,1,0)\| = \sqrt{2}.$$

20. $W = \{a(1,-2,0) + b(0,0,1)\}$. The set $\{(1,-2,0), (0,0,1)\}$ is an orthogonal basis. The vectors $\left(\frac{1}{\sqrt{5}}, \frac{-2}{\sqrt{5}}, 0 \right)$ and $(0,0,1)$ are therefore an orthonormal basis.

$$\text{proj}_W \mathbf{x} = \left((2,4,-1) \cdot \left(\frac{1}{\sqrt{5}}, \frac{-2}{\sqrt{5}}, 0 \right) \right) \left(\frac{1}{\sqrt{5}}, \frac{-2}{\sqrt{5}}, 0 \right) + ((2,4,-1) \cdot (0,0,1))(0,0,1)$$

$$= \frac{-6}{5}(1,-2,0) + (-1)(0,0,1) = \left(\frac{-6}{5}, \frac{12}{5}, -1 \right), \text{ so}$$

$$d(\mathbf{x}, W) = \|\mathbf{x} - \text{proj}_W \mathbf{x}\| = \left\| (2,4,-1) - \left(\frac{-6}{5}, \frac{12}{5}, -1 \right) \right\| = \left\| \left(\frac{16}{5}, \frac{8}{5}, 0 \right) \right\| = \frac{8}{5}\sqrt{5}.$$

21. $W = \{a(1,2,3)\}$. The vector $\left(\frac{1}{\sqrt{14}}, \frac{2}{\sqrt{14}}, \frac{3}{\sqrt{14}} \right)$ is an orthonormal basis for W.

$$\text{proj}_W \mathbf{x} = \left((1,3,-2) \cdot \left(\frac{1}{\sqrt{14}}, \frac{2}{\sqrt{14}}, \frac{3}{\sqrt{14}} \right) \right) \left(\frac{1}{\sqrt{14}}, \frac{2}{\sqrt{14}}, \frac{3}{\sqrt{14}} \right) = \left(\frac{1}{14}, \frac{2}{14}, \frac{3}{14} \right), \text{ so}$$

$$d(\mathbf{x}, W) = \|\mathbf{x} - \text{proj}_W \mathbf{x}\| = \left\| (1,3,-2) - \left(\frac{1}{14}, \frac{2}{14}, \frac{3}{14} \right) \right\| = \left\| \left(\frac{13}{14}, \frac{40}{14}, \frac{-31}{14} \right) \right\| = \frac{\sqrt{2730}}{14}$$

Section 5.7

22. $(1,2,-2)$ and $(6,1,4)$ are orthogonal. If the vector (a,b,c) is orthogonal to both, then $(1,2,-2)\cdot(a,b,c) = a + 2b - 2c = 0$ and $(6,1,4)\cdot(a,b,c) = 6a + b + 4c = 0$. One solution to this system of equations is $a = -10$, $b = 16$, and $c = 11$. The set $\{(1,2,-2), (6,1,4), (-10,16,11)\}$ is therefore an orthogonal basis for $\mathbf{R}^3$.

23. $\|\mathbf{v}\| = \sqrt{\mathbf{v}\cdot\mathbf{v}}$ so we need to show that $\mathbf{v}\cdot\mathbf{v} = (\mathbf{v}\cdot\mathbf{u}_1)^2 + (\mathbf{v}\cdot\mathbf{u}_2)^2 + \ldots + (\mathbf{v}\cdot\mathbf{u}_n)^2$.
$\mathbf{v} = (\mathbf{v}\cdot\mathbf{u}_1)\mathbf{u}_1 + (\mathbf{v}\cdot\mathbf{u}_2)\mathbf{u}_2 + \ldots + (\mathbf{v}\cdot\mathbf{u}_n)\mathbf{u}_n$, so
$\mathbf{v}\cdot\mathbf{v} = ((\mathbf{v}\cdot\mathbf{u}_1)\mathbf{u}_1 + (\mathbf{v}\cdot\mathbf{u}_2)\mathbf{u}_2 + \ldots + (\mathbf{v}\cdot\mathbf{u}_n)\mathbf{u}_n)\cdot((\mathbf{v}\cdot\mathbf{u}_1)\mathbf{u}_1 + (\mathbf{v}\cdot\mathbf{u}_2)\mathbf{u}_2 + \ldots + (\mathbf{v}\cdot\mathbf{u}_n)\mathbf{u}_n)$
$= (\mathbf{v}\cdot\mathbf{u}_1)\mathbf{u}_1\cdot(\mathbf{v}\cdot\mathbf{u}_1)\mathbf{u}_1 + (\mathbf{v}\cdot\mathbf{u}_2)\mathbf{u}_2\cdot(\mathbf{v}\cdot\mathbf{u}_2)\mathbf{u}_2 + \ldots + (\mathbf{v}\cdot\mathbf{u}_n)\mathbf{u}_n\cdot(\mathbf{v}\cdot\mathbf{u}_n)\mathbf{u}_n$. All the other terms are zero because $\mathbf{u}_i\cdot\mathbf{u}_j = 0$ for $i \neq j$, and since $\mathbf{u}_i\cdot\mathbf{u}_i = 1$ we have
$\mathbf{v}\cdot\mathbf{v} = (\mathbf{v}\cdot\mathbf{u}_1)^2 + (\mathbf{v}\cdot\mathbf{u}_2)^2 + \ldots + (\mathbf{v}\cdot\mathbf{u}_n)^2$.

24. $\{\mathbf{u}_1, \mathbf{u}_2, \ldots, \mathbf{u}_n\}$ is a set of n linearly independent vectors in a vector space with dimension n. Therefore by Theorem 4.11, $\{\mathbf{u}_1, \mathbf{u}_2, \ldots, \mathbf{u}_n\}$ is a basis for the vector space.

25. $\text{proj}_\mathbf{u}\mathbf{v} = \dfrac{\mathbf{v}\cdot\mathbf{u}}{\mathbf{u}\cdot\mathbf{u}}\mathbf{u}$, so $(\mathbf{v} - \text{proj}_\mathbf{u}\mathbf{v})\cdot\mathbf{u} = (\mathbf{v} - \dfrac{\mathbf{v}\cdot\mathbf{u}}{\mathbf{u}\cdot\mathbf{u}}\mathbf{u})\cdot\mathbf{u} = \mathbf{v}\cdot\mathbf{u} - \dfrac{\mathbf{v}\cdot\mathbf{u}}{\mathbf{u}\cdot\mathbf{u}}\mathbf{u}\cdot\mathbf{u} = 0$.

26. We show that the column vectors are unit vectors and that they are mutually orthogonal.

    (a) $\mathbf{a}_1 = \begin{bmatrix} 1 \\ 0 \end{bmatrix}$ and $\mathbf{a}_2 = \begin{bmatrix} 0 \\ 1 \end{bmatrix}$. $\|\mathbf{a}_1\| = \|\mathbf{a}_2\| = 1$ and $\mathbf{a}_1\cdot\mathbf{a}_2 = 0$.

    (b) $\mathbf{a}_1 = \begin{bmatrix} 0 \\ 1 \end{bmatrix}$ and $\mathbf{a}_2 = \begin{bmatrix} -1 \\ 0 \end{bmatrix}$. $\|\mathbf{a}_1\| = \|\mathbf{a}_2\| = 1$ and $\mathbf{a}_1\cdot\mathbf{a}_2 = 0$.

    (c) $\mathbf{a}_1 = \begin{bmatrix} \frac{\sqrt{3}}{2} \\ \frac{-1}{2} \end{bmatrix}$ and $\mathbf{a}_2 = \begin{bmatrix} \frac{1}{2} \\ \frac{\sqrt{3}}{2} \end{bmatrix}$. $\|\mathbf{a}_1\|^2 = \frac{3}{4} + \frac{1}{4} = 1$ and $\|\mathbf{a}_2\|^2 = \frac{1}{4} + \frac{3}{4} = 1$, so $\|a_1\| = \|a_2\| = 1$, and $a_1 \cdot a_2 = \frac{\sqrt{3}}{2} \times \frac{1}{2} - \frac{1}{2} \times \frac{\sqrt{3}}{2} = 0$.

(d) $a_1 = \begin{bmatrix} 1 \\ 0 \\ 0 \end{bmatrix}, a_2 = \begin{bmatrix} 0 \\ 0 \\ 1 \end{bmatrix}$, and $a_3 = \begin{bmatrix} 0 \\ -1 \\ 0 \end{bmatrix}$.

$\|a_1\| = \|a_2\| = \|a_3\| = 1$ and $a_1 \cdot a_2 = a_1 \cdot a_3 = a_2 \cdot a_3 = 0$.

(e) $a_1 = \begin{bmatrix} 0 \\ \frac{-2}{\sqrt{6}} \\ \frac{1}{\sqrt{3}} \end{bmatrix}, a_2 = \begin{bmatrix} \frac{1}{\sqrt{2}} \\ \frac{1}{\sqrt{6}} \\ \frac{1}{\sqrt{3}} \end{bmatrix}$, and $a_3 = \begin{bmatrix} \frac{-1}{\sqrt{2}} \\ \frac{1}{\sqrt{6}} \\ \frac{1}{\sqrt{3}} \end{bmatrix}$. $\|a_1\|^2 = \frac{4}{6} + \frac{1}{3} = 1$,

$\|a_2\|^2 = \frac{1}{2} + \frac{1}{6} + \frac{1}{3} = 1$, and $\|a_3\|^2 = \frac{1}{2} + \frac{1}{6} + \frac{1}{3} = 1$, so $\|a_1\| = \|a_2\| = \|a_3\| = 1$, and $a_1 \cdot a_2 = a_1 \cdot a_3 = 0 + \frac{-2}{6} + \frac{1}{3} = 0$, $a_2 \cdot a_3 = \frac{-1}{2} + \frac{1}{6} + \frac{1}{3} = 0$.

27. (a) The row vectors are $r_1 = \begin{bmatrix} \frac{1}{\sqrt{2}} & \frac{1}{\sqrt{2}} \end{bmatrix}$ and $r_2 = \begin{bmatrix} \frac{-1}{\sqrt{2}} & \frac{1}{\sqrt{2}} \end{bmatrix}$.

$\|r_1\|^2 = \|r_2\|^2 = \frac{1}{2} + \frac{1}{2} = 1$, so $\|r_1\| = \|r_2\| = 1$, and $r_1 \cdot r_2 = \frac{1}{\sqrt{2}} \times \frac{-1}{\sqrt{2}} + \frac{1}{\sqrt{2}} \times \frac{1}{\sqrt{2}} = 0$. Thus the rows are orthonormal vectors.

$AA^t = \begin{bmatrix} \frac{1}{\sqrt{2}} & \frac{1}{\sqrt{2}} \\ \frac{-1}{\sqrt{2}} & \frac{1}{\sqrt{2}} \end{bmatrix} \begin{bmatrix} \frac{1}{\sqrt{2}} & \frac{-1}{\sqrt{2}} \\ \frac{1}{\sqrt{2}} & \frac{1}{\sqrt{2}} \end{bmatrix} = \begin{bmatrix} 1 & 0 \\ 0 & 1 \end{bmatrix}$, so $A^{-1} = A^t$.

The rows of A, which have been shown to be orthonormal, are the columns of $A^{-1} = A^t$, so A^{-1} is orthogonal. $|A| = \frac{1}{\sqrt{2}} \times \frac{1}{\sqrt{2}} - \frac{1}{\sqrt{2}} \times \frac{-1}{\sqrt{2}} = 1$.

(b) The row vectors are $r_1 = \begin{bmatrix} \frac{\sqrt{3}}{2} & \frac{1}{2} \end{bmatrix}$ and $r_2 = \begin{bmatrix} \frac{1}{2} & \frac{-\sqrt{3}}{2} \end{bmatrix}$. $\|r_1\|^2 = \frac{3}{4} + \frac{1}{4} = 1$ and $\|r_2\|^2 = \frac{1}{4} + \frac{3}{4} = 1$, so $\|r_1\| = \|r_2\| = 1$, and $r_1 \cdot r_2 = \frac{\sqrt{3}}{2} \times \frac{1}{2} - \frac{1}{2} \times \frac{\sqrt{3}}{2} = 0$. Thus the rows are orthonormal vectors.

$$AA^t = \begin{bmatrix} \frac{\sqrt{3}}{2} & \frac{1}{2} \\ \frac{1}{2} & \frac{-\sqrt{3}}{2} \end{bmatrix} \begin{bmatrix} \frac{\sqrt{3}}{2} & \frac{1}{2} \\ \frac{1}{2} & \frac{-\sqrt{3}}{2} \end{bmatrix} = \begin{bmatrix} 1 & 0 \\ 0 & 1 \end{bmatrix}, \text{ so } A^{-1} = A^t.$$

The rows of A, which have been shown to be orthonormal, are the columns of $A^{-1} = A^t$, so A^{-1} is orthogonal. $|A| = \frac{\sqrt{3}}{2} \times \frac{-\sqrt{3}}{2} - \frac{1}{2} \times \frac{1}{2} = -1$.

28. It is proved in the text that if A is orthogonal, then $A^{-1} = A^t$, so we need to show that if $A^{-1} = A^t$ then A is orthogonal. The (i,j)th element of $AA^{-1} = AA^t = I$ is $a_{1i} a_{1j} + a_{2i} a_{2j} + \ldots + a_{ni} a_{nj}$, the dot product of the ith and jth columns of A. If $i \neq j$, this sum is zero because the nondiagonal elements of I are zero. If $i = j$, this sum is 1 because the diagonal elements of I are 1. Thus we have proved that the columns of A are mutually orthogonal and that they are all unit vectors, so A is an orthogonal matrix.

29. Let $C = AB$. Then $C^t = (AB)^t = B^t A^t = B^{-1} A^{-1} = (AB)^{-1} = C^{-1}$. Thus by Exercise 28, C is orthogonal.

30. $AA^t = (I - 2uu^t)(I - 2uu^t)^t = (I - 2uu^t)(I - 2(u^t)^t u^t) = (I - 2uu^t)(I - 2uu^t)$
    $= I - 4uu^t + 4uu^t uu^t = I - 4uu^t + 4u(u^t u)u^t = I - 4uu^t + 4u1u^t = I.$
    Thus $AA^t = I$. Multiply both sides by A^{-1}. $A^{-1}AA^t = A^{-1}I$, $A = A^{-1}$. Thus A is orthogonal.

31. To show that A^{-1} is unitary it is necessary to show that $(A^{-1})^{-1} = (\overline{A^{-1}})^t$.
    Since A is unitary $A^{-1} = \overline{A}^t$. Thus $(A^{-1})^t = (\overline{A}^t)^t = \overline{A}$, so that $(\overline{A^{-1}})^t = A = (A^{-1})^{-1}$.

32. The (i,j)th element of $A\overline{A}^t$ is $a_{i1} \overline{a}_{j1} + a_{i2} \overline{a}_{j2} + \ldots + a_{in} \overline{a}_{jn}$. If $A\overline{A}^t = I$, then $a_{i1} \overline{a}_{j1} + a_{i2} \overline{a}_{j2} + \ldots + a_{in} \overline{a}_{jn} = 1$ if $i = j$ and zero if $i \neq j$, so that the columns of A are a set of mutually orthogonal unit vectors in $\mathbb{C}^n$. Conversely, if the columns of A are a set of mutually orthogonal unit vectors in $\mathbb{C}^n$, then $a_{i1} \overline{a}_{j1} + a_{i2} \overline{a}_{j2} + \ldots + a_{in} \overline{a}_{jn} = 1$ if $i = j$ and 0 if $i \neq j$, so that $A\overline{A}^t = I$.

Chapter 5 Review Exercises

1. (a) $(a, b, a-2) + (c, d, c-2) = (a+c, b+d, a+c-4)$, so the sum of two such vectors is not in the set. Thus the set is not a subspace of $\mathbf{R}^3$.

   (b) $(a,-2a,3a) + (b,-2b,3b) = (a+b, -2(a+b), 3(a+b))$ and $c(a, -2a, 3a) = (ca, -2ca, 3ca)$, so the sum and scalar product of vectors in the set is in the set. Thus the set is a subspace of $\mathbf{R}^3$.

   (c) $(a, b, 2a-3b) + (e, f, 2e-3f) = (a+e, b+f, 2(a+e)-3(b+f))$ and $c(a, b, 2a-3b) = (ca, cb, 2ca-3cb)$, so the sum and scalar product of vectors in the set is in the set. Thus the set is a subspace of $\mathbf{R}^3$.

   (d) $(a,2,b) + (c,2,d) = (a+c, 4, b+d)$, so the sum of vectors in the set is not in the set. Thus the set is not a subspace of $\mathbf{R}^3$.

2. Only the subset (a) is a subspace of $\mathbf{R}^3$. None of the other subsets is closed under scalar multiplication.

3. (a) Not a subspace: $\begin{bmatrix} 1 & 2 \\ 3 & 4 \end{bmatrix}$ is in the subset. $0 \begin{bmatrix} 1 & 2 \\ 3 & 4 \end{bmatrix} = \begin{bmatrix} 0 & 0 \\ 0 & 0 \end{bmatrix}$ is not in the subset. Not closed under scalar multiplication.

   (b) A subspace: Let $\begin{bmatrix} a & b \\ c & d \end{bmatrix}$ and $\begin{bmatrix} p & q \\ r & s \end{bmatrix}$ be in the subset. Thus $a+b+c+d = 0$ and $p+q+r+s = 0$.

   $\begin{bmatrix} a & b \\ c & d \end{bmatrix} + \begin{bmatrix} p & q \\ r & s \end{bmatrix} = \begin{bmatrix} a+p & b+q \\ c+r & d+s \end{bmatrix}$, and $a+p+b+q+c+r+d+s = 0$.

   $k \begin{bmatrix} a & b \\ c & d \end{bmatrix} = \begin{bmatrix} ka & kb \\ kc & kd \end{bmatrix}$, and $ka+kb+kc+kd = k(a+b+c+d) = 0$.

   Closed under addition and under scalar multiplication.

   (c) A subspace: $\begin{bmatrix} a & b \\ -b & c \end{bmatrix} + \begin{bmatrix} p & q \\ -q & r \end{bmatrix} = \begin{bmatrix} a+p & b+q \\ -(b+q) & c+r \end{bmatrix}$.

   $k \begin{bmatrix} a & b \\ -b & c \end{bmatrix} = \begin{bmatrix} ka & kb \\ -kb & kc \end{bmatrix}$. Closed under addition and under scalar multiplication.

   (d) Not a subspace: $\begin{bmatrix} a & 3 \\ b & c \end{bmatrix} + \begin{bmatrix} a & 3 \\ b & c \end{bmatrix} = \begin{bmatrix} 2a & 6 \\ 2b & 2c \end{bmatrix}$ and $2 \begin{bmatrix} a & 3 \\ b & c \end{bmatrix} = \begin{bmatrix} 2a & 6 \\ b & c \end{bmatrix}$.

Chapter 5 Review Exercises

(2, 2) element becomes 6, not closed under addition or multiplication.

(e) Not a subspace: Consider $A=\begin{bmatrix} 1 & 1 \\ 1 & 1 \end{bmatrix}$ and $B=\begin{bmatrix} 1 & 2 \\ 3 & 6 \end{bmatrix}$. |A|=0 and |B|=0.
$A+B=\begin{bmatrix} 2 & 3 \\ 4 & 7 \end{bmatrix}$. |A+B|=2≠0. Not closed under addition.

4. $(f + g)(x) = 3x - 1 + 2x^2 + 3 = 2x^2 + 3x + 2$, $3f(x) = 3(3x - 1) = 9x - 3$, and
$(2f - 3g)(x) = 2(3x - 1) - 3(2x^2 + 3) = -6x^2 + 6x - 11$.

5. $3x^2 + ax - b + 3x^2 + cx - d = 6x^2 + (a+c)x - (b+d)$ is not in S, so the set is not closed under addition and therefore is not a subspace of P_2.

6. Necessary: Let **v** be in the subspace and 0 be the zero scalar. Then $0\mathbf{v} = \mathbf{0}$, the zero vector (Theorem 5.1(a)). The subspace is closed under scalar multiplication. Thus **0** is in the subspace.
Not sufficient: The subset of $\mathbf{R}^2$ consisting of vectors of the form (a, a^2) contains the zero vector. It is not closed under addition, thus not a subspace.

7. (a) $(3,15,-4) = 2(1,2,-1) + 3(2,4,0) - (5,1,2)$.

    (b) $(-3,-4,7) = a(5,-1,3) + b(2,0,3) + c(4,1,0)$ if and only if $-3 = 5a + 2b + 4c$, $-4 = -a + c$, and $7 = 3a + 3b$. This system of equations has the unique solution $a = 25/21$, $b = 24/21$, $c = -59/21$, so $(-3,-4,7)$ is a linear combination of $(5,-1,3)$, $(2,0,3)$, and $(4,1,0)$.

8. No: $2A+3B-C = 2\begin{bmatrix} 1 & 2 \\ 4 & 3 \end{bmatrix} + 3\begin{bmatrix} 0 & -2 \\ 1 & 5 \end{bmatrix} - \begin{bmatrix} 1 & 1 \\ 1 & 1 \end{bmatrix} = \begin{bmatrix} 1 & -3 \\ 10 & 20 \end{bmatrix}$.

9. (a) $(x_1,x_2,x_3) = a(1,2,3) + b(-2,4,1) + c(0,6,4)$ if and only if $x_1 = a - 2b$, $x_2 = 2a + 4b + 6c$, and $x_3 = 3a + b + 4c$. The coefficient matrix for this system of equations has nonzero determinant, so the system has a unique solution. Thus each vector in $\mathbf{R}^3$ is a linear combination of the given vectors.

    (b) $(x_1,x_2,x_3) = a(-2,1,0) + b(1,2,-1) + c(-1,8,-3)$ if and only if $x_1 = -2a + b - c$, $x_2 = a + 2b + 8c$, $x_3 = -b - 3c$. The coefficient matrix for this system of equations has determinant zero, so the coefficient matrix is row equivalent to a 3x3 matrix having at least one row of zeros. Thus the augmented matrix is row equivalent to a 3x4 matrix having 0 0 0 N for its third row, where N is not zero for some values of

197

Chapter 5 Review Exercises

x_1, x_2, and x_3. So for some values of x_1, x_2, and x_3, the system of equations has no solution. The given set of vectors does not span $\mathbf{R}^3$.

A shorter argument is that the given vectors are not linearly independent and, in fact, span a two-dimensional subspace of $\mathbf{R}^3$, so they certainly do not span $\mathbf{R}^3$.

10. $13x^2 + 8x - 21 = 2(2x^2 + x - 3) - 3(-3x^2 - 2x + 5)$.

11. (a) $a(1,-2,0) + b(0,1,3) + c(2,0,12) = (0,0,0)$ if and only if $a + 2c = 0$, $-2a + b = 0$, and $3b + 12c = 0$. The coefficient matrix for this homogeneous system of equations has determinant zero, so there are solutions with a, b, and c not all zero. Thus these vectors are linearly dependent.

   (b) $a(-1,18,7) + b(-1,4,1) + c(1,3,2) = (0,0,0)$ if and only if $-a - b + c = 0$, $18a + 4b + 3c = 0$, and $7a + b + 2c = 0$. The coefficient matrix for this homogeneous system of equations has determinant zero, so there are solutions with a, b, and c not all zero. Thus these vectors are linearly dependent.

   (c) $a(5,-1,3) + b(2,1,0) + c(3,-2,2) = (0,0,0)$ if and only if $5a + 2b + 3c = 0$, $-a + b - 2c = 0$, and $3a + 2c = 0$. The coefficient matrix for this homogeneous system of equations has nonzero determinant, so the solution $a = b = c = 0$ is unique. The vectors are therefore linearly independent.

12. (a) and (b) In each case the set consists of two linearly independent vectors. By Theorem 5.11 they are therefore a basis for $\mathbf{R}^2$.

    (c) and (d) In each case the set consists of three linearly independent vectors. By Theorem 5.11 they are therefore a basis for $\mathbf{R}^3$.

13. $(10,9,8) = 2(-1,3,1) + 3(4,1,2)$.

14. $ax^2 + bx + c = -a(-x^2) + (b/3)(3x) + (c/2)2$.

15. Any of the sets $\{(1,-2,3), (4,1,-1), (1,0,0)\}$, $\{(1,-2,3), (4,1,-1), (0,1,0)\}$, $\{(1,-2,3), (4,1,-1), (0,0,1)\}$ is a basis for $\mathbf{R}^3$.

Chapter 5 Review Exercises

16. $(a,b,c,a-2b+3c) = a(1,0,0,1) + b(0,1,0,-2) + c(0,0,1,3)$. The linearly independent set $\{(1,0,0,1), (0,1,0,-2), (0,0,1,3)\}$ is a basis for the subspace.

17. $\begin{bmatrix} 1 & 0 & 0 \\ 0 & 0 & 0 \\ 0 & 0 & 0 \end{bmatrix}, \begin{bmatrix} 0 & 1 & 0 \\ 0 & 0 & 0 \\ 0 & 0 & 0 \end{bmatrix}, \begin{bmatrix} 0 & 0 & 1 \\ 0 & 0 & 0 \\ 0 & 0 & 0 \end{bmatrix}, \begin{bmatrix} 0 & 0 & 0 \\ 0 & 1 & 0 \\ 0 & 0 & 0 \end{bmatrix}, \begin{bmatrix} 0 & 0 & 0 \\ 0 & 0 & 1 \\ 0 & 0 & 0 \end{bmatrix}$, and $\begin{bmatrix} 0 & 0 & 0 \\ 0 & 0 & 0 \\ 0 & 0 & 1 \end{bmatrix}$ are a basis for the vector space of upper triangular 3x3 matrices.

18. $a(x^2 + 2x - 3) + b(3x^2 + x - 1) + c(4x^2 + 3x - 3) = 0$ if and only if $a + 3b + 4c = 0$, $2a + b + 3c = 0$, and $-3a - b - 3c = 0$. This system of homogeneous equations has the unique solution $a = b = c = 0$. Thus the given functions are linearly independent. The dimension of P_2 is 3, so the three functions are a basis.

19. (a) $\begin{bmatrix} 1 & 2 & -1 \\ -1 & 3 & 4 \\ 0 & 5 & 3 \end{bmatrix} \approx \begin{bmatrix} 1 & 2 & -1 \\ 0 & 5 & 3 \\ 0 & 5 & 3 \end{bmatrix} \approx \begin{bmatrix} 1 & 2 & -1 \\ 0 & 1 & 3/5 \\ 0 & 0 & 0 \end{bmatrix}$, so the rank of the matrix is 2.

(b) $\begin{bmatrix} 2 & 1 & 4 \\ -2 & 0 & -1 \\ 3 & 2 & 7 \end{bmatrix} \approx \begin{bmatrix} 1 & 1/2 & 2 \\ 0 & 1 & 3 \\ 0 & 1/2 & 1 \end{bmatrix} \approx \begin{bmatrix} 1 & 1/2 & 2 \\ 0 & 1 & 3 \\ 0 & 0 & 1 \end{bmatrix}$, so the rank of the matrix is 3.

(c) $\begin{bmatrix} -2 & 4 & 8 \\ 1 & -2 & 4 \\ 4 & -8 & 16 \end{bmatrix} \approx \begin{bmatrix} 1 & -2 & -4 \\ 0 & 0 & 1 \\ 0 & 0 & 0 \end{bmatrix}$, so the rank of the matrix is 2.

20. $\begin{bmatrix} 1 & -2 & 3 & 4 \\ -1 & 3 & 1 & -2 \\ 2 & -3 & 10 & 10 \end{bmatrix} \approx \begin{bmatrix} 1 & -2 & 3 & 4 \\ 0 & 1 & 4 & 2 \\ 0 & 1 & 4 & 2 \end{bmatrix} \approx \begin{bmatrix} 1 & -2 & 3 & 4 \\ 0 & 1 & 4 & 2 \\ 0 & 0 & 0 & 0 \end{bmatrix}$, so the vectors

$(1,-2,3,4)$ and $(0,1,4,2)$ are a basis for the subspace.

21. $\mathbf{v} = a\mathbf{v}_1 + b\mathbf{v}_2 = a\mathbf{v}_1 + b\mathbf{v}_2 + 0\mathbf{v}_3$.

199

Chapter 5 Review Exercises

22. If $a\mathbf{v}_1 + b\mathbf{v}_2 = \mathbf{0}$ then $a\mathbf{v}_1 + b\mathbf{v}_2 + 0\mathbf{v}_3 = \mathbf{0}$ and since the set $\{\mathbf{v}_1, \mathbf{v}_2, \mathbf{v}_3\}$ is linearly independent this means $a = b = 0$, so $\{\mathbf{v}_1, \mathbf{v}_2\}$ must be linearly independent.

    If $a\mathbf{v}_1 + c\mathbf{v}_3 = \mathbf{0}$ then $a\mathbf{v}_1 + 0\mathbf{v}_2 + c\mathbf{v}_3 = \mathbf{0}$ and since the set $\{\mathbf{v}_1, \mathbf{v}_2, \mathbf{v}_3\}$ is linearly independent this means $a = c = 0$, so $\{\mathbf{v}_1, \mathbf{v}_3\}$ must be linearly independent.

    If $b\mathbf{v}_2 + c\mathbf{v}_3 = \mathbf{0}$ then $0\mathbf{v}_1 + b\mathbf{v}_2 + c\mathbf{v}_3 = \mathbf{0}$ and since the set $\{\mathbf{v}_1, \mathbf{v}_2, \mathbf{v}_3\}$ is linearly independent this means $b = c = 0$, so $\{\mathbf{v}_2, \mathbf{v}_3\}$ must be linearly independent.

    Since $\{\mathbf{v}_1, \mathbf{v}_2, \mathbf{v}_3\}$ is linearly independent, none of the three vectors can be the zero vector. Thus $a\mathbf{v}_1 = \mathbf{0}$ (or $b\mathbf{v}_2 = \mathbf{0}$ or $c\mathbf{v}_3 = \mathbf{0}$) only if $a = 0$ (or $b = 0$ or $c = 0$).

23. $a(\mathbf{v}_1 + 2\mathbf{v}_2) + b(3\mathbf{v}_1 - \mathbf{v}_2) = (a + 3b)\mathbf{v}_1 + (2a - b)\mathbf{v}_2$. There are scalars c and d, not both zero, with $c\mathbf{v}_1 + d\mathbf{v}_2 = \mathbf{0}$. Solve the system $c = a + 3b$, $d = 2a - b$: $a = \frac{3d+c}{7}$, $b = \frac{2c-d}{7}$. $3d+c$ and $2c-d$ cannot both be zero unless both c and d are zero, so at least one of a and b is nonzero and $a(\mathbf{v}_1 + 2\mathbf{v}_2) + b(3\mathbf{v}_1 - \mathbf{v}_2) = c\mathbf{v}_1 + d\mathbf{v}_2 = \mathbf{0}$. So $\mathbf{v}_1 + 2\mathbf{v}_2$ and $3\mathbf{v}_1 - \mathbf{v}_2$ are linearly dependent.

24. (a) True. The vectors are both in the subspace which has dimension 2, and they are linearly independent.

    (b) False. The dimension of $\mathbf{R}^2$ is 2. No set of more than two vectors can be linearly independent.

    (c) True. This is a direct result of Theorem 4.11 and the definitions.

    (d) True. If two vectors in $\mathbf{R}^2$ are not a basis, they are linearly dependent and therefore collinear.

    (e) False. For the vectors to be linearly dependent all three would have to lie on the same line or in the same plane. It is much more likely that the first two vectors would not lie on the same line and that the third vector would not lie in the plane of the first two. Reasoning in terms of matrices, if you write the vectors as rows of a matrix, the determinant of the matrix would have to be zero for the vectors to be linearly dependent. Zero is only one of an infinite number of possible values for the determinant of a 3x3 matrix.

25. (a) false   (b) true   (c) true   (d) true   (e) false

Chapter 5 Review Exercises

26. If rank(A) = n, the row space of A is **R**^{n}, so the reduced echelon form of A must have n linearly independent rows; i.e., it must be I$_n$. If A is row equivalent to I$_n$, the rows of I$_n$ are linear combinations of the rows of A. But the rows of I$_n$ span **R**^{n}, so the rows of A span **R**^{n} and rank(A) = n.

27. $\text{proj}_\mathbf{u}\mathbf{v} = \dfrac{\mathbf{v}\cdot\mathbf{u}}{\mathbf{u}\cdot\mathbf{u}}\,\mathbf{u}$.

   (a) $\text{proj}_\mathbf{u}\mathbf{v} = \dfrac{(1,3)\cdot(2,4)}{(2,4)\cdot(2,4)}(2,4) = \dfrac{7}{10}(2,4) = \left(\dfrac{7}{5},\dfrac{14}{5}\right)$.

   (b) $\text{proj}_\mathbf{u}\mathbf{v} = \dfrac{(-1,3,4)\cdot(-1,2,4)}{(-1,2,4)\cdot(-1,2,4)}(-1,2,4) = \dfrac{23}{21}(-1,2,4) = \left(-\dfrac{23}{21},\dfrac{46}{21},\dfrac{92}{21}\right)$.

28. $\mathbf{u}_1 = (1,2,3,-1)$, $\mathbf{u}_2 = (2,0,-1,1) - \dfrac{(2,0,-1,1)\cdot(1,2,3,-1)}{(1,2,3,-1)\cdot(1,2,3,-1)}(1,2,3,-1)$

   $= (2,0,-1,1) - \dfrac{-2}{15}(1,2,3,-1) = \left(\dfrac{32}{15},\dfrac{4}{15},\dfrac{-9}{15},\dfrac{13}{15}\right)$, and

   $\mathbf{u}_3 = (3,2,0,1) - \dfrac{(3,2,0,1)\cdot(1,2,3,-1)}{(1,2,3,-1)\cdot(1,2,3,-1)}(1,2,3,-1) - \dfrac{(3,2,0,1)\cdot(32,4,-9,13)}{(32,4,-9,13)\cdot(32,4,-9,13)}(32,4,-9,13)$

   $= (3,2,0,1) - \dfrac{2}{5}(1,2,3,-1) - \dfrac{39}{430}(32,4,-9,13)$

   $= \left(\dfrac{13}{5},\dfrac{6}{5},\dfrac{-6}{5},\dfrac{7}{5}\right) - \dfrac{39}{86}\left(\dfrac{32}{5},\dfrac{4}{5},\dfrac{-9}{5},\dfrac{13}{5}\right) = \left(\dfrac{-26}{86},\dfrac{72}{86},\dfrac{-33}{86},\dfrac{19}{86}\right)$

   are an orthogonal basis for the subspace of **R**^{4}. $\|\mathbf{u}_1\| = \sqrt{15}$, $\|\mathbf{u}_2\| = \dfrac{1}{15}\sqrt{1290}$,

   and $\|\mathbf{u}_3\| = \dfrac{1}{86}\sqrt{7310}$, so the set $\left\{\left(\dfrac{1}{\sqrt{15}},\dfrac{2}{\sqrt{15}},\dfrac{3}{\sqrt{15}},\dfrac{-1}{\sqrt{15}}\right),\right.$

   $\left(\dfrac{32}{\sqrt{1290}},\dfrac{4}{\sqrt{1290}},\dfrac{-9}{\sqrt{1290}},\dfrac{13}{\sqrt{1290}}\right),\left.\left(\dfrac{-26}{\sqrt{7310}},\dfrac{72}{\sqrt{7310}},\dfrac{-33}{\sqrt{7310}},\dfrac{19}{\sqrt{7310}}\right)\right\}$

   is an orthonormal basis for the subspace.

29. $(x,y,x+2y) = x(1,0,1) + y(0,1,2)$. The vectors $(1,0,1)$ and $(0,1,2)$ span the subspace and

Chapter 5 Review Exercises

are linearly independent so they are a basis. $\mathbf{u}_1 = (1,0,1)$ and

$\mathbf{u}_2 = (0,1,2) - \frac{(0,1,2)\cdot(1,0,1)}{(1,0,1)\cdot(1,0,1)}(1,0,1) = (0,1,2) - (1,0,1) = (-1,1,1)$ are an orthogonal

basis, and $\|\mathbf{u}_1\| = \sqrt{2}$ and $\|\mathbf{u}_2\| = \sqrt{3}$, so the set $\left\{\left(\frac{1}{\sqrt{2}}, 0, \frac{1}{\sqrt{2}}\right), \left(\frac{-1}{\sqrt{3}}, \frac{1}{\sqrt{3}}, \frac{1}{\sqrt{3}}\right)\right\}$

is an orthonormal basis for the subspace.

30. $\mathbf{u}_1 = (2,1,1)$ and $\mathbf{u}_2 = (1,-1,3) - \frac{(1,-1,3)\cdot(2,1,1)}{(2,1,1)\cdot(2,1,1)}(2,1,1) = (1,-1,3) - \frac{2}{3}(2,1,1)$

$= \left(\frac{-1}{3}, \frac{-5}{3}, \frac{7}{3}\right)$ are an orthogonal basis for W. $\|\mathbf{u}_1\| = \sqrt{6}$ and $\|\mathbf{u}_2\| = \frac{5}{3}\sqrt{3}$,

so $\left(\frac{2}{\sqrt{6}}, \frac{1}{\sqrt{6}}, \frac{1}{\sqrt{6}}\right)$ and $\left(\frac{-1}{5\sqrt{3}}, \frac{-5}{5\sqrt{3}}, \frac{7}{5\sqrt{3}}\right)$ are an orthonormal basis.

$\text{proj}_W(3,1,-2) = \left((3,1,-2)\cdot\left(\frac{2}{\sqrt{6}}, \frac{1}{\sqrt{6}}, \frac{1}{\sqrt{6}}\right)\right)\left(\frac{2}{\sqrt{6}}, \frac{1}{\sqrt{6}}, \frac{1}{\sqrt{6}}\right)$

$+ \left((3,1,-2)\cdot\left(\frac{-1}{5\sqrt{3}}, \frac{-5}{5\sqrt{3}}, \frac{7}{5\sqrt{3}}\right)\right)\left(\frac{-1}{5\sqrt{3}}, \frac{-5}{5\sqrt{3}}, \frac{7}{5\sqrt{3}}\right)$

$= \frac{5}{6}(2,1,1) + \frac{-22}{75}(-1,-5,7) = \left(\frac{294}{150}, \frac{345}{150}, \frac{-183}{150}\right)$.

31. $W = \{a(1,3,0) + b(0,0,1)\}$. $(1,3,0)$ and $(0,0,1)$ are an orthogonal basis for the subspace,
so $\left(\frac{1}{\sqrt{10}}, \frac{3}{\sqrt{10}}, 0\right)$ and $(0,0,1)$ are an orthonormal basis. Let $\mathbf{x} = (1,2,-4)$.

$\text{proj}_W \mathbf{x} = \left((1,2,-4)\cdot\left(\frac{1}{\sqrt{10}}, \frac{3}{\sqrt{10}}, 0\right)\right)\left(\frac{1}{\sqrt{10}}, \frac{3}{\sqrt{10}}, 0\right) + ((1,2,-4)\cdot(0,0,1))(0,0,1)$

$= \frac{7}{10}(1,3,0) - 4(0,0,1) = \left(\frac{7}{10}, \frac{21}{10}, -4\right)$. Thus

$d(\mathbf{x}, W) = \|\mathbf{x} - \text{proj}_W \mathbf{x}\| = \left\|(1,2,-4) - \left(\frac{7}{10}, \frac{21}{10}, -4\right)\right\| = \left\|\left(\frac{3}{10}, \frac{-1}{10}, 0\right)\right\| = \frac{1}{10}\sqrt{10}$.

32. If A is orthogonal, the rows of A form an orthonormal set. The rows of A are the columns of A^t, so from the definition of orthogonal matrix, A^t is orthogonal. Interchange A and A^t in the argument above to show that if A^t is orthogonal then A is orthogonal.

33. $W = \{a(1,0,1) + b(0,1,-2)\}$. $\mathbf{u}_1 = (1,0,1)$ and $\mathbf{u}_2 = (0,1,-2) - \dfrac{(0,1,-2)\cdot(1,0,1)}{(1,0,1)\cdot(1,0,1)}(1,0,1)$

$= (0,1,-2) - (-1)(1,0,1) = (1,1,-1)$ are an orthogonal basis. The vectors

$\left(\dfrac{1}{\sqrt{2}},0,\dfrac{1}{\sqrt{2}}\right)$ and $\left(\dfrac{1}{\sqrt{3}},\dfrac{1}{\sqrt{3}},\dfrac{-1}{\sqrt{3}}\right)$ are therefore an orthonormal basis.

$\mathbf{v} = (1,3,-1) = \mathbf{w} + \mathbf{w}_\perp$, where

$\mathbf{w} = \mathrm{proj}_W \mathbf{v} = \left((1,3,-1)\cdot\left(\dfrac{1}{\sqrt{2}},0,\dfrac{1}{\sqrt{2}}\right)\right)\left(\dfrac{1}{\sqrt{2}},0,\dfrac{1}{\sqrt{2}}\right)$

$\quad + \left((1,3,-1)\cdot\left(\dfrac{1}{\sqrt{3}},\dfrac{1}{\sqrt{3}},\dfrac{-1}{\sqrt{3}}\right)\right)\left(\dfrac{1}{\sqrt{3}},\dfrac{1}{\sqrt{3}},\dfrac{-1}{\sqrt{3}}\right) = \dfrac{5}{3}(1,1,-1) = \left(\dfrac{5}{3},\dfrac{5}{3},\dfrac{-5}{3}\right)$ and

$\mathbf{w}_\perp = \mathbf{v} - \mathrm{proj}_W \mathbf{v} = (1,3,-1) - \left(\dfrac{5}{3},\dfrac{5}{3},\dfrac{-5}{3}\right) = \left(\dfrac{-2}{3},\dfrac{4}{3},\dfrac{2}{3}\right)$.

34. Suppose $\mathbf{u}$ and $\mathbf{v}$ are orthogonal and that $a\mathbf{u} + b\mathbf{v} = \mathbf{0}$, so that $a\mathbf{u} = -b\mathbf{v}$. Then $a\mathbf{u}\cdot a\mathbf{u} = a\mathbf{u}\cdot(-b\mathbf{v}) = (-ab)\mathbf{u}\cdot\mathbf{v} = 0$, so $a\mathbf{u} = \mathbf{0}$. Thus $a = 0$ since $\mathbf{u} \neq \mathbf{0}$. In the same way $b = 0$, and so $\mathbf{u}$ and $\mathbf{v}$ are linearly independent.

35. If $\mathbf{u}\cdot\mathbf{v} = 0$ and $\mathbf{u}\cdot\mathbf{w} = 0$ then $\mathbf{u}\cdot(a\mathbf{v} + b\mathbf{w}) = \mathbf{u}\cdot(a\mathbf{v}) + \mathbf{u}\cdot(b\mathbf{w}) = a(\mathbf{u}\cdot\mathbf{v}) + b(\mathbf{u}\cdot\mathbf{w}) = 0$.

Chapter 6

Exercise Set 6.1

1. $\begin{vmatrix} 5-\lambda & 4 \\ 1 & 2-\lambda \end{vmatrix} = (5-\lambda)(2-\lambda) - 4 = \lambda^2 - 7\lambda + 6 = (\lambda-6)(\lambda-1)$, so the eigenvalues are $\lambda = 6$ and $\lambda = 1$. For $\lambda = 6$, the eigenvectors are the solutions of $\begin{bmatrix} -1 & 4 \\ 1 & -4 \end{bmatrix} \begin{bmatrix} x_1 \\ x_2 \end{bmatrix} = 0$, so the eigenvectors are vectors of the form $r \begin{bmatrix} 4 \\ 1 \end{bmatrix}$. For $\lambda = 1$, the eigenvectors are the solutions of $\begin{bmatrix} 4 & 4 \\ 1 & 1 \end{bmatrix} \begin{bmatrix} x_1 \\ x_2 \end{bmatrix} = 0$, so the eigenvectors are vectors of the form $s \begin{bmatrix} -1 \\ 1 \end{bmatrix}$.

2. $\begin{vmatrix} 1-\lambda & -2 \\ 1 & 4-\lambda \end{vmatrix} = (1-\lambda)(4-\lambda) + 2 = \lambda^2 - 5\lambda + 6 = (\lambda-2)(\lambda-3)$, so the eigenvalues are $\lambda = 2$ and $\lambda = 3$. For $\lambda = 2$, the eigenvectors are the solutions of $\begin{bmatrix} -1 & -2 \\ 1 & 2 \end{bmatrix} \begin{bmatrix} x_1 \\ x_2 \end{bmatrix} = 0$, so the eigenvectors are vectors of the form $r \begin{bmatrix} -2 \\ 1 \end{bmatrix}$. For $\lambda = 3$, the eigenvectors are the solutions of $\begin{bmatrix} -2 & -2 \\ 1 & 1 \end{bmatrix} \begin{bmatrix} x_1 \\ x_2 \end{bmatrix} = 0$, so the eigenvectors are vectors of the form $s \begin{bmatrix} -1 \\ 1 \end{bmatrix}$.

3. $\begin{vmatrix} 5-\lambda & 6 \\ -2 & -2-\lambda \end{vmatrix} = (5-\lambda)(-2-\lambda) + 12 = \lambda^2 - 3\lambda + 2 = (\lambda-2)(\lambda-1)$, so the eigenvalues are $\lambda = 2$ and $\lambda = 1$. For $\lambda = 2$, the eigenvectors are the solutions of $\begin{bmatrix} 3 & 6 \\ -2 & -4 \end{bmatrix} \begin{bmatrix} x_1 \\ x_2 \end{bmatrix} = 0$, so the eigenvectors are vectors of the form $r \begin{bmatrix} -2 \\ 1 \end{bmatrix}$. For $\lambda = 1$, the eigenvectors are the

solutions of $\begin{bmatrix} 4 & 6 \\ -2 & -3 \end{bmatrix} \begin{bmatrix} x_1 \\ x_2 \end{bmatrix} = \mathbf{0}$, so the eigenvectors are vectors of the form $s \begin{bmatrix} -3 \\ 2 \end{bmatrix}$.

4. $\begin{vmatrix} 5-\lambda & 2 \\ -8 & -3-\lambda \end{vmatrix} = (5-\lambda)(-3-\lambda) + 16 = \lambda^2 - 2\lambda + 1 = (\lambda-1)(\lambda-1)$, so the only eigenvalue is $\lambda = 1$. The eigenvectors are the solutions of $\begin{bmatrix} 4 & 2 \\ -8 & -4 \end{bmatrix} \begin{bmatrix} x_1 \\ x_2 \end{bmatrix} = \mathbf{0}$, so the eigenvectors are vectors of the form $r \begin{bmatrix} 1 \\ -2 \end{bmatrix}$.

5. $\begin{vmatrix} 1-\lambda & 2 \\ 2 & 1-\lambda \end{vmatrix} = (1-\lambda)(1-\lambda) - 4 = \lambda^2 - 2\lambda - 3 = (\lambda-3)(\lambda+1)$, so the eigenvalues are $\lambda = 3$ and $\lambda = -1$. For $\lambda = 3$, the eigenvectors are the solutions of $\begin{bmatrix} -2 & 2 \\ 2 & -2 \end{bmatrix} \begin{bmatrix} x_1 \\ x_2 \end{bmatrix} = \mathbf{0}$, so the eigenvectors are vectors of the form $r \begin{bmatrix} 1 \\ 1 \end{bmatrix}$. For $\lambda = -1$, the eigenvectors are the solutions of $\begin{bmatrix} 2 & 2 \\ 2 & 2 \end{bmatrix} \begin{bmatrix} x_1 \\ x_2 \end{bmatrix} = \mathbf{0}$, so the eigenvectors are vectors of the form $s \begin{bmatrix} -1 \\ 1 \end{bmatrix}$.

6. $\begin{vmatrix} 2-\lambda & 1 \\ -1 & 4-\lambda \end{vmatrix} = (2-\lambda)(4-\lambda) + 1 = \lambda^2 - 6\lambda + 9 = (\lambda-3)(\lambda-3)$, so the only eigenvalue is $\lambda = 3$. The eigenvectors are the solutions of $\begin{bmatrix} -1 & 1 \\ -1 & 1 \end{bmatrix} \begin{bmatrix} x_1 \\ x_2 \end{bmatrix} = \mathbf{0}$, so the eigenvectors are vectors of the form $r \begin{bmatrix} 1 \\ 1 \end{bmatrix}$.

7. $\begin{vmatrix} 3-\lambda & -1 \\ 2 & -\lambda \end{vmatrix} = (3-\lambda)(-\lambda) + 2 = \lambda^2 - 3\lambda + 2 = (\lambda-2)(\lambda-1)$, so the eigenvalues are $\lambda = 2$ and $\lambda = 1$. For $\lambda = 2$, the eigenvectors are the solutions of $\begin{bmatrix} 1 & -1 \\ 2 & -2 \end{bmatrix} \begin{bmatrix} x_1 \\ x_2 \end{bmatrix} = \mathbf{0}$, so the eigenvectors are vectors of the form $r \begin{bmatrix} 1 \\ 1 \end{bmatrix}$. For $\lambda = 1$, the eigenvectors are the solutions of $\begin{bmatrix} 2 & -1 \\ 2 & -1 \end{bmatrix} \begin{bmatrix} x_1 \\ x_2 \end{bmatrix} = \mathbf{0}$, so the eigenvectors are vectors of the form $s \begin{bmatrix} 1 \\ 2 \end{bmatrix}$.

8. $\begin{vmatrix} 2-\lambda & -4 \\ -1 & 2-\lambda \end{vmatrix} = (2-\lambda)(2-\lambda) - 4 = \lambda^2 - 4\lambda = \lambda(\lambda-4)$, so the eigenvalues are $\lambda = 0$ and $\lambda = 4$. For $\lambda = 0$, the eigenvectors are the solutions of $\begin{bmatrix} 2 & -4 \\ -1 & 2 \end{bmatrix} \begin{bmatrix} x_1 \\ x_2 \end{bmatrix} = \mathbf{0}$, so the eigenvectors are vectors of the form $r \begin{bmatrix} 2 \\ 1 \end{bmatrix}$. For $\lambda = 4$, the eigenvectors are the solutions of $\begin{bmatrix} -2 & -4 \\ -1 & -2 \end{bmatrix} \begin{bmatrix} x_1 \\ x_2 \end{bmatrix} = \mathbf{0}$, so the eigenvectors are vectors of the form $s \begin{bmatrix} -2 \\ 1 \end{bmatrix}$.

9. $\begin{vmatrix} 3-\lambda & 2 & -2 \\ -3 & -1-\lambda & 3 \\ 1 & 2 & -\lambda \end{vmatrix} = (3-\lambda)(-1-\lambda)(-\lambda) + 18 + 2(-1-\lambda) - 6(3-\lambda) - 6\lambda = -\lambda^3 + 2\lambda^2 + \lambda - 2$

$= (1-\lambda^2)(\lambda-2)$, so the eigenvalues are $\lambda = 1$, $\lambda = -1$, and $\lambda = 2$. For $\lambda = 1$, the eigenvectors are the solutions of $\begin{bmatrix} 2 & 2 & -2 \\ -3 & -2 & 3 \\ 1 & 2 & -1 \end{bmatrix} \begin{bmatrix} x_1 \\ x_2 \\ x_3 \end{bmatrix} = \mathbf{0}$, so the eigenvectors are vectors of the form $r \begin{bmatrix} 1 \\ 0 \\ 1 \end{bmatrix}$. For $\lambda = -1$, the eigenvectors are the solutions of

$\begin{bmatrix} 4 & 2 & -2 \\ -3 & 0 & 3 \\ 1 & 2 & 1 \end{bmatrix} \begin{bmatrix} x_1 \\ x_2 \\ x_3 \end{bmatrix} = \mathbf{0}$, so the eigenvectors are vectors of the form $s \begin{bmatrix} 1 \\ -1 \\ 1 \end{bmatrix}$.

For $\lambda = 2$, the eigenvectors are the solutions of $\begin{bmatrix} 1 & 2 & -2 \\ -3 & -3 & 3 \\ 1 & 2 & -2 \end{bmatrix} \begin{bmatrix} x_1 \\ x_2 \\ x_3 \end{bmatrix} = \mathbf{0}$, so the

eigenvectors are vectors of the form $t \begin{bmatrix} 0 \\ 1 \\ 1 \end{bmatrix}$.

10. $\begin{vmatrix} 1-\lambda & -2 & 2 \\ -2 & 1-\lambda & 2 \\ -2 & 0 & 3-\lambda \end{vmatrix} = (1-\lambda)^2(3-\lambda)$, so the eigenvalues are $\lambda = 1$ and $\lambda = 3$. For $\lambda = 1$, the

eigenvectors are the solutions of $\begin{bmatrix} 0 & -2 & 2 \\ -2 & 0 & 2 \\ -2 & 0 & 2 \end{bmatrix} \begin{bmatrix} x_1 \\ x_2 \\ x_3 \end{bmatrix} = \mathbf{0}$, so the eigenvectors are

vectors of the form $r \begin{bmatrix} 1 \\ 1 \\ 1 \end{bmatrix}$. For $\lambda = 3$, the eigenvectors are the solutions of

$\begin{bmatrix} -2 & -2 & 2 \\ -2 & -2 & 2 \\ -2 & 0 & 0 \end{bmatrix} \begin{bmatrix} x_1 \\ x_2 \\ x_3 \end{bmatrix} = \mathbf{0}$, so the eigenvectors are vectors of the form $s \begin{bmatrix} 0 \\ 1 \\ 1 \end{bmatrix}$.

11. $\begin{vmatrix} 1-\lambda & 0 & 0 \\ -2 & 1-\lambda & 2 \\ -2 & 0 & 3-\lambda \end{vmatrix} = (1-\lambda)^2(3-\lambda)$, so the eigenvalues are $\lambda = 1$ and $\lambda = 3$. For $\lambda = 1$, the

eigenvectors are the solutions of $\begin{bmatrix} 0 & 0 & 0 \\ -2 & 0 & 2 \\ -2 & 0 & 2 \end{bmatrix} \begin{bmatrix} x_1 \\ x_2 \\ x_3 \end{bmatrix} = \mathbf{0}$, so the eigenvectors are

vectors of the form $r \begin{bmatrix} 1 \\ 0 \\ 1 \end{bmatrix} + s \begin{bmatrix} 0 \\ 1 \\ 0 \end{bmatrix}$. For $\lambda = 3$, the eigenvectors are the solutions of

$\begin{bmatrix} -2 & 0 & 0 \\ -2 & -2 & 2 \\ -2 & 0 & 0 \end{bmatrix} \begin{bmatrix} x_1 \\ x_2 \\ x_3 \end{bmatrix} = \mathbf{0}$, so the eigenvectors are vectors of the form $t \begin{bmatrix} 0 \\ 1 \\ 1 \end{bmatrix}$.

12. $\begin{vmatrix} 1-\lambda & 0 & 0 \\ -2 & 5-\lambda & -2 \\ -2 & 4 & -1-\lambda \end{vmatrix} = (1-\lambda)(5-\lambda)(-1-\lambda) + 8(1-\lambda) = (1-\lambda)^2(3-\lambda)$, so the eigenvalues are

$\lambda = 1$ and $\lambda = 3$. For $\lambda = 1$, the eigenvectors are the solutions of $\begin{bmatrix} 0 & 0 & 0 \\ -2 & 4 & -2 \\ -2 & 4 & -2 \end{bmatrix} \begin{bmatrix} x_1 \\ x_2 \\ x_3 \end{bmatrix} = \mathbf{0}$,

so the eigenvectors are vectors of the form $r \begin{bmatrix} -1 \\ 0 \\ 1 \end{bmatrix} + s \begin{bmatrix} 1 \\ 1 \\ 1 \end{bmatrix}$. For $\lambda = 3$, the eigenvectors

are the solutions of $\begin{bmatrix} -2 & 0 & 0 \\ -2 & 2 & -2 \\ -2 & 4 & -4 \end{bmatrix} \begin{bmatrix} x_1 \\ x_2 \\ x_3 \end{bmatrix} = \mathbf{0}$, so the eigenvectors are vectors of the form

$t \begin{bmatrix} 0 \\ 1 \\ 1 \end{bmatrix}$.

Section 6.1

13. $\begin{vmatrix} 15-\lambda & 7 & -7 \\ -1 & 1-\lambda & 1 \\ 13 & 7 & -5-\lambda \end{vmatrix} = (1-\lambda)(16-10\lambda+\lambda^2) = (1-\lambda)(2-\lambda)(8-\lambda)$, so the eigenvalues are

$\lambda = 1$, $\lambda = 2$, and $\lambda = 8$. For $\lambda = 1$, the eigenvectors are the solutions of

$\begin{bmatrix} 14 & 7 & -7 \\ -1 & 0 & 1 \\ 13 & 7 & -6 \end{bmatrix} \begin{bmatrix} x_1 \\ x_2 \\ x_3 \end{bmatrix} = \mathbf{0}$, so the eigenvectors are vectors of the form $r \begin{bmatrix} 1 \\ -1 \\ 1 \end{bmatrix}$. For $\lambda = 2$,

the eigenvectors are the solutions of $\begin{bmatrix} 13 & 7 & -7 \\ -1 & -1 & 1 \\ 13 & 7 & -7 \end{bmatrix} \begin{bmatrix} x_1 \\ x_2 \\ x_3 \end{bmatrix} = \mathbf{0}$, so the eigenvectors are

vectors of the form $s \begin{bmatrix} 0 \\ 1 \\ 1 \end{bmatrix}$. For $\lambda = 8$, the eigenvectors are the solutions of

$\begin{bmatrix} 7 & 7 & -7 \\ -1 & -7 & 1 \\ 13 & 7 & -13 \end{bmatrix} \begin{bmatrix} x_1 \\ x_2 \\ x_3 \end{bmatrix} = \mathbf{0}$, so the eigenvectors are vectors of the form $t \begin{bmatrix} 1 \\ 0 \\ 1 \end{bmatrix}$.

14. $\begin{vmatrix} 5-\lambda & -2 & 2 \\ 4 & -3-\lambda & 4 \\ 4 & -6 & 7-\lambda \end{vmatrix} = (5-\lambda)(3-\lambda)(1-\lambda)$, so the eigenvalues are $\lambda = 5$, $\lambda = 3$, and $\lambda = 1$.

For $\lambda = 5$, the eigenvectors are the solutions of $\begin{bmatrix} 0 & -2 & 2 \\ 4 & -8 & 4 \\ 4 & -6 & 2 \end{bmatrix} \begin{bmatrix} x_1 \\ x_2 \\ x_3 \end{bmatrix} = \mathbf{0}$, so the

eigenvectors are vectors of the form $r \begin{bmatrix} 1 \\ 1 \\ 1 \end{bmatrix}$. For $\lambda = 3$, the eigenvectors are the solutions of $\begin{bmatrix} 2 & -2 & 2 \\ 4 & -6 & 4 \\ 4 & -6 & 4 \end{bmatrix} \begin{bmatrix} x_1 \\ x_2 \\ x_3 \end{bmatrix} = \mathbf{0}$, so the eigenvectors are vectors of the form $s \begin{bmatrix} 1 \\ 0 \\ -1 \end{bmatrix}$. For $\lambda = 1$,

the eigenvectors are the solutions of $\begin{bmatrix} 4 & -2 & 2 \\ 4 & -4 & 4 \\ 4 & -6 & 6 \end{bmatrix} \begin{bmatrix} x_1 \\ x_2 \\ x_3 \end{bmatrix} = \mathbf{0}$, so the eigenvectors are

vectors of the form $t \begin{bmatrix} 0 \\ 1 \\ 1 \end{bmatrix}$.

15. $\begin{vmatrix} 4-\lambda & 2 & -2 & 2 \\ 1 & 3-\lambda & 1 & -1 \\ 0 & 0 & 2-\lambda & 0 \\ 1 & 1 & -3 & 5-\lambda \end{vmatrix} = (2-\lambda) \begin{vmatrix} 4-\lambda & 2 & 2 \\ 1 & 3-\lambda & -1 \\ 1 & 1 & 5-\lambda \end{vmatrix} = (2-\lambda)(4-\lambda)(2-\lambda)(6-\lambda)$, so the

eigenvalues are $\lambda = 2$, $\lambda = 4$, and $\lambda = 6$. For $\lambda = 2$, the eigenvectors are the solutions of

$\begin{bmatrix} 2 & 2 & -2 & 2 \\ 1 & 1 & 1 & -1 \\ 0 & 0 & 0 & 0 \\ 1 & 1 & -3 & 3 \end{bmatrix} \begin{bmatrix} x_1 \\ x_2 \\ x_3 \\ x_4 \end{bmatrix} = \mathbf{0}$, so the eigenvectors are vectors of the form

$r \begin{bmatrix} 1 \\ -1 \\ 0 \\ 0 \end{bmatrix} + s \begin{bmatrix} 0 \\ 0 \\ 1 \\ 1 \end{bmatrix}$. For $\lambda = 4$, the eigenvectors are the solutions of

$\begin{bmatrix} 0 & 2 & -2 & 2 \\ 1 & -1 & 1 & -1 \\ 0 & 0 & -2 & 0 \\ 1 & 1 & -3 & 1 \end{bmatrix} \begin{bmatrix} x_1 \\ x_2 \\ x_3 \\ x_4 \end{bmatrix} = \mathbf{0}$, so the eigenvectors are vectors of the form $t \begin{bmatrix} 0 \\ 1 \\ 0 \\ -1 \end{bmatrix}$.

Section 6.1

For $\lambda = 6$, the eigenvectors are the solutions of $\begin{bmatrix} -2 & 2 & -2 & 2 \\ 1 & -3 & 1 & -1 \\ 0 & 0 & -4 & 0 \\ 1 & 1 & -3 & -1 \end{bmatrix} \begin{bmatrix} x_1 \\ x_2 \\ x_3 \\ x_4 \end{bmatrix} = \mathbf{0}$, so the

eigenvectors are vectors of the form $p \begin{bmatrix} 1 \\ 0 \\ 0 \\ 1 \end{bmatrix}$.

16. $\begin{vmatrix} 3-\lambda & 5 & -5 & 5 \\ 3 & 1-\lambda & 3 & -3 \\ -2 & 2 & -\lambda & 2 \\ 0 & 4 & -6 & 8-\lambda \end{vmatrix} = \begin{vmatrix} 3-\lambda & 5 & -5 & 8-\lambda \\ 3 & 1-\lambda & 3 & 0 \\ -2 & 2 & -\lambda & 0 \\ 0 & 4 & -6 & 8-\lambda \end{vmatrix} = \begin{vmatrix} 3-\lambda & 1 & 1 & 0 \\ 3 & 1-\lambda & 3 & 0 \\ -2 & 2 & -\lambda & 0 \\ 0 & 4 & -6 & 8-\lambda \end{vmatrix}$

$= (8-\lambda) \begin{vmatrix} 3-\lambda & 1 & 1 \\ 3 & 1-\lambda & 3 \\ -2 & 2 & -\lambda \end{vmatrix} = (8-\lambda)(4-\lambda)(2-\lambda)(2+\lambda)$, so the eigenvalues are $\lambda = 8$, $\lambda = 4$,

$\lambda = 2$, and $\lambda = -2$. For $\lambda = 2$, the eigenvectors are the solutions of

$\begin{bmatrix} 1 & 5 & -5 & 5 \\ 3 & -1 & 3 & -3 \\ -2 & 2 & -2 & 2 \\ 0 & 4 & -6 & 6 \end{bmatrix} \begin{bmatrix} x_1 \\ x_2 \\ x_3 \\ x_4 \end{bmatrix} = \mathbf{0}$, so the eigenvectors are vectors of the form $r \begin{bmatrix} 0 \\ 0 \\ 1 \\ 1 \end{bmatrix}$. For

$\lambda = 4$, the eigenvectors are the solutions of $\begin{bmatrix} -1 & 5 & -5 & 5 \\ 3 & -3 & 3 & -3 \\ -2 & 2 & -4 & 2 \\ 0 & 4 & -6 & 4 \end{bmatrix} \begin{bmatrix} x_1 \\ x_2 \\ x_3 \\ x_4 \end{bmatrix} = \mathbf{0}$, so the

eigenvectors are vectors of the form $s \begin{bmatrix} 0 \\ 1 \\ 0 \\ -1 \end{bmatrix}$. For $\lambda = 8$, the eigenvectors are the solutions of $\begin{bmatrix} -5 & 5 & -5 & 5 \\ 3 & -7 & 3 & -3 \\ -2 & 2 & -8 & 2 \\ 0 & 4 & -6 & 0 \end{bmatrix} \begin{bmatrix} x_1 \\ x_2 \\ x_3 \\ x_4 \end{bmatrix} = \mathbf{0}$, so the eigenvectors are vectors of the form

$t \begin{bmatrix} 1 \\ 0 \\ 0 \\ 1 \end{bmatrix}$. For $\lambda = -2$, the eigenvectors are the solutions of $\begin{bmatrix} 5 & 5 & -5 & 5 \\ 3 & 3 & 3 & -3 \\ -2 & 2 & 2 & 2 \\ 0 & 4 & -6 & 10 \end{bmatrix} \begin{bmatrix} x_1 \\ x_2 \\ x_3 \\ x_4 \end{bmatrix} = \mathbf{0}$,

so the eigenvectors are vectors of the form $p \begin{bmatrix} 1 \\ -1 \\ 1 \\ 1 \end{bmatrix}$.

17. $\begin{vmatrix} 1-\lambda & 0 \\ 0 & 1-\lambda \end{vmatrix} = (1-\lambda)^2$, so the only eigenvalue is $\lambda = 1$. The eigenvectors are the

solutions of $\begin{bmatrix} 0 & 0 \\ 0 & 0 \end{bmatrix} \begin{bmatrix} x_1 \\ x_2 \end{bmatrix} = \mathbf{0}$, so the eigenvectors are all the vectors in $\mathbf{R}^2$. The

transformation represented by the identity matrix is the identity transformation that maps each vector in $\mathbf{R}^2$ into itself.

18. $\begin{vmatrix} 3-\lambda & 0 \\ 0 & 3-\lambda \end{vmatrix} = (3-\lambda)^2$, so the only eigenvalue is $\lambda = 3$. The eigenvectors are the

solutions of $\begin{bmatrix} 0 & 0 \\ 0 & 0 \end{bmatrix} \begin{bmatrix} x_1 \\ x_2 \end{bmatrix} = \mathbf{0}$, so the eigenvectors are all the vectors in $\mathbf{R}^2$. The

transformation represented by the given matrix is an expansion transformation that maps each vector $\mathbf{v}$ in $\mathbf{R}^2$ into the vector $3\mathbf{v}$. Thus each vector in $\mathbf{R}^2$ has the same direction as its image.

Section 6.1

19. $\begin{vmatrix} -2-\lambda & 0 \\ 0 & -2-\lambda \end{vmatrix} = (-2-\lambda)^2$, so the only eigenvalue is $\lambda = -2$. The eigenvectors are the

    solutions of $\begin{bmatrix} 0 & 0 \\ 0 & 0 \end{bmatrix} \begin{bmatrix} x_1 \\ x_2 \end{bmatrix} = \mathbf{0}$, so the eigenvectors are all the vectors in $\mathbf{R}^2$. The

    transformation represented by the given matrix maps each vector $\mathbf{v}$ in $\mathbf{R}^2$ into the vector $-2\mathbf{v}$. Thus each vector in $\mathbf{R}^2$ has the direction opposite the direction of its image.

20. $\begin{vmatrix} 1-\lambda & 1 \\ -2 & -1-\lambda \end{vmatrix} = (1-\lambda)(-1-\lambda) + 2 = \lambda^2 + 1 \neq 0$ for any real value of λ, so there are no

    real eigenvalues.

21. $\begin{vmatrix} -\lambda & -1 \\ 1 & -\lambda \end{vmatrix} = \lambda^2 + 1 \neq 0$ for any real value of λ, so there are no real eigenvalues.

    The given matrix is a rotation matrix that rotates each vector in $\mathbf{R}^2$ through a 90° angle. Thus no vector has the same or opposite direction as its image.

22. If A is a diagonal matrix with diagonal elements a_{ii}, then $A - \lambda I_n$ is also a diagonal matrix with diagonal elements $a_{ii} - \lambda$. Thus $|A - \lambda I_n|$ is the product of the terms $a_{ii} - \lambda$, and the solutions of the equation $|A - \lambda I_n| = 0$ are the values $\lambda = a_{ii}$, the diagonal elements of A.

23. If A is an upper triangular matrix with diagonal elements a_{ii}, then $A - \lambda I_n$ is also an upper triangular matrix with diagonal elements $a_{ii} - \lambda$. $|A - \lambda I_n|$ is the product of the terms $a_{ii} - \lambda$, and the solutions of the equation $|A - \lambda I_n| = 0$ are the values $\lambda = a_{ii}$, the diagonal elements of A.

24. $(A - \lambda I_n)^t = A^t - (\lambda I_n)^t = A^t - \lambda I_n$, so $|A - \lambda I_n| = |(A - \lambda I_n)^t| = |A^t - \lambda I_n|$, that is, A and A^t have the same characteristic polynomial and therefore the same eigenvalues.

Section 6.1

25. Zero is an eigenvalue of a matrix A if and only if the equation $A\mathbf{x} = 0\mathbf{x} = \mathbf{0}$ has a nonzero solution $\mathbf{x}$. $A\mathbf{x} = \mathbf{0}$ has a nonzero solution if and only if A is singular.

26. If λ is an eigenvalue of A and X is an eigenvector corresponding to λ, then $AX = \lambda X$, $A^2X = A(AX) = A(\lambda X) = \lambda AX = \lambda(\lambda X) = \lambda^2 X$ (i.e., λ^2 is an eigenvalue of A^2 with corresponding eigenvector X), and $A^m X = A^{m-1}(AX) = A^{m-1}(\lambda X) = \lambda A^{m-1}X = \lambda A^{m-2}(AX)$ $= \lambda A^{m-2}(\lambda X) = \lambda^2 A^{m-2}X = \lambda^2 A^{m-3}(AX) = \ldots = \lambda^{m-1}A(AX) = \lambda^{m-1}A(\lambda X) = \lambda^m X$, so that λ^m is an eigenvalue of A^m with corresponding eigenvector X.

27. $|A - 0I_n| = |A|$, so $|A - 0I_n| = 0$ if and only if $|A| = 0$.

28. The characteristic polynomial for the nxn zero matrix O_n is λ^n. Thus the only eigenvalue of O_n is $\lambda = 0$. However if λ is an eigenvalue of A then λ^k is an eigenvalue of $A^k = O_n$ (see Exercise 26). Thus $\lambda^k = 0$ so $\lambda = 0$.

29. The characteristic polynomial of A is $|A - \lambda I_n| = \lambda^n + c_{n-1}\lambda^{n-1} + \ldots + c_1\lambda + c_0$. Substituting $\lambda = 0$, this equation becomes $|A| = c_0$.

30. $\begin{vmatrix} 1-\lambda & 2 \\ 1 & -\lambda \end{vmatrix} = (1-\lambda)(-\lambda) - 2 = \lambda^2 - \lambda - 2 = (\lambda-2)(\lambda+1)$, so the eigenvalues are $\lambda = 2$ and $\lambda = -1$. The corresponding eigenvectors (written as row vectors) are [2 1] and [1 −1].

$\begin{vmatrix} 2-\lambda & 3 \\ 1 & -\lambda \end{vmatrix} = (2-\lambda)(-\lambda) - 3 = \lambda^2 - 2\lambda - 3 = (\lambda-3)(\lambda+1)$, so the eigenvalues are $\lambda = 3$ and $\lambda = -1$. The corresponding eigenvectors are [3 1] and [1 −1].

$$\begin{vmatrix} 3-\lambda & 4 \\ 1 & -\lambda \end{vmatrix} = (3-\lambda)(-\lambda) - 4 = \lambda^2 - 3\lambda - 4 = (\lambda-4)(\lambda+1),$$ so the eigenvalues are $\lambda = 4$

and $\lambda = -1$. The corresponding eigenvectors are [4 1] and [1 −1].

$$\begin{vmatrix} 4-\lambda & 5 \\ 1 & -\lambda \end{vmatrix} = (4-\lambda)(-\lambda) - 5 = \lambda^2 - 4\lambda - 5 = (\lambda-5)(\lambda+1),$$ so the eigenvalues are $\lambda = 5$

and $\lambda = -1$. The corresponding eigenvectors are [5 1] and [1 −1].

Conjecture: The matrix $\begin{bmatrix} a & a+1 \\ 1 & 0 \end{bmatrix}$ has eigenvalues $\lambda = a+1$ and $\lambda = -1$ and corresponding eigenvectors [a+1 1] and [1 −1].

Proof: $\begin{bmatrix} a & a+1 \\ 1 & 0 \end{bmatrix} \begin{bmatrix} a+1 \\ 1 \end{bmatrix} = (a+1) \begin{bmatrix} a+1 \\ 1 \end{bmatrix}$ and $\begin{bmatrix} a & a+1 \\ 1 & 0 \end{bmatrix} \begin{bmatrix} 1 \\ -1 \end{bmatrix} = \begin{bmatrix} -1 \\ 1 \end{bmatrix} = (-1) \begin{bmatrix} 1 \\ -1 \end{bmatrix}.$

31. (a) $\begin{vmatrix} -\lambda & 2 \\ -1 & 3-\lambda \end{vmatrix} = (-\lambda)(3-\lambda) + 2 = \lambda^2 - 3\lambda + 2.$

$\begin{bmatrix} 0 & 2 \\ -1 & 3 \end{bmatrix}^2 - 3 \begin{bmatrix} 0 & 2 \\ -1 & 3 \end{bmatrix} + 2 \begin{bmatrix} 1 & 0 \\ 0 & 1 \end{bmatrix} = \begin{bmatrix} -2 & 6 \\ -3 & 7 \end{bmatrix} + \begin{bmatrix} 2 & -6 \\ 3 & -7 \end{bmatrix} = \begin{bmatrix} 0 & 0 \\ 0 & 0 \end{bmatrix}.$

(b) $\begin{vmatrix} 8-\lambda & -10 \\ 5 & -7-\lambda \end{vmatrix} = (8-\lambda)(-7-\lambda) + 50 = \lambda^2 - \lambda - 6.$

$\begin{bmatrix} 8 & -10 \\ 5 & -7 \end{bmatrix}^2 - \begin{bmatrix} 8 & -10 \\ 5 & -7 \end{bmatrix} - 6 \begin{bmatrix} 1 & 0 \\ 0 & 1 \end{bmatrix} = \begin{bmatrix} 14 & -10 \\ 5 & -1 \end{bmatrix} + \begin{bmatrix} -14 & 10 \\ -5 & 1 \end{bmatrix} = \begin{bmatrix} 0 & 0 \\ 0 & 0 \end{bmatrix}.$

(c) $\begin{vmatrix} 6-\lambda & -8 \\ 4 & -6-\lambda \end{vmatrix} = (6-\lambda)(-6-\lambda) + 32 = \lambda^2 - 4.$

$\begin{bmatrix} 6 & -8 \\ 4 & -6 \end{bmatrix}^2 - 4 \begin{bmatrix} 1 & 0 \\ 0 & 1 \end{bmatrix} = \begin{bmatrix} 4 & 0 \\ 0 & 4 \end{bmatrix} - \begin{bmatrix} 4 & 0 \\ 0 & 4 \end{bmatrix} = \begin{bmatrix} 0 & 0 \\ 0 & 0 \end{bmatrix}.$

Section 6.2

(d) $\begin{vmatrix} -1-\lambda & 5 \\ -10 & 14-\lambda \end{vmatrix} = (-1-\lambda)(14-\lambda) + 50 = \lambda^2 - 13\lambda + 36.$

$\begin{bmatrix} -1 & 5 \\ -10 & 14 \end{bmatrix}^2 - 13 \begin{bmatrix} -1 & 5 \\ -10 & 14 \end{bmatrix} + 36 \begin{bmatrix} 1 & 0 \\ 0 & 1 \end{bmatrix} = \begin{bmatrix} -49 & 65 \\ -130 & 146 \end{bmatrix} + \begin{bmatrix} 49 & -65 \\ 130 & -146 \end{bmatrix}$

$= \begin{bmatrix} 0 & 0 \\ 0 & 0 \end{bmatrix}.$

Exercise Set 6.2

1. The eigenvectors of $\lambda = 1$ are vectors of the form $r \begin{bmatrix} 2 \\ 1 \end{bmatrix}$. If there is no change in total population $2r + r = 200 + 60 = 260$, so $r = 260/3$. Thus the long-term prediction is that population in metropolitan areas will be $2r = 173.3333$ million and population in nonmetropolitan areas will be $r = 86.6667$ million.

2. The eigenvectors of $\lambda = 1$ are vectors of the form $r \begin{bmatrix} .7 \\ 1.3 \\ 1 \end{bmatrix}$. If there is no change in total population $.7r + 1.3r + r = 58 + 142 + 60 = 260$, so $r = 260/3$. Thus the long-term prediction is that population in the cities will be $.7r = 60.6667$ million, population in the suburbs will be $1.3r = 104$ million, and population in nonmetropolitan areas will be $r = 112.6667$ million.

3. $P^2 = \begin{bmatrix} .375 & .25 & .125 \\ .5 & .5 & .5 \\ .125 & .25 & .375 \end{bmatrix}$, and since all terms are positive, P is regular. The eigenvectors of $\lambda = 1$ are vectors of the form $r \begin{bmatrix} 1 \\ 2 \\ 1 \end{bmatrix}$. The powers of P approach the stochastic matrix $Q = \begin{bmatrix} s & s & s \\ 2s & 2s & 2s \\ s & s & s \end{bmatrix}$, so $s = .25$ and $Q = \begin{bmatrix} .25 & .25 & .25 \\ .5 & .5 & .5 \\ .25 & .25 & .25 \end{bmatrix}$.

The columns of Q indicate that when guinea pigs are bred with hybrids, only the long-term distribution of types AA, Aa, and aa will be 1:2:1. That is, the long-term probabilities

Section 6.2

of Types AA, Aa, and aa are .25, .5, and .25.

4. $$P = \begin{bmatrix} .65 & .23 \\ .35 & .77 \end{bmatrix} \begin{matrix} \text{wet} \\ \text{dry} \end{matrix}$$ with columns labeled wet, dry

(a) $P^2 = \begin{bmatrix} .5 & .33 \\ .5 & .67 \end{bmatrix}$. If Thursday is dry, the probability that Saturday will also be dry is .67, the (2,2) term in P^2.

(b) The eigenvectors of P corresponding to $\lambda = 1$ are vectors of the form $r\begin{bmatrix} 23 \\ 35 \end{bmatrix}$.

Thus the powers of P approach the stochastic matrix $Q = \begin{bmatrix} 23s & 23s \\ 35s & 35s \end{bmatrix}$, so

$(23+35)s = 1$ and $s = \frac{1}{58}$. $\frac{23}{58} = .4$, $\frac{35}{58} = .6$, and $Q = \begin{bmatrix} .4 & .4 \\ .6 & .6 \end{bmatrix}$, so the long-term probability for a wet day in December is .4 and for a dry day is .6.

5. P is regular since all terms of P^2 are positive. The eigenvectors of P corresponding to $\lambda = 1$ are vectors of the form $r\begin{bmatrix} 3 \\ 3 \\ 2 \end{bmatrix}$; thus the distribution of rats in rooms 1, 2, and 3 is

3:3:2. The powers of P approach the stochastic matrix $Q = \begin{bmatrix} 3s & 3s & 3s \\ 3s & 3s & 3s \\ 2s & 2s & 2s \end{bmatrix}$, so

$3s + 3s + 2s = 1$ and $s = 1/8$. The long-term probability that a given rat will be in room 2 is therefore 3/8.

6. room 1   2   3   4
$$P = \begin{bmatrix} 0 & 1/3 & 0 & 1/4 \\ 1/2 & 0 & 1/3 & 1/4 \\ 0 & 1/3 & 0 & 1/2 \\ 1/2 & 1/3 & 2/3 & 0 \end{bmatrix} \begin{matrix} 1 \\ 2 \\ 3 \\ 4 \end{matrix}$$

P is regular since every term in P^2 is positive.

The eigenvectors of P corresponding to $\lambda = 1$ are vectors of the form

$r \begin{bmatrix} 2 \\ 3 \\ 3 \\ 4 \end{bmatrix}$; thus the distribution of rats in rooms 1, 2, 3, and 4 is 2:3:3:4. The powers of P

approach the stochastic matrix $Q = \begin{bmatrix} 2s & 2s & 2s & 2s \\ 3s & 3s & 3s & 3s \\ 3s & 3s & 3s & 3s \\ 4s & 4s & 4s & 4s \end{bmatrix}$, so $2s + 3s + 3s + 4s = 1$

and $s = 1/12$. The long-term probability that a given rat will be in room 4 is therefore $4/12 = 1/3$.

7. $P = \begin{bmatrix} .75 & .20 \\ .25 & .80 \end{bmatrix}$. The eigenvectors of $\lambda = 1$ are vectors of the form $r \begin{bmatrix} 4 \\ 5 \end{bmatrix}$. The powers of

P approach $Q = \begin{bmatrix} 4s & 4s \\ 5s & 5s \end{bmatrix}$, so $4s + 5s = 1$ and $s = 1/9$. If current trends continue, the eventual distribution will be $4/9 = 44.4\%$ using company A and $5/9 = 55.6\%$ using company B.

8. $P = \begin{bmatrix} .8 & .2 & .05 \\ .05 & .75 & .05 \\ .15 & .05 & .9 \end{bmatrix}$. The eigenvectors of $\lambda = 1$ are vectors of the form $r \begin{bmatrix} 9 \\ 5 \\ 16 \end{bmatrix}$. The

powers of P approach $Q = \begin{bmatrix} 9s & 9s & 9s \\ 5s & 5s & 5s \\ 16s & 16s & 16s \end{bmatrix}$, so $9s + 5s + 16s = 1$ and $s = \dfrac{1}{30}$. If

Section 6.3

current buying patterns continue, the eventual distribution will be 9/30 = 30% using product I, 5/30 = $16\frac{2}{3}$% using product II, and 16/30 = $53\frac{1}{3}$% using product III.

9. The sum of the terms in each column of a stochastic matrix A is 1, so the sum of the terms in each column of A − I is zero. It has previously been proved (Exercise 15, Section 3.2) that if the sum of the terms in each column of a matrix is zero, the determinant of the matrix is zero. Thus |A − 1I| = |A − I| = 0, and 1 is an eigenvalue of A.

Exercise Set 6.3

1. (a) $C^{-1}AC = \begin{bmatrix} 3 & -5 \\ -1 & 2 \end{bmatrix}\begin{bmatrix} 1 & 2 \\ -1 & 3 \end{bmatrix}\begin{bmatrix} 2 & 5 \\ 1 & 3 \end{bmatrix} = \begin{bmatrix} 7 & 13 \\ -2 & -3 \end{bmatrix}$.

   (b) $C^{-1}AC = \begin{bmatrix} -1 & 2 \\ 2 & -3 \end{bmatrix}\begin{bmatrix} -8 & 18 \\ -6 & 13 \end{bmatrix}\begin{bmatrix} 3 & 2 \\ 2 & 1 \end{bmatrix} = \begin{bmatrix} 4 & 0 \\ 0 & 1 \end{bmatrix}$.

   (c) $C^{-1}AC = \begin{bmatrix} 4 & -1 \\ -7 & 2 \end{bmatrix}\begin{bmatrix} 0 & 4 \\ 3 & 2 \end{bmatrix}\begin{bmatrix} 2 & 1 \\ 7 & 4 \end{bmatrix} = \begin{bmatrix} 92 & 53 \\ -156 & -90 \end{bmatrix}$.

2. (a) $C^{-1}AC = \begin{bmatrix} -1 & 2 & 2 \\ 0 & 1 & 1 \\ 2 & 0 & -1 \end{bmatrix}\begin{bmatrix} 2 & 0 & 0 \\ -2 & 2 & 1 \\ 2 & 0 & 1 \end{bmatrix}\begin{bmatrix} -1 & 2 & 0 \\ 2 & -3 & 1 \\ -2 & 4 & -1 \end{bmatrix} = \begin{bmatrix} 2 & 0 & 0 \\ 0 & 2 & 0 \\ 0 & 0 & 1 \end{bmatrix}$.

   (b) $C^{-1}AC = \begin{bmatrix} -1 & -3 & 2 \\ 1 & 2 & -1 \\ 1 & 1 & 1 \end{bmatrix}\begin{bmatrix} 1 & 2 & 3 \\ 0 & -1 & 2 \\ 1 & 1 & 0 \end{bmatrix}\begin{bmatrix} 3 & 5 & -1 \\ -2 & -3 & 1 \\ -1 & -2 & 1 \end{bmatrix} = \begin{bmatrix} 6 & 14 & -7 \\ -5 & -11 & 6 \\ -3 & -6 & 5 \end{bmatrix}$.

3. (a) The eigenvalues and eigenvectors of this matrix were found in Exercise 1 in

Section 6.1. They are $\lambda = 6$ with eigenvectors $r\begin{bmatrix} 4 \\ 1 \end{bmatrix}$ and $\lambda = 1$ with

eigenvectors $s\begin{bmatrix} -1 \\ 1 \end{bmatrix}$. $\frac{1}{5}\begin{bmatrix} 1 & 1 \\ -1 & 4 \end{bmatrix}\begin{bmatrix} 5 & 4 \\ 1 & 2 \end{bmatrix}\begin{bmatrix} 4 & -1 \\ 1 & 1 \end{bmatrix} = \begin{bmatrix} 6 & 0 \\ 0 & 1 \end{bmatrix}$.

(b) $\begin{vmatrix} 2-\lambda & 1 \\ 2 & 3-\lambda \end{vmatrix} = (2-\lambda)(3-\lambda) - 2 = \lambda^2 - 5\lambda + 4 = (\lambda-4)(\lambda-1)$, so the eigenvalues are

$\lambda = 4$ and $\lambda = 1$. For $\lambda = 4$, the eigenvectors are the solutions of

$\begin{bmatrix} -2 & 1 \\ 2 & -1 \end{bmatrix}\begin{bmatrix} x_1 \\ x_2 \end{bmatrix} = \mathbf{0}$, so the eigenvectors are vectors of the form $r\begin{bmatrix} 1 \\ 2 \end{bmatrix}$. For $\lambda = 1$,

the eigenvectors are the solutions of $\begin{bmatrix} 1 & 1 \\ 2 & 2 \end{bmatrix}\begin{bmatrix} x_1 \\ x_2 \end{bmatrix} = \mathbf{0}$, so the eigenvectors are

vectors of the form $s\begin{bmatrix} -1 \\ 1 \end{bmatrix}$. $\frac{1}{3}\begin{bmatrix} 1 & 1 \\ -2 & 1 \end{bmatrix}\begin{bmatrix} 2 & 1 \\ 2 & 3 \end{bmatrix}\begin{bmatrix} 1 & -1 \\ 2 & 1 \end{bmatrix} = \begin{bmatrix} 4 & 0 \\ 0 & 1 \end{bmatrix}$.

(c) $\begin{vmatrix} 1-\lambda & 1 \\ 0 & 1-\lambda \end{vmatrix} = (1-\lambda)(1-\lambda)$, so the only eigenvalue is $\lambda = 1$. The eigenvectors are

the solutions of $\begin{bmatrix} 0 & 1 \\ 0 & 0 \end{bmatrix}\begin{bmatrix} x_1 \\ x_2 \end{bmatrix} = \mathbf{0}$, so the eigenvectors are vectors of the form $r\begin{bmatrix} 1 \\ 0 \end{bmatrix}$.

There are not two linearly independent eigenvectors, so the matrix cannot be diagonalized.

(d) $\begin{vmatrix} 4-\lambda & -1 \\ 2 & 1-\lambda \end{vmatrix} = (4-\lambda)(1-\lambda) + 2 = \lambda^2 - 5\lambda + 6 = (\lambda-2)(\lambda-3)$, so the eigenvalues are

$\lambda = 2$ and $\lambda = 3$. For $\lambda = 2$, the eigenvectors are the solutions of

$\begin{bmatrix} 2 & -1 \\ 2 & -1 \end{bmatrix}\begin{bmatrix} x_1 \\ x_2 \end{bmatrix} = \mathbf{0}$, so the eigenvectors are vectors of the form $r\begin{bmatrix} 1 \\ 2 \end{bmatrix}$. For $\lambda = 3$, the eigenvectors are the solutions of $\begin{bmatrix} 1 & -1 \\ 2 & -2 \end{bmatrix}\begin{bmatrix} x_1 \\ x_2 \end{bmatrix} = \mathbf{0}$, so the eigenvectors are

vectors of the form $s\begin{bmatrix} 1 \\ 1 \end{bmatrix}$. $\begin{bmatrix} -1 & 1 \\ 2 & -1 \end{bmatrix}\begin{bmatrix} 4 & -1 \\ 2 & 1 \end{bmatrix}\begin{bmatrix} 1 & 1 \\ 2 & 1 \end{bmatrix} = \begin{bmatrix} 2 & 0 \\ 0 & 3 \end{bmatrix}$.

(e) $\begin{vmatrix} 4-\lambda & 1 \\ -1 & 2-\lambda \end{vmatrix} = (4-\lambda)(2-\lambda) + 1 = \lambda^2 - 6\lambda + 9 = (\lambda-3)(\lambda-3)$, so the only eigenvalue

is $\lambda = 3$. The eigenvectors are the solutions of $\begin{bmatrix} 1 & 1 \\ -1 & -1 \end{bmatrix}\begin{bmatrix} x_1 \\ x_2 \end{bmatrix} = \mathbf{0}$, so the

eigenvectors are vectors of the form $r\begin{bmatrix} 1 \\ -1 \end{bmatrix}$. Since there are not two linearly

independent eigenvectors, the matrix cannot be diagonalized.

4. (a) $\begin{vmatrix} -7-\lambda & 10 \\ -5 & 8-\lambda \end{vmatrix} = (-7-\lambda)(8-\lambda) + 50 = \lambda^2 - \lambda - 6 = (\lambda+2)(\lambda-3)$, so the eigenvalues are

$\lambda = -2$ and $\lambda = 3$. For $\lambda = -2$, the eigenvectors are the solutions of

$\begin{bmatrix} -5 & 10 \\ -5 & 10 \end{bmatrix}\begin{bmatrix} x_1 \\ x_2 \end{bmatrix} = \mathbf{0}$, so the eigenvectors are vectors of the form $r\begin{bmatrix} 2 \\ 1 \end{bmatrix}$. For $\lambda = 3$,

the eigenvectors are the solutions of $\begin{bmatrix} -10 & 10 \\ -5 & 5 \end{bmatrix}\begin{bmatrix} x_1 \\ x_2 \end{bmatrix} = \mathbf{0}$, so the eigenvectors

are vectors of the form $s\begin{bmatrix} 1 \\ 1 \end{bmatrix}$. $\begin{bmatrix} 1 & -1 \\ -1 & 2 \end{bmatrix}\begin{bmatrix} -7 & 10 \\ -5 & 8 \end{bmatrix}\begin{bmatrix} 2 & 1 \\ 1 & 1 \end{bmatrix} = \begin{bmatrix} -2 & 0 \\ 0 & 3 \end{bmatrix}$.

(b) $\begin{vmatrix} 7-\lambda & 4 \\ -8 & -5-\lambda \end{vmatrix} = (7-\lambda)(-5-\lambda) + 32 = \lambda^2 - 2\lambda - 3 = (\lambda+1)(\lambda-3)$, so the eigenvalues

are $\lambda = -1$ and $\lambda = 3$. For $\lambda = -1$, the eigenvectors are the solutions of

$\begin{bmatrix} 8 & 4 \\ -8 & -4 \end{bmatrix}\begin{bmatrix} x_1 \\ x_2 \end{bmatrix} = \mathbf{0}$, so the eigenvectors are vectors of the form $r\begin{bmatrix} 1 \\ -2 \end{bmatrix}$. For $\lambda = 3$, the eigenvectors are the solutions of $\begin{bmatrix} 4 & 4 \\ -8 & -8 \end{bmatrix}\begin{bmatrix} x_1 \\ x_2 \end{bmatrix} = \mathbf{0}$, so the eigenvectors are vectors of the form $s\begin{bmatrix} 1 \\ -1 \end{bmatrix}$. $\begin{bmatrix} -1 & -1 \\ 2 & 1 \end{bmatrix}\begin{bmatrix} 7 & 4 \\ -8 & -5 \end{bmatrix}\begin{bmatrix} 1 & 1 \\ -2 & -1 \end{bmatrix} = \begin{bmatrix} -1 & 0 \\ 0 & 3 \end{bmatrix}$.

(c) $\begin{vmatrix} 1-\lambda & -2 \\ 2 & -3-\lambda \end{vmatrix} = (1-\lambda)(-3-\lambda) + 4 = \lambda^2 + 2\lambda + 1 = (\lambda+1)(\lambda+1)$, so the only eigenvalue is $\lambda = -1$. The eigenvectors are the solutions of $\begin{bmatrix} 2 & -2 \\ 2 & -2 \end{bmatrix}\begin{bmatrix} x_1 \\ x_2 \end{bmatrix} = \mathbf{0}$, so the eigenvectors are vectors of the form $r\begin{bmatrix} 1 \\ -1 \end{bmatrix}$. Since there are not two linearly independent eigenvectors, the matrix cannot be diagonalized.

(d) $\begin{vmatrix} 7-\lambda & -4 \\ 1 & 3-\lambda \end{vmatrix} = (7-\lambda)(3-\lambda) + 4 = \lambda^2 - 10\lambda + 25 = (\lambda-5)(\lambda-5)$, so the only eigenvalue is $\lambda = 5$. The eigenvectors are the solutions of $\begin{bmatrix} 2 & -4 \\ 1 & -2 \end{bmatrix}\begin{bmatrix} x_1 \\ x_2 \end{bmatrix} = \mathbf{0}$, so the eigenvectors are vectors of the form $r\begin{bmatrix} 2 \\ 1 \end{bmatrix}$. Since there are not two linearly independent eigenvectors, the matrix cannot be diagonalized.

(e) $\begin{vmatrix} a-\lambda & b \\ 0 & a-\lambda \end{vmatrix} = (a-\lambda)(a-\lambda)$, so the only eigenvalue is $\lambda = a$. The eigenvectors are the solutions of $\begin{bmatrix} 0 & b \\ 0 & 0 \end{bmatrix}\begin{bmatrix} x_1 \\ x_2 \end{bmatrix} = \mathbf{0}$, so the eigenvectors are vectors of the form $r\begin{bmatrix} 1 \\ 0 \end{bmatrix}$.

Section 6.3

There are not two linearly independent eigenvectors, so the matrix cannot be diagonalized.

5. (a) The eigenvalues and eigenvectors of this matrix were found in Exercise 13 in Section 8.1. They are $\lambda = 1, \lambda = 2, \lambda = 8$, with corresponding eigenvectors

$$r\begin{bmatrix} 1 \\ -1 \\ 1 \end{bmatrix}, s\begin{bmatrix} 0 \\ 1 \\ 1 \end{bmatrix}, t\begin{bmatrix} 1 \\ 0 \\ 1 \end{bmatrix}. \quad \begin{bmatrix} -1 & -1 & 1 \\ -1 & 0 & 1 \\ 2 & 1 & -1 \end{bmatrix}\begin{bmatrix} 15 & 7 & -7 \\ -1 & 1 & 1 \\ 13 & 7 & -5 \end{bmatrix}\begin{bmatrix} 1 & 0 & 1 \\ -1 & 1 & 0 \\ 1 & 1 & 1 \end{bmatrix} = \begin{bmatrix} 1 & 0 & 0 \\ 0 & 2 & 0 \\ 0 & 0 & 8 \end{bmatrix}.$$

(b) The eigenvalues and eigenvectors of this matrix were found in Exercise 14 in Section 4.7. They are $\lambda = 5, \lambda = 3, \lambda = 1$, with corresponding eigenvectors

$$r\begin{bmatrix} 1 \\ 1 \\ 1 \end{bmatrix}, s\begin{bmatrix} 1 \\ 0 \\ -1 \end{bmatrix}, t\begin{bmatrix} 0 \\ 1 \\ 1 \end{bmatrix}. \quad \begin{bmatrix} 1 & -1 & 1 \\ 0 & 1 & -1 \\ -1 & 2 & -1 \end{bmatrix}\begin{bmatrix} 5 & -2 & 2 \\ 4 & -3 & 4 \\ 4 & -6 & 7 \end{bmatrix}\begin{bmatrix} 1 & 1 & 0 \\ 1 & 0 & 1 \\ 1 & -1 & 1 \end{bmatrix} = \begin{bmatrix} 5 & 0 & 0 \\ 0 & 3 & 0 \\ 0 & 0 & 1 \end{bmatrix}.$$

(c) $\begin{vmatrix} 1-\lambda & 0 & 0 \\ -2 & 1-\lambda & 2 \\ -2 & 0 & 3-\lambda \end{vmatrix} = (1-\lambda)^2(3-\lambda)$, so the eigenvalues are $\lambda = 1$ and $\lambda = 3$.

For $\lambda = 1$, the eigenvectors are $r\begin{bmatrix} 1 \\ 0 \\ 1 \end{bmatrix} + s\begin{bmatrix} 0 \\ 1 \\ 0 \end{bmatrix}$. For $\lambda = 3$, the eigenvectors are

$$t\begin{bmatrix} 0 \\ 1 \\ 1 \end{bmatrix}. \quad \begin{bmatrix} 1 & 0 & 0 \\ 1 & 1 & -1 \\ -1 & 0 & 1 \end{bmatrix}\begin{bmatrix} 1 & 0 & 0 \\ -2 & 1 & 2 \\ -2 & 0 & 3 \end{bmatrix}\begin{bmatrix} 1 & 0 & 0 \\ 0 & 1 & 1 \\ 1 & 0 & 1 \end{bmatrix} = \begin{bmatrix} 1 & 0 & 0 \\ 0 & 1 & 0 \\ 0 & 0 & 3 \end{bmatrix}.$$

(d) The eigenvalues are $\lambda = 3, \lambda = 2, \lambda = -4$, with corresponding eigenvectors

$$r\begin{bmatrix} 1 \\ 1 \\ 0 \end{bmatrix}, s\begin{bmatrix} 0 \\ 1 \\ 0 \end{bmatrix}, t\begin{bmatrix} 0 \\ 0 \\ 1 \end{bmatrix}. \quad \begin{bmatrix} 1 & 0 & 0 \\ -1 & 1 & 0 \\ 0 & 0 & 1 \end{bmatrix}\begin{bmatrix} 3 & 0 & 0 \\ 1 & 2 & 0 \\ 0 & 0 & -4 \end{bmatrix}\begin{bmatrix} 1 & 0 & 0 \\ 1 & 1 & 0 \\ 0 & 0 & 1 \end{bmatrix} = \begin{bmatrix} 3 & 0 & 0 \\ 0 & 2 & 0 \\ 0 & 0 & -4 \end{bmatrix}.$$

Section 6.3

6. (a) The eigenvalues (found in exercise 5, Section 8.1) are $\lambda = 3$ and $\lambda = -1$, with corresponding eigenvectors $r \begin{bmatrix} 1 \\ 1 \end{bmatrix}$ and $s \begin{bmatrix} -1 \\ 1 \end{bmatrix}$. The set $\left\{ \begin{bmatrix} \frac{1}{\sqrt{2}} \\ \frac{1}{\sqrt{2}} \end{bmatrix}, \begin{bmatrix} \frac{-1}{\sqrt{2}} \\ \frac{1}{\sqrt{2}} \end{bmatrix} \right\}$ is an orthonormal basis.

$$\begin{bmatrix} \frac{1}{\sqrt{2}} & \frac{1}{\sqrt{2}} \\ \frac{-1}{\sqrt{2}} & \frac{1}{\sqrt{2}} \end{bmatrix} \begin{bmatrix} 1 & 2 \\ 2 & 1 \end{bmatrix} \begin{bmatrix} \frac{1}{\sqrt{2}} & \frac{-1}{\sqrt{2}} \\ \frac{1}{\sqrt{2}} & \frac{1}{\sqrt{2}} \end{bmatrix} = \begin{bmatrix} 3 & 0 \\ 0 & -1 \end{bmatrix}.$$

(b) $\begin{vmatrix} 11-\lambda & 2 \\ 2 & 14-\lambda \end{vmatrix} = (11-\lambda)(14-\lambda) - 4 = \lambda^2 - 25\lambda + 150 = (\lambda-15)(\lambda-10)$, so the eigenvalues are $\lambda = 15$ and $\lambda = 10$, with corresponding eigenvectors $r \begin{bmatrix} 1 \\ 2 \end{bmatrix}$ and $s \begin{bmatrix} -2 \\ 1 \end{bmatrix}$. The set $\left\{ \begin{bmatrix} \frac{1}{\sqrt{5}} \\ \frac{2}{\sqrt{5}} \end{bmatrix}, \begin{bmatrix} \frac{-2}{\sqrt{5}} \\ \frac{1}{\sqrt{5}} \end{bmatrix} \right\}$ is an orthonormal basis.

$$\begin{bmatrix} \frac{1}{\sqrt{5}} & \frac{2}{\sqrt{5}} \\ \frac{-2}{\sqrt{5}} & \frac{1}{\sqrt{5}} \end{bmatrix} \begin{bmatrix} 11 & 2 \\ 2 & 14 \end{bmatrix} \begin{bmatrix} \frac{1}{\sqrt{5}} & \frac{-2}{\sqrt{5}} \\ \frac{2}{\sqrt{5}} & \frac{1}{\sqrt{5}} \end{bmatrix} = \begin{bmatrix} 15 & 0 \\ 0 & 10 \end{bmatrix}.$$

(c) $\begin{vmatrix} 3-\lambda & 1 \\ 1 & 3-\lambda \end{vmatrix} = (3-\lambda)(3-\lambda) - 1 = \lambda^2 - 6\lambda + 8 = (\lambda-4)(\lambda-2)$, so the eigenvalues are $\lambda = 4$ and $\lambda = 2$, with corresponding eigenvectors $r \begin{bmatrix} 1 \\ 1 \end{bmatrix}$ and $s \begin{bmatrix} -1 \\ 1 \end{bmatrix}$. The set $\left\{ \begin{bmatrix} \frac{1}{\sqrt{2}} \\ \frac{1}{\sqrt{2}} \end{bmatrix}, \begin{bmatrix} \frac{-1}{\sqrt{2}} \\ \frac{1}{\sqrt{2}} \end{bmatrix} \right\}$ is an orthonormal basis.

$$\begin{bmatrix} \frac{1}{\sqrt{2}} & \frac{1}{\sqrt{2}} \\ \frac{-1}{\sqrt{2}} & \frac{1}{\sqrt{2}} \end{bmatrix} \begin{bmatrix} 3 & 1 \\ 1 & 3 \end{bmatrix} \begin{bmatrix} \frac{1}{\sqrt{2}} & \frac{-1}{\sqrt{2}} \\ \frac{1}{\sqrt{2}} & \frac{1}{\sqrt{2}} \end{bmatrix} = \begin{bmatrix} 4 & 0 \\ 0 & 2 \end{bmatrix}.$$

(d) $\begin{vmatrix} -1-\lambda & -8 \\ -8 & 11-\lambda \end{vmatrix} = (-1-\lambda)(11-\lambda) - 64 = \lambda^2 - 10\lambda - 75 = (\lambda-15)(\lambda+5)$, so the

eigenvalues are $\lambda = -5$ and $\lambda = 15$, with corresponding eigenvectors

$r\begin{bmatrix} 2 \\ 1 \end{bmatrix}$ and $s\begin{bmatrix} -1 \\ 2 \end{bmatrix}$. The set $\left\{ \begin{bmatrix} \frac{2}{\sqrt{5}} \\ \frac{1}{\sqrt{5}} \end{bmatrix}, \begin{bmatrix} \frac{-1}{\sqrt{5}} \\ \frac{2}{\sqrt{5}} \end{bmatrix} \right\}$ is an orthonormal basis.

$$\begin{bmatrix} \frac{2}{\sqrt{5}} & \frac{1}{\sqrt{5}} \\ \frac{-1}{\sqrt{5}} & \frac{2}{\sqrt{5}} \end{bmatrix} \begin{bmatrix} -1 & -8 \\ -8 & 11 \end{bmatrix} \begin{bmatrix} \frac{2}{\sqrt{5}} & \frac{-1}{\sqrt{5}} \\ \frac{1}{\sqrt{5}} & \frac{2}{\sqrt{5}} \end{bmatrix} = \begin{bmatrix} -5 & 0 \\ 0 & 15 \end{bmatrix}.$$

7. (a) $\begin{vmatrix} 1-\lambda & 5 \\ 5 & 1-\lambda \end{vmatrix} = (1-\lambda)(1-\lambda) - 25 = \lambda^2 - 2\lambda - 24 = (\lambda-6)(\lambda+4)$, so the

eigenvalues are $\lambda = 6$ and $\lambda = -4$, with corresponding eigenvectors

$r\begin{bmatrix} 1 \\ 1 \end{bmatrix}$ and $s\begin{bmatrix} -1 \\ 1 \end{bmatrix}$. The set $\left\{ \begin{bmatrix} \frac{1}{\sqrt{2}} \\ \frac{1}{\sqrt{2}} \end{bmatrix}, \begin{bmatrix} \frac{-1}{\sqrt{2}} \\ \frac{1}{\sqrt{2}} \end{bmatrix} \right\}$ is an orthonormal basis.

$$\begin{bmatrix} \frac{1}{\sqrt{2}} & \frac{1}{\sqrt{2}} \\ \frac{-1}{\sqrt{2}} & \frac{1}{\sqrt{2}} \end{bmatrix} \begin{bmatrix} 1 & 5 \\ 5 & 1 \end{bmatrix} \begin{bmatrix} \frac{1}{\sqrt{2}} & \frac{-1}{\sqrt{2}} \\ \frac{1}{\sqrt{2}} & \frac{1}{\sqrt{2}} \end{bmatrix} = \begin{bmatrix} 6 & 0 \\ 0 & -4 \end{bmatrix}.$$

(b) $\begin{vmatrix} 9-\lambda & 2 \\ 2 & 6-\lambda \end{vmatrix} = (9-\lambda)(6-\lambda) - 4 = \lambda^2 - 15\lambda + 50 = (\lambda-10)(\lambda-5)$, so the eigenvalues are $\lambda = 10$ and $\lambda = 5$, with corresponding eigenvectors

$r\begin{bmatrix} 2 \\ 1 \end{bmatrix}$ and $s\begin{bmatrix} -1 \\ 2 \end{bmatrix}$. The set $\left\{ \begin{bmatrix} \frac{2}{\sqrt{5}} \\ \frac{1}{\sqrt{5}} \end{bmatrix}, \begin{bmatrix} \frac{-1}{\sqrt{5}} \\ \frac{2}{\sqrt{5}} \end{bmatrix} \right\}$ is an orthonormal basis.

$$\begin{bmatrix} \frac{2}{\sqrt{5}} & \frac{1}{\sqrt{5}} \\ \frac{-1}{\sqrt{5}} & \frac{2}{\sqrt{5}} \end{bmatrix} \begin{bmatrix} 9 & 2 \\ 2 & 6 \end{bmatrix} \begin{bmatrix} \frac{2}{\sqrt{5}} & \frac{-1}{\sqrt{5}} \\ \frac{1}{\sqrt{5}} & \frac{2}{\sqrt{5}} \end{bmatrix} = \begin{bmatrix} 10 & 0 \\ 0 & 5 \end{bmatrix}.$$

(c) $\begin{vmatrix} 1-\lambda & 3 \\ 3 & 9-\lambda \end{vmatrix} = (1-\lambda)(9-\lambda) - 9 = \lambda^2 - 10\lambda = (\lambda-10)\lambda$, so the eigenvalues

are $\lambda = 10$ and $\lambda = 0$, with corresponding eigenvectors $r\begin{bmatrix} 1 \\ 3 \end{bmatrix}$ and $s\begin{bmatrix} -3 \\ 1 \end{bmatrix}$.

The set $\left\{ \begin{bmatrix} \frac{1}{\sqrt{10}} \\ \frac{3}{\sqrt{10}} \end{bmatrix}, \begin{bmatrix} \frac{-3}{\sqrt{10}} \\ \frac{1}{\sqrt{10}} \end{bmatrix} \right\}$ is an orthonormal basis.

$$\begin{bmatrix} \frac{1}{\sqrt{10}} & \frac{3}{\sqrt{10}} \\ \frac{-3}{\sqrt{10}} & \frac{1}{\sqrt{10}} \end{bmatrix} \begin{bmatrix} 1 & 3 \\ 3 & 9 \end{bmatrix} \begin{bmatrix} \frac{1}{\sqrt{10}} & \frac{-3}{\sqrt{10}} \\ \frac{3}{\sqrt{10}} & \frac{1}{\sqrt{10}} \end{bmatrix} = \begin{bmatrix} 10 & 0 \\ 0 & 0 \end{bmatrix}.$$

(d) $\begin{vmatrix} 1.5-\lambda & -.5 \\ -.5 & 1.5-\lambda \end{vmatrix} = (1.5-\lambda)(1.5-\lambda) - .25 = \lambda^2 - 3\lambda + 2 = (\lambda-1)(\lambda-2)$, so the

eigenvalues are $\lambda = 1$ and $\lambda = 2$, with corresponding eigenvectors

$r\begin{bmatrix} 1 \\ 1 \end{bmatrix}$ and $s\begin{bmatrix} -1 \\ 1 \end{bmatrix}$. The set $\left\{ \begin{bmatrix} \frac{1}{\sqrt{2}} \\ \frac{1}{\sqrt{2}} \end{bmatrix}, \begin{bmatrix} \frac{-1}{\sqrt{2}} \\ \frac{1}{\sqrt{2}} \end{bmatrix} \right\}$ is an orthonormal basis.

$$\begin{bmatrix} \frac{1}{\sqrt{2}} & \frac{1}{\sqrt{2}} \\ \frac{-1}{\sqrt{2}} & \frac{1}{\sqrt{2}} \end{bmatrix} \begin{bmatrix} 1.5 & -.5 \\ -.5 & 1.5 \end{bmatrix} \begin{bmatrix} \frac{1}{\sqrt{2}} & \frac{-1}{\sqrt{2}} \\ \frac{1}{\sqrt{2}} & \frac{1}{\sqrt{2}} \end{bmatrix} = \begin{bmatrix} 1 & 0 \\ 0 & 2 \end{bmatrix}.$$

8. (a) $\begin{vmatrix} -\lambda & 2 & 0 \\ 2 & -\lambda & 0 \\ 0 & 0 & 1-\lambda \end{vmatrix} = \lambda^2(1-\lambda) - 4(1-\lambda)$, so the eigenvalues are $\lambda = 1$ and $\lambda = \pm 2$.

For $\lambda = 1$, the eigenvectors are $r\begin{bmatrix} 0 \\ 0 \\ 1 \end{bmatrix}$, for $\lambda = 2$, the eigenvectors are $s\begin{bmatrix} 1 \\ 1 \\ 0 \end{bmatrix}$, and for

$\lambda = -2$, the eigenvectors are $t\begin{bmatrix} 1 \\ -1 \\ 0 \end{bmatrix}$.

$$\begin{bmatrix} 0 & 0 & 1 \\ 1/\sqrt{2} & 1/\sqrt{2} & 0 \\ 1/\sqrt{2} & -1/\sqrt{2} & 0 \end{bmatrix} \begin{bmatrix} 0 & 2 & 0 \\ 2 & 0 & 0 \\ 0 & 0 & 1 \end{bmatrix} \begin{bmatrix} 0 & 1/\sqrt{2} & 1/\sqrt{2} \\ 0 & 1/\sqrt{2} & -1/\sqrt{2} \\ 1 & 0 & 0 \end{bmatrix} = \begin{bmatrix} 1 & 0 & 0 \\ 0 & 2 & 0 \\ 0 & 0 & -2 \end{bmatrix}.$$

(b) $\begin{vmatrix} 9-\lambda & -3 & 3 \\ -3 & 6-\lambda & -6 \\ 3 & -6 & 6-\lambda \end{vmatrix} = (9-\lambda)(6-\lambda)^2 + 54\lambda - 324 = -\lambda(15-\lambda)(6-\lambda)$, so the

eigenvalues are $\lambda = 0$, $\lambda = 15$, and $\lambda = 6$. For $\lambda = 0$, the eigenvectors are $r\begin{bmatrix} 0 \\ 1 \\ 1 \end{bmatrix}$,

for $\lambda = 15$, the eigenvectors are $s\begin{bmatrix} 1 \\ -1 \\ 1 \end{bmatrix}$, and for $\lambda = 6$, the eigenvectors are $t\begin{bmatrix} 2 \\ 1 \\ -1 \end{bmatrix}$.

$$\begin{bmatrix} 0 & 1/\sqrt{2} & 1/\sqrt{2} \\ 1/\sqrt{3} & -1/\sqrt{3} & 1/\sqrt{3} \\ 2/\sqrt{6} & 1/\sqrt{6} & -1/\sqrt{6} \end{bmatrix} \begin{bmatrix} 9 & -3 & 3 \\ -3 & 6 & -6 \\ 3 & -6 & 6 \end{bmatrix} \begin{bmatrix} 0 & 1/\sqrt{3} & 2/\sqrt{6} \\ 1/\sqrt{2} & -1/\sqrt{3} & 1/\sqrt{6} \\ 1/\sqrt{2} & 1/\sqrt{3} & -1/\sqrt{6} \end{bmatrix}$$

$$= \begin{bmatrix} 0 & 0 & 0 \\ 0 & 15 & 0 \\ 0 & 0 & 6 \end{bmatrix}.$$

(c) $\begin{vmatrix} 1-\lambda & 2 & -2 \\ 2 & 4-\lambda & -4 \\ -2 & -4 & 4-\lambda \end{vmatrix} = (1-\lambda)(4-\lambda)^2 + 24\lambda - 16 = \lambda^2(9-\lambda)$, so the eigenvalues are

$\lambda = 0$ and $\lambda = 9$. For $\lambda = 0$, the eigenvectors are $r\begin{bmatrix} 0 \\ 1 \\ 1 \end{bmatrix} + s\begin{bmatrix} 2 \\ -1 \\ 0 \end{bmatrix}$, and for $\lambda = 9$, the

eigenvectors are $t\begin{bmatrix} 1 \\ 2 \\ -2 \end{bmatrix}$.

$$\begin{bmatrix} 0 & 1/\sqrt{2} & 1/\sqrt{2} \\ 2/\sqrt{5} & -1/\sqrt{5} & 0 \\ 1/3 & 2/3 & -2/3 \end{bmatrix} \begin{bmatrix} 1 & 2 & -2 \\ 2 & 4 & -4 \\ -2 & -4 & 4 \end{bmatrix} \begin{bmatrix} 0 & 2/\sqrt{5} & 1/3 \\ 1/\sqrt{2} & -1/\sqrt{5} & 2/3 \\ 1/\sqrt{2} & 0 & -2/3 \end{bmatrix} = \begin{bmatrix} 0 & 0 & 0 \\ 0 & 0 & 0 \\ 0 & 0 & 9 \end{bmatrix}.$$

9. (a) $\begin{bmatrix} 1 & 5 \\ 5 & 1 \end{bmatrix}^8 = \begin{bmatrix} \frac{1}{\sqrt{2}} & \frac{-1}{\sqrt{2}} \\ \frac{1}{\sqrt{2}} & \frac{1}{\sqrt{2}} \end{bmatrix} \begin{bmatrix} 6 & 0 \\ 0 & -4 \end{bmatrix}^8 \begin{bmatrix} \frac{1}{\sqrt{2}} & \frac{1}{\sqrt{2}} \\ \frac{-1}{\sqrt{2}} & \frac{1}{\sqrt{2}} \end{bmatrix}$

$= \frac{1}{2} \begin{bmatrix} 1 & -1 \\ 1 & 1 \end{bmatrix} \begin{bmatrix} 1679616 & 0 \\ 0 & 65536 \end{bmatrix} \begin{bmatrix} 1 & 1 \\ -1 & 1 \end{bmatrix} = \begin{bmatrix} 872576 & 807040 \\ 807040 & 872576 \end{bmatrix}.$

(b) $\begin{bmatrix} 9 & 2 \\ 2 & 6 \end{bmatrix}^5 = \begin{bmatrix} \frac{2}{\sqrt{5}} & \frac{-1}{\sqrt{5}} \\ \frac{1}{\sqrt{5}} & \frac{2}{\sqrt{5}} \end{bmatrix} \begin{bmatrix} 10 & 0 \\ 0 & 5 \end{bmatrix}^5 \begin{bmatrix} \frac{2}{\sqrt{5}} & \frac{1}{\sqrt{5}} \\ \frac{-1}{\sqrt{5}} & \frac{2}{\sqrt{5}} \end{bmatrix}$

$= \frac{1}{5} \begin{bmatrix} 2 & -1 \\ 1 & 2 \end{bmatrix} \begin{bmatrix} 100000 & 0 \\ 0 & 3125 \end{bmatrix} \begin{bmatrix} 2 & 1 \\ -1 & 2 \end{bmatrix} = \begin{bmatrix} 80625 & 38750 \\ 38750 & 22500 \end{bmatrix}.$

(c) $\begin{bmatrix} 1 & 3 \\ 3 & 9 \end{bmatrix}^6 = \begin{bmatrix} \frac{1}{\sqrt{10}} & \frac{-3}{\sqrt{10}} \\ \frac{3}{\sqrt{10}} & \frac{1}{\sqrt{10}} \end{bmatrix} \begin{bmatrix} 10 & 0 \\ 0 & 0 \end{bmatrix}^6 \begin{bmatrix} \frac{1}{\sqrt{10}} & \frac{3}{\sqrt{10}} \\ \frac{-3}{\sqrt{10}} & \frac{1}{\sqrt{10}} \end{bmatrix}$

$= \frac{1}{10} \begin{bmatrix} 1 & -3 \\ 3 & 1 \end{bmatrix} \begin{bmatrix} 1000000 & 0 \\ 0 & 0 \end{bmatrix} \begin{bmatrix} 1 & 3 \\ -3 & 1 \end{bmatrix} = \begin{bmatrix} 100000 & 300000 \\ 300000 & 900000 \end{bmatrix}.$

(d) $\begin{bmatrix} 1.5 & -.5 \\ -.5 & 1.5 \end{bmatrix}^{16} = \begin{bmatrix} \frac{1}{\sqrt{2}} & \frac{-1}{\sqrt{2}} \\ \frac{1}{\sqrt{2}} & \frac{1}{\sqrt{2}} \end{bmatrix} \begin{bmatrix} 1 & 0 \\ 0 & 2 \end{bmatrix}^{16} \begin{bmatrix} \frac{1}{\sqrt{2}} & \frac{1}{\sqrt{2}} \\ \frac{-1}{\sqrt{2}} & \frac{1}{\sqrt{2}} \end{bmatrix}$

$= \frac{1}{2} \begin{bmatrix} 1 & -1 \\ 1 & 1 \end{bmatrix} \begin{bmatrix} 1 & 0 \\ 0 & 65536 \end{bmatrix} \begin{bmatrix} 1 & 1 \\ -1 & 1 \end{bmatrix} = \begin{bmatrix} 32768.5 & -32767.5 \\ -32767.5 & 32768.5 \end{bmatrix}.$

10. (a) $\begin{bmatrix} 0 & 2 & 0 \\ 2 & 0 & 0 \\ 0 & 0 & 1 \end{bmatrix}^6 = \begin{bmatrix} 0 & 1/\sqrt{2} & 1/\sqrt{2} \\ 0 & 1/\sqrt{2} & -1/\sqrt{2} \\ 1 & 0 & 0 \end{bmatrix} \begin{bmatrix} 1 & 0 & 0 \\ 0 & 2 & 0 \\ 0 & 0 & -2 \end{bmatrix}^6 \begin{bmatrix} 0 & 0 & 1 \\ 1/\sqrt{2} & 1/\sqrt{2} & 0 \\ 1/\sqrt{2} & -1/\sqrt{2} & 0 \end{bmatrix}$

$= \frac{1}{2} \begin{bmatrix} 0 & 1 & 1 \\ 0 & 1 & -1 \\ \sqrt{2} & 0 & 0 \end{bmatrix} \begin{bmatrix} 1 & 0 & 0 \\ 0 & 64 & 0 \\ 0 & 0 & 64 \end{bmatrix} \begin{bmatrix} 0 & 0 & \sqrt{2} \\ 1 & 1 & 0 \\ 1 & -1 & 0 \end{bmatrix} = \begin{bmatrix} 64 & 0 & 0 \\ 0 & 64 & 0 \\ 0 & 0 & 1 \end{bmatrix}.$

(b) $\begin{bmatrix} 9 & -3 & 3 \\ -3 & 6 & -6 \\ 3 & -6 & 6 \end{bmatrix}^5 = \begin{bmatrix} 0 & 1/\sqrt{3} & 2/\sqrt{6} \\ 1/\sqrt{2} & -1/\sqrt{3} & 1/\sqrt{6} \\ 1/\sqrt{2} & 1/\sqrt{3} & -1/\sqrt{6} \end{bmatrix} \begin{bmatrix} 0 & 0 & 0 \\ 0 & 15 & 0 \\ 0 & 0 & 6 \end{bmatrix}^5 \begin{bmatrix} 0 & 1/\sqrt{2} & 1/\sqrt{2} \\ 1/\sqrt{3} & -1/\sqrt{3} & 1/\sqrt{3} \\ 2/\sqrt{6} & 1/\sqrt{6} & -1/\sqrt{6} \end{bmatrix}$

$= \frac{1}{6} \begin{bmatrix} 0 & \sqrt{2} & 2 \\ \sqrt{3} & -\sqrt{2} & 1 \\ \sqrt{3} & \sqrt{2} & -1 \end{bmatrix} \begin{bmatrix} 0 & 0 & 0 \\ 0 & 759375 & 0 \\ 0 & 0 & 7776 \end{bmatrix} \begin{bmatrix} 0 & \sqrt{3} & \sqrt{3} \\ \sqrt{2} & -\sqrt{2} & \sqrt{2} \\ 2 & 1 & -1 \end{bmatrix}$

$$= \begin{bmatrix} 258309 & -250533 & 250533 \\ -250533 & 254421 & -254421 \\ 250533 & -254421 & 254421 \end{bmatrix}.$$

(c) $\begin{bmatrix} 1 & 2 & -2 \\ 2 & 4 & -4 \\ -2 & -4 & 4 \end{bmatrix}^4 = \begin{bmatrix} 0 & 2/\sqrt{5} & 1/3 \\ 1/\sqrt{2} & -1/\sqrt{5} & 2/3 \\ 1/\sqrt{2} & 0 & -2/3 \end{bmatrix} \begin{bmatrix} 0 & 0 & 0 \\ 0 & 0 & 0 \\ 0 & 0 & 9 \end{bmatrix}^4 \begin{bmatrix} 0 & 1/\sqrt{2} & 1/\sqrt{2} \\ 2/\sqrt{5} & -1/\sqrt{5} & 0 \\ 1/3 & 2/3 & -2/3 \end{bmatrix}$

$= \begin{bmatrix} 0 & 2/\sqrt{5} & 1/3 \\ 1/\sqrt{2} & -1/\sqrt{5} & 2/3 \\ 1/\sqrt{2} & 0 & -2/3 \end{bmatrix} \begin{bmatrix} 0 & 0 & 0 \\ 0 & 0 & 0 \\ 0 & 0 & 6561 \end{bmatrix} \begin{bmatrix} 0 & 1/\sqrt{2} & 1/\sqrt{2} \\ 2/\sqrt{5} & -1/\sqrt{5} & 0 \\ 1/3 & 2/3 & -2/3 \end{bmatrix}$

$= \begin{bmatrix} 927 & 1458 & -1458 \\ 1458 & 2916 & -2916 \\ -1458 & -2916 & 2916 \end{bmatrix}.$

11. (a) If A and B are similar, then $B = C^{-1}AC$, so that $|B| = |C^{-1}AC| = |C^{-1}||A||C| = |A|$.

(b) If A and B are similar, they represent the same linear operator T with respect to two different bases on the vector space. Thus rank(A) = dim range(T) and rank(B) = dim range(T), so rank(A) = rank(B).

(c) We know from Exercise 17, Section 2.3 that for nxn matrices E and F, tr(EF) = tr(FE). Thus if $B = C^{-1}AC$, then $tr(B) = tr(C^{-1}(AC)) = tr((AC)C^{-1}) = tr(A(CC^{-1}))$ = tr(A).

(d) If $B = C^{-1}AC$ then $B^t = (C^{-1}AC)^t = (AC)^t(C^{-1})^t = C^tA^t(C^{-1})^t = C^tA^t(C^t)^{-1}$. Thus $B^t = E^{-1}A^tE$, where $E = (C^{-1})^t$, so B^t and A^t are similar.

(e) If A and B are similar, then A is nonsingular if and only if $|A| \neq 0$ if and only if $|B| \neq 0$ (since $|A| = |B|$) if and only if B is nonsingular.

(f) If $B = C^{-1}AC$ and A is nonsingular, then B is also nonsingular and $B^{-1} = (C^{-1}AC)^{-1} = C^{-1}A^{-1}C$, so B^{-1} is similar to A^{-1}.

12. Let A be a symmetric matrix and λ_1 and λ_2 distinct eigenvalues of A with corresponding

eigenvectors X and Y. $X \cdot Y = X^t Y = Y^t X$, $AX = \lambda_1 X$, and $AY = \lambda_2 Y$. Thus $X^t AY = X^t \lambda_2 Y = \lambda_2 X^t Y = \lambda_2 X \cdot Y$ and $Y^t AX = Y^t \lambda_1 X = \lambda_1 Y^t X = \lambda_1 X \cdot Y$. However, since $Y^t AX$ is a real number, $Y^t AX = (Y^t AX)^t = X^t AY$, so $\lambda_1 X \cdot Y = \lambda_2 X \cdot Y$. Since $\lambda_1 \neq \lambda_2$, we conclude that $X \cdot Y = 0$.

13. (a) Let $A = C^{-1}BC$ and let D be the diagonal matrix $D = E^{-1}AE$. Then $D = E^{-1}(C^{-1}BC)E = (CE)^{-1}B(CE)$, so B is diagonalizable to the same diagonal matrix D.

    (b) If $A = C^{-1}BC$ then $A + kI = C^{-1}BC + kC^{-1}IC = C^{-1}BC + C^{-1}kIC = C^{-1}(B+kI)C$, so $B + kI$ and $A + kI$ are similar for any scalar k.

14. Let $D = C^{-1}AC$. Then $D^2 = (C^{-1}AC)(C^{-1}AC) = C^{-1}A^2C$ and $D^n = \underbrace{(C^{-1}AC)(C^{-1}AC) \ldots (C^{-1}AC)(C^{-1}AC)}_{n \text{ terms}} = C^{-1}A^nC$.

15. The matrix $\begin{bmatrix} 4 & -1 \\ 2 & 1 \end{bmatrix}$ has eigenvectors $\lambda = 2$ and $\lambda = 3$ with corresponding eigenvectors $r\begin{bmatrix} 1 \\ 2 \end{bmatrix}$ and $s\begin{bmatrix} 1 \\ 1 \end{bmatrix}$. If $C = \begin{bmatrix} 1 & 1 \\ 2 & 1 \end{bmatrix}$, then $C^{-1}AC = \begin{bmatrix} 2 & 0 \\ 0 & 3 \end{bmatrix}$, but if $C = \begin{bmatrix} 1 & 1 \\ 1 & 2 \end{bmatrix}$, then $C^{-1}AC = \begin{bmatrix} 3 & 0 \\ 0 & 2 \end{bmatrix}$. Thus the diagonal matrix is not unique. An eigenvalue λ will occupy the ith diagonal position in the diagonal matrix if the ith column of C is an eigenvector corresponding to λ.

16. Let $D = C^{-1}AC$ be the diagonalization of A. If the eigenvalues of A are $\lambda_1, \lambda_2, \ldots, \lambda_n$, then $|D| = \lambda_1 \lambda_2 \ldots \lambda_n$. However $|A| = |D|$, so $|A| = \lambda_1 \lambda_2 \ldots \lambda_n$.

17. If $B = C^{-1}AC = C^tAC$, then $B^t = (C^tAC)^t = C^tA^t(C^t)^t = C^tA^tC = C^tAC = B$. Thus B is symmetric.

18. $B = E^tAE$ and $C = F^tBF$, so $C = F^t(E^tAE)F = (EF)^tA(EF)$. The product of orthogonal matrices is orthogonal, so C and A are orthogonally similar.

Exercise Set 6.4

1. (a) $x^2 + 4xy + 2y^2 = [x\ y] \begin{bmatrix} 1 & 2 \\ 2 & 2 \end{bmatrix} \begin{bmatrix} x \\ y \end{bmatrix}.$

   (b) $3x^2 + 2xy - 4y^2 = [x\ y] \begin{bmatrix} 3 & 1 \\ 1 & -4 \end{bmatrix} \begin{bmatrix} x \\ y \end{bmatrix}.$

   (c) $7x^2 - 6xy - y^2 = [x\ y] \begin{bmatrix} 7 & -3 \\ -3 & -1 \end{bmatrix} \begin{bmatrix} x \\ y \end{bmatrix}.$

   (d) $2x^2 + 5xy + 3y^2 = [x\ y] \begin{bmatrix} 2 & 5/2 \\ 5/2 & 3 \end{bmatrix} \begin{bmatrix} x \\ y \end{bmatrix}.$

   (e) $-3x^2 - 7xy + 4y^2 = [x\ y] \begin{bmatrix} -3 & -7/2 \\ -7/2 & 4 \end{bmatrix} \begin{bmatrix} x \\ y \end{bmatrix}.$

   (f) $5x^2 + 3xy - 2y^2 = [x\ y] \begin{bmatrix} 5 & 3/2 \\ 3/2 & -2 \end{bmatrix} \begin{bmatrix} x \\ y \end{bmatrix}.$

2. (a) $11x^2 + 4xy + 14y^2 - 60 = [x\ y] \begin{bmatrix} 11 & 2 \\ 2 & 14 \end{bmatrix} \begin{bmatrix} x \\ y \end{bmatrix} - 60.$ From Exercise 6(b) in Section 6.3, $C^t \begin{bmatrix} 11 & 2 \\ 2 & 14 \end{bmatrix} C = \begin{bmatrix} 15 & 0 \\ 0 & 10 \end{bmatrix}$, where $C = \begin{bmatrix} \frac{1}{\sqrt{5}} & \frac{-2}{\sqrt{5}} \\ \frac{2}{\sqrt{5}} & \frac{1}{\sqrt{5}} \end{bmatrix}$, so the given equation becomes $[x\ y]\, C \begin{bmatrix} 15 & 0 \\ 0 & 10 \end{bmatrix} C^t \begin{bmatrix} x \\ y \end{bmatrix} - 60 = 0$, or $[x'\ y'] \begin{bmatrix} 15 & 0 \\ 0 & 10 \end{bmatrix} \begin{bmatrix} x' \\ y' \end{bmatrix} - 60 = 0$, where $[x'\ y'] = [x\ y]\, C$. Thus $15x'^2 + 10y'^2 = 60$; i.e., $\frac{x'^2}{4} + \frac{y'^2}{6} = 1$. The graph is an ellipse with the lines $y = 2x$ and $x = -2y$ as axes.

Section 6.4

Figure for 2(a): $a = 2, b = \sqrt{6}$

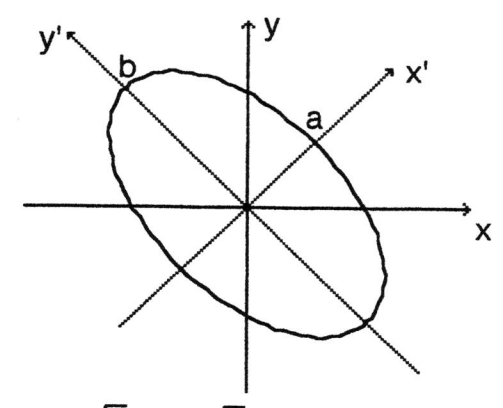

Figure for 2(b): $a = \sqrt{3}, b = \sqrt{6}$

(b) $3x^2 + 2xy + 3y^2 - 12 = \begin{bmatrix} x & y \end{bmatrix} \begin{bmatrix} 3 & 1 \\ 1 & 3 \end{bmatrix} \begin{bmatrix} x \\ y \end{bmatrix} - 12$. From Exercise 6(c) in Section 6.3, $C^t \begin{bmatrix} 3 & 1 \\ 1 & 3 \end{bmatrix} C = \begin{bmatrix} 4 & 0 \\ 0 & 2 \end{bmatrix}$, where $C = \begin{bmatrix} \frac{1}{\sqrt{2}} & \frac{-1}{\sqrt{2}} \\ \frac{1}{\sqrt{2}} & \frac{1}{\sqrt{2}} \end{bmatrix}$, so the given equation becomes $\begin{bmatrix} x & y \end{bmatrix} C \begin{bmatrix} 4 & 0 \\ 0 & 2 \end{bmatrix} C^t \begin{bmatrix} x \\ y \end{bmatrix} - 12 = 0$, or $\begin{bmatrix} x' & y' \end{bmatrix} \begin{bmatrix} 4 & 0 \\ 0 & 2 \end{bmatrix} \begin{bmatrix} x' \\ y' \end{bmatrix} - 12 = 0$, where $\begin{bmatrix} x' & y' \end{bmatrix} = \begin{bmatrix} x & y \end{bmatrix} C$. Thus $4x'^2 + 2y'^2 = 12$; i.e., $\frac{x'^2}{3} + \frac{y'^2}{6} = 1$. The graph is an ellipse with the lines $y = x$ and $y = -x$ as axes.

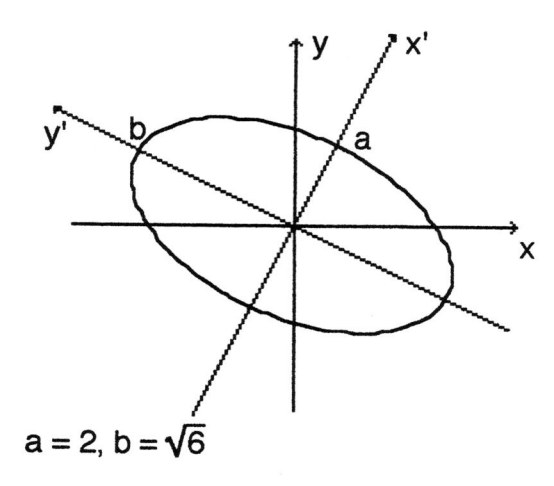

(c) $x^2 - 6xy + y^2 - 8 = \begin{bmatrix} x & y \end{bmatrix} \begin{bmatrix} 1 & -3 \\ -3 & 1 \end{bmatrix} \begin{bmatrix} x \\ y \end{bmatrix} - 8$. The symmetric matrix has eigenvalues $\lambda = -2$ and $\lambda = 4$ with corresponding orthonormal eigenvectors $\begin{bmatrix} \frac{1}{\sqrt{2}} \\ \frac{1}{\sqrt{2}} \end{bmatrix}$ and $\begin{bmatrix} \frac{-1}{\sqrt{2}} \\ \frac{1}{\sqrt{2}} \end{bmatrix}$, so $C^t \begin{bmatrix} 1 & -3 \\ -3 & 1 \end{bmatrix} C = \begin{bmatrix} -2 & 0 \\ 0 & 4 \end{bmatrix}$, where $C = \begin{bmatrix} \frac{1}{\sqrt{2}} & \frac{-1}{\sqrt{2}} \\ \frac{1}{\sqrt{2}} & \frac{1}{\sqrt{2}} \end{bmatrix}$, so the given equation becomes $\begin{bmatrix} x & y \end{bmatrix} C \begin{bmatrix} -2 & 0 \\ 0 & 4 \end{bmatrix} C^t \begin{bmatrix} x \\ y \end{bmatrix} - 8 = 0$, or

Section 6.4

$[x'\ y'] \begin{bmatrix} -2 & 0 \\ 0 & 4 \end{bmatrix} \begin{bmatrix} x' \\ y' \end{bmatrix} - 8 = 0$, where $[x'\ y'] = [x\ y]\, C$. Thus $-2x'^2 + 4y'^2 = 8$;

i.e., $\dfrac{-x'^2}{4} + \dfrac{y'^2}{2} = 1$. The graph is a hyperbola with the lines $y = x$ and $y = -x$ as axes.

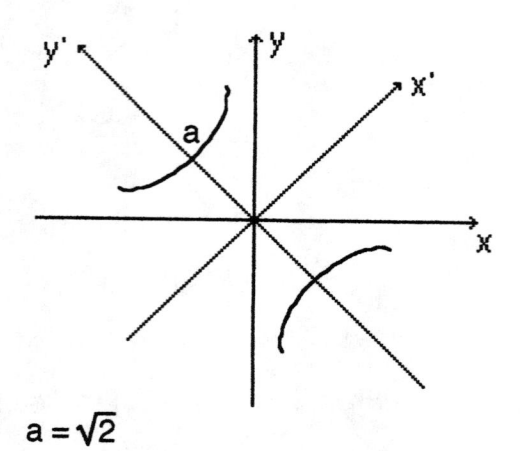

$a = \sqrt{2}$

Figure for 2(c)

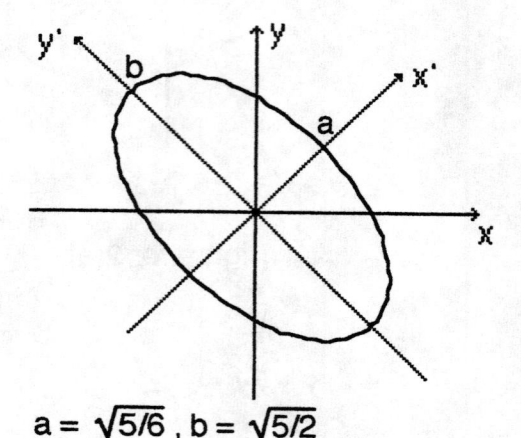

$a = \sqrt{5/6},\ b = \sqrt{5/2}$

Figure for 2(d)

(d) $4x^2 + 4xy + 4y^2 - 5 = [x\ y] \begin{bmatrix} 4 & 2 \\ 2 & 4 \end{bmatrix} \begin{bmatrix} x \\ y \end{bmatrix} - 5$. The symmetric matrix has

eigenvalues $\lambda = 6$ and $\lambda = 2$ with corresponding orthonormal eigenvectors

$\begin{bmatrix} \frac{1}{\sqrt{2}} \\ \frac{1}{\sqrt{2}} \end{bmatrix}$ and $\begin{bmatrix} \frac{-1}{\sqrt{2}} \\ \frac{1}{\sqrt{2}} \end{bmatrix}$, so $C^t \begin{bmatrix} 4 & 2 \\ 2 & 4 \end{bmatrix} C = \begin{bmatrix} 6 & 0 \\ 0 & 2 \end{bmatrix}$, where $C = \begin{bmatrix} \frac{1}{\sqrt{2}} & \frac{-1}{\sqrt{2}} \\ \frac{1}{\sqrt{2}} & \frac{1}{\sqrt{2}} \end{bmatrix}$, so

the given equation becomes $[x\ y]\, C \begin{bmatrix} 6 & 0 \\ 0 & 2 \end{bmatrix} C^t \begin{bmatrix} x \\ y \end{bmatrix} - 5 = 0$, or

$[x'\ y'] \begin{bmatrix} 6 & 0 \\ 0 & 2 \end{bmatrix} \begin{bmatrix} x' \\ y' \end{bmatrix} - 5 = 0$, where $[x'\ y'] = [x\ y]\, C$. Thus $6x'^2 + 2y'^2 = 5$,

i.e. $\dfrac{6x'^2}{5} + \dfrac{2y'^2}{5} = 1$. The graph is an ellipse with the lines $y = x$ and $y = -x$ as axes.

Section 6.4

3. (a) $a_n = a_{n-1} + 2a_{n-2}$, $a_1 = 1$, and $a_2 = 2$. Let $b_n = a_{n-1}$. Thus $\begin{bmatrix} a_n \\ b_n \end{bmatrix} = \begin{bmatrix} 1 & 2 \\ 1 & 0 \end{bmatrix} \begin{bmatrix} a_{n-1} \\ b_{n-1} \end{bmatrix}$.

The matrix has eigenvalues $\lambda = -1$ and $\lambda = 2$ with corresponding eigenvectors

$\begin{bmatrix} 1 \\ -1 \end{bmatrix}$ and $\begin{bmatrix} 2 \\ 1 \end{bmatrix}$. Let $C = \begin{bmatrix} 1 & 2 \\ -1 & 1 \end{bmatrix}$. $C^{-1} = \frac{1}{3}\begin{bmatrix} 1 & -2 \\ 1 & 1 \end{bmatrix}$, and

$\begin{bmatrix} a_n \\ b_n \end{bmatrix} = C\begin{bmatrix} (-1)^{n-2} & 0 \\ 0 & 2^{n-2} \end{bmatrix} C^{-1}\begin{bmatrix} 2 \\ 1 \end{bmatrix} = \frac{1}{3}\begin{bmatrix} 2^n + 2^{n-1} \\ 2^{n-1} + 2^{n-2} \end{bmatrix}$, so $a_n = \frac{1}{3}(2^n + 2^{n-1}) = 2^{n-1}$.

$a_{10} = 2^9 = 512$.

(b) $a_n = 2a_{n-1} + 3a_{n-2}$, $a_1 = 1$, and $a_2 = 3$. Let $b_n = a_{n-1}$. Thus $\begin{bmatrix} a_n \\ b_n \end{bmatrix} = \begin{bmatrix} 2 & 3 \\ 1 & 0 \end{bmatrix}\begin{bmatrix} a_{n-1} \\ b_{n-1} \end{bmatrix}$.

The matrix has eigenvalues $\lambda = -1$ and $\lambda = 3$ with corresponding eigenvectors

$\begin{bmatrix} 1 \\ -1 \end{bmatrix}$ and $\begin{bmatrix} 3 \\ 1 \end{bmatrix}$. Let $C = \begin{bmatrix} 1 & 3 \\ -1 & 1 \end{bmatrix}$. $C^{-1} = \frac{1}{4}\begin{bmatrix} 1 & -3 \\ 1 & 1 \end{bmatrix}$, and

$\begin{bmatrix} a_n \\ b_n \end{bmatrix} = C\begin{bmatrix} (-1)^{n-2} & 0 \\ 0 & 3^{n-2} \end{bmatrix} C^{-1}\begin{bmatrix} 3 \\ 1 \end{bmatrix} = \frac{1}{4}\begin{bmatrix} 3^n + 3^{n-1} \\ 3^{n-1} + 3^{n-2} \end{bmatrix}$, so $a_n = \frac{1}{4}(3^n + 3^{n-1}) = 3^{n-1}$.

$a_9 = 3^8 = 6561$.

(c) $a_n = 3a_{n-1} + 4a_{n-2}$, $a_1 = 1$, and $a_2 = -1$. Let $b_n = a_{n-1}$. Thus $\begin{bmatrix} a_n \\ b_n \end{bmatrix} = \begin{bmatrix} 3 & 4 \\ 1 & 0 \end{bmatrix}\begin{bmatrix} a_{n-1} \\ b_{n-1} \end{bmatrix}$.

The matrix has eigenvalues $\lambda = -1$ and $\lambda = 4$ with corresponding eigenvectors

$\begin{bmatrix} 1 \\ -1 \end{bmatrix}$ and $\begin{bmatrix} 4 \\ 1 \end{bmatrix}$. Let $C = \begin{bmatrix} 1 & 4 \\ -1 & 1 \end{bmatrix}$. $C^{-1} = \frac{1}{5}\begin{bmatrix} 1 & -4 \\ 1 & 1 \end{bmatrix}$, and

$\begin{bmatrix} a_n \\ b_n \end{bmatrix} = C\begin{bmatrix} (-1)^{n-2} & 0 \\ 0 & 4^{n-2} \end{bmatrix} C^{-1}\begin{bmatrix} -1 \\ 1 \end{bmatrix} = \begin{bmatrix} (-1)^{n-1} \\ (-1)^{n-2} \end{bmatrix}$, so $a_n = (-1)^{n-1}$.

$a_{12} = (-1)^{11} = -1$.

Section 6.4

4. $a_n = a_{n-1} + a_{n-2}$, $a_1 = 1$, and $a_2 = 1$. Let $b_n = a_{n-1}$. Thus $\begin{bmatrix} a_n \\ b_n \end{bmatrix} = \begin{bmatrix} 1 & 1 \\ 1 & 0 \end{bmatrix} \begin{bmatrix} a_{n-1} \\ b_{n-1} \end{bmatrix}$.

The matrix has characteristic polynomial $\lambda^2 - \lambda - 1$, so the eigenvalues are $\lambda = \dfrac{1 \pm \sqrt{5}}{2}$

with corresponding eigenvectors $\begin{bmatrix} \frac{1+\sqrt{5}}{2} \\ 1 \end{bmatrix}$ and $\begin{bmatrix} \frac{1-\sqrt{5}}{2} \\ 1 \end{bmatrix}$. Let $C = \begin{bmatrix} \frac{1+\sqrt{5}}{2} & \frac{1-\sqrt{5}}{2} \\ 1 & 1 \end{bmatrix}$.

$C^{-1} = \dfrac{1}{\sqrt{5}} \begin{bmatrix} 1 & \frac{-1+\sqrt{5}}{2} \\ -1 & \frac{1-\sqrt{5}}{2} \end{bmatrix}$, and $\begin{bmatrix} a_n \\ b_n \end{bmatrix} = C \begin{bmatrix} \left(\frac{1+\sqrt{5}}{2}\right)^{n-2} & 0 \\ 0 & \left(\frac{1-\sqrt{5}}{2}\right)^{n-2} \end{bmatrix} C^{-1} \begin{bmatrix} 1 \\ 1 \end{bmatrix}$

$= \dfrac{1}{\sqrt{5}} \begin{bmatrix} \left(\frac{1+\sqrt{5}}{2}\right)^n - \left(\frac{1-\sqrt{5}}{2}\right)^n \\ \left(\frac{1+\sqrt{5}}{2}\right)^{n-1} - \left(\frac{1-\sqrt{5}}{2}\right)^{n-1} \end{bmatrix}$, so $a_n = \dfrac{1}{\sqrt{5}} \left(\left(\frac{1+\sqrt{5}}{2}\right)^n - \left(\frac{1-\sqrt{5}}{2}\right)^n \right)$ and

$\dfrac{a_n}{a_{n-1}} = \dfrac{\left(\frac{1+\sqrt{5}}{2}\right)^n - \left(\frac{1-\sqrt{5}}{2}\right)^n}{\left(\frac{1+\sqrt{5}}{2}\right)^{n-1} - \left(\frac{1-\sqrt{5}}{2}\right)^{n-1}}$. We multiply the numerator by $\dfrac{\left(\frac{1+\sqrt{5}}{2}\right)}{\left(\frac{1+\sqrt{5}}{2}\right)^n}$

and the denominator by $\dfrac{1}{\left(\frac{1+\sqrt{5}}{2}\right)^{n-1}}$ to get $\dfrac{a_n}{a_{n-1}} = \dfrac{1+\sqrt{5}}{2} \cdot \dfrac{1 - \left(\frac{-3+\sqrt{5}}{2}\right)^n}{1 - \left(\frac{-3+\sqrt{5}}{2}\right)^{n-1}}$.

As n increases, $\left(\dfrac{-3+\sqrt{5}}{2}\right)^n$ approaches zero, so $\dfrac{a_n}{a_{n-1}}$ approaches $\dfrac{1+\sqrt{5}}{2}$.

5. $m\ddot{x}_1 = \dfrac{1}{a}(-2Tx_1 + Tx_2)$ and $m\ddot{x}_2 = \dfrac{1}{a}(Tx_1 - 2Tx_2)$, so $\begin{bmatrix} \ddot{x}_1 \\ \ddot{x}_2 \end{bmatrix} = \dfrac{T}{ma} \begin{bmatrix} -2 & 1 \\ 1 & -2 \end{bmatrix} \begin{bmatrix} x_1 \\ x_2 \end{bmatrix}$.

236

Section 6.4

The matrix has eigenvalues $\lambda = -3$ and $\lambda = -1$ with corresponding eigenvectors $\begin{bmatrix} 1 \\ -1 \end{bmatrix}$

and $\begin{bmatrix} 1 \\ 1 \end{bmatrix}$. $C^{-1}AC = \begin{bmatrix} -3 & 0 \\ 0 & -1 \end{bmatrix}$, where $C = \begin{bmatrix} 1 & 1 \\ -1 & 1 \end{bmatrix}$. Let $\begin{bmatrix} x_1' \\ x_2' \end{bmatrix} = C^{-1}\begin{bmatrix} x_1 \\ x_2 \end{bmatrix}$.

$\begin{bmatrix} \ddot{x}_1' \\ \ddot{x}_2' \end{bmatrix} = \frac{T}{ma}\begin{bmatrix} -3 & 0 \\ 0 & -1 \end{bmatrix}\begin{bmatrix} x_1' \\ x_2' \end{bmatrix}$, so that $\ddot{x}_1' = \frac{-3T}{ma} x_1'$ and $\ddot{x}_2' = \frac{-T}{ma} x_2'$.

Solutions of these equations are $x_1' = b_1 \cos(\alpha_1 t + \gamma_1)$ and $x_2' = b_2 \cos(\alpha_2 t + \gamma_2)$, where $\alpha_1 = (3T/ma)^{1/2}$ and $\alpha_2 = (T/ma)^{1/2}$. We now must solve for x_1 and x_2.

$\begin{bmatrix} x_1 \\ x_2 \end{bmatrix} = C\begin{bmatrix} x_1' \\ x_2' \end{bmatrix} = \begin{bmatrix} 1 & 1 \\ -1 & 1 \end{bmatrix}\begin{bmatrix} x_1' \\ x_2' \end{bmatrix} = x_1'\begin{bmatrix} 1 \\ -1 \end{bmatrix} + x_2'\begin{bmatrix} 1 \\ 1 \end{bmatrix}$

$= b_1 \cos(\alpha_1 t + \gamma_1)\begin{bmatrix} 1 \\ -1 \end{bmatrix} + b_2 \cos(\alpha_2 t + \gamma_2)\begin{bmatrix} 1 \\ 1 \end{bmatrix}$. Thus the normal modes are

mode 1: $\begin{bmatrix} x_1 \\ x_2 \end{bmatrix} = \cos(\alpha_1 t + \gamma_1)\begin{bmatrix} 1 \\ -1 \end{bmatrix}$, where $\alpha_1 = \left(\frac{3T}{ma}\right)^{1/2}$, and

mode 2: $\begin{bmatrix} x_1 \\ x_2 \end{bmatrix} = \cos(\alpha_2 t + \gamma_2)\begin{bmatrix} 1 \\ 1 \end{bmatrix}$, where $\alpha_2 = \left(\frac{T}{ma}\right)^{1/2}$.

6. $M\ddot{x}_1 = -5x_1 + 2x_2$ and $M\ddot{x}_2 = 2x_1 - 2x_2$, so $\begin{bmatrix} \ddot{x}_1 \\ \ddot{x}_2 \end{bmatrix} = \frac{1}{M}\begin{bmatrix} -5 & 2 \\ 2 & -2 \end{bmatrix}\begin{bmatrix} x_1 \\ x_2 \end{bmatrix}$.

The matrix has eigenvalues $\lambda = -1$ and $\lambda = -6$ with corresponding eigenvectors $\begin{bmatrix} 1 \\ 2 \end{bmatrix}$

and $\begin{bmatrix} 2 \\ -1 \end{bmatrix}$. $C^{-1}AC = \begin{bmatrix} -1 & 0 \\ 0 & -6 \end{bmatrix}$, where $C = \begin{bmatrix} 1 & 2 \\ 2 & -1 \end{bmatrix}$. Let $\begin{bmatrix} x_1' \\ x_2' \end{bmatrix} = C^{-1}\begin{bmatrix} x_1 \\ x_2 \end{bmatrix}$.

$\begin{bmatrix} \ddot{x}_1' \\ \ddot{x}_2' \end{bmatrix} = \frac{1}{M}\begin{bmatrix} -1 & 0 \\ 0 & -6 \end{bmatrix}\begin{bmatrix} x_1' \\ x_2' \end{bmatrix}$, so that $\ddot{x}_1' = \frac{-1}{M} x_1'$ and $\ddot{x}_2' = \frac{-6}{M} x_2'$.

Solutions of these equations are $x_1' = b_1 \cos(\alpha_1 t + \gamma_1)$ and $x_2' = b_2 \cos(\alpha_2 t + \gamma_2)$, where $\alpha_1 = (1/M)^{1/2}$ and $\alpha_2 = (6/M)^{1/2}$. We now must solve for x_1 and x_2.

$$\begin{bmatrix} x_1 \\ x_2 \end{bmatrix} = C \begin{bmatrix} x_1' \\ x_2' \end{bmatrix} = \begin{bmatrix} 1 & 2 \\ 2 & -1 \end{bmatrix} \begin{bmatrix} x_1' \\ x_2' \end{bmatrix} = x_1' \begin{bmatrix} 1 \\ 2 \end{bmatrix} + x_2' \begin{bmatrix} 2 \\ -1 \end{bmatrix}$$

$$= b_1 \cos(\alpha_1 t + \gamma_1) \begin{bmatrix} 1 \\ 2 \end{bmatrix} + b_2 \cos(\alpha_2 t + \gamma_2) \begin{bmatrix} 2 \\ -1 \end{bmatrix}.$$

7. $m\ddot{x}_1 = -3x_1 + 2x_2$ and $m\ddot{x}_2 = 2x_1 - 5x_2$, so $\begin{bmatrix} \ddot{x}_1 \\ \ddot{x}_2 \end{bmatrix} = \frac{1}{m} \begin{bmatrix} -3 & 2 \\ 2 & -5 \end{bmatrix} \begin{bmatrix} x_1 \\ x_2 \end{bmatrix}$.

The matrix has eigenvalues $\lambda = -4+\sqrt{5}$ and $\lambda = -4-\sqrt{5}$ with corresponding

eigenvectors $\begin{bmatrix} 1 \\ \frac{-1+\sqrt{5}}{2} \end{bmatrix}$ and $\begin{bmatrix} 1 \\ \frac{-1-\sqrt{5}}{2} \end{bmatrix}$. $C^{-1}AC = \begin{bmatrix} -4+\sqrt{5} & 0 \\ 0 & -4-\sqrt{5} \end{bmatrix}$, where

$C = \begin{bmatrix} 1 & 1 \\ \frac{-1+\sqrt{5}}{2} & \frac{-1-\sqrt{5}}{2} \end{bmatrix}$. Let $\begin{bmatrix} x_1' \\ x_2' \end{bmatrix} = C^{-1} \begin{bmatrix} x_1 \\ x_2 \end{bmatrix}$.

$\begin{bmatrix} \ddot{x}_1' \\ \ddot{x}_2' \end{bmatrix} = \frac{1}{m} \begin{bmatrix} -4+\sqrt{5} & 0 \\ 0 & -4-\sqrt{5} \end{bmatrix} \begin{bmatrix} x_1' \\ x_2' \end{bmatrix}$, so that $\ddot{x}_1' = \frac{-4+\sqrt{5}}{m} x_1'$ and $\ddot{x}_2' = \frac{-4-\sqrt{5}}{m} x_2'$.

Solutions of these equations are $x_1' = b_1 \cos(\alpha_1 t + \gamma_1)$ and $x_2' = b_2 \cos(\alpha_2 t + \gamma_2)$, where $\alpha_1 = \left(\frac{4-\sqrt{5}}{m}\right)^{1/2}$ and $\alpha_2 = \left(\frac{4+\sqrt{5}}{m}\right)^{1/2}$. We now must solve for x_1 and x_2.

$$\begin{bmatrix} x_1 \\ x_2 \end{bmatrix} = C \begin{bmatrix} x_1' \\ x_2' \end{bmatrix} = \begin{bmatrix} 1 & 1 \\ \frac{-1+\sqrt{5}}{2} & \frac{-1-\sqrt{5}}{2} \end{bmatrix} \begin{bmatrix} x_1' \\ x_2' \end{bmatrix} = x_1' \begin{bmatrix} 1 \\ \frac{-1+\sqrt{5}}{2} \end{bmatrix} + x_2' \begin{bmatrix} 1 \\ \frac{-1-\sqrt{5}}{2} \end{bmatrix}$$

$$= b_1 \cos(\alpha_1 t + \gamma_1) \begin{bmatrix} 1 \\ \frac{-1+\sqrt{5}}{2} \end{bmatrix} + b_2 \cos(\alpha_2 t + \gamma_2) \begin{bmatrix} 1 \\ \frac{-1-\sqrt{5}}{2} \end{bmatrix}.$$

8. $x' = OA' = OC + CA' = OC + DP = OB\sin\theta + BP\cos\theta = y\sin\theta + x\cos\theta$
    $y' = OB' = PA' = BC - BD = OB\cos\theta - BP\sin\theta = y\cos\theta - x\sin\theta$

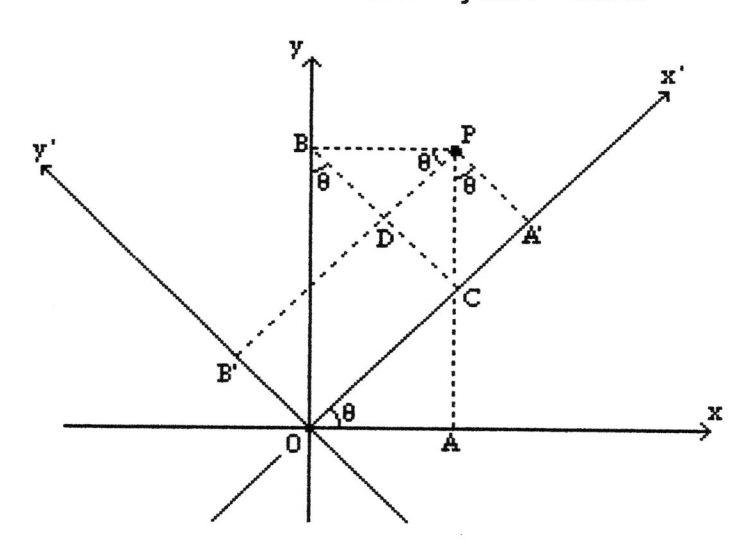

Chapter 6 Review Exercises

1. $\begin{vmatrix} 5-\lambda & -7 & 7 \\ 4 & -3-\lambda & 4 \\ 4 & -1 & 2-\lambda \end{vmatrix} = (5-\lambda)(1-\lambda)(-2-\lambda)$, so the eigenvalues are $\lambda = 5$, $\lambda = 1$, and $\lambda = -2$. For $\lambda = 5$, the eigenvectors are the solutions of $\begin{bmatrix} 0 & -7 & 7 \\ 4 & -8 & 4 \\ 4 & -1 & -3 \end{bmatrix} \begin{bmatrix} x_1 \\ x_2 \\ x_3 \end{bmatrix} = \mathbf{0}$, so the eigenvectors are vectors of the form $r\begin{bmatrix} 1 \\ 1 \\ 1 \end{bmatrix}$. For $\lambda = 1$, the eigenvectors are the solutions of $\begin{bmatrix} 4 & -7 & 7 \\ 4 & -4 & 4 \\ 4 & -1 & 1 \end{bmatrix} \begin{bmatrix} x_1 \\ x_2 \\ x_3 \end{bmatrix} = \mathbf{0}$, so the eigenvectors are vectors of the form $s\begin{bmatrix} 0 \\ 1 \\ 1 \end{bmatrix}$. For

$\lambda = -2$, the eigenvectors are the solutions of $\begin{bmatrix} 7 & -7 & 7 \\ 4 & -1 & 4 \\ 4 & -1 & 4 \end{bmatrix} \begin{bmatrix} x_1 \\ x_2 \\ x_3 \end{bmatrix} = 0$, so the eigenvectors are vectors of the form $t \begin{bmatrix} 1 \\ 0 \\ -1 \end{bmatrix}$.

2. Let λ be an eigenvalue of A with eigenvector $\mathbf{x}$. A is invertible so $\lambda \neq 0$. $A\mathbf{x} = \lambda \mathbf{x}$, so $\mathbf{x} = A^{-1}A\mathbf{x} = A^{-1}\lambda\mathbf{x} = \lambda A^{-1}\mathbf{x}$. Thus $\frac{1}{\lambda}\mathbf{x} = A^{-1}\mathbf{x}$, so the eigenvalues for A^{-1} are the inverses of the eigenvalues for A and the corresponding eigenvectors are the same.

3. $A\mathbf{x} = \lambda\mathbf{x}$, so $A\mathbf{x} - kI\mathbf{x} = \lambda\mathbf{x} - kI\mathbf{x} = \lambda\mathbf{x} - k\mathbf{x}$, so $(A - kI)\mathbf{x} = (\lambda - k)\mathbf{x}$. Thus $\lambda - k$ is an eigenvalue for $A - kI$ with corresponding eigenvector $\mathbf{x}$.

4. $C^{-1}AC = \begin{bmatrix} 1 & -1 \\ -1 & 2 \end{bmatrix} \begin{bmatrix} 4 & -2 \\ 1 & 1 \end{bmatrix} \begin{bmatrix} 2 & 1 \\ 1 & 1 \end{bmatrix} = \begin{bmatrix} 3 & 0 \\ 0 & 2 \end{bmatrix}$.

5. $\begin{vmatrix} 1-\lambda & 1 \\ -2 & 4-\lambda \end{vmatrix} = (3-\lambda)(2-\lambda)$, so the eigenvalues are $\lambda = 3$ and $\lambda = 2$. For $\lambda = 3$, the eigenvectors are $r \begin{bmatrix} 1 \\ 2 \end{bmatrix}$ and for $\lambda = 2$, the eigenvectors are $s \begin{bmatrix} 1 \\ 1 \end{bmatrix}$. Let $C = \begin{bmatrix} 1 & 1 \\ 2 & 1 \end{bmatrix}$.

$C^{-1}AC = \begin{bmatrix} -1 & 1 \\ 2 & -1 \end{bmatrix} \begin{bmatrix} 1 & 1 \\ -2 & 4 \end{bmatrix} \begin{bmatrix} 1 & 1 \\ 2 & 1 \end{bmatrix} = \begin{bmatrix} 3 & 0 \\ 0 & 2 \end{bmatrix}$.

6. $\begin{vmatrix} 7-\lambda & -2 & 1 \\ -2 & 10-\lambda & -2 \\ 1 & -2 & 7-\lambda \end{vmatrix} = (6-\lambda)(6-\lambda)(12-\lambda)$, so the eigenvalues are $\lambda = 6$ and $\lambda = 12$.

For $\lambda = 6$, the eigenvectors are vectors of the form $r \begin{bmatrix} 1 \\ 0 \\ -1 \end{bmatrix} + s \begin{bmatrix} 1 \\ 1 \\ 1 \end{bmatrix}$. For $\lambda = 12$, the eigenvectors are vectors of the form $t\begin{bmatrix} 1 \\ -2 \\ 1 \end{bmatrix}$. Orthonormal eigenvectors are

$\begin{bmatrix} 1/\sqrt{2} \\ 0 \\ -1/\sqrt{2} \end{bmatrix}, \begin{bmatrix} 1/\sqrt{3} \\ 1/\sqrt{3} \\ 1/\sqrt{3} \end{bmatrix}, \begin{bmatrix} 1/\sqrt{6} \\ -2/\sqrt{6} \\ 1/\sqrt{6} \end{bmatrix}$. Let $C = \begin{bmatrix} 1/\sqrt{2} & 1/\sqrt{3} & 1/\sqrt{6} \\ 0 & 1/\sqrt{3} & -2/\sqrt{6} \\ -1/\sqrt{2} & 1/\sqrt{3} & 1/\sqrt{6} \end{bmatrix}$.

$\begin{bmatrix} 1/\sqrt{2} & 0 & -1/\sqrt{2} \\ 1/\sqrt{3} & 1/\sqrt{3} & 1/\sqrt{3} \\ 1/\sqrt{6} & -2/\sqrt{6} & 1/\sqrt{6} \end{bmatrix} \begin{bmatrix} 7 & -2 & 1 \\ -2 & 10 & -2 \\ 1 & -2 & 7 \end{bmatrix} \begin{bmatrix} 1/\sqrt{2} & 1/\sqrt{3} & 1/\sqrt{6} \\ 0 & 1/\sqrt{3} & -2/\sqrt{6} \\ -1/\sqrt{2} & 1/\sqrt{3} & 1/\sqrt{6} \end{bmatrix} = \begin{bmatrix} 6 & 0 & 0 \\ 0 & 6 & 0 \\ 0 & 0 & 12 \end{bmatrix}$.

7. $\begin{vmatrix} a-\lambda & b \\ b & c-\lambda \end{vmatrix} = (a-\lambda)(c-\lambda) - b^2 = \lambda^2 - (a+c)\lambda + ac - b^2$. The characteristic equation

$\lambda^2 - (a+c)\lambda + ac - b^2 = 0$ has roots $\lambda = \frac{1}{2}(a+c \pm \sqrt{D})$ where $D = (a-c)^2 + 4b^2$ (from the quadratic formula). D is nonnegative for all values of a, b, and c, so the roots are real.

8. If A is symmetric then, A can be diagonalized, $D = C^{-1}AC$, where the diagonal elements of D are the eigenvalues of A. If A has only one eigenvalue, λ, then $D = \lambda I$, so $\lambda I = C^{-1}AC$. Thus $A = CC^{-1}ACC^{-1} = C\lambda IC^{-1} = \lambda CIC^{-1} = \lambda CC^{-1} = \lambda I$.

9. $-x^2 - 16xy + 11y^2 - 30 = [x\ y]\begin{bmatrix} -1 & -8 \\ -8 & 11 \end{bmatrix}\begin{bmatrix} x \\ y \end{bmatrix} - 30$. The symmetric matrix has

eigenvalues $\lambda = 15$ and $\lambda = -5$ with corresponding orthonormal eigenvectors

$\begin{bmatrix} \frac{1}{\sqrt{5}} \\ \frac{-2}{\sqrt{5}} \end{bmatrix}$ and $\begin{bmatrix} \frac{2}{\sqrt{5}} \\ \frac{1}{\sqrt{5}} \end{bmatrix}$, so $C^t \begin{bmatrix} -1 & -8 \\ -8 & 11 \end{bmatrix} C = \begin{bmatrix} 15 & 0 \\ 0 & -5 \end{bmatrix}$, where $C = \begin{bmatrix} \frac{1}{\sqrt{5}} & \frac{2}{\sqrt{5}} \\ \frac{-2}{\sqrt{5}} & \frac{1}{\sqrt{5}} \end{bmatrix}$, so the

given equation becomes $[x\ y] C \begin{bmatrix} 15 & 0 \\ 0 & -5 \end{bmatrix} C^t \begin{bmatrix} x \\ y \end{bmatrix} - 30 = 0$, or

$[x'\ y'] \begin{bmatrix} 15 & 0 \\ 0 & -5 \end{bmatrix} \begin{bmatrix} x' \\ y' \end{bmatrix} - 30 = 0$, where $[x'\ y'] = [x\ y] C$. Thus $15x'^2 - 5y'^2 = 30$;

i.e., $\dfrac{x'^2}{2} - \dfrac{y'^2}{6} = 1$. The graph is a hyperbola with axes $y = -2x$ and $x = 2y$.

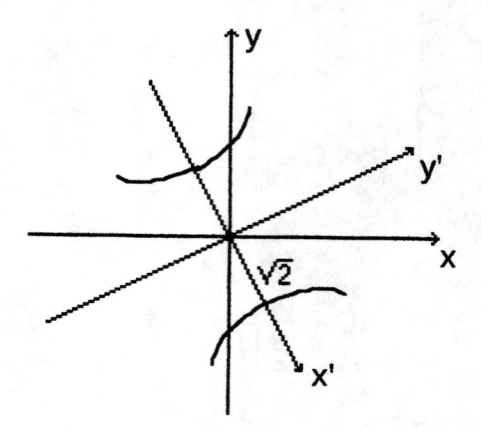

10. $a_n = 4a_{n-1} + 5a_{n-2}$, $a_1 = 3$, and $a_2 = 2$. Let $b_n = a_{n-1}$. Thus $\begin{bmatrix} a_n \\ b_n \end{bmatrix} = \begin{bmatrix} 4 & 5 \\ 1 & 0 \end{bmatrix} \begin{bmatrix} a_{n-1} \\ b_{n-1} \end{bmatrix}$.

The matrix has eigenvalues $\lambda = -1$ and $\lambda = 5$ with corresponding eigenvectors

$\begin{bmatrix} 1 \\ -1 \end{bmatrix}$ and $\begin{bmatrix} 5 \\ 1 \end{bmatrix}$. Let $C = \begin{bmatrix} 1 & 5 \\ -1 & 1 \end{bmatrix}$. $C^{-1} = \dfrac{1}{6} \begin{bmatrix} 1 & -5 \\ 1 & 1 \end{bmatrix}$, and

$\begin{bmatrix} a_n \\ b_n \end{bmatrix} = C \begin{bmatrix} (-1)^{n-2} & 0 \\ 0 & 5^{n-2} \end{bmatrix} C^{-1} \begin{bmatrix} 2 \\ 3 \end{bmatrix} = \dfrac{1}{6} \begin{bmatrix} 13(-1)^{n-1} + 5^n \\ 13(-1)^{n-2} + 5^{n-1} \end{bmatrix}$, so

$a_n = \dfrac{1}{6}(13(-1)^{n-1} + 5^n)$. $a_{12} = \dfrac{1}{6}(13(-1)^{11} + 5^{12}) = \dfrac{1}{6}(-13 + 244140625)$

$= 40{,}690{,}102$.

Chapter 7

Exercise Set 7.1

1. $T((x_1,y_1)+(x_2,y_2)) = T(x_1+x_2, y_1+y_2) = (3(x_1+x_2), x_1+x_2+y_1+y_2) = (3x_1+3x_2, x_1+y_1+x_2+y_2)$

   $= (3x_1, x_1+y_1) + (3x_2, x_2+y_2) = T(x_1,y_1) + T(x_2,y_2)$ and $T(c(x,y)) = T(cx,cy) = (3cx, cx+cy)$.

   $= c(3x, x+y) = cT(x,y)$. Thus T is linear.

   $T(1,3) = (3(1), 1+3) = (3,4)$. $T(-1,2) = (3(-1), -1+2) = (-3,1)$.

2. $T((x_1,y_1)+(x_2,y_2)) = T(x_1+x_2, y_1+y_2) = (x_1+x_2+2(y_1+y_2), 3(y_1+y_2), x_1+x_2-(y_1+y_2)) =$

   $(x_1+x_2+2y_1+2y_2, 3y_1+3y_2, x_1+x_2-y_1-y_2) = (x_1+2y_1, 3y_1, x_1-y_1) + (x_2+2y_2, 3y_2, x_2-y_2)$

   $= T(x_1,y_1) + T(x_2,y_2)$. $T(c(x,y)) = T(cx,cy) = (cx+2cy, 3cy, cx-cy) = c(x+2y, 3y, x-y)$

   $= cT(x,y)$, so T is linear.

   $T(0,4) = (0+2(4), 3(4), 0-4) = (8,12,-4)$. $T(1,1) = (1+2(1), 3(1), 1-1) = (3,9,0)$.

3. (a) $T(c(x,y)) = T(cx,cy) = (cy)^2 = c^2y^2$. But $cT(x,y) = cy^2$. $T(c(x,y)) \neq cT(x,y)$.
   Thus T is not linear.

   (b) $T(c(x,y)) = T(cx,cy) = cx-3$. But $cT(x,y) = c(x-3) = cx-3c$. $T(c(x,y)) \neq cT(x,y)$. Not linear.

4. $T((ax^2 + bx + c)+(px^2 + qx + r)) = T((a+p)x^2 + (b+q)x + c+r) = (c+r)x^2 + (a+p) =$

   $(cx^2 + a) + (rx^2 + p) = T(ax^2 + bx + c) + T(px^2 + qx + r)$.

   $T(k(ax^2 + bx + c)) = T(kax^2 + kbx + kc) = kcx^2 + ka = k(cx^2 + a) = kT(ax^2 + bx + c)$.

   T is linear. $T(3x^2-x+2) = 2x^2 + 3$.

5. $T((ax^2 + bx + c)+(px^2 + qx + r)) = T((a+p)x^2 + (b+q)x + c+r) = (b+q)x + c + r =$

243

Section 7.1

$(bx+c) + (qx+r) = T((ax^2 + bx + c) + T((ax^2 + bx + c)$.

$T(k(ax^2 + bx + c)) = T(kax^2 + kbx + kc) = kbx + kc = k(bx+c) = kT(ax^2 + bx + c)$.

T is linear. $T(3x^2 - x + 2) = -x+2$. $T(ax^2 - x + 2) = x-2$ for any a.

6. $T(k(ax + b)) = T(kax + kb) = x + ka$. But $kT(ax + b) = k(x+a) = kx + ka$.

   $T(k(x,y)) \neq kT(x,y)$. Thus T is not linear.

7. (a) $\begin{bmatrix} 1 & 2 \\ 3 & 0 \end{bmatrix} \begin{bmatrix} x \\ y \end{bmatrix} = \begin{bmatrix} x+2y \\ 3x \end{bmatrix}$; thus (x,y) is in the kernel of T if $x+2y = 0$ and $3x = 0$.

   The only solution of this system of homogeneous equations is $x = 0, y = 0$, so only the zero vector is in the kernel of T. The columns of the matrix are linearly independent, so the range is $\mathbf{R}^2$. Dim ker(T) = 0, dim range(T) = 2, and dim domain(T) = 2, so dim ker(T) + dim range(T) = dim domain(T).

   (b) $\begin{bmatrix} 2 & 0 \\ 3 & 0 \end{bmatrix} \begin{bmatrix} x \\ y \end{bmatrix} = \begin{bmatrix} 2x \\ 3x \end{bmatrix}$; thus (x,y) is in the kernel of T if $2x = 0$ and $3x = 0$, i.e., if $x = 0$.

   The kernel of T is the set {(0,r)} and the range is the set {(2r,3r)}. Dim ker(T) = 1, dim range(T) = 1, and dim domain(T) = 2, so dim ker(T) + dim range(T) = dim domain(T).

   (c) $\begin{bmatrix} 2 & 4 \\ 4 & 8 \end{bmatrix} \begin{bmatrix} x \\ y \end{bmatrix} = \begin{bmatrix} 2x+4y \\ 4x+8y \end{bmatrix}$; thus (x,y) is in the kernel of T if $2x+4y = 0$ and $4x+8y = 0$,

   i.e., if $x = -2y$. The kernel of T is the set {(-2r,r)} and the range is the set {(r,2r)}. Dim ker(T) = 1, dim range(T) = 1, and dim domain(T) = 2, so dim ker(T) + dim range(T) = dim domain(T).

   (d) $\begin{bmatrix} 1 & 2 \\ -1 & 3 \end{bmatrix} \begin{bmatrix} x \\ y \end{bmatrix} = \begin{bmatrix} x+2y \\ -x+3y \end{bmatrix}$; thus (x,y) is in the kernel of T if $x+2y = 0$ and $-x+3y = 0$.

   The only solution of this system of homogeneous equations is $x = 0, y = 0$, so only the zero vector is in the kernel of T. The range is $\mathbf{R}^2$. Dim ker(T) = 0, dim range(T) = 2, and dim domain(T) = 2, so dim ker(T) + dim range(T) = dim domain(T).

   (e) $\begin{bmatrix} 1 & 2 & 3 \\ 0 & 1 & 2 \end{bmatrix} \begin{bmatrix} x \\ y \\ z \end{bmatrix} = \begin{bmatrix} x+2y+3z \\ y+2z \end{bmatrix}$; thus (x,y,z) is in the kernel of T if $x+2y+3z = 0$ and

$y+2z = 0$. The general solution to this system of equations is $x = r$, $y = -2r$, and $z = r$, so the kernel of T is the set $\{(r,-2r,r)\}$. The range is $\mathbf{R}^2$. Dim ker(T) = 1, dim range(T) = 2, and dim domain(T) = 3, so dim ker(T) + dim range(T) = dim domain(T).

(f) $\begin{bmatrix} 1 & 0 & 0 \\ 0 & 2 & 0 \\ 0 & 0 & 3 \end{bmatrix} \begin{bmatrix} x \\ y \\ z \end{bmatrix} = \begin{bmatrix} x \\ 2y \\ 3z \end{bmatrix}$; thus (x,y,z) is in the kernel of T if $x = 0$, $2y = 0$, and $3z = 0$,

i.e., if $x = y = z = 0$. The kernel of T is the zero vector. The range is $\mathbf{R}^3$.
Dim ker(T) = 0, dim range(T) = 3, and dim domain(T) = 3,
so dim ker(T) + dim range(T) = dim domain(T).

(g) $\begin{bmatrix} 0 & 1 & 0 \\ 0 & 2 & 0 \\ 0 & 0 & 4 \end{bmatrix} \begin{bmatrix} x \\ y \\ z \end{bmatrix} = \begin{bmatrix} y \\ 2y \\ 4z \end{bmatrix}$; thus (x,y,z) is in the kernel of T if $y = 0$, $2y = 0$, and $4z = 0$,

i.e., if $y = z = 0$. The kernel of T is the set $\{(r,0,0)\}$. The range is the set $\{(a,2a,b)\}$.
Dim ker(T) = 1, dim range(T) = 2, and dim domain(T) = 3,
so dim ker(T) + dim range(T) = dim domain(T).

(h) $\begin{bmatrix} 1 & 2 & 1 \\ -1 & -2 & 0 \\ 2 & 4 & 1 \end{bmatrix} \begin{bmatrix} x \\ y \\ z \end{bmatrix} = \begin{bmatrix} x+2y+z \\ -x-2y \\ 2x+4y+z \end{bmatrix}$; thus (x,y,z) is in the kernel of T if $x+2y+z = 0$,

$-x-2y = 0$ and $2x+4y+z = 0$, i.e., if $x = -2y$ and $z = 0$. The kernel of T is the set $\{(-2r,r,0)\}$ and the range is the set $\{(a+b,-a,2a+b)\}$, using the first and third columns of the matrix as a basis. Dim ker(T) = 1, dim range(T) = 2, and dim domain(T) = 3, so dim ker(T) + dim range(T) = dim domain(T).

(i) $\begin{bmatrix} 1 & 1 & 5 \\ 0 & 1 & 3 \\ 2 & 1 & 7 \end{bmatrix} \begin{bmatrix} x \\ y \\ z \end{bmatrix} = \begin{bmatrix} x+y+5z \\ y+3z \\ 2x+y+7z \end{bmatrix}$; thus (x,y,z) is in the kernel of T if $x+y+5z = 0$,

$y+3z = 0$, and $2x+y+7z = 0$, i.e., if $x = -2z$ and $y = -3z$. The kernel of T is the set $\{(-2r,-3r,r)\}$ and the range is the set $\{(a+b,b,2a+b)\}$, using the first two columns of the matrix as a basis. Dim ker(T) = 1, dim range(T) = 2, and dim domain(T) = 3, so dim ker(T) + dim range(T) = dim domain(T).

8. (a) The kernel is the set $\{(0,r,s)\}$ and the range is the set $\{(a,0,0)\}$. Dim ker(T) = 2, dim range(T) = 1, and dim domain(T) = 3, so dim ker(T) + dim range(T) = dim domain(T).

(b) The kernel is the set $\{(r,-r,0)\}$ and the range is $\mathbf{R}^2$. Dim ker(T) = 1, dim range(T) = 2, and dim domain(T) = 3, so dim ker(T) + dim range(T) = dim domain(T).

- (c) The kernel is the set $\{(-r-s,r,s)\}$ and the range is **R**. Dim ker(T) = 2, dim range(T) = 1, and dim domain(T) = 3, so dim ker(T) + dim range(T) = dim domain(T).

- (d) The kernel is the set $\{(0,r)\}$ and the range is the set $\{(a,2a,3a)\}$. Dim ker(T) = 1, dim range(T) = 1, and dim domain(T) = 2, so dim ker(T) + dim range(T) = dim domain(T).

- (e) The kernel is the zero vector and the range is the set $\{(3a,a-b,b)\}$. Dim ker(T) = 0, dim range(T) = 2, and dim domain(T) = 2, so dim ker(T) + dim range(T) = dim domain(T).

9. (a) T(**u**+**w**) = 5(**u**+**w**) = 5**u** + 5**w** = T(**u**) + T(**w**) and T(c**u**) = 5(c**u**) = 5c**u** = c(5**u**) = cT(**u**), so T is linear. ker(T) is the zero vector and range(T) is U (since for any vector **u** in U, **u** = T(.2**u**)).

   (b) T(c**u**) = 2(c**u**) + 3**v** ≠ c(2**u** + 3**v**) = cT(**u**), so T is not linear.

   (c) T(**u**+**w**) = **u**+**w** = T(**u**) + T(**w**) and T(c**u**) = c**u** = cT(**u**), so T is linear. ker(T) is the zero vector and range(T) is U.

   (d) T(**u**+**w**) = **0** = **0** + **0** = T(**u**) + T(**w**) and T(c**u**) = **0** = c**0** = cT(**u**), so T is linear. ker(T) is U and range(T) is **0**.

10. (a) The dimension of the range of the transformation is the rank of the matrix = 2. The domain is $\mathbf{R}^3$, which has dimension 3, so the dimension of the kernel is 1. This transformation is not one-to-one.

    (b) The dimension of the range of the transformation is the rank of the matrix = 3. The domain is $\mathbf{R}^3$, which has dimension 3, so the dimension of the kernel is 0. This transformation is one-to-one.

    (c) The dimension of the range of the transformation is the rank of the matrix = 3. The domain is $\mathbf{R}^4$, which has dimension 4, so the dimension of the kernel is 1. This transformation is not one-to-one.

    (d) The dimension of the range of the transformation is the rank of the matrix = 2. The domain is $\mathbf{R}^3$, which has dimension 3, so the dimension of the kernel is 1. This transformation is not one-to-one.

    (e) The dimension of the range of the transformation is the rank of the matrix = 2. The domain is $\mathbf{R}^4$, which has dimension 4, so the dimension of the kernel is 2. This transformation is not one-to-one.

- (f) The dimension of the range of the transformation is the rank of the matrix = 3. The domain is **R**^{3}, which has dimension 3, so the dimension of the kernel is 0. This transformation is one-to-one.

11. (a) |A| = 12 ≠ 0, so the transformation is nonsingular and therefore one-to-one.

    (b) |B| = 0, so the transformation is not one-to-one.

    (c) |C| = 0, so the transformation is not one-to-one.

    (d) |D| = 0, so the transformation is not one-to-one.

    (e) |E| = –24, so the transformation is one-to-one.

    (f) |F| = 212, so the transformation is one-to-one.

12. (a) The set of fixed points is the set of all points for which (x,y) = (x,3y). This is the set {(r,0)}.

    (b) The set of fixed points is the set of all points for which (x,y) = (x,2). This is the set {(r,2)}.

    (c) This transformation has no fixed points since there are no solutions to the equation y = y + 1.

    (d) All points in U are fixed points, since T maps each point to itself.

    (e) The set of fixed points is the set of all points for which (x,y) = (y,x). This is the set {(r,r)}.

    (f) The set of fixed points is the set of all points for which (x,y) = (x+y,x–y). This set contains only the zero vector.

    (g) If **u** and **v** are fixed points of a linear transformation T: U → U, then T(**u**) = **u** and T(**v**) = **v**, so that T(**u**+**v**) = T(**u**) + T(**v**) = **u** + **v** and T(c**u**) = cT(**u**) = c**u**. Thus both **u**+**v** and c**u** are fixed points of T, and the set of fixed points is therefore a subspace of U.

13. (a) Let **u**$_1$, **u**$_2$, ..., **u**$_n$ be a basis for U. The vectors T(**u**$_1$), T(**u**$_2$), ..., T(**u**$_n$) span range(T), for if **v** is a vector in range(T), then **v** = T(**u**) for some vector **u** = a$_1$**u**$_1$ + a$_2$**u**$_2$ + ... + a$_n$**u**$_n$ in U, and so **v** = T(**u**) = T(a$_1$**u**$_1$ + a$_2$**u**$_2$ + ... + a$_n$**u**$_n$) = a$_1$T(**u**$_1$) + a$_2$T(**u**$_2$) + ... + a$_n$T(**u**$_n$). Also the vectors T(**u**$_1$), T(**u**$_2$), ..., T(**u**$_n$) are linearly independent, because if **0** = a$_1$T(**u**$_1$) + a$_2$T(**u**$_2$) + ... + a$_n$T(**u**$_n$)

$= T(a_1\mathbf{u}_1 + a_2\mathbf{u}_2 + \ldots + a_n\mathbf{u}_n)$ then $a_1\mathbf{u}_1 + a_2\mathbf{u}_2 + \ldots + a_n\mathbf{u}_n = \mathbf{0}$ since $\ker(T) = \mathbf{0}$, and $a_1 = a_2 = \ldots = a_n = 0$ since $\mathbf{u}_1, \mathbf{u}_2, \ldots, \mathbf{u}_n$ are linearly independent. Thus $T(\mathbf{u}_1), T(\mathbf{u}_2), \ldots, T(\mathbf{u}_n)$ are a basis for range(T). So dim range(T) = dim domain(T), and since dim ker(T) = 0, the equality is proved.

(b) If ker(T) = U, then range(T) = **0**, since all vectors in U are mapped into **0**. dim ker(T) = dim U = dim domain(T), and since dim range(T) = 0, the equality is proved.

14. (a) If $\ker(T) = \mathbf{0}$ and $\mathbf{u}_1, \mathbf{u}_2, \ldots, \mathbf{u}_n$ are a basis for U, then $T(\mathbf{u}_1), T(\mathbf{u}_2), \ldots, T(\mathbf{u}_n)$ are a basis for range(T) (see Exercise 7(a)). Thus $\{T(\mathbf{u}_1), T(\mathbf{u}_2), \ldots, T(\mathbf{u}_n)\}$ is a set of n linearly independent vectors in V. If dim(V) = dim(U) = n, this set of vectors is a basis for V. Since V and range(T) have the same basis vectors, they are the same vector space.

If $\mathbf{u}_1, \mathbf{u}_2, \ldots, \mathbf{u}_n$ are a basis for U, then $T(\mathbf{u}_1), T(\mathbf{u}_2), \ldots, T(\mathbf{u}_n)$ span range(T). If range(T) = V then dim range(T) = n and $T(\mathbf{u}_1), T(\mathbf{u}_2), \ldots, T(\mathbf{u}_n)$ are a basis for range(T) = V. If $\mathbf{0} = T(\mathbf{u}) = T(a_1\mathbf{u}_1 + a_2\mathbf{u}_2 + \ldots + a_n\mathbf{u}_n)$
$= a_1 T(\mathbf{u}_1) + a_2 T(\mathbf{u}_2) + \ldots + a_n T(\mathbf{u}_n)$ then $a_1 = a_2 = \ldots = a_n = 0$ since $T(\mathbf{u}_1), T(\mathbf{u}_2), \ldots, T(\mathbf{u}_n)$ are a basis. Thus $\mathbf{u} = \mathbf{0}$ and $\ker(T) = \mathbf{0}$.

(b) If range(T) = **0**, then T(**u**) = **0** for all vectors **u** in U. Thus ker(T) = U.

If ker(T) = U, then T(**u**) = **0** for all vectors **u** in U. Thus the only vector in range(T) is **0**.

15. dim ker(T) + dim range(T) = dim domain(T). dim ker(T) ≥ 0, so dim range(T) ≤ dim domain(T).

16. dim range(T) = rank A = rank A^t = dim range(T^t).

17. $T(x,y,z) = (x,y,0) = (1,2,0)$ if x = 1, y = 2, z = r. Thus the set of vectors mapped by T into (1,2,0) is the set {(1,2,r)}.

Section 7.1

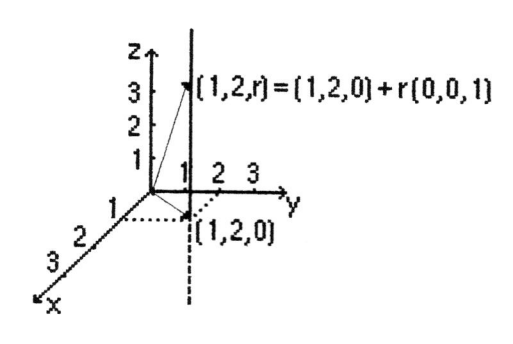

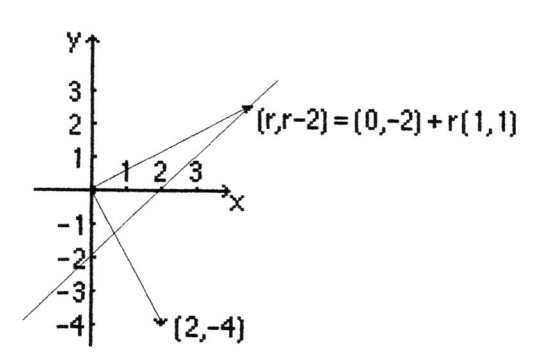

Figure for exercise 17

Figure for exercise 18

18. $T(x,y) = (x-y, 2y-2x) = (2,-4)$ if $x-y = 2$ and $2y-2x = -4$, i.e. if $y = x-2$. Thus the set of vectors mapped by T into $(2,-4)$ is the set $\{(r, r-2)\}$.

19. $T(x,y) = (2x, 3x) = (4,6)$ if $2x = 4$ and $3x = 6$, i.e. if $x = 2$. Thus the set of vectors mapped by T into $(4,6)$ is the set $\{(2, r)\}$. This set is not a subspace of $\mathbf{R}^2$. It does not contain the zero vector.

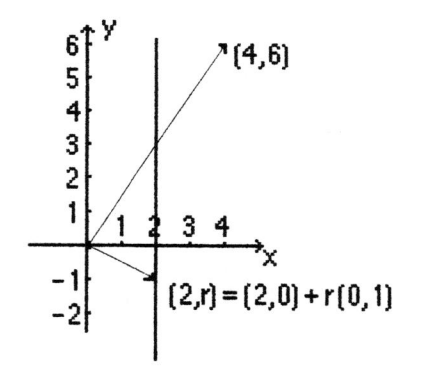

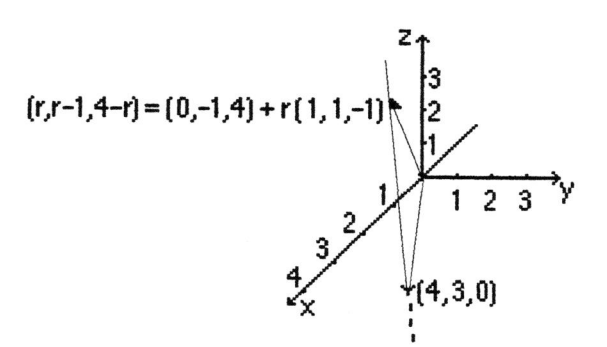

Figure for exercise 19

Figure for exercise 20

20. $T(x,y,z) = (x-y, x+z) = (1,4)$ if $x-y = 1$ and $x+z = 4$. Thus the set of vectors mapped by T into $(1,4)$ is the set $\{(r, r-1, 4-r)\}$. This set is not a subspace of $\mathbf{R}^3$. It does not contain the zero vector.

21. This set is not a subspace because it does not contain the zero vector.

Section 7.1

22. $T(a_2x^2 + a_1x + a_0) + T(b_2x^2 + b_1x + b_0) = (a_2 + a_1)x^2 + a_1x + 2a_0 + (b_2 + b_1)x^2 + b_1x + 2b_0$

$= (a_2 + a_1 + b_2 + b_1)x^2 + (a_1 + b_1)x + 2(a_0 + b_0) = T((a_2 + b_2)x^2 + (a_1 + b_1)x + a_0 + b_0)$
$= T(a_2x^2 + a_1x + a_0 + b_2x^2 + b_1x + b_0)$ and $T(c(a_2x^2 + a_1x + a_0)) = T(ca_2x^2 + ca_1x + ca_0)$
$= (ca_2 + ca_1)x^2 + ca_1x + 2ca_0 = c((a_2 + a_1)x^2 + a_1x + 2a_0) = cT(a_2x^2 + a_1x + a_0)$, so
T is linear.

$T(a_2x^2 + a_1x + a_0) = 0$ if $a_2 = a_1 = a_0 = 0$, so ker(T) is the zero polynomial and range(T) is P_2. A basis for P_2 is the set $\{1, x, x^2\}$.

23. $T(a_3x^3 + a_2x^2 + a_1x + a_0) + T(b_3x^3 + b_2x^2 + b_1x + b_0) = a_3x^2 - a_0 + b_3x^2 - b_0$
$= (a_3 + b_3)x^2 - (a_0 + b_0) = T((a_3 + b_3)x^3 + (a_2 + b_2)x^2 + (a_1 + b_1)x + a_0 + b_0)$
$= T(a_3x^3 + a_2x^2 + a_1x + a_0 + b_3x^3 + b_2x^2 + b_1x + b_0)$ and $T(c(a_3x^3 + a_2x^2 + a_1x + a_0))$
$= T(ca_3x^3 + ca_2x^2 + ca_1x + ca_0) = ca_3x^2 - ca_0 = c(a_3x^2 - a_0) = cT(a_3x^3 + a_2x^2 + a_1x + a_0)$,
so T is linear.

$T(a_3x^3 + a_2x^2 + a_1x + a_0) = 0$ if $a_3 = a_0 = 0$, thus ker(T) = $\{a_2x^2 + a_1x\}$ (basis = $\{x, x^2\}$) and range(T) = $\{a_3x^2 - a_0\}$ (basis = $\{1, x^2\}$).

24. $g(a_2x^2 + a_1x + a_0) + g(b_2x^2 + b_1x + b_0) = 2a_2x^3 + a_1x + 3a_0 + 2b_2x^3 + b_1x + 3b_0$
$= 2(a_2 + b_2)x^3 + (a_1 + b_1)x + 3(a_0 + b_0) = g((a_2 + b_2)x^2 + (a_1 + b_1)x + a_0 + b_0)$
$= g(a_2x^2 + a_1x + a_0 + b_2x^2 + b_1x + b_0)$ and $g(c(a_2x^2 + a_1x + a_0))$
$= g(ca_2x^2 + ca_1x + ca_0) = 2ca_2x^3 + ca_1x + 3ca_0 = c(2a_2x^3 + a_1x + 3a_0)$
$= cg(a_2x^2 + a_1x + a_0)$, so g is linear.

$g(a_2x^2 + a_1x + a_0) = 0$ if $a_2 = a_1 = a_0 = 0$, so ker(g) is the zero polynomial and range(g) = $\{a_3x^3 + a_1x + a_0\}$. A basis for range(g) is the set $\{1, x, x^3\}$.

25. $D(x^3 - 3x^2 + 2x + 1) = 3x^2 - 6x + 2$.

$D(a_nx^n + \ldots + a_1x + a_0) = 0$ if $a_n = \ldots = a_1 = 0$, i.e. ker(D) = the set of constant polynomials. Range(D) = P_{n-1}, because every polynomial of degree less than or equal

Section 7.1

to n−1 is the derivative of a polynomial of degree one larger than its own degree, and no polynomial of degree n is the derivative of a polynomial of degree n or less.

26. (a) $D^2(2x^3 + 3x^2 - 5x + 4) = D(6x^2 + 6x - 5) = 12x + 6$.

$D^2(a_n x^n + \ldots + a_1 x + a_0) = D(na_n x^{n-1} + \ldots + 2a_2 x + a_1)$
$= n(n-1)a_n x^{n-2} + \ldots + 6a_3 x + 2a_2$.

$D^2(a_n x^n + \ldots + a_1 x + a_0) + D^2(b_n x^n + \ldots + b_1 x + b_0)$
$= n(n-1)a_n x^{n-2} + \ldots + 6a_3 x + 2a_2 + n(n-1)b_n x^{n-2} + \ldots + 6b_3 x + 2b_2$
$= n(n-1)(a_n + b_n) x^{n-2} + \ldots + 6(a_3 + b_3)x + 2(a_2 + b_2)$
$= D^2((a_n + b_n) x^n + \ldots + (a_1 + b_1)x + a_0 + b_0)$ and
$D^2(c(a_n x^n + \ldots + a_1 x + a_0)) = D^2(ca_n x^n + \ldots + ca_1 x + ca_0)$
$= n(n-1)ca_n x^{n-2} + \ldots + 6ca_3 x + 2ca_2 = c(n(n-1)a_n x^{n-2} + \ldots + 6a_3 x + 2a_2)$
$= cD^2(a_n x^n + \ldots + a_1 x + a_0)$, so D^2 is linear.

$D^2(a_n x^n + \ldots + a_1 x + a_0) = n(n-1)a_n x^{n-2} + \ldots + 6a_3 x + 2a_2 = 0$ if
$a_n = \ldots = a_3 = a_2 = 0$. Thus $\ker(D^2) = \{a_1 x + a_0\} = P_1$. $\text{Range}(D^2) = P_{n-2}$ because every polynomial of degree less than or equal to n−2 is the second derivative of a polynomial of degree two larger than its own degree, and no polynomial of degree n−1 or larger is the second derivative of a polynomial of degree n or less.

(b) $(D^2 + D + 3)(x^3 - 2x^2 + 6x + 1) = 6x - 4 + 3x^2 - 4x + 6 + 3(x^3 - 2x^2 + 6x + 1)$
$= 3x^3 - 3x^2 + 20x + 5$.

$(D^2 + D + 3)(a_n x^n + \ldots + a_1 x + a_0) = n(n-1)a_n x^{n-2} + \ldots + 6a_3 x + 2a_2$
$+ na_n x^{n-1} + \ldots + 2a_2 x + a_1 + 3(a_n x^n + \ldots + a_1 x + a_0)$.

$(D^2 + D + 3)(a_n x^n + \ldots + a_1 x + a_0) + (D^2 + D + 3)(b_n x^n + \ldots + b_1 x + b_0)$
$= n(n-1)a_n x^{n-2} + \ldots + 6a_3 x + 2a_2 + na_n x^{n-1} + \ldots + 2a_2 x + a_1$
$+ 3(a_n x^n + \ldots + a_1 x + a_0) + n(n-1)b_n x^{n-2} + \ldots + 6b_3 x + 2b_2$
$+ nb_n x^{n-1} + \ldots + 2b_2 x + b_1 + 3(b_n x^n + \ldots + b_1 x + b_0)$
$= n(n-1)(a_n + b_n) x^{n-2} + \ldots + 6(a_3 + b_3)x + 2(a_2 + b_2)$
$+ n(a_n + b_n) x^{n-1} + \ldots + 2(a_2 + b_2)x + a_1 + b_1$
$+ 3((a_n + b_n) x^n + \ldots + (a_1 + b_1)x + a_0 + b_0)$

$= (D^2 + D + 3)((a_n + b_n)x^n + \ldots + (a_1 + b_1)x + a_0 + b_0)$ and
$(D^2 + D + 3)(c(a_n x^n + \ldots + a_1 x + a_0)) = (D^2 + D + 3)(ca_n x^n + \ldots + ca_1 x + ca_0)$
$= n(n-1)ca_n x^{n-2} + \ldots + 6ca_3 x + 2ca_2 + nca_n x^{n-1} + \ldots + 2ca_2 x + ca_1$
$+ 3(ca_n x^n + \ldots + ca_1 x + ca_0) = c(n(n-1)a_n x^{n-2} + \ldots + 6a_3 x + 2a_2$
$+ na_n x^{n-1} + \ldots + 2a_2 x + a_1 + 3(a_n x^n + \ldots + a_1 x + a_0))$.
$= c(D^2 + D + 3)(a_n x^n + \ldots + a_1 x + a_0)$, so $(D^2 + D + 3)$ is linear.

$(D^2 + D + 3)(a_n x^n + \ldots + a_1 x + a_0) = 3a_n x^n + (3a_{n-1} + na_n)x^{n-1}$
$+ (3a_{n-2} + (n-1)a_{n-1} + n(n-1)a_n)x^{n-2} + \ldots + (3a_1 + 2a_2 + 6a_3)x + (3a_0 + a_1 + 2a_2)$.
Thus if $(D^2 + D + 3)(a_n x^n + \ldots + a_1 x + a_0) = 0$, then $a_n = a_{n-1} = \ldots = a_2 = a_1 = a_0 = 0$,
so that $\ker(D^2 + D + 3)$ is the zero polynomial and range $(D^2 + D + 3) = P_n$.

27. $D(a_n x^n + \ldots + a_1 x + a_0) = na_n x^{n-1} + \ldots + 3a_3 x^2 + 2a_2 x + a_1 = 3x^2 - 4x + 7$
if $a_n = a_{n-1} = \ldots = a_4 = 0$, $a_3 = 1$, $a_2 = -2$, $a_1 = 7$, and $a_0 = r$, where r is any real number. Thus the set of polynomials mapped into $3x^2 - 4x + 7$ is the set $\{x^3 - 2x^2 + 7x + r\}$.

28. $D^2(a_n x^n + \ldots + a_1 x + a_0) = n(n-1)a_n x^{n-2} + \ldots + 20a_5 x^3 + 12a_4 x^2 + 6a_3 x + 2a_2$
$= 4x^3 + 6x^2 - 2x + 3$ if $a_n = a_{n-1} = \ldots = a_6 = 0$, $a_5 = 1/5$, $a_4 = 1/2$, $a_3 = -1/3$, $a_2 = 3/2$, $a_1 = r$, and $a_0 = s$, where r and s are any real numbers. Thus the set of polynomials mapped into $4x^3 + 6x^2 - 2x + 3$ is the set $\{x^5/5 + x^4/2 - x^3/3 + 3x^2/2 + rx + s\}$.

29. $J(f) = \int_0^1 f(x)\,dx$. $J(f+g) = \int_0^1 (f(x)+g(x))\,dx = \int_0^1 f(x)\,dx + \int_0^1 g(x)\,dx = J(f) + J(g)$ and
$J(cf) = \int_0^1 cf(x)\,dx = c\int_0^1 f(x)\,dx = cJ(f)$, so J is a linear mapping of P_n to $\mathbf{R}$.

30. $\int_0^1 (8x^3 + 6x^2 + 4x + 1)\,dx = [2x^4 + 2x^3 + 2x^2 + x]_0^1 = 7$.

$\int_0^1 (a_n x^n + a_{n-1} x^{n-1} + \ldots + a_1 x + a_0)\,dx = \frac{a_n}{n+1} + \frac{a_{n-1}}{n} + \ldots + \frac{a_1}{2} + a_0$,

Section 7.1

so ker(T) is all polynomials $a_n x^n + a_{n-1} x^{n-1} + \ldots + a_1 x + a_0$ for which

$\frac{a_n}{n+1} + \frac{a_{n-1}}{n} + \ldots + \frac{a_1}{2} = -a_0$. For any real number c, let p(x) = c. Then

$\int_0^1 p(x)\, dx = \int_0^1 c\, dx = [cx]_0^1 = c$, so range(T) = **R**. The set of elements mapped into

2 is the set of all polynomials $a_n x^n + a_{n-1} x^{n-1} + \ldots + a_1 x + a_0$ for which

$\frac{a_n}{n+1} + \frac{a_{n-1}}{n} + \ldots + \frac{a_1}{2} + a_0 = 2$.

31. (a) $T(A+C) = (A+C)^t = A^t + C^t = T(A) + T(C)$ and $T(cA) = (cA)^t = cA^t = cT(A)$, so T is linear. $A^t = O$ if and only if $A = O$, so ker(T) = O and range(T) = U.

(b) $T(A+C) = |A+C|$ and $T(A) + T(C) = |A| + |C| \neq |A+C|$, so T is not linear.

(c) $T(A+C) = tr(A+C)$ and $T(A) + T(C) = tr(A) + tr(C) \neq tr(A+C)$, so T is not linear.

For both (b) and (c), the matrices $A = \begin{bmatrix} 2 & 0 \\ 0 & 1 \end{bmatrix}$ and $C = \begin{bmatrix} -2 & 0 \\ 0 & -1 \end{bmatrix}$ demonstrate the inequalities. $tr(A) + tr(C) = |A| + |C| = 2 + 2 = 4$ and $tr(A+C) = |A+C| = 0$.

(d) $T(A+C) = (A+C)^2 \neq A^2 + C^2 = T(A) + T(C)$, so T is not linear.

(e) $T(A+C) = A+C+B \neq (A+B) + (C+B) = T(A) + T(C)$, so T is not linear.

(f) $T(A+C) = a_{11} + c_{11} = T(A) + T(C)$ and $T(cA) = ca_{11} = cT(A)$, so T is linear.

$\text{Ker}(T) = \left\{ \begin{bmatrix} 0 & a \\ b & c \end{bmatrix} \right\}$ and range(T) = **R**.

(g) $T(A+C) = 0 = 0 + 0 = T(A) + T(C)$ and $T(cA) = 0 = c \times 0 = cT(A)$, so T is linear.
Ker(T) = U and range(T) = 0.

253

(h) $T(A+C) = I_2 \neq I_2 + I_2 = T(A) + T(C)$, so T is not linear.

(i) $T(A+C) = A+C + (A+C)^t = A+C + A^t + C^t = A + A^t + C + C^t = T(A) + T(C)$ and $T(cA) = cA + (cA)^t = cA + cA^t = c(A + A^t) = cT(A)$, so T is linear.

$$\begin{bmatrix} a & b \\ c & d \end{bmatrix} + \begin{bmatrix} a & c \\ b & d \end{bmatrix} = \begin{bmatrix} 2a & b+c \\ b+c & 2d \end{bmatrix}, \text{ so } \ker(T) = \left\{ \begin{bmatrix} 0 & r \\ -r & 0 \end{bmatrix} \right\}$$ and range(T) is the

set of all symmetric matrices.

(j) If $A = \begin{bmatrix} 1 & 0 \\ 0 & 1 \end{bmatrix}$ and $C = \begin{bmatrix} 1 & 0 \\ 0 & 1 \end{bmatrix}$ then $A+C = \begin{bmatrix} 2 & 0 \\ 0 & 2 \end{bmatrix}$ and

$T(A+C) = I_2 \neq I_2 + I_2 = T(A) + T(C)$, so T is not linear.

32. T is one-to-one if and only if ker(T) is the zero vector if and only if dim ker(T) = 0 if and only if dim range(T) = dim domain(T).

33. Let T be a linear transformation from $\mathbf{R}^3$ to $\mathbf{R}^2$. Dim domain(T) = 3 and dim range(T) ≤ 2, so dim ker(T) = dim domain(T) – dim range(T) ≥ 1. Thus ker(T) is not the zero vector and T is not one-to-one.

In general, if dim(U) > dim(V) and T: U → V is a linear transformation, then dim ker(T) = dim domain(T) – dim range(T) ≥ dim(U) – dim(V) ≥ 1. Thus ker(T) is not the zero vector and T is not one-to-one.

34. Suppose T preserves linear independence, and suppose T(**u**) = T(**w**). Then $\mathbf{0} = T(\mathbf{u}) - T(\mathbf{w}) = T(\mathbf{u} - \mathbf{w})$. $\mathbf{u} - \mathbf{w} = a_1 \mathbf{u}_1 + a_2 \mathbf{u}_2 + \ldots + a_n \mathbf{u}_n$, where $\{\mathbf{u}_1, \mathbf{u}_2, \ldots, \mathbf{u}_n\}$ is a basis for U, so $\mathbf{0} = T(\mathbf{u} - \mathbf{w}) = T(a_1 \mathbf{u}_1 + a_2 \mathbf{u}_2 + \ldots + a_n \mathbf{u}_n)$ $= a_1 T(\mathbf{u}_1) + a_2 T(\mathbf{u}_2) + \ldots + a_n T(\mathbf{u}_n)$. But $\{T(\mathbf{u}_1), T(\mathbf{u}_2), \ldots, T(\mathbf{u}_n)\}$ is a linearly independent set in V so $a_1 = a_2 = \ldots = a_n = 0$. Therefore $\mathbf{u} - \mathbf{w} = \mathbf{0}$, and T is one-to-one.

Now suppose T is one-to-one, and **u** and **w** are linearly independent vectors in U. If $\mathbf{0} = aT(\mathbf{u}) + bT(\mathbf{w}) = T(a\mathbf{u} + b\mathbf{w})$, then $a\mathbf{u} + b\mathbf{w} = \mathbf{0}$, but since **u** and **w** are linearly independent this means that a = b = 0, so that T(**u**) and T(**w**) are linearly independent, so T preserves linear independence.

Section 7.1

35. $T(a_1v_1) = a_1T(v_1)$ and $T(a_1v_1+a_2v_2) = T(a_1v_1)+T(a_2v_2) = a_1T(v_1)+a_2T(v_2)$. Thus $T(a_1v_1+a_2v_2+a_3v_3) = T(a_1v_1+a_2v_2)+T(a_3v_3) = a_1T(v_1)+a_2T(v_2)+a_3T(v_3)$. Continuing in this way, $T(a_1v_1+\ldots+a_{m-1}v_{m-1}+a_mv_m) = T(a_1v_1+\ldots+a_{m-1}v_{m-1})+T(a_mv_m)$ $= a_1T(v_1) +\ldots+a_{m-1}T(v_{m-1})+a_mT(v_m)$.

Proof by induction: $T(a_1v_1) = a_1T(v_1)$ from the definition of linear transformation.

If $T(a_1v_1+\ldots+a_{m-1}v_{m-1}) = a_1T(v_1)+\ldots+a_{m-1}T(v_{m-1})$, then $T(a_1v_1+\ldots+a_{m-1}v_{m-1}+a_mv_m)$ $= T(a_1v_1+\ldots+a_{m-1}v_{m-1})+T(a_mv_m) = a_1T(v_1)+\ldots+a_{m-1}T(v_{m-1})+a_mT(v_m)$.

36. $T(x_1+x_2) = (x_1+x_2)\cdot y = x_1\cdot y+x_2\cdot y = T(x_1)+T(x_2)$ and $T(cx) = cx\cdot y = c(x\cdot y) = cT(x)$, so T is a linear transformation from $\mathbf{R}^n$ to $\mathbf{R}$.

37. (a) $(T_1+T_2)(u+v) = T_1(u+v) +T_2(u+v) = T_1(u) +T_1(v) +T_2(u)+T_2(v)$
$= T_1(u) +T_2(u) +T_1(v)+T_2(v) = (T_1+T_2)(u) +(T_1+T_2)(v)$ and $(T_1+T_2)(cu)$
$T_1(cu) + T_2(cu) = cT_1(u) + cT_2(u) = c(T_1(u) + T_2(u)) = c(T_1+T_2)(u)$, so T_1+T_2 is linear.

$(cT_1)(u+v) = cT_1(u+v) = c(T_1(u)+T_1(v)) = cT_1(u)+cT_1(v) = (cT_1)(u)+(cT_1)(v)$ and
$(cT_1)(ku) = cT_1(ku) = c(kT_1)(u) = ckT_1(u) = kcT_1(u) = k(cT_1)(u)$, so cT_1 is linear.

(b) If $T_1(x) = A_1x$ and $T_2(x) = A_2x$, then $(T_1+T_2)(x) = T_1(x) + T_2(x) = A_1x + A_2x$
$= (A_1 + A_2)x$ and $(cT_1)(x) = cT_1(x) = c(A_1x) = cA_1x$.

(c) T_1+T_2 and cT_1 are both linear transformations, so the closure conditions (axioms 1 and 2) for a vector space are satisfied.

Axiom 3 $(T_1+T_2)(u) = T_1(u) + T_2(u) = T_2(u) + T_1(u) = (T_2+T_1)(u)$, so

255

$$T_1+T_2 = T_2+T_1.$$

Axiom 4 $((T_1+T_2)+T_3)(\mathbf{u}) = (T_1+T_2)(\mathbf{u})+T_3(\mathbf{u}) = (T_1(\mathbf{u})+T_2(\mathbf{u}))+T_3(\mathbf{u})$

$= T_1(\mathbf{u}) + (T_2(\mathbf{u})+T_3(\mathbf{u})) = T_1(\mathbf{u}) + (T_2+T_3)(\mathbf{u}) = (T_1 + (T_2+T_3))(\mathbf{u}),$

so $(T_1+T_2)+T_3 = T_1 + (T_2+T_3)$.

Axiom 5 Let O be the linear transformation defined by $O(\mathbf{u}) = \mathbf{0}$. Then

$(T_1+O)(\mathbf{u}) = T_1(\mathbf{u}) + O(\mathbf{u}) = T_1(\mathbf{u}) + \mathbf{0} = T_1(\mathbf{u})$, so $T_1+O = T_1$.

Axiom 6 $-T_1$ is the linear transformation defined by $(-T_1)(\mathbf{u}) = -T_1(\mathbf{u})$.

$(T_1+(-T_1))(\mathbf{u}) = T_1(\mathbf{u}) + (-T_1)(\mathbf{u}) = \mathbf{0}$, so $T_1+(-T_1) = O$.

Axiom 7 $(c(T_1+T_2))(\mathbf{u}) = c(T_1+T_2)(\mathbf{u}) = c(T_1(\mathbf{u}) + T_2(\mathbf{u})) = cT_1(\mathbf{u}) + cT_2(\mathbf{u})$

$= (cT_1)(\mathbf{u}) + (cT_2)(\mathbf{u})$, so $c(T_1+T_2) = cT_1 + cT_2$.

Axiom 8 $((c+d)T_1)(\mathbf{u}) = (c+d)T_1(\mathbf{u}) = cT_1(\mathbf{u}) + dT_1(\mathbf{u}) = (cT_1)(\mathbf{u}) + (dT_1)(\mathbf{u})$

$= (cT_1 + dT_1)(\mathbf{u})$, so $(c+d)T_1 = cT_1 + dT_1$.

Axiom 9 $((cd)T_1)(\mathbf{u}) = (cd)T_1(\mathbf{u}) = c(dT_1)(\mathbf{u})$, so $(cd)T_1 = c(dT_1)$.

Axiom 10 $(1T_1)(\mathbf{u}) = 1T_1(\mathbf{u}) = T_1(\mathbf{u})$, so $1T_1 = T_1$.

Exercise Set 7.2

1. $(r, r, 1) = r(1,1,0) + (0,0,1)$.

2. $(0,1,2) = r(0,0,0) + (0,1,2)$.

3. $(r+1, 2r, r) = r(1,2,1) + (1,0,0)$.

4. $(r-s+3, r, s) = r(1,1,0) + s(-1,0,1) + (3,0,0)$.

Section 7.2

5. $(-4r+3s-2, -5r+5s-6, r, s) = r(-4,-5,1,0) + s(3,5,0,1) + (-2,-6,0,0)$.

6. $(-32r-23s-19, 13r+9s+8, -5r-3s-3, r, s)$
   $= r(-32,13, -5,1,0) + s(-23,9,-3, 0,1) + (-19,8,-3,0,0)$.

7. The system $A\mathbf{x} = \mathbf{y}$ has solutions if the vector $\mathbf{y}$ is in range(T), i.e., if the coordinates of $\mathbf{y} = (y_1, y_2, y_3)$ satisfy the condition $y_3 = y_1 + y_2$.

   (a) $2 = 1 + 1$, so solutions exist.
   (b) $3 \neq -1 + 2$, so there are no solutions.
   (c) $5 = 3 + 2$, so solutions exist.
   (d) $5 \neq 2 + 4$, so there are no solutions.

8. Any particular solution of the system will do in place of $\mathbf{x}_1$. $(2,3,1)$ is another particular solution (obtained from the general solution by letting $r = 1$).

9. $(0,-2,-1,0)$ is another particular solution (obtained from the general solution by letting $r = -1$ and $s = 0$).

10. The coefficient matrix is A. Since $\mathbf{x}_1 = (1,-1,4)$ is a solution, substitute $x_1 = 1$, $x_2 = -1$, $x_3 = 4$ in the left side of each equation to find the number on the right.

    $$x_1 + 2x_2 + x_3 = 3$$
    $$x_2 + 2x_3 = 7$$
    $$x_1 + x_2 - x_3 = -4$$

11. The coefficient matrix is A. Since $\mathbf{x}_1 = (2,0,3)$ is a solution, substitute $x_1 = 2$, $x_2 = 0$, $x_3 = 3$ in the left side of each equation to find the number on the right.

    $$2x_1 + x_2 = 4$$
    $$3x_1 + 3x_2 + x_3 = 9$$
    $$x_2 + x_3 = 3$$

Section 7.2

12. Let $T(\mathbf{x}) = A\mathbf{x}$. First suppose $\ker(T) = \mathbf{0}$. If $A\mathbf{x}_1 = \mathbf{y}$ and $A\mathbf{x}_2 = \mathbf{y}$ then $A(\mathbf{x}_1 - \mathbf{x}_2) = \mathbf{0}$. Thus $\mathbf{x}_1 - \mathbf{x}_2$ is in $\ker(T)$, so $\mathbf{x}_1 - \mathbf{x}_2 = \mathbf{0}$; that is, $\mathbf{x}_1 = \mathbf{x}_2$, so there is exactly one solution to $A\mathbf{x} = \mathbf{y}$.

Now suppose $A\mathbf{x} = \mathbf{y}$ has $\mathbf{x}_1$ as its unique solution. If $\mathbf{z}$ is in $\ker(T)$, then $A\mathbf{z} = \mathbf{0}$, so $A(\mathbf{x}_1 + \mathbf{z}) = \mathbf{y} + \mathbf{0} = \mathbf{y}$. Thus $\mathbf{x}_1 + \mathbf{z}$ is a solution to $A\mathbf{x} = \mathbf{y}$. That means $\mathbf{x}_1 + \mathbf{z} = \mathbf{x}_1$, so $\mathbf{z} = \mathbf{0}$, and $\ker(T)$ is the zero vector.

13. (a) $(D^2+D-2)e^{mx} = 0$, so $(m^2+m-2)e^{mx} = 0$, which gives $(m+2)(m-1) = 0$, so that $m = -2$ or $m = 1$. Thus a basis for the kernel is the set $\{e^{-2x}, e^x\}$.
$\ker(D^2+D-2) = \{ae^{-2x} + be^x\}$.

(b) $(D^2+4D+3)e^{mx} = 0$, so $(m^2+4m+3)e^{mx} = 0$, which gives $(m+3)(m+1) = 0$, so that $m = -3$ or $m = -1$. Thus a basis for the kernel is the set $\{e^{-3x}, e^{-x}\}$.
$\ker(D^2+4D+3) = \{ae^{-3x} + be^{-x}\}$.

(c) $(D^2+2D-8)e^{mx} = 0$, so $(m^2+2m-8)e^{mx} = 0$, which gives $(m+4)(m-2) = 0$, so that $m = -4$ or $m = 2$. Thus a basis for the kernel is the set $\{e^{-4x}, e^{2x}\}$.
$\ker(D^2+2D-8) = \{ae^{-4x} + be^{2x}\}$.

14. (a) $(D^2+5D+6)e^{mx} = 0$ gives $(m^2+5m+6)e^{mx} = 0$, which gives $(m+3)(m+2) = 0$, so that $m = -3$ or $m = -2$. Thus a basis for the kernel is the set $\{e^{-3x}, e^{-2x}\}$.

A particular solution is given by $6y = 8$ or $y = 4/3$, so the general solution is $y = re^{-3x} + se^{-2x} + 4/3$.

(b) $(D^2-3D)e^{mx} = 0$ gives $(m^2-3m)e^{mx} = 0$, which gives $(m-3)m = 0$, so that $m = 3$ or $m = 0$. Thus a basis for the kernel is the set $\{e^{3x}, 1\}$.

A particular solution is given by $-3dy/dx = 8$ or $y = -8x/3$, so the general solution is $y = re^{3x} + s - 8x/3$.

(c) $(D^2-7D+12)e^{mx} = 0$ gives $(m^2-7m+12)e^{mx} = 0$, which gives $(m-3)(m-4) = 0$, so that $m = 3$ or $m = 4$. Thus a basis for the kernel is the set $\{e^{3x}, e^{4x}\}$.

Section 7.3

A particular solution is given by $12y = 24$ or $y = 2$, so the general solution is $y = re^{3x} + se^{4x} + 2$.

(d) $(D^2+8D)e^{mx} = 0$ gives $(m^2+8m)e^{mx} = 0$, which gives $(m+8)m = 0$, so that $m = -8$ or $m = 0$. Thus a basis for the kernel is the set $\{e^{-8x}, 1\}$.

A particular solution is of the form $y = ke^{2x}$. We determine k. $dy/dx = 2ke^{2x}$ and $d^2y/dx^2 = 4ke^{2x}$, so that $d^2y/dx^2 + 8dy/dx = 20ke^{2x} = 3e^{2x}$. Thus $k = 3/20$, and the general solution is $y = re^{-8x} + s + 3e^{2x}/20$.

Exercise Set 7.3

1. $(2,-3) = 2(1,0) + -3(0,1)$, so $u_B = \begin{bmatrix} 2 \\ -3 \end{bmatrix}$.

2. $(8,-1) = a(3,-1) + b(2,1) = (3a+2b,-a+b)$, so $8 = 3a+2b$ and $-1 = -a+b$. This system of equations has the unique solution $a = 2$, $b = 1$, so $u_B = \begin{bmatrix} 2 \\ 1 \end{bmatrix}$.

3. $(5,1) = 5(1,0) + 1(0,1)$, so $u_B = \begin{bmatrix} 5 \\ 1 \end{bmatrix}$.

4. $(-7,-5) = a(1,-1) + b(3,1) = (a+3b,-a+b)$, so $-7 = a+3b$ and $-5 = -a+b$. This system of equations has the unique solution $a = 2$, $b = -3$, so $u_B = \begin{bmatrix} 2 \\ -3 \end{bmatrix}$.

5. $(4,0,-2) = 4(1,0,0) + 0(0,1,0) + -2(0,0,1)$, so $u_B = \begin{bmatrix} 4 \\ 0 \\ -2 \end{bmatrix}$.

6. $(6,-3,1) = a(1,-1,0) + b(2,1,-1) + c(2,0,0)$, so $6 = a+2b+2c$, $-3 = -a+b$, and $1 = -b$. This

system of equations has the unique solution a = 2, b = -1, c = 3, so $u_B = \begin{bmatrix} 2 \\ -1 \\ 3 \end{bmatrix}$.

7. $(3,-1,7) = a(2,0,-1) + b(0,1,3) + c(1,1,1)$, so $3 = 2a+c$, $-1 = b+c$, and $7 = -a+3b+c$. This system of equations has the unique solution a = 16/3, b = 20/3, and c = -23/3, so

$u_B = \begin{bmatrix} 16/3 \\ 20/3 \\ -23/3 \end{bmatrix}$.

8. $(1,-6,-8) = a(1,2,3) + b(1,-1,0) + c(0,1,-2)$, so $1 = a+b$, $-6 = 2a-b+c$, and $-8 = 3a-2c$. This system of equations has the unique solution a = -2, b = 3, c = 1, so $u_B = \begin{bmatrix} -2 \\ 3 \\ 1 \end{bmatrix}$.

9. $7x - 3 = 7(x) + -3(1)$, so $u_B = \begin{bmatrix} 7 \\ -3 \end{bmatrix}$.

10. $2x - 6 = a(3x - 5) + b(x - 1)$, so $2 = 3a+b$ and $-6 = -5a-b$. This system of equations has the unique solution a = 2, b = -4, so $u_B = \begin{bmatrix} 2 \\ -4 \end{bmatrix}$.

11. $4x^2 - 5x + 2 = 4(x^2) + -5(x) + 2(1)$, so $u_B = \begin{bmatrix} 4 \\ -5 \\ 2 \end{bmatrix}$.

12. $3x^2 - 6x - 2 = a(x^2) + b(x - 1) + c(2x)$, so $3 = a$, $-6 = b+2c$, and $-2 = -b$. This system of equations has the unique solution a = 3, b = 2, c = -4, so $u_B = \begin{bmatrix} 3 \\ 2 \\ -4 \end{bmatrix}$.

13. $\mathbf{u}\cdot(0,-1,0) = 0$, $\mathbf{u}\cdot(3/5,0,-4/5) = 11$, $\mathbf{u}\cdot(4/5,0,3/5) = -2$, so $\mathbf{u}_B = \begin{bmatrix} 0 \\ 11 \\ -2 \end{bmatrix}$.

14. $\mathbf{u}\cdot(1/3,2/3,2/3) = 11/3$, $\mathbf{u}\cdot(2/3,-2/3,1/3) = 1/3$, $\mathbf{u}\cdot(2/3,1/3,-2/3) = -2/3$, so $\mathbf{u}_B = \begin{bmatrix} 11/3 \\ 1/3 \\ -2/3 \end{bmatrix}$.

15. $\mathbf{u}\cdot(1,0,0) = 2$, $\mathbf{u}\cdot(0,1/\sqrt{2},1/\sqrt{2}) = 5\sqrt{2}$, $\mathbf{u}\cdot(0,1/\sqrt{2},-1/\sqrt{2}) = -3\sqrt{2}$, so $\mathbf{u}_B = \begin{bmatrix} 2 \\ 5\sqrt{2} \\ -3\sqrt{2} \end{bmatrix}$.

16. $P = \begin{bmatrix} 2 & 1 \\ 3 & 2 \end{bmatrix}$, and $\mathbf{u}_{B'} = P\mathbf{u}_B = \begin{bmatrix} 4 \\ 7 \end{bmatrix}$, $\mathbf{v}_{B'} = P\mathbf{v}_B = \begin{bmatrix} 5 \\ 7 \end{bmatrix}$, and $\mathbf{w}_{B'} = P\mathbf{w}_B = \begin{bmatrix} 8 \\ 14 \end{bmatrix}$.

17. $P = \begin{bmatrix} 4 & 3 \\ 1 & -1 \end{bmatrix}$, and $\mathbf{u}_{B'} = P\mathbf{u}_B = \begin{bmatrix} 12 \\ 3 \end{bmatrix}$, $\mathbf{v}_{B'} = P\mathbf{v}_B = \begin{bmatrix} 11 \\ 1 \end{bmatrix}$, and $\mathbf{w}_{B'} = P\mathbf{w}_B = \begin{bmatrix} -5 \\ -3 \end{bmatrix}$.

18. $P = \begin{bmatrix} 1 & 2 \\ 1 & -3 \end{bmatrix}$, and $\mathbf{u}_{B'} = P\mathbf{u}_B = \begin{bmatrix} 2 \\ -3 \end{bmatrix}$, $\mathbf{v}_{B'} = P\mathbf{v}_B = \begin{bmatrix} -3 \\ 7 \end{bmatrix}$, and $\mathbf{w}_{B'} = P\mathbf{w}_B = \begin{bmatrix} 5 \\ 0 \end{bmatrix}$.

19. $P = \begin{bmatrix} 3 & 2 \\ 2 & 1 \end{bmatrix}$, and $\mathbf{u}_{B'} = P\mathbf{u}_B = \begin{bmatrix} 4 \\ 3 \end{bmatrix}$, $\mathbf{v}_{B'} = P\mathbf{v}_B = \begin{bmatrix} -2 \\ -1 \end{bmatrix}$, and $\mathbf{w}_{B'} = P\mathbf{w}_B = \begin{bmatrix} 21 \\ 13 \end{bmatrix}$.

20. $P = \begin{bmatrix} 2 & -3 \\ -3 & 4 \end{bmatrix}$, and $\mathbf{u}_{B'} = P\mathbf{u}_B = \begin{bmatrix} -1 \\ 1 \end{bmatrix}$, $\mathbf{v}_{B'} = P\mathbf{v}_B = \begin{bmatrix} 6 \\ -9 \end{bmatrix}$, and $\mathbf{w}_{B'} = P\mathbf{w}_B = \begin{bmatrix} 2 \\ -4 \end{bmatrix}$.

21. The transition matrix from B' to B is $P = \begin{bmatrix} 5 & 3 \\ 3 & 2 \end{bmatrix}$, so the transition matrix from B to B' is

$P^{-1} = \begin{bmatrix} 2 & -3 \\ -3 & 5 \end{bmatrix}$, and $\mathbf{u}_{B'} = P^{-1}\mathbf{u}_B = \begin{bmatrix} -19 \\ 31 \end{bmatrix}$.

Section 7.3

22. The transition matrix from B' to B is $P = \begin{bmatrix} 1 & -1 \\ 2 & -1 \end{bmatrix}$, so the transition matrix from B to B' is

$P^{-1} = \begin{bmatrix} -1 & 1 \\ -2 & 1 \end{bmatrix}$, and $\mathbf{u}_{B'} = P^{-1}\mathbf{u}_B = \begin{bmatrix} -5 \\ -13 \end{bmatrix}$.

23. The transition matrix from B to the standard basis is $R = \begin{bmatrix} 1 & 3 \\ 2 & 0 \end{bmatrix}$ and the transition matrix from B' to the standard basis is $Q = \begin{bmatrix} 2 & 3 \\ 1 & 2 \end{bmatrix}$. $P = Q^{-1}R = \begin{bmatrix} 2 & -3 \\ -1 & 2 \end{bmatrix}\begin{bmatrix} 1 & 3 \\ 2 & 0 \end{bmatrix}$

$= \begin{bmatrix} -4 & 6 \\ 3 & -3 \end{bmatrix}$, and $\mathbf{u}_{B'} = P\mathbf{u}_B = \begin{bmatrix} 24 \\ -15 \end{bmatrix}$.

24. The transition matrix from B to the standard basis is $R = \begin{bmatrix} 3 & 1 \\ -1 & 1 \end{bmatrix}$ and the transition matrix from B' to the standard basis is $Q = \begin{bmatrix} 2 & 1 \\ 5 & 2 \end{bmatrix}$. $P = Q^{-1}R = \begin{bmatrix} 2 & -1 \\ -5 & 2 \end{bmatrix}\begin{bmatrix} 3 & 1 \\ -1 & 1 \end{bmatrix}$

$= \begin{bmatrix} -7 & -1 \\ 17 & 3 \end{bmatrix}$, and $\mathbf{u}_{B'} = P\mathbf{u}_B = \begin{bmatrix} -47 \\ 113 \end{bmatrix}$.

25. $3x^2 = 3(x^2) + 0(x) + 0(1)$, $x - 1 = 0(x^2) + 1(x) + -1(1)$, and $4 = 0(x^2) + 0(x) + 4(1)$, so the transition matrix from B' to B is $P = \begin{bmatrix} 3 & 0 & 0 \\ 0 & 1 & 0 \\ 0 & -1 & 4 \end{bmatrix}$. The transition matrix from B to B' is

$P^{-1} = \begin{bmatrix} 1/3 & 0 & 0 \\ 0 & 1 & 0 \\ 0 & 1/4 & 1/4 \end{bmatrix}$. The coordinate vectors relative to B' of $3x^2 + 4x + 8$, $6x^2 + 4$,

$8x + 12$, and $3x^2 + 4x + 4$ are respectively $P^{-1}\begin{bmatrix} 3 \\ 4 \\ 8 \end{bmatrix} = \begin{bmatrix} 1 \\ 4 \\ 3 \end{bmatrix}$, $P^{-1}\begin{bmatrix} 6 \\ 0 \\ 4 \end{bmatrix} = \begin{bmatrix} 2 \\ 0 \\ 1 \end{bmatrix}$,

Section 7.3

$$P^{-1}\begin{bmatrix} 0 \\ 8 \\ 12 \end{bmatrix} = \begin{bmatrix} 0 \\ 8 \\ 5 \end{bmatrix}, \text{ and } P^{-1}\begin{bmatrix} 3 \\ 4 \\ 4 \end{bmatrix} = \begin{bmatrix} 1 \\ 4 \\ 2 \end{bmatrix}.$$

26. $x + 2 = 1(x) + 2(1)$ and $3 = 0(x) + 3(1)$, so the transition matrix from B' to B is

    $P = \begin{bmatrix} 1 & 0 \\ 2 & 3 \end{bmatrix}$ and the transition matrix from B to B' is $P^{-1} = \frac{1}{3}\begin{bmatrix} 3 & 0 \\ -2 & 1 \end{bmatrix}$. The coordinate

    vectors relative to B' of $3x + 3$, $6x$, $6x + 9$, and $12x - 3$ are respectively $P^{-1}\begin{bmatrix} 3 \\ 3 \end{bmatrix} = \begin{bmatrix} 3 \\ -1 \end{bmatrix}$,

    $P^{-1}\begin{bmatrix} 6 \\ 0 \end{bmatrix} = \begin{bmatrix} 6 \\ -4 \end{bmatrix}$, $P^{-1}\begin{bmatrix} 6 \\ 9 \end{bmatrix} = \begin{bmatrix} 6 \\ -1 \end{bmatrix}$, and $P^{-1}\begin{bmatrix} 12 \\ -3 \end{bmatrix} = \begin{bmatrix} 12 \\ -9 \end{bmatrix}$.

27. Let $T\left(\begin{bmatrix} a & 0 \\ 0 & b \end{bmatrix}\right) = (a,b)$. $T\left(\begin{bmatrix} a & 0 \\ 0 & b \end{bmatrix} + \begin{bmatrix} d & 0 \\ 0 & e \end{bmatrix}\right) = T\left(\begin{bmatrix} a+d & 0 \\ 0 & b+e \end{bmatrix}\right) = (a+d, b+e)$

    $= (a,b) + (d,e) = T\left(\begin{bmatrix} a & 0 \\ 0 & b \end{bmatrix}\right) + T\left(\begin{bmatrix} d & 0 \\ 0 & e \end{bmatrix}\right)$ and $T\left(c\begin{bmatrix} a & 0 \\ 0 & b \end{bmatrix}\right) = T\left(\begin{bmatrix} ca & 0 \\ 0 & cb \end{bmatrix}\right)$

    $= (ca, cb) = c(a,b) = cT\left(\begin{bmatrix} a & 0 \\ 0 & b \end{bmatrix}\right)$, so T is linear. T is one-to-one because if

    $T\left(\begin{bmatrix} x & 0 \\ 0 & y \end{bmatrix}\right) = (a,b)$, then $x = a$ and $y = b$, and T is onto because if (a,b) is an element of

    $\mathbf{R}^2$ then $T\left(\begin{bmatrix} a & 0 \\ 0 & b \end{bmatrix}\right) = (a,b)$. Thus T is an isomorphism.

28. Let $T\left(\begin{bmatrix} a & b \\ b & c \end{bmatrix}\right) = (a,b,c)$. $T\left(\begin{bmatrix} a & b \\ b & c \end{bmatrix} + \begin{bmatrix} f & g \\ g & h \end{bmatrix}\right) = T\left(\begin{bmatrix} a+f & b+g \\ b+g & c+h \end{bmatrix}\right) = (a+f, b+g, c+h)$

    $= (a,b,c) + (f,g,h) = T\left(\begin{bmatrix} a & b \\ b & c \end{bmatrix}\right) + T\left(\begin{bmatrix} f & g \\ g & h \end{bmatrix}\right)$ and $T\left(d\begin{bmatrix} a & b \\ b & c \end{bmatrix}\right) = T\left(\begin{bmatrix} da & db \\ db & dc \end{bmatrix}\right)$

    $= (da, db, dc) = d(a,b,c) = dT\left(\begin{bmatrix} a & b \\ b & c \end{bmatrix}\right)$, so T is linear. T is one-to-one because if

$T\left(\begin{bmatrix} x & y \\ y & z \end{bmatrix}\right) = (a,b,c)$, then $x = a$, $y = b$, and $z = c$, and T is onto because if (a,b,c) is an element of $\mathbf{R}^2$, then $T\left(\begin{bmatrix} a & b \\ b & c \end{bmatrix}\right) = (a,b,c)$. Thus T is an isomorphism.

29. T is a linear transformation from $\mathbf{R}^n$ to $\mathbf{R}^n$. If T is an isomorphism, then T is onto $\mathbf{R}^n$. The range of T is $\mathbf{R}^n$, and since the columns of A are a basis for the range of T, this means the rank of A is n. Thus A is nonsingular.

If A is nonsingular, the rank of A is n, so the columns of A are linearly independent and therefore are a basis for $\mathbf{R}^n$, so the range of A is $\mathbf{R}^n$; i.e., T is onto. Dim ker(T) = 0, since dim range(T) + dim ker(T) = dim domain(T) and n = dim range(T) = dim domain(T), so t is one-to-one. Thus the linear transformation T is an isomorphism.

30. A is nonsingular so A^{-1} is also nonsingular. Thus, from Exercise 29, the linear transformation of $\mathbf{R}^n$ to $\mathbf{R}^n$ defined by A^{-1} is an isomorphism.

Exercise Set 7.4

1. $T(0,1,-1) = T(0(1,0,0) + 1(0,1,0) - 1(0,0,1)) = T(0,1,0) - T(0,0,1) = (0,-2) - (-1,1)$
   $= (1,-3)$. Alternatively, the matrix of T with respect to the standard basis is

   $A = \begin{bmatrix} 2 & 0 & -1 \\ 1 & -2 & 1 \end{bmatrix}$. $A\begin{bmatrix} 0 \\ 1 \\ -1 \end{bmatrix} = \begin{bmatrix} 1 \\ -3 \end{bmatrix}$, so $T(0,1,-1) = (1,-3)$.

2. $T(3,-2) = T(3(1,0) - 2(0,1)) = 3T(1,0) - 2T(0,1) = 3(4) - 2(-3) = 18$. Alternatively, the matrix of T with respect to the standard basis is $[\,4\ -3\,]$. $[\,4\ -3\,]\begin{bmatrix} 3 \\ -2 \end{bmatrix} = 18$, so $T(3,-2) = 18$.

3. $T(2,1) = T(2(1,0) + (0,1)) = 2T(1,0) + T(0,1) = 2(2,5) + (1,-3) = (5,7)$. Alternatively, the

matrix of T with respect to the standard basis is A = $\begin{bmatrix} 2 & 1 \\ 5 & -3 \end{bmatrix}$. $A\begin{bmatrix} 2 \\ 1 \end{bmatrix} = \begin{bmatrix} 5 \\ 7 \end{bmatrix}$, so T(2,1) = (5,7).

4. $T(3x^2 - 2x + 1) = T(3(x^2) - 2(x) + 1(1)) = 3T(x^2) - 2T(x) + T(1) = 3(3x+1) - 2(2) + 2x - 5$
= 9x + 3 - 4 + 2x - 5 = 11x - 6. Alternatively, the matrix of T with respect to the standard

basis is A = $\begin{bmatrix} 3 & 0 & 2 \\ 1 & 2 & -5 \end{bmatrix}$. $A\begin{bmatrix} 3 \\ -2 \\ 1 \end{bmatrix} = \begin{bmatrix} 11 \\ -6 \end{bmatrix}$, so $T(3x^2 - 2x + 1) = 11x - 6$.

5. $T(x^2 + 3x - 2) = T(x^2 + 3(x) - 2(1)) = T(x^2) + 3T(x) - 2T(1) = x^2 + 3 + 3(2x^2 + 4x - 1) - 2(3x - 1)$
$= x^2 + 3 + 6x^2 + 12x - 3 - 6x + 2 = 7x^2 + 6x + 2$. Alternatively, the matrix of T with

respect to the standard basis is A = $\begin{bmatrix} 1 & 2 & 0 \\ 0 & 4 & 3 \\ 3 & -1 & -1 \end{bmatrix}$. $A\begin{bmatrix} 1 \\ 3 \\ -2 \end{bmatrix} = \begin{bmatrix} 7 \\ 6 \\ 2 \end{bmatrix}$, so

$T(x^2 + 3x - 2) = 7x^2 + 6x + 2$.

6. The coordinate vectors of $T(u_1)$ and $T(u_2)$ relative to the basis $\{v_1, v_2\}$ are $\begin{bmatrix} 2 \\ 3 \end{bmatrix}$ and $\begin{bmatrix} 4 \\ -1 \end{bmatrix}$, so A = $\begin{bmatrix} 2 & 4 \\ 3 & -1 \end{bmatrix}$. The coordinate vector of u relative to the basis $\{u_1, u_2\}$ is $\begin{bmatrix} 2 \\ 5 \end{bmatrix}$, so the coordinate vector of T(u) relative to $\{v_1, v_2\}$ is $A\begin{bmatrix} 2 \\ 5 \end{bmatrix} = \begin{bmatrix} 24 \\ 1 \end{bmatrix}$. Thus T(u) $= 24v_1 + v_2$.

7. The coordinate vectors of $T(u_1)$ and $T(u_2)$ relative to the basis $\{v_1, v_2, v_3\}$ are $\begin{bmatrix} 2 \\ 1 \\ -3 \end{bmatrix}$ and $\begin{bmatrix} 1 \\ -2 \\ 3 \end{bmatrix}$, so A = $\begin{bmatrix} 2 & 1 \\ 1 & -2 \\ -3 & 1 \end{bmatrix}$. The coordinate vector of u relative to the basis $\{u_1, u_2\}$ is

Section 7.4

$\begin{bmatrix} 4 \\ -7 \end{bmatrix}$, so the coordinate vector of T(u) relative to $\{v_1, v_2, v_3\}$ is $A \begin{bmatrix} 4 \\ -7 \end{bmatrix} = \begin{bmatrix} 1 \\ 18 \\ -19 \end{bmatrix}$. Thus

$T(u) = v_1 + 18v_2 - 19v_3$.

8. The coordinate vectors of $T(u_1)$, $T(u_2)$, and $T(u_3)$ relative to the basis $\{v_1, v_2, v_3\}$ are $\begin{bmatrix} 1 \\ 1 \\ 1 \end{bmatrix}$, $\begin{bmatrix} 3 \\ -2 \\ 0 \end{bmatrix}$, and $\begin{bmatrix} 1 \\ 2 \\ -1 \end{bmatrix}$, so $A = \begin{bmatrix} 1 & 3 & 1 \\ 1 & -2 & 2 \\ 1 & 0 & -1 \end{bmatrix}$. The coordinate vector of u relative to the basis $\{u_1, u_2, u_3\}$ is $\begin{bmatrix} 3 \\ 2 \\ -5 \end{bmatrix}$, so the coordinate vector of T(u) relative to $\{v_1, v_2, v_3\}$ is $A \begin{bmatrix} 3 \\ 2 \\ -5 \end{bmatrix}$

$= \begin{bmatrix} 4 \\ -11 \\ 8 \end{bmatrix}$. Thus $T(u) = 4v_1 - 11v_2 + 8v_3$.

9. (a) $T(1,0,0) = (1,0) = 1(1,0) + 0(0,1)$, $T(0,1,0) = (0,0) = 0(1,0) + 0(0,1)$, and $T(0,0,1)$

$= (0,1) = 0(1,0) + 1(0,1)$, so $A = \begin{bmatrix} 1 & 0 & 0 \\ 0 & 0 & 1 \end{bmatrix}$. $(1,2,3) = 1(1,0,0) + 2(0,1,0) + 3(0,0,1)$,

so the coordinate vector of T(1,2,3) relative to the standard basis of R^2 is $A \begin{bmatrix} 1 \\ 2 \\ 3 \end{bmatrix}$

$= \begin{bmatrix} 1 \\ 3 \end{bmatrix}$. Thus $T(1,2,3) = 1(1,0) + 3(0,1) = (1,3)$.

(b) $T(1,0,0) = (3,0) = 3(1,0) + 0(0,1)$, $T(0,1,0) = (0,1) = 0(1,0) + 1(0,1)$, and $T(0,0,1)$

$= (0,1) = 0(1,0) + 1(0,1)$, so $A = \begin{bmatrix} 3 & 0 & 0 \\ 0 & 1 & 1 \end{bmatrix}$. $(1,2,3) = 1(1,0,0) + 2(0,1,0) + 3(0,0,1)$, so the coordinate vector of T(1,2,3) relative to the standard basis of $\mathbf{R}^2$ is $A \begin{bmatrix} 1 \\ 2 \\ 3 \end{bmatrix}$

$= \begin{bmatrix} 3 \\ 5 \end{bmatrix}$. Thus T(1,2,3) = 3(1,0) + 5(0,1) = (3,5).

(c) T(1,0,0) = (1,2) = 1(1,0) + 2(0,1), T(0,1,0) = (1,−1) = 1(1,0) − 1(0,1), and T(0,0,1)

= (0,0) = 0(1,0) + 0(0,1), so $A = \begin{bmatrix} 1 & 1 & 0 \\ 2 & -1 & 0 \end{bmatrix}$. (1,2,3) = 1(1,0,0) + 2(0,1,0) + 3(0,0,1),

so the coordinate vector of T(1,2,3) relative to the standard basis of $\mathbf{R}^2$ is $A \begin{bmatrix} 1 \\ 2 \\ 3 \end{bmatrix}$

$= \begin{bmatrix} 3 \\ 0 \end{bmatrix}$. Thus T(1,2,3) = 3(1,0) + 0(0,1) = (3,0).

10. (a) T(1,0,0) = (1,0,0) = 1(1,0,0) + 0(0,1,0) + 0(0,0,1), T(0,1,0) = (0,2,0) = 0(1,0,0)

+ 2(0,1,0), and T(0,0,1) = (0,0,3) = 0(1,0,0) + 0(0,1,0) + 3(0,0,1), so $A = \begin{bmatrix} 1 & 0 & 0 \\ 0 & 2 & 0 \\ 0 & 0 & 3 \end{bmatrix}$.

(−1,5,2) = −1(1,0,0) + 5(0,1,0) + 2(0,0,1), so the coordinate vector of T(−1,5,2)

relative to the standard basis of $\mathbf{R}^3$ is $A \begin{bmatrix} -1 \\ 5 \\ 2 \end{bmatrix} = \begin{bmatrix} -1 \\ 10 \\ 6 \end{bmatrix}$. Thus T(−1,5,2) = −1(1,0,0)

+ 10(0,1,0) + 6(0,0,1) = (−1,10,6).

(b) T(1,0,0) = (1,0,0), T(0,1,0) = (0,1,0), and T(0,0,1) = (0,0,1), so $A = I_3$. The

coordinate vector of T(−1,5,2) relative to the standard basis of $\mathbf{R}^3$ is $A \begin{bmatrix} -1 \\ 5 \\ 2 \end{bmatrix}$

$= \begin{bmatrix} -1 \\ 5 \\ 2 \end{bmatrix}$, so $T(-1,5,2) = (-1,5,2)$.

(c) $T(1,0,0) = (1,0,0)$, $T(0,1,0) = (0,0,0)$, and $T(0,0,1) = (0,0,0)$, so $A = \begin{bmatrix} 1 & 0 & 0 \\ 0 & 0 & 0 \\ 0 & 0 & 0 \end{bmatrix}$.

The coordinate vector of $T(-1,5,2)$ relative to the standard basis of $\mathbf{R}^3$ is $A \begin{bmatrix} -1 \\ 5 \\ 2 \end{bmatrix}$

$= \begin{bmatrix} -1 \\ 0 \\ 0 \end{bmatrix}$, so $T(-1,5,2) = (-1,0,0)$.

(d) $T(1,0,0) = (1,0,1)$, $T(0,1,0) = (1,3,2)$, and $T(0,0,1) = (0,0,-4)$, so $A = \begin{bmatrix} 1 & 1 & 0 \\ 0 & 3 & 0 \\ 1 & 2 & -4 \end{bmatrix}$.

The coordinate vector of $T(-1,5,2)$ relative to the standard basis of $\mathbf{R}^3$ is $A \begin{bmatrix} -1 \\ 5 \\ 2 \end{bmatrix}$

$= \begin{bmatrix} 4 \\ 15 \\ 1 \end{bmatrix}$, so $T(-1,5,2) = (4,15,1)$.

11. $T(\mathbf{u}_1) = (2,1) = -2\mathbf{u}_1' + 1\mathbf{u}_2'$, $T(\mathbf{u}_2) = (2,3) = -2\mathbf{u}_1' + 3\mathbf{u}_2'$, and $T(\mathbf{u}_3) = (-1,2) = 1\mathbf{u}_1' + 2\mathbf{u}_2'$,

so $A = \begin{bmatrix} -2 & -2 & 1 \\ 1 & 3 & 2 \end{bmatrix}$. $\mathbf{u} = (3,-4,0) = 2\mathbf{u}_1 + \mathbf{u}_2 - \mathbf{u}_3$, so the coordinate vector of $\mathbf{u}$

relative to the basis $\{\mathbf{u}_1, \mathbf{u}_2, \mathbf{u}_3\}$ is $\begin{bmatrix} 2 \\ 1 \\ -1 \end{bmatrix}$ and the coordinate vector of $T(\mathbf{u})$ relative to the

basis $\{\mathbf{u}_1', \mathbf{u}_2'\}$ is $A \begin{bmatrix} 2 \\ 1 \\ -1 \end{bmatrix} = \begin{bmatrix} -7 \\ 3 \end{bmatrix}$. Thus $T(\mathbf{u}) = -7\mathbf{u}_1' + 3\mathbf{u}_2' = (7,3)$.

12. $T(\mathbf{u}_1) = (1,4,6) = 1\mathbf{u}_1' + 2\mathbf{u}_2' - 6\mathbf{u}_3'$ and $T(\mathbf{u}_2) = (4,3,-2) = 4\mathbf{u}_1' + 3/2\mathbf{u}_2' + 2\mathbf{u}_3'$, so

$A = \begin{bmatrix} 1 & 4 \\ 2 & 3/2 \\ -6 & 2 \end{bmatrix}$. $\mathbf{u} = (9,1) = 1\mathbf{u}_1 + 2\mathbf{u}_2$, so the coordinate vector of $\mathbf{u}$ relative to the

basis $\{\mathbf{u}_1, \mathbf{u}_2\}$ is $\begin{bmatrix} 1 \\ 2 \end{bmatrix}$ and the coordinate vector of $T(\mathbf{u})$ relative to the basis $\{\mathbf{u}_1', \mathbf{u}_2', \mathbf{u}_3'\}$ is

$A \begin{bmatrix} 1 \\ 2 \end{bmatrix} = \begin{bmatrix} 9 \\ 5 \\ -2 \end{bmatrix}$. Thus $T(\mathbf{u}) = 9\mathbf{u}_1' + 5\mathbf{u}_2' - 2\mathbf{u}_3' = (9,10,2)$.

13. $T(\mathbf{u}_1) = T(1,2) = (2,3) = 2\mathbf{u}_1 + 1\mathbf{u}_2$ and $T(\mathbf{u}_2) = T(0,-1) = (0,-1) = 0\mathbf{u}_1 + 1\mathbf{u}_2$, so

$A = \begin{bmatrix} 2 & 0 \\ 1 & 1 \end{bmatrix}$. $\mathbf{u} = (-1,3) = -1\mathbf{u}_1 - 5\mathbf{u}_2$, so the coordinate vector of $\mathbf{u}$ relative to the

basis $\{\mathbf{u}_1, \mathbf{u}_2\}$ is $\begin{bmatrix} -1 \\ -5 \end{bmatrix}$ and the coordinate vector of $T(\mathbf{u})$ relative to the basis $\{\mathbf{u}_1, \mathbf{u}_2\}$ is

$A \begin{bmatrix} -1 \\ -5 \end{bmatrix} = \begin{bmatrix} -2 \\ -6 \end{bmatrix}$. Thus $T(\mathbf{u}) = -2\mathbf{u}_1 - 6\mathbf{u}_2 = (-2,2)$.

14. $D(2x^2) = 4x = 0(2x^2) + 4(x) + 0(-1)$, $D(x) = 1 = 0(2x^2) + 0(x) - (-1)$, and $D(-1) = 0$

$= 0(2x^2) + 0(x) + 0(-1)$, so $A = \begin{bmatrix} 0 & 0 & 0 \\ 4 & 0 & 0 \\ 0 & -1 & 0 \end{bmatrix}$. $3x^2 - 2x + 4 = \frac{3}{2}(2x^2) - 2(x) - 4(-1)$, so

the coordinate vector of $3x^2 - 2x + 4$ relative to the given basis is $\begin{bmatrix} 3/2 \\ -2 \\ -4 \end{bmatrix}$, and the

coordinate vector of $T(3x^2 - 2x + 4)$ is $A \begin{bmatrix} 3/2 \\ -2 \\ -4 \end{bmatrix} = \begin{bmatrix} 0 \\ 6 \\ 2 \end{bmatrix}$. Thus $T(3x^2 - 2x + 4)$

$= 0(2x^2) + 6(x) + 2(-1) = 6x - 2$.

Section 7.4

15. $(D^2 + 2D + 1)(\sin x) = 2\cos x = 0(\sin x) + 2(\cos x)$ and $(D^2 + 2D + 1)(\cos x) = -2\sin x$

$= -2(\sin x) + 0(\cos x)$, so $A = \begin{bmatrix} 0 & -2 \\ 2 & 0 \end{bmatrix}$. $3\sin x + \cos x$ has coordinate vector $\begin{bmatrix} 3 \\ 1 \end{bmatrix}$, so

the coordinate vector of $(D^2 + 2D + 1)(3\sin x + \cos x)$ is $A\begin{bmatrix} 3 \\ 1 \end{bmatrix} = \begin{bmatrix} -2 \\ 6 \end{bmatrix}$. Thus

$(D^2 + 2D + 1)(3\sin x + \cos x) = -2\sin x + 6\cos x$.

16. (a) $T(x^2) = 0 = 0(x^2) + 0(x) + 0(1)$, $T(x) = x^2 + x = 1(x^2) + 1(x) + 0(1)$, and $T(1) = x^2 - x$

$= 1(x^2) - 1(x) + 0(1)$, so $A = \begin{bmatrix} 0 & 1 & 1 \\ 0 & 1 & -1 \\ 0 & 0 & 0 \end{bmatrix}$.

(b) $T(x) = x = 0(x^2) + 1(x) + 0(1)$ and $T(1) = x^2 + 1 = 1(x^2) + 0(x) + 1(1)$, so $A = \begin{bmatrix} 0 & 1 \\ 1 & 0 \\ 0 & 1 \end{bmatrix}$.

(c) $T(x^2) = -1 = 0(x) - 1(1)$, $T(x) = 1 = 0(x) + 1(1)$, and $T(1) = 2x = 2(x) + 1(1)$, so

$A = \begin{bmatrix} 0 & 0 & 2 \\ -1 & 1 & 0 \end{bmatrix}$.

17. $T(x^2 + x) = x = 1(x) + 0(1)$, $T(x) = 0 = 0(x) + 0(1)$, and $T(1) = 1 = 0(x) + 1(1)$, so

$A = \begin{bmatrix} 1 & 0 & 0 \\ 0 & 0 & 1 \end{bmatrix}$. $3x^2 + 2x - 1 = 3(x^2 + x) - 1(x) - 1(1)$, so the coordinate vector for

$3x^2 + 2x - 1$ relative to the basis $\{x^2 + x, x, 1\}$ is $\begin{bmatrix} 3 \\ -1 \\ -1 \end{bmatrix}$, and the coordinate vector for

Section 7.4

$T(3x^2 + 2x - 1)$ relative to the basis $\{x, 1\}$ is $A \begin{bmatrix} 3 \\ -1 \\ -1 \end{bmatrix} = \begin{bmatrix} 3 \\ -1 \end{bmatrix}$. Thus $T(3x^2 + 2x - 1)$

$= 3(x) - 1(1) = 3x - 1$.

18. $T(x+1) = x - 1 = 1(x+1) - 1(2)$ and $T(2) = 2x = 2(x+1) - 1(2)$, so $A = \begin{bmatrix} 1 & 2 \\ -1 & -1 \end{bmatrix}$.

$4x - 3 = 4(x+1) - 7/2(2)$, so the coordinate vector for $4x - 3$ relative to the given basis

is $\begin{bmatrix} 4 \\ -7/2 \end{bmatrix}$ and the coordinate vector for $T(4x - 3)$ is $A \begin{bmatrix} 4 \\ -7/2 \end{bmatrix} = \begin{bmatrix} -3 \\ -1/2 \end{bmatrix}$. Thus $T(4x - 3)$

$= -3(x+1) - 1/2(2) = -3x - 4$.

19. $T(x) = x = 1(x) + 0(1)$ and $T(1) = x - 1 = 1(x) - 1(1)$, so $A = \begin{bmatrix} 1 & 1 \\ 0 & -1 \end{bmatrix}$.

$x + 1 = 1(x) + 1(1)$ and $x - 1 = 1(x) - 1(1)$, so $P = \begin{bmatrix} 1 & 1 \\ 1 & -1 \end{bmatrix}$ and $P^{-1} = \frac{1}{2} \begin{bmatrix} 1 & 1 \\ 1 & -1 \end{bmatrix}$.

$P^{-1}AP = \frac{1}{2} \begin{bmatrix} 1 & 1 \\ 3 & -1 \end{bmatrix}$, which is the matrix of T with respect to the basis $\{x+1, x-1\}$.

20. $T(1,0) = (2,1) = 2(1,0) + 1(0,1)$ and $T(0,1) = (0,1) = 0(1,0) + 1(0,1)$, so $A = \begin{bmatrix} 2 & 0 \\ 1 & 1 \end{bmatrix}$.

$(1,1) = 1(1,0) + 1(0,1)$ and $(2,1) = 2(1,0) + 1(0,1)$, so $P = \begin{bmatrix} 1 & 2 \\ 1 & 1 \end{bmatrix}$ and $P^{-1} = \begin{bmatrix} -1 & 2 \\ 1 & -1 \end{bmatrix}$.

$P^{-1}AP = \begin{bmatrix} 2 & 2 \\ 0 & 1 \end{bmatrix}$, which is the matrix of T with respect to the basis $\{(1,1), (2,1)\}$.

21. $T(1,0) = (1,1) = 1(1,0) + 1(0,1)$ and $T(0,1) = (-1,1) = -1(1,0) + 1(0,1)$, so $A = \begin{bmatrix} 1 & -1 \\ 1 & 1 \end{bmatrix}$.

$(-2,1) = -2(1,0) + 1(0,1)$ and $(1,2) = 1(1,0) + 2(0,1)$, so $P = \begin{bmatrix} -2 & 1 \\ 1 & 2 \end{bmatrix}$ and P^{-1}

$= \frac{1}{5} \begin{bmatrix} -2 & 1 \\ 1 & 2 \end{bmatrix}$. $P^{-1}AP = \begin{bmatrix} 1 & 1 \\ -1 & 1 \end{bmatrix}$, which is the matrix of T with respect to the basis

$\{(-2,1), (1,2)\}$.

22. (a) Let $\{u_1, u_2, \ldots, u_m\}$ be a basis for U, and let $v_{m+1}, \ldots, v_n$ be additional linearly independent vectors in V so that $\{u_1, u_2, \ldots, u_m, v_{m+1}, \ldots, v_n\}$ is a basis for V. Let T be a linear transformation that maps each of $u_1, u_2, \ldots, u_m$ into the zero vector of W. Then each element of U will be in the kernel of T. If $\dim(W) \geq n-m$, then $v_{m+1}, \ldots, v_n$ can be mapped by T into linearly independent vectors in W, and $U = \ker(T)$. If $\dim(W) < n-m$, then $T(v_{m+1}), \ldots, T(v_n)$ will be linearly dependent and there will be constants $a_{m+1}, \ldots, a_n$, not all zero, such that $T(a_{m+1}v_{m+1} + \ldots + a_n v_n)$ $= a_{m+1}T(v_{m+1}) + \ldots + a_n T(v_n) = 0$. The vector $a_{m+1}v_{m+1} + \ldots + a_n v_n$ is not in U, for if it were, there would be constants $b_1, b_2, \ldots, b_m$ such that $b_1 u_1 + b_2 u_2 + \ldots + b_m u_m$ $= a_{m+1}v_{m+1} + \ldots + a_n v_n$; i.e., $b_1 u_1 + b_2 u_2 + \ldots + b_m u_m - a_{m+1}v_{m+1} - \ldots - a_n v_n = 0$. This cannot be because the vectors $u_1, u_2, \ldots, u_m, v_{m+1}, \ldots, v_n$ are linearly independent. Thus in this case U is contained in but not equal to $\ker(T)$.

(b) The set $\{(1,3,-1), (1,0,0), (0,1,0)\}$ is a basis for $\mathbf{R}^3$. Let T be the linear transformation given by $T(1,3,-1) = (0,0)$, $T(1,0,0) = (1,0)$ and $T(0,1,0) = (0,1)$. Let $u = a_1(1,3,-1) + a_2(1,0,0) + a_3(0,1,0)$ and suppose $T(u) = (0,0)$. Thus $a_1 T(1,3,-1) + a_2 T(1,0,0) + a_3 T(0,1,0) = a_1(0,0) + a_2(1,0) + a_3(0,1) = (0,0)$, so that $a_2(1,0) + a_3(0,1) = (0,0)$. Since $(1,0)$ and $(0,1)$ are linearly independent, a_2 and a_3 must be zero and thus u is a multiple of $(1,3,-1)$.

(c) $\{(r+s, 2r, -s)\} = \{r(1,2,0) + s(1,0,-1)\}$. The set $\{(1,2,0), (1,0,-1)\}$ is a basis for this subspace of $\mathbf{R}^3$ and $\{(1,2,0), (1,0,-1), (0,1,0)\}$ is a basis for $\mathbf{R}^3$. Let T be the linear transformation given by $T(1,2,0) = (0,0)$, $T(1,0,-1) = (0,0)$, and $T(0,1,0) = (1,0)$. Let $u = a_1(1,2,0) + a_2(1,0,-1) + a_3(0,1,0)$ and suppose $T(u) = (0,0)$. Thus $a_1 T(1,2,0) + a_2 T(1,0,-1) + a_3 T(0,1,0) = a_1(0,0) + a_2(0,0) + a_3(1,0) = (0,0)$, so that $a_3(1,0) = (0,0)$, and therefore $a_3 = 0$. Thus $u = a_1(1,2,0) + a_2(1,0,-1)$.

23. The set $\{(2,-1), (1,0)\}$ is a basis for $\mathbf{R}^2$. Let T be the linear transformation given by $T(2,-1) = (0,0)$ and $T(1,0) = (1,0)$. Let $u = a_1(2,-1) + a_2(1,0)$ and suppose $T(u) = (0,0)$. Thus $a_1 T(2,-1) + a_2 T(1,0) = a_1(0,0) + a_2(1,0) = (0,0)$, so that $a_2(1,0) = (0,0)$, and

therefore $a_2 = 0$. Thus **u** is a multiple of $(2,-1)$.

24. The set $\{(4,1), (1,0)\}$ is a basis for $\mathbf{R}^2$. Let T be the linear transformation given by $T(4,1) = (0,0,0)$ and $T(1,0) = (1,0,0)$. Let $\mathbf{u} = a_1(4,1) + a_2(1,0)$ and suppose $T(\mathbf{u}) = (0,0,0)$. Thus $a_1 T(4,1) + a_2 T(1,0) = a_1(0,0,0) + a_2(1,0,0) = (0,0,0)$, so that $a_2(1,0,0) = (0,0,0)$, and therefore $a_2 = 0$. Thus **u** is a multiple of $(4,1)$.

25. $T(\mathbf{u} + \mathbf{v}) = \mathbf{u} + \mathbf{v} = T(\mathbf{u}) + T(\mathbf{v})$ and $T(c\mathbf{u}) = c\mathbf{u} = cT(\mathbf{u})$, so T is linear. Let $B = \{\mathbf{u}_1, \mathbf{u}_2, \ldots, \mathbf{u}_n\}$ be a basis for U. $T(\mathbf{u}_i) = \mathbf{u}_i$, so if A is the matrix of T with respect to B, its ith column will have a 1 in the ith position and zeros everywhere else. Thus $A = I_n$.

26. $T(\mathbf{u} + \mathbf{v}) = \mathbf{0} = \mathbf{0} + \mathbf{0} = T(\mathbf{u}) + T(\mathbf{v})$ and $T(c\mathbf{u}) = \mathbf{0} = c\mathbf{0} = cT(\mathbf{u})$, so T is linear. Let $B = \{\mathbf{u}_1, \mathbf{u}_2, \ldots, \mathbf{u}_n\}$. $T(\mathbf{u}_i) = \mathbf{0}$, so if A is the matrix of T with respect to B, its ith column will consist of all zeros. Thus $A = O_n$, the zero matrix.

27. If **u** is an element of U and $T(\mathbf{u}) = \mathbf{v}$ and $L(T(\mathbf{u})) = \mathbf{w}$, and if $\mathbf{u}_B$, $\mathbf{v}_{B'}$, and $\mathbf{w}_{B''}$ are the coordinate vectors of **u**, **v**, and **w** relative to the bases B, B', and B'', then $P\mathbf{u}_B = \mathbf{v}_{B'}$ and $Q\mathbf{v}_{B'} = \mathbf{w}_{B''}$. Thus $QP\mathbf{u}_B = Q\mathbf{v}_{B'} = \mathbf{w}_{B''}$ and QP is the matrix of L∘T with respect to B and B''.

28. If **u** is an element of U and $\mathbf{u}_B$ is the coordinate vector of **u** relative to basis B, then the coordinate vector of $T(\mathbf{u})$ relative to the basis B' is given by $A\mathbf{u}_B$, where A is the matrix of T with respect to the bases B and B'. If A is also the matrix of L with respect to the same two bases, then the coordinate vector of $L(\mathbf{u})$ relative to the basis B' will also be given by $A\mathbf{u}_B$. But this means that the coordinate vector of $L(\mathbf{u})$ relative to the basis B' is the same as the coordinate vector of $T(\mathbf{u})$ relative to the basis B'. Thus $L(\mathbf{u}) = T(\mathbf{u})$ for every vector **u** in U and therefore $L = T$.

29. (a) $T(1,0) = (4,2) = 4(1,0) + 2(0,1)$ and $T(0,1) = (2,4) = 2(1,0) + 4(0,1)$, so the matrix representation of T relative to the standard basis is $A = \begin{bmatrix} 4 & 2 \\ 2 & 4 \end{bmatrix}$.

$\begin{vmatrix} 4-\lambda & 2 \\ 2 & 4-\lambda \end{vmatrix} = (4-\lambda)(4-\lambda) - 4 = \lambda^2 - 8\lambda + 12 = (\lambda-6)(\lambda-2)$, so the eigenvalues are

$\lambda = 6$ and $\lambda = 2$ with corresponding eigenvectors (written as row vectors) r [1 1] and s [-1 1]. Orthonormal eigenvectors are $\left[\frac{1}{\sqrt{2}} \ \frac{1}{\sqrt{2}}\right]$ and $\left[\frac{-1}{\sqrt{2}} \ \frac{1}{\sqrt{2}}\right]$. Let B' be the basis $\left\{\left(\frac{1}{\sqrt{2}}, \frac{1}{\sqrt{2}}\right), \left(\frac{-1}{\sqrt{2}}, \frac{1}{\sqrt{2}}\right)\right\}$. The matrix representation of T relative to B' is

$A' = \begin{bmatrix} 6 & 0 \\ 0 & 2 \end{bmatrix}$. The transition matrix from the standard basis to B' is $C = \begin{bmatrix} \frac{1}{\sqrt{2}} & \frac{-1}{\sqrt{2}} \\ \frac{1}{\sqrt{2}} & \frac{1}{\sqrt{2}} \end{bmatrix}$

and $C^{-1}AC = A'$.

The standard basis defines an xy coordinate system and the basis B' defines an x'y' coordinate system, rotated 45° counterclockwise from the xy system. The transformation T is a scaling in the x'y' system with factor 6 in the x' direction and factor 2 in the y' direction.

(b) $T(1,0) = (5,3) = 5(1,0) + 3(0,1)$ and $T(0,1) = (3,5) = 3(1,0) + 5(0,1)$, so the matrix representation of T relative to the standard basis is $A = \begin{bmatrix} 5 & 3 \\ 3 & 5 \end{bmatrix}$.

$\begin{vmatrix} 5-\lambda & 3 \\ 3 & 5-\lambda \end{vmatrix} = (5-\lambda)(5-\lambda) - 9 = \lambda^2 - 10\lambda + 16 = (\lambda-8)(\lambda-2)$, so the eigenvalues are

$\lambda = 8$ and $\lambda = 2$ with corresponding eigenvectors (written as row vectors) r [1 1] and s [-1 1]. Orthonormal eigenvectors are $\left[\frac{1}{\sqrt{2}} \ \frac{1}{\sqrt{2}}\right]$ and $\left[\frac{-1}{\sqrt{2}} \ \frac{1}{\sqrt{2}}\right]$. Let B' be the basis $\left\{\left(\frac{1}{\sqrt{2}}, \frac{1}{\sqrt{2}}\right), \left(\frac{-1}{\sqrt{2}}, \frac{1}{\sqrt{2}}\right)\right\}$. The matrix representation of T relative to B' is

$A' = \begin{bmatrix} 8 & 0 \\ 0 & 2 \end{bmatrix}$. The transition matrix from the standard basis to B' is $C = \begin{bmatrix} \frac{1}{\sqrt{2}} & \frac{-1}{\sqrt{2}} \\ \frac{1}{\sqrt{2}} & \frac{1}{\sqrt{2}} \end{bmatrix}$

and $C^{-1}AC = A'$.

Section 7.4

The standard basis defines an xy coordinate system and the basis B' defines an x'y' coordinate system, rotated 45° counterclockwise from the xy system. The transformation T is a scaling in the x'y' system with factor 8 in the x' direction and factor 2 in the y' direction.

(c) $T(1,0) = (9,2) = 9(1,0) + 2(0,1)$ and $T(0,1) = (2,4) = 2(1,0) + 6(0,1)$, so the matrix representation of T relative to the standard basis is $A = \begin{bmatrix} 9 & 2 \\ 2 & 6 \end{bmatrix}$.

A has eigenvalues $\lambda = 10$ and $\lambda = 5$ with corresponding eigenvectors r [2 1] and s [-1 2]. Orthonormal eigenvectors are $\left[\frac{2}{\sqrt{5}} \;\; \frac{1}{\sqrt{5}}\right]$ and $\left[\frac{-1}{\sqrt{5}} \;\; \frac{2}{\sqrt{5}}\right]$.

Let B' be the basis $\left\{ \left(\frac{2}{\sqrt{5}}, \frac{1}{\sqrt{5}}\right), \left(\frac{-1}{\sqrt{5}}, \frac{2}{\sqrt{5}}\right) \right\}$. The matrix representation of T relative to B' is $A' = \begin{bmatrix} 10 & 0 \\ 0 & 5 \end{bmatrix}$. The transition matrix from the standard basis to B' is

$$C = \begin{bmatrix} \frac{2}{\sqrt{5}} & \frac{-1}{\sqrt{5}} \\ \frac{1}{\sqrt{5}} & \frac{2}{\sqrt{5}} \end{bmatrix} \text{ and } C^{-1}AC = A'.$$

The standard basis defines an xy coordinate system and the basis B' defines an x'y' coordinate system, rotated counterclockwise through an angle θ from the xy system, where $\cos\theta = 2/\sqrt{5}$ and $\sin\theta = 1/\sqrt{5}$. The transformation T is a scaling in the x'y' system with factor 10 in the x' direction and factor 5 in the y' direction.

(d) $T(1,0) = (14,2) = 14(1,0) + 2(0,1)$ and $T(0,1) = (2,11) = 2(1,0) + 11(0,1)$, so the matrix representation of T relative to the standard basis is $A = \begin{bmatrix} 14 & 2 \\ 2 & 11 \end{bmatrix}$.

$$\begin{vmatrix} 14-\lambda & 2 \\ 2 & 11-\lambda \end{vmatrix} = (14-\lambda)(11-\lambda) - 4 = \lambda^2 - 25\lambda + 150 = (\lambda-15)(\lambda-10),$$ so the eigenvalues are $\lambda = 15$ and $\lambda = 10$ with corresponding eigenvectors r [2 1]

and s [–1 2]. Orthonormal eigenvectors are $\left[\frac{2}{\sqrt{5}} \ \frac{1}{\sqrt{5}}\right]$ and $\left[\frac{-1}{\sqrt{5}} \ \frac{2}{\sqrt{5}}\right]$.

Let B' be the basis $\left\{\left(\frac{2}{\sqrt{5}}, \frac{1}{\sqrt{5}}\right), \left(\frac{-1}{\sqrt{5}}, \frac{2}{\sqrt{5}}\right)\right\}$. The matrix representation of T relative to B' is $A' = \begin{bmatrix} 15 & 0 \\ 0 & 10 \end{bmatrix}$. The transition matrix from the standard basis to B' is

$$C = \begin{bmatrix} \frac{2}{\sqrt{5}} & \frac{-1}{\sqrt{5}} \\ \frac{1}{\sqrt{5}} & \frac{2}{\sqrt{5}} \end{bmatrix} \text{ and } C^{-1}AC = A'.$$

The standard basis defines an xy coordinate system and the basis B' defines an x'y' coordinate system, rotated counterclockwise through an angle θ from the xy system, where $\cos θ = 2/\sqrt{5}$ and $\sin θ = 1/\sqrt{5}$. The transformation T is a scaling in the x'y' system with factor 15 in the x' direction and factor 10 in the y' direction.

30. (a) T(1,0) = (8,9) = 8(1,0) + 9(0,1) and T(0,1) = (–6,–7) = –6(1,0) – 7(0,1), so the matrix representation of T relative to the standard basis is $A = \begin{bmatrix} 8 & -6 \\ 9 & -7 \end{bmatrix}$.

$$\begin{vmatrix} 8-λ & -6 \\ 9 & -7-λ \end{vmatrix} = (8-λ)(-7-λ) + 54 = λ^2 - λ - 2 = (λ-2)(λ+1), \text{ so the}$$

eigenvalues are λ = 2 and λ = –1 with corresponding eigenvectors r [1 1] and s [2 3]. Let B' be the basis {(1,1), (2,3)}. The matrix representation of T relative to B' is $A' = \begin{bmatrix} 2 & 0 \\ 0 & -1 \end{bmatrix}$. The transition matrix from the standard basis to B' is

$$C = \begin{bmatrix} 1 & 2 \\ 1 & 3 \end{bmatrix} \text{ and } C^{-1}AC = A'.$$

The standard basis defines an xy coordinate system and the basis B' defines an x'y' coordinate system, which is not rectangular. The transformation T is a scaling in the x'y' system with factor 2 in the x' direction and factor 1 in the y' direction followed by a reflection about the x' axis.

Section 7.4

(b) $T(1,0) = (-2,-10) = -2(1,0) - 10(0,1)$ and $T(0,1) = (2,7) = 2(1,0) + 7(0,1)$, so the matrix representation of T relative to the standard basis is $A = \begin{bmatrix} -2 & 2 \\ -10 & 7 \end{bmatrix}$.

$\begin{vmatrix} -2-\lambda & 2 \\ -10 & 7-\lambda \end{vmatrix} = (-2-\lambda)(7-\lambda) + 20 = \lambda^2 - 5\lambda + 6 = (\lambda-2)(\lambda-3)$, so the eigenvalues are $\lambda = 2$ and $\lambda = 3$ with corresponding eigenvectors r [1 2] and s [2 5]. Let B' be the basis $\{(1,2), (2,5)\}$. The matrix representation of T relative to B' is $A' = \begin{bmatrix} 2 & 0 \\ 0 & 3 \end{bmatrix}$. The transition matrix from the standard basis to B' is $C = \begin{bmatrix} 1 & 2 \\ 2 & 5 \end{bmatrix}$ and $C^{-1}AC = A'$.

The standard basis defines an xy coordinate system and the basis B' defines an x'y' coordinate system, which is not rectangular. The transformation T is a scaling in the x'y' system with factor 2 in the x' direction and factor 3 in the y' direction.

(c) $T(1,0) = (3,2) = 3(1,0) + 2(0,1)$ and $T(0,1) = (-4,-3) = -4(1,0) - 3(0,1)$, so the matrix representation of T relative to the standard basis is $A = \begin{bmatrix} 3 & -4 \\ 2 & -3 \end{bmatrix}$.

$\begin{vmatrix} 3-\lambda & -4 \\ 2 & -3-\lambda \end{vmatrix} = (3-\lambda)(-3-\lambda) + 8 = \lambda^2 - 1 = (\lambda-1)(\lambda+1)$, so the eigenvalues are $\lambda = 1$ and $\lambda = -1$ with corresponding eigenvectors r [2 1] and s [1 1]. Let B' be the basis $\{(2,1), (1,1)\}$. The matrix representation of T relative to B' is $A' = \begin{bmatrix} 1 & 0 \\ 0 & -1 \end{bmatrix}$. The transition matrix from the standard basis to B' is $C = \begin{bmatrix} 2 & 1 \\ 1 & 1 \end{bmatrix}$ and $C^{-1}AC = A'$.

The standard basis defines an xy coordinate system and the basis B' defines an x'y' coordinate system, which is not rectangular. The transformation T is a reflection about the x' axis.

(d) $T(1,0) = (7,-10) = 7(1,0) - 10(0,1)$ and $T(0,1) = (5,-8) = 5(1,0) - 8(0,1)$, so the matrix representation of T relative to the standard basis is $A = \begin{bmatrix} 7 & 5 \\ -10 & -8 \end{bmatrix}$.

$\begin{vmatrix} 7-\lambda & 5 \\ -10 & -8-\lambda \end{vmatrix} = (7-\lambda)(-8-\lambda) + 50 = \lambda^2 + \lambda - 6 = (\lambda-2)(\lambda+3)$, so the eigenvalues are $\lambda = 2$ and $\lambda = -3$ with corresponding eigenvectors r [-1 1] and s [1 -2]. Let B' be the basis $\{(-1,1), (1,2)\}$. The matrix representation of T relative to B' is $A' = \begin{bmatrix} 2 & 0 \\ 0 & -3 \end{bmatrix}$. The transition matrix from the standard basis to B' is

$C = \begin{bmatrix} -1 & 1 \\ 1 & -2 \end{bmatrix}$ and $C^{-1}AC = A'$.

The standard basis defines an xy coordinate system and the basis B' defines an x'y' coordinate system, which is not rectangular. The transformation T is a scaling in the x'y' system with factor 2 in the x' direction and factor 3 in the y' direction followed by a reflection about the x' axis.

Chapter 7 Review Exercises

1. $\begin{bmatrix} 6 & 4 \\ 3 & 2 \end{bmatrix} \begin{bmatrix} x \\ y \end{bmatrix} = \begin{bmatrix} 6x+4y \\ 3x+2y \end{bmatrix}$, thus (x,y) is in the kernel of T if $3x+2y = 0$, i.e. if $y = -3x/2$.

   The kernel of T is the set $\{(r,-3r/2)\}$ and the range is the set $\{(2r,r)\}$. Dim ker(T) = 1, dim range(T) = 1 and dim domain(T) = 2, so dim ker(T) + dim range(T) = dim domain(T).

2. $\begin{bmatrix} 1 & 1 & 1 \\ 0 & 1 & -1 \\ 2 & 3 & 1 \end{bmatrix} \begin{bmatrix} x \\ y \\ z \end{bmatrix} = \begin{bmatrix} x+y+z \\ y-z \\ 2x+3y+z \end{bmatrix}$; thus (x,y,z) is in the kernel of T if $x+y+z = 0$, $y-z = 0$, and $2x+3y+z = 0$, i.e., if $x = -2y$ and $z = y$. The kernel of T is the set $\{(-2r,r,r)\}$ with basis $\{(-2,1,1)\}$, and a basis for the range is the set $\{(1,0,2), (1,1,3)\}$.

3. (a) The dimension of the range of the transformation is the rank of the matrix = 2. The domain is $\mathbf{R}^4$, which has dimension 4, so the dimension of the kernel is 2. This transformation is not one-to-one.

   (b) $|B| \neq 0$, so the transformation is one-to-one.

4. The kernel is the set $\{(0,r,r)\}$ with basis $\{(0,1,1)\}$ and the range is the set $\{(a,2a,b)\}$ with basis $\{(1,2,0), (0,0,1)\}$.

Chapter 7 Review Exercises

5. $g(a_2x^2 + a_1x + a_0) + g(b_2x^2 + b_1x + b_0)$

    $= (a_2 - a_1)x^3 - a_1x + 2a_0 + (b_2 - b_1)x^3 - b_1x + 2b_0$

    $= (a_2 + b_2 - a_1 - b_1)x^3 - (a_1 + b_1)x + 2(a_0 + b_0)$

    $= g((a_2 + b_2)x^2 + (a_1 + b_1)x + a_0 + b_0)$

    $= g(a_2x^2 + a_1x + a_0 + b_2x^2 + b_1x + b_0)$.

Addition is preserved.

$g(c(a_2x^2 + a_1x + a_0))$

$= g(ca_2x^2 + ca_1x + ca_0)$

$= (ca_2 - ca_1)x^3 - ca_1x + 2ca_0$

$= c((a_2 - a_1)x^3 - a_1x + 2a_0)$

$= cg(a_2x^2 + a_1x + a_0)$.

Scalar multiplication is preserved.

Thus g is linear.

Ker(g) is the set of all polynomials $a_2x^2 + a_1x + a_0$ with $a_2 - a_1 = 0$, $a_1 = 0$, and $a_0 = 0$, i.e., $a_2 = a_1 = a_0 = 0$, so ker(g) is the zero polynomial. Range(g) is the set of all polynomials $ax^3 + bx + c$. The set $\{x^3, x, 1\}$ is a basis for range(g).

6. $(D^2 - 2D + 1)(a_nx^n + \ldots + a_1x + a_0) =$

    $n(n-1)a_nx^{n-2} + \ldots + 6a_3x + 2a_2 - 2(na_nx^{n-1} + \ldots + 2a_2x + a_1) + (a_nx^n + \ldots + a_1x + a_0)$.

Thus,

$$(D^2 - 2D + 1)(a_n x^n + \ldots + a_1 x + a_0) + (D^2 - 2D + 1)(b_n x^n + \ldots + b_1 x + b_0)$$
$$= n(n-1)a_n x^{n-2} + \ldots + 6a_3 x + 2a_2 - 2(na_n x^{n-1} + \ldots + 2a_2 x + a_1)$$
$$+ (a_n x^n + \ldots + a_1 x + a_0) + n(n-1)b_n x^{n-2} + \ldots + 6b_3 x + 2b_2$$
$$- 2(nb_n x^{n-1} + \ldots + 2b_2 x + b_1) + (b_n x^n + \ldots + b_1 x + b_0)$$
$$= n(n-1)(a_n + b_n)x^{n-2} + \ldots + 6(a_3 + b_3)x + 2(a_2 + b_2)$$
$$- 2(n(a_n + b_n)x^{n-1} + \ldots + 2(a_2 + b_2)x + a_1 + b_1) + ((a_n + b_n)x^n + \ldots$$
$$+ (a_1 + b_1)x + a_0 + b_0)$$
$$= (D^2 - 2D + 1)((a_n + b_n)x^n + \ldots + (a_1 + b_1)x + a_0 + b_0).$$

Addition is preserved.

$$(D^2 - 2D + 1)(c(a_n x^n + \ldots + a_1 x + a_0))$$
$$= (D^2 - 2D + 1)(ca_n x^n + \ldots + ca_1 x + ca_0)$$
$$= n(n-1)ca_n x^{n-2} + \ldots + 6ca_3 x + 2ca_2 - 2(nca_n x^{n-1} + \ldots + 2ca_2 x + ca_1)$$
$$+ (ca_n x^n + \ldots + ca_1 x + ca_0) = c(n(n-1)a_n x^{n-2} + \ldots + 6a_3 x + 2a_2$$
$$- 2(na_n x^{n-1} + \ldots + 2a_2 x + a_1) + (a_n x^n + \ldots + a_1 x + a_0))$$
$$= c(D^2 - 2D + 1)(a_n x^n + \ldots + a_1 x + a_0).$$

Scalar multiplication is preserved.

Thus $(D^2 - 2D + 1)$ is linear.

$(D^2 - 2D + 1)(a_n x^n + \ldots + a_1 x + a_0) = a_n x^n + (a_{n-1} - 2na_n)x^{n-1}$
$+ (a_{n-2} - 2(n-1)a_{n-1} + n(n-1)a_n)x^{n-2} + \ldots + (a_1 - 4a_2 + 6a_3)x + (a_0 - 2a_1 + 2a_2)$. Thus if $(D^2 - 2D + 1)(a_n x^n + \ldots + a_1 x + a_0) = 12x - 4$, then $a_n = a_{n-1} = \ldots = a_2 = 0$, $a_1 = 12$, and $a_0 = 20$. Thus the only polynomial mapped into $12x - 4$ is the polynomial $12x + 20$.

Chapter 7 Review Exercises

7. $(-1, 18) = a(1,3) + b(-1,4)$. Thus, $-1 = a - b$ and $18 = 3a + 4b$.

   This system of equations has the unique solution $a = 2$, $b = 3$,

   so the coordinate vector is $\begin{bmatrix} a \\ b \end{bmatrix} = \begin{bmatrix} 2 \\ 3 \end{bmatrix}$.

8. $3x^2 + 2x - 13 = a(x^2 + 1) + b(x + 2) + c(x - 3)$.

   Thus, $3 = a$, $2 = b + c$, and $-13 = a + 2b - 3c$.

   This system of equations has the unique solution $a = 3$, $b = -2$, $c = 4$,

   so the coordinate vector is $\begin{bmatrix} a \\ b \\ c \end{bmatrix} = \begin{bmatrix} 3 \\ -2 \\ 4 \end{bmatrix}$.

9. $(0,5,-15) \cdot (0,1,0) = 5$, $(0,5,-15) \cdot (-3/5, 0, 4/5) = -12$.

   $(0,5,-15) \cdot (4/5, 0, 3/5) = -9$.

   Thus the coordinate vector is $\begin{bmatrix} 5 \\ -12 \\ -9 \end{bmatrix}$.

10. $P = \begin{bmatrix} 1 & 5 \\ 3 & 2 \end{bmatrix}$, and $\mathbf{u}_{B'} = P\mathbf{u}_B = \begin{bmatrix} 8 \\ 11 \end{bmatrix}$,

    $\mathbf{v}_{B'} = P\mathbf{v}_B = \begin{bmatrix} -5 \\ 11 \end{bmatrix}$, and $\mathbf{w}_{B'} = P\mathbf{w}_B = \begin{bmatrix} 9 \\ 14 \end{bmatrix}$.

Chapter 7 Review Exercises

11. The transition matrix from B to the standard basis is $R = \begin{bmatrix} -1 & 2 \\ 2 & 1 \end{bmatrix}$ and the transition matrix from B' to the standard basis is $Q = \begin{bmatrix} 4 & -3 \\ 3 & 2 \end{bmatrix}$.

$$P = Q^{-1}R = \frac{1}{17}\begin{bmatrix} 2 & 3 \\ -3 & 4 \end{bmatrix}\begin{bmatrix} -1 & 2 \\ 2 & 1 \end{bmatrix} = \frac{1}{17}\begin{bmatrix} 4 & 7 \\ 11 & -2 \end{bmatrix}.$$

$$u_{B'} = Pu_B = \frac{1}{17}\begin{bmatrix} 4 & 7 \\ 11 & -2 \end{bmatrix}\begin{bmatrix} 4 \\ 1 \end{bmatrix} = \frac{1}{17}\begin{bmatrix} 23 \\ 42 \end{bmatrix}.$$

12. The matrix with respect to the standard basis is $A = \begin{bmatrix} 3 & -1 \\ 2 & 4 \end{bmatrix}$.

$A\begin{bmatrix} 2 \\ 7 \end{bmatrix} = \begin{bmatrix} -1 \\ 32 \end{bmatrix}$. Thus, $T(2,7) = (-1,32)$.

13. The matrix of T with respect to the given bases is $A = \begin{bmatrix} 1 & 3 \\ 5 & -1 \\ -2 & 2 \end{bmatrix}$.

$A\begin{bmatrix} 2 \\ -3 \end{bmatrix} = \begin{bmatrix} -7 \\ 13 \\ -10 \end{bmatrix}$. Thus, $T(u) = -7v_1 + 13v_2 - 10v_3$.

14. $T(1,0,0) = (2,0)$, $T(0,1,0) = (0,-3)$, and $T(0,0,1) = (0,0)$.

The matrix of T with respect to the standard bases is $A = \begin{bmatrix} 2 & 0 & 0 \\ 0 & -3 & 0 \end{bmatrix}$. $A\begin{bmatrix} 1 \\ 2 \\ 3 \end{bmatrix} = \begin{bmatrix} 2 \\ -6 \end{bmatrix}$.

Thus, $T(1,2,3) = (2,-6)$.

15. $T(x^2) = x^2$, $T(x) = -x^2$, and $T(1) = 2x$, so $A = \begin{bmatrix} 1 & -1 & 0 \\ 0 & 0 & 2 \\ 0 & 0 & 0 \end{bmatrix}$. $A \begin{bmatrix} 2 \\ -1 \\ 3 \end{bmatrix} = \begin{bmatrix} 3 \\ 6 \\ 0 \end{bmatrix}$.

Thus, $T(2x^2 - x + 3) = 3x^2 + 6x$.

16. $T(1,0) = (3,1)$ and $T(0,1) = (0,-1)$, so the matrix of T with respect to the standard basis is

$A = \begin{bmatrix} 3 & 0 \\ 1 & -1 \end{bmatrix}$.

The transition matrix from the basis $\{(1,2), (2,3)\}$ to the standard basis

is $P = \begin{bmatrix} 1 & 2 \\ 2 & 3 \end{bmatrix}$.

Thus $B = P^{-1}AP = \begin{bmatrix} -3 & 2 \\ 2 & -1 \end{bmatrix} \begin{bmatrix} 3 & 0 \\ 1 & -1 \end{bmatrix} \begin{bmatrix} 1 & 2 \\ 2 & 3 \end{bmatrix} = \begin{bmatrix} -11 & -20 \\ 7 & 13 \end{bmatrix}$ is the

matrix of T with respect to the basis $\{(1,2), (2,3)\}$.

Chapter 8

Exercise Set 8.1

1. $\mathbf{u} = (x_1, x_2)$, $\mathbf{v} = (y_1, y_2)$, $\mathbf{w} = (z_1, z_2)$, and c is a scalar.

    $\langle \mathbf{u}, \mathbf{v} \rangle = 4x_1 y_1 + 9x_2 y_2 = 4y_1 x_1 + 9y_2 x_2 = \langle \mathbf{v}, \mathbf{u} \rangle$,

    $\langle \mathbf{u}+\mathbf{v}, \mathbf{w} \rangle = 4(x_1 + y_1)z_1 + 9(x_2 + y_2)z_2 = 4x_1 z_1 + 4y_1 z_1 + 9x_2 z_2 + 9y_2 z_2$

    $= 4x_1 z_1 + 9x_2 z_2 + 4x_2 z_2 + 9y_2 z_2 = \langle \mathbf{u}, \mathbf{w} \rangle + \langle \mathbf{v}, \mathbf{w} \rangle$,

    $\langle c\mathbf{u}, \mathbf{v} \rangle = 4cx_1 y_1 + 9cx_2 y_2 = c(4x_1 y_1 + 9x_2 y_2) = c\langle \mathbf{u}, \mathbf{v} \rangle$, and

    $\langle \mathbf{u}, \mathbf{u} \rangle = 4x_1 x_1 + 9x_2 x_2 = 4x_1^2 + 9x_2^2 \geq 0$ and equality holds if and only if $4x_1^2 = 0$ and $9x_2^2 = 0$, i.e., if and only if $x_1 = 0$ and $x_2 = 0$. Thus the given function is an inner product on $\mathbf{R}^2$.

2. $\mathbf{u} = (x_1, x_2, x_3)$, $\mathbf{v} = (y_1, y_2, y_3)$ $\mathbf{w} = (z_1, z_2, z_3)$, and c is a scalar.

    $\langle \mathbf{u}, \mathbf{v} \rangle = x_1 y_1 + 2x_2 y_2 + 4x_3 y_3 = y_1 x_1 + 2y_2 x_2 + 4y_3 x_3 = \langle \mathbf{v}, \mathbf{u} \rangle$,

    $\langle \mathbf{u}+\mathbf{v}, \mathbf{w} \rangle = (x_1 + y_1)z_1 + 2(x_2 + y_2)z_2 + 4(x_3 + y_3)z_3$

    $= x_1 z_1 + y_1 z_1 + 2x_2 z_2 + 2y_2 z_2 + 4x_3 z_3 + 4y_3 z_3$

    $= x_1 z_1 + 2x_2 z_2 + 4x_3 z_3 + y_1 z_1 + 2y_2 z_2 + 4y_3 z_3 = \langle \mathbf{u}, \mathbf{w} \rangle + \langle \mathbf{v}, \mathbf{w} \rangle$,

    $\langle c\mathbf{u}, \mathbf{v} \rangle = cx_1 y_1 + 2cx_2 y_2 + 4cx_3 y_3 = c(x_1 y_1 + 2x_2 y_2 + 4x_3 y_3) = c\langle \mathbf{u}, \mathbf{v} \rangle$, and

    $\langle \mathbf{u}, \mathbf{u} \rangle = x_1 x_1 + 2x_2 x_2 + 4x_3 x_3 = x_1^2 + 2x_2^2 + 4x_3^2 \geq 0$ and equality holds if and only if $4x_1^2 = 0$, $2x_2^2 = 0$, and $4x_3^2 = 0$, i.e. if and only if $x_1 = 0$, $x_2 = 0$, and $x_3 = 0$. Thus the given function is an inner product on $\mathbf{R}^3$.

3. $\langle u,u \rangle = 2x_1 x_1 - x_2 x_2 = 2x_1^2 - x_2^2 < 0$ for the vector $u = (1,2)$, so condition 4 of the definition is not satisfied.

4. $u = \begin{bmatrix} a & b \\ c & d \end{bmatrix}$, $v = \begin{bmatrix} e & f \\ g & h \end{bmatrix}$, and $w = \begin{bmatrix} j & k \\ l & m \end{bmatrix}$.

   $\langle u+v, w \rangle = (a+e)j + (b+f)k + (c+g)l + (d+h)m = aj + ej + bk + fk + cl + gl + dm + hm$

   $= aj + bk + cl + dm + ej + fk + gl + hm = \langle u,w \rangle + \langle v,w \rangle$, so axiom 2 is satisfied.

   $\langle u,u \rangle = aa + bb + cc + dd = a^2 + b^2 + c^2 + d^2 \geq 0$ and equality holds if and only if $a = b = c = d = 0$, so axiom 4 is satisfied.

5. $u = \begin{bmatrix} a & b \\ c & d \end{bmatrix}$, $v = \begin{bmatrix} e & f \\ g & h \end{bmatrix}$, $w = \begin{bmatrix} j & k \\ l & m \end{bmatrix}$, and s is a scalar.

   $\langle u,v \rangle = ae + 2bf + 3cg + 4dh = ea + 2fb + 3gc + 4hd = \langle v,u \rangle$,

   $\langle u+v, w \rangle = (a+e)j + 2(b+f)k + 3(c+g)l + 4(d+h)m$

   $= aj + ej + 2bk + 2fk + 3cl + 3gl + 4dm + 4hm$

   $= aj + 2bk + 3cl + 4dm + ej + 2fk + 3gl + 4hm = \langle u,w \rangle + \langle v,w \rangle$,

   $\langle su,v \rangle = sae + 2sbf + 3scg + 4sdh = s(ea + 2fb + 3gc + 4hd) = s\langle u,v \rangle$, and

   $\langle u,u \rangle = aa + 2bb + 3cc + 4dd = a^2 + 2b^2 + 3c^2 + 4d^2 \geq 0$ and equality holds if and only if $a = b = c = d = 0$. Thus the given function is an inner product on M_{22}.

   (a) $\langle u,v \rangle = 1 \times 4 + 2(2 \times 1) + 3(0 \times -3) + 4(-3 \times 2) = 4 + 4 + 0 - 24 = -16$.

   (b) $\langle u,v \rangle = -2 \times 5 + 2(4 \times -2) + 3(1 \times 0) + 4(0 \times -3) = -10 - 16 + 0 + 0 = -26$.

6. $\langle cf, g \rangle = \int_0^1 cf(x)g(x)dx = c\int_0^1 f(x)g(x)dx = c\langle f,g \rangle$, so axiom 3 is satisfied.

   $\langle f,f \rangle = \int_0^1 f(x)f(x)dx = \int_0^1 f(x)^2 dx \geq 0$ and equality holds if and only if $f(x)$ is the zero function, so axiom 4 is satisfied.

Section 8.1

7. $\langle f,g\rangle = \int_a^b f(x)g(x)dx = \int_a^b g(x)f(x)dx = \langle g,f\rangle,$

    $\langle f+g,h\rangle = \int_a^b (f(x)+g(x))h(x)dx = \int_a^b f(x)h(x)dx + \int_a^b g(x)h(x)dx = \langle f,h\rangle + \langle g,h\rangle,$

    $\langle cf,g\rangle = \int_a^b cf(x)g(x)dx = c\int_a^b f(x)g(x)dx = c\langle f,g\rangle,$ and

    $\langle f,f\rangle = \int_a^b f(x)f(x)dx = \int_a^b f(x)^2 dx \geq 0$ and equality holds if and only if $f(x)$ is the zero function. Thus the given function with $a < b$ defines an inner product on P_n.

    If $a = b$, the integral is zero for all functions f and g, violating axiom 4.

8. (a) $\langle f,g\rangle = \int_0^1 (2x+1)(3x-2)\, dx = \int_0^1 (6x^2-x-2)\, dx = \left[2x^3 - \frac{1}{2}x^2 - 2x\right]_0^1 = -\frac{1}{2}.$

    (b) $\langle f,g\rangle = \int_0^1 (x^2+2)(3)\, dx = \int_0^1 (3x^2+6)\, dx = [x^3 + 6x]_0^1 = 7.$

    (c) $\langle f,g\rangle = \int_0^1 (x^2+3x-2)(x+1)\, dx = \int_0^1 (x^3+4x^2+x-2)\, dx = \left[\frac{1}{4}x^4 + \frac{4}{3}x^3 + \frac{1}{2}x^2 - 2x\right]_0^1$

    $= \frac{1}{12}.$

    (d) $\langle f,g\rangle = \int_0^1 (x^3+2x-1)(3x^2-4x+2)\, dx = \int_0^1 (3x^5-4x^4+8x^3-11x^2+8x-2)\, dx$

    $= \left[\frac{1}{2}x^6 - \frac{4}{5}x^5 + 2x^4 - \frac{11}{3}x^3 + 4x^2 - 2x\right]_0^1 = \frac{1}{30}.$

9. (a) $\|f\|^2 = \int_0^1 (4x-2)^2\, dx = \int_0^1 (16x^2-16x+4)\, dx = \left[\frac{16}{3}x^3 - 8x^2 + 4x\right]_0^1 = \frac{4}{3},$ so $\|f\| = \frac{2}{\sqrt{3}}.$

    (b) $\|f\|^2 = \int_0^1 (7x^3)^2\, dx = \int_0^1 (49x^6)\, dx = [7x^7]_0^1 = 7,$ so $\|f\| = \sqrt{7}.$

Section 8.1

(c) $\|f\|^2 = \int_0^1 (3x^2+2)^2 \, dx = \int_0^1 (9x^4+12x^2+4) \, dx = \left[\frac{9}{5}x^5+4x^3+4x\right]_0^1 = \frac{49}{5}$, so $\|f\| = \frac{7}{\sqrt{5}}$.

(d) $\|g\|^2 = \int_0^1 (x^2+x+1)^2 \, dx = \int_0^1 (x^4+2x^3+3x^2+2x+1) \, dx = \left[\frac{1}{5}x^5+\frac{1}{2}x^4+x^3+x^2+x\right]_0^1$

$= \frac{37}{10}$, so $\|g\| = \sqrt{\frac{37}{10}}$.

10. $\langle f,g \rangle = \int_0^1 (x^2)(4x-3) \, dx = \int_0^1 (4x^3-3x^2) \, dx = [x^4-x^3]_0^1 = 0$, so the functions are orthogonal.

11. $\langle f,g \rangle = \int_0^1 (1)\left(\frac{1}{2}-x\right) dx = \int_0^1 \left(\frac{1}{2}-x\right) dx = \left[\frac{1}{2}x-\frac{1}{2}x^2\right]_0^1 = 0$, so the functions are orthogonal.

12. $\langle f,g \rangle = \int_0^1 (5x^2)(9x) \, dx = \int_0^1 (45x^3) \, dx = \left[\frac{45}{4}x^4\right]_0^1 = \frac{45}{4}$,

$\|f\|^2 = \int_0^1 (5x^2)^2 \, dx = \int_0^1 (25x^4) \, dx = [5x^5]_0^1 = 5$, so $\|f\| = \sqrt{5}$, and

$\|g\|^2 = \int_0^1 (9x)^2 \, dx = \int_0^1 (81x^2) \, dx = [27x^3]_0^1 = 27$, so $\|g\| = 3\sqrt{3}$. Thus

$\cos \theta = \frac{45}{4\sqrt{5}\, 3\sqrt{3}} = \frac{3 \times 15}{4 \times 3\sqrt{15}} = \frac{\sqrt{15}}{4}$.

13. $\langle f,g \rangle = \int_0^1 (6x+12)(ax+b) \, dx = \int_0^1 (6ax^2+12ax+6bx+12b) \, dx$

$= [2ax^3+6ax^2+3bx^2+12bx]_0^1 = 2a+6a+3b+12b$, so any choice of a and b that makes

$8a+15b = 0$ will do. Let $a = 15$ and $b = -8$. $g(x) = 15x - 8$ is orthogonal to $f(x)$.

14. $d(f,g) = \|f-g\| = \|2x+4\|$. $\|2x+4\|^2 = \int_0^1 (2x+4)^2 \, dx = \int_0^1 (4x^2+16x+16) \, dx$

$= \left[\frac{4}{3}x^3 + 8x^2 + 16x\right]_0^1 = \frac{4}{3} + 8 + 16 = \frac{76}{3}$, so $d(f,g) = \sqrt{\frac{76}{3}}$.

15. $d(f,g)^2 = \|f-g\|^2 = \|-2x+3\|^2 = \int_0^1 (-2x+3)^2 \, dx = \int_0^1 (4x^2-12x+9) \, dx$

$= \left[\frac{4}{3}x^3 - 6x^2 + 9x\right]_0^1 = \frac{4}{3} - 6 + 9 = \frac{13}{3}$, and $d(f,h)^2 = \|f-h\|^2 = \|3x-4\|^2 = \int_0^1 (3x-4)^2 \, dx$

$= \int_0^1 (9x^2 - 24x + 16) \, dx = [3x^3 - 12x^2 + 16x]_0^1 = 7 > \frac{13}{3}$, so g is closer to f.

16. (a) $\left\langle \begin{bmatrix} 1 & 2 \\ 3 & 4 \end{bmatrix}, \begin{bmatrix} -2 & 0 \\ -3 & 5 \end{bmatrix} \right\rangle = 1 \times -2 + 2 \times 0 + 3 \times -3 + 4 \times 5 = -2 + 0 - 9 + 20 = 9.$

(b) $\left\langle \begin{bmatrix} 0 & -3 \\ 2 & 5 \end{bmatrix}, \begin{bmatrix} 3 & 6 \\ -2 & -7 \end{bmatrix} \right\rangle = 0 \times 3 + -3 \times 6 + 2 \times -2 + 5 \times -7 = 0 - 18 - 4 - 35 = -57.$

17. (a) $\left\| \begin{bmatrix} 1 & 2 \\ 3 & 4 \end{bmatrix} \right\|^2 = 1^2 + 2^2 + 3^2 + 4^2 = 1 + 4 + 9 + 16 = 30$, so $\left\| \begin{bmatrix} 1 & 2 \\ 3 & 4 \end{bmatrix} \right\| = \sqrt{30}$.

(b) $\left\| \begin{bmatrix} 0 & 1 \\ -1 & 3 \end{bmatrix} \right\|^2 = 0 + 1^2 + (-1)^2 + 3^2 = 0 + 1 + 1 + 9 = 11$, so $\left\| \begin{bmatrix} 0 & 1 \\ -1 & 3 \end{bmatrix} \right\| = \sqrt{11}$.

(c) $\left\| \begin{bmatrix} 5 & -2 \\ -1 & 6 \end{bmatrix} \right\|^2 = 5^2 + (-2)^2 + (-1)^2 + 6^2 = 25 + 4 + 1 + 36 = 66$, so

$\left\| \begin{bmatrix} 5 & -2 \\ -1 & 6 \end{bmatrix} \right\| = \sqrt{66}$.

Section 8.1

(d) $\left\|\begin{bmatrix} 4 & -2 \\ -1 & -3 \end{bmatrix}\right\|^2 = 4^2 + (-2)^2 + (-1)^2 + (-3)^2 = 16 + 4 + 1 + 9 = 30$, so

$\left\|\begin{bmatrix} 4 & -2 \\ -1 & -3 \end{bmatrix}\right\| = \sqrt{30}$.

18. (a) $\left\langle \begin{bmatrix} 1 & 2 \\ -1 & 1 \end{bmatrix}, \begin{bmatrix} 2 & 4 \\ 3 & -7 \end{bmatrix} \right\rangle = 1 \times 2 + 2 \times 4 + -1 \times 3 + 1 \times -7 = 2 + 8 - 3 - 7 = 0$, so the matrices are orthogonal.

(b) $\left\langle \begin{bmatrix} 5 & 2 \\ -3 & 2 \end{bmatrix}, \begin{bmatrix} -1 & 6 \\ 1 & -2 \end{bmatrix} \right\rangle = 5 \times -1 + 2 \times 6 + -3 \times 1 + 2 \times -2 = -5 + 12 - 3 - 4 = 0$, so the matrices are orthogonal.

19. $\left\langle \begin{bmatrix} 1 & 2 \\ 3 & 4 \end{bmatrix}, \begin{bmatrix} a & b \\ c & d \end{bmatrix} \right\rangle = a + 2b + 3c + 4d$, so any choice of a,b,c, and d satisfying the condition $a + 2b + 3c + 4d = 0$ will do. One such choice is $a = -3$, $b = 2$, $c = 1$, $d = -1$.

20. (a) $d\left(\begin{bmatrix} 4 & 0 \\ -1 & 3 \end{bmatrix}, \begin{bmatrix} 1 & 1 \\ 1 & 1 \end{bmatrix} \right) = \left\| \begin{bmatrix} 3 & -1 \\ -2 & 2 \end{bmatrix} \right\|$. $\left\| \begin{bmatrix} 3 & -1 \\ -2 & 2 \end{bmatrix} \right\|^2 = 18$, so

$d\left(\begin{bmatrix} 4 & 0 \\ -1 & 3 \end{bmatrix}, \begin{bmatrix} 1 & 1 \\ 1 & 1 \end{bmatrix} \right) = \sqrt{18} = 3\sqrt{2}$.

(b) $d\left(\begin{bmatrix} 2 & -3 \\ -1 & 4 \end{bmatrix}, \begin{bmatrix} -3 & 2 \\ 1 & 0 \end{bmatrix} \right) = \left\| \begin{bmatrix} 5 & -5 \\ -2 & 4 \end{bmatrix} \right\|$. $\left\| \begin{bmatrix} 5 & -5 \\ -2 & 4 \end{bmatrix} \right\|^2 = 70$, so

$d\left(\begin{bmatrix} 2 & -3 \\ -1 & 4 \end{bmatrix}, \begin{bmatrix} -3 & 2 \\ 1 & 0 \end{bmatrix} \right) = \sqrt{70}$.

21. (a) $\langle u, v \rangle = (2-i)(3+2i) + (3+2i)(2-i) = 2(6 + 2 - 3i + 4i) = 16 + 2i$.

$\|u\|^2 = (2-i)(2+i) + (3+2i)(3-2i) = 4 + 1 + 9 + 4 = 18$, so $\|u\| = \sqrt{18} = 3\sqrt{2}$.

$\|v\|^2 = (3-2i)(3+2i) + (2+i)(2-i) = 18$, so $\|v\| = \sqrt{18} = 3\sqrt{2}$.

290

Section 8.1

$d(\mathbf{u},\mathbf{v})^2 = \|\mathbf{u}-\mathbf{v}\|^2 = \|(-1+i, 1+i)\|^2 = (-1+i)(-1-i) + (1+i)(1-i) = 4$, so $d(\mathbf{u},\mathbf{v}) = 2$.
u and **v** are not orthogonal since $\langle\mathbf{u},\mathbf{v}\rangle \neq 0$.

(b) $\langle\mathbf{u},\mathbf{v}\rangle = (4+3i)(2-i) + (1-i)(4+5i) = 8 + 3 - 4i + 6i + 4 + 5 + 5i - 4i = 20 + 3i$.

$\|\mathbf{u}\|^2 = (4+3i)(4-3i) + (1-i)(1+i) = 16 + 9 + 1 + 1 = 27$, so $\|\mathbf{u}\| = \sqrt{27} = 3\sqrt{3}$.

$\|\mathbf{v}\|^2 = (2+i)(2-i) + (4-5i)(4+5i) = 4 + 1 + 16 + 25 = 46$, so $\|\mathbf{v}\| = \sqrt{46}$.

$d(\mathbf{u},\mathbf{v})^2 = \|\mathbf{u}-\mathbf{v}\|^2 = \|(2+2i, -3+4i)\|^2 = (2+2i)(2-2i) + (-3+4i)(-3-4i) = 33$, so
$d(\mathbf{u},\mathbf{v}) = \sqrt{33}$. **u** and **v** are not orthogonal since $\langle\mathbf{u},\mathbf{v}\rangle \neq 0$.

(c) $\langle\mathbf{u},\mathbf{v}\rangle = (2+3i)(i) + (-1)(3-2i) = 2i - 3 - 3 + 2i = -6 + 4i$.

$\|\mathbf{u}\|^2 = (2+3i)(2-3i) + (-1)(-1) = 4 + 9 + 1 = 14$, so $\|\mathbf{u}\| = \sqrt{14}$.

$\|\mathbf{v}\|^2 = (-i)(i) + (3+2i)(3-2i) = 1 + 9 + 4 = 14$, so $\|\mathbf{v}\| = \sqrt{14}$.

$d(\mathbf{u},\mathbf{v})^2 = \|\mathbf{u}-\mathbf{v}\|^2 = \|(2+4i, -4-2i)\|^2 = (2+4i)(2-4i) + (-4-2i)(-4+2i) = 40$, so
$d(\mathbf{u},\mathbf{v}) = \sqrt{40} = 2\sqrt{10}$. **u** and **v** are not orthogonal since $\langle\mathbf{u},\mathbf{v}\rangle \neq 0$.

(d) $\langle\mathbf{u},\mathbf{v}\rangle = (2-3i)(1) + (-2+3i)(1) = 2 - 3i - 2 + 3i = 0$.

$\|\mathbf{u}\|^2 = (2-3i)(2+3i) + (-2+3i)(-2-3i) = 4 + 9 + 4 + 9 = 26$, so $\|\mathbf{u}\| = \sqrt{26}$.

$\|\mathbf{v}\|^2 = (1)(1) + (1)(1) = 2$, so $\|\mathbf{v}\| = \sqrt{2}$.

$d(\mathbf{u},\mathbf{v})^2 = \|\mathbf{u}-\mathbf{v}\|^2 = \|(1-3i, -3+3i)\|^2 = (1-3i)(1+3i) + (-3+3i)(-3-3i) = 28$, so
$d(\mathbf{u},\mathbf{v}) = \sqrt{28} = 2\sqrt{7}$. **u** and **v** are orthogonal since $\langle\mathbf{u},\mathbf{v}\rangle = 0$.

22. (a) $\langle\mathbf{u},\mathbf{v}\rangle = (1+4i)(1-i) + (1+i)(-4-i) = 1 + 4 - i + 4i - 4 + 1 - i - 4i = 2 - 2i$.

$\|\mathbf{u}\|^2 = (1+4i)(1-4i) + (1+i)(1-i) = 1 + 16 + 1 + 1 = 19$, so $\|\mathbf{u}\| = \sqrt{19}$.

$\|\mathbf{v}\|^2 = (1+i)(1-i) + (-4+i)(-4-i) = 1 + 1 + 16 + 1 = 19$, so $\|\mathbf{v}\| = \sqrt{19}$.

$d(\mathbf{u},\mathbf{v})^2 = \|\mathbf{u}-\mathbf{v}\|^2 = \|(3i, 5)\|^2 = (3i)(-3i) + (5)(5) = 34$, so

$d(u,v) = \sqrt{34}$. **u** and **v** are not orthogonal since $\langle u,v \rangle \neq 0$.

(b) $\langle u,v \rangle = (2+7i)(3+4i) + (1+i)(2-5i) = 6 - 28 + 21i + 8i + 2 + 5 + 2i - 5i = -15+26i$.

$\|u\|^2 = (2+7i)(2-7i) + (1+i)(1-i) = 4 + 49 + 1 + 1 = 55$, so $\|u\| = \sqrt{55}$.

$\|v\|^2 = (3-4i)(3+4i) + (2+5i)(2-5i) = 9 + 16 + 4 + 25 = 54$, so $\|v\| = \sqrt{54} = 3\sqrt{6}$.

$d(u,v)^2 = \|u-v\|^2 = \|(-1+11i, -1-4i)\|^2 = (-1+11i)(-1-11i) + (-1-4i)(-1+4i) = 139$,

so $d(u,v) = \sqrt{139}$. **u** and **v** are not orthogonal since $\langle u,v \rangle \neq 0$.

(c) $\langle u,v \rangle = (1-3i)(2+i) + (1+i)(-5i) = 2 + 3 - 6i + i + 5 - 5i = 10-10i$.

$\|u\|^2 = (1-3i)(1+3i) + (1+i)(1-i) = 1 + 9 + 1 + 1 = 12$, so $\|u\| = \sqrt{12} = 2\sqrt{3}$.

$\|v\|^2 = (2-i)(2+i) + (5i)(-5i) = 4 + 1 + 25 = 30$, so $\|v\| = \sqrt{30}$.

$d(u,v)^2 = \|u-v\|^2 = \|(-1-2i, 1-4i)\|^2 = (-1-2i)(-1+2i) + (1-4i)(1+4i) = 22$,

so $d(u,v) = \sqrt{22}$. **u** and **v** are not orthogonal since $\langle u,v \rangle \neq 0$.

(d) $\langle u,v \rangle = (3+i)(2-i) + (2+2i)\left(-2-\frac{3}{2}i\right) = 6 + 1 - 3i + 2i - 4 + 3 - 4i - 3i = 6-8i$.

$\|u\|^2 = (3+i)(3-i) + (2+2i)(2-2i) = 9 + 1 + 4 + 4 = 18$, so $\|u\| = \sqrt{18} = 3\sqrt{2}$.

$\|v\|^2 = (2+i)(2-i) + \left(-2+\frac{3}{2}i\right)\left(-2-\frac{3}{2}i\right) = 4 + 1 + 4 + \frac{9}{4} = \frac{45}{4}$, so $\|v\| = \frac{3}{2}\sqrt{5}$.

$d(u,v)^2 = \|u-v\|^2 = \left\|\left(1, 4+\frac{1}{2}i\right)\right\|^2 = (1)(1) + \left(4+\frac{1}{2}i\right)\left(4-\frac{1}{2}i\right) = \frac{69}{4}$, so

$d(u,v) = \frac{1}{2}\sqrt{69}$. **u** and **v** are not orthogonal since $\langle u,v \rangle \neq 0$.

23. $\mathbf{u} = (x_1, \ldots, x_n)$, $\mathbf{v} = (y_1, \ldots, y_n)$, $\mathbf{w} = (z_1, \ldots, z_n)$, and c is a scalar.

$\langle u,v\rangle = x_1\bar{y}_1 + \ldots + x_n\bar{y}_n = \bar{y}_1 x_1 + \ldots + \bar{y}_n x_n = \overline{y_1\bar{x}_1} + \ldots + \overline{y_n \bar{x}_n} = \overline{\langle v,u\rangle} = \langle v,u\rangle$,

$\langle u+v,w\rangle = (x_1+y_1)\bar{z}_1 + \ldots + (x_n+y_n)\bar{z}_n = x_1\bar{z}_1 + y_1\bar{z}_1 + \ldots + x_n\bar{z}_n + y_n\bar{z}_n$
$= x_1\bar{z}_1 + \ldots + x_n\bar{z}_n + y_1\bar{z}_1 + \ldots + y_n\bar{z}_n = \langle u,w\rangle + \langle v,w\rangle$,

$\langle cu,v\rangle = cx_1\bar{y}_1 + \ldots + cx_n\bar{y}_n = c(x_1\bar{y}_1 + \ldots + x_n\bar{y}_n) = c\langle u,v\rangle$, and

$\langle u,u\rangle = x_1\bar{x}_1 + \ldots + x_n\bar{x}_n \geq 0$ since all $x_i\bar{x}_i \geq 0$, and equality holds if and only if all $x_i = 0$.

Thus the given function is an inner product on $\mathbf{C}^n$.

24. $\langle u,kv\rangle = x_1\overline{k}\,\bar{y}_1 + \ldots + x_n\overline{k}\,\bar{y}_n = \overline{k}x_1\bar{y}_1 + \ldots + \overline{k}x_n\bar{y}_n = \overline{k}(x_1\bar{y}_1 + \ldots + x_n\bar{y}_n)$

$= \overline{k}\langle u,v\rangle$.

25. (a) Let **u** be any nonzero vector in the inner product space. Then using axiom 1, the fact that $0\mathbf{u} = \mathbf{0}$, axiom 3, and the fact that zero times any real number is zero we have $\langle v,0\rangle = \langle 0,v\rangle = \langle 0u,v\rangle = 0\langle u,v\rangle = 0$.

(b) From axiom 1, axiom 2, and axiom 1 again $\langle u,v+w\rangle = \langle v+w,u\rangle = \langle v,u\rangle + \langle w,u\rangle$
$= \langle u,v\rangle + \langle u,w\rangle$.

(c) From axiom 1, axiom 3, and axiom 1 again $\langle u,cv\rangle = \langle cv,u\rangle = c\langle v,u\rangle = c\langle u,v\rangle$.

26. (a) $\langle u,v\rangle = uAv^t = (uA)\cdot v = v\cdot(uA) = v(uA)^t = vAu^t = \langle v,u\rangle$,

$\langle u+v,w\rangle = (u+v)Aw^t = uAw^t + vAw^t = \langle u,w\rangle + \langle v,w\rangle$,

$\langle cu,v\rangle = (cu)Av^t = c(uAv^t) = c\langle u,v\rangle$, and

$\langle u,u\rangle = uAu^t > 0$ if $u \neq 0$ and $= 0$ if $u = 0$, so this function is an inner product on $\mathbf{R}^n$.

(b) Let $u = (x_1, \ldots, x_n)$ and $v = (y_1, \ldots, y_n)$. $u\cdot v = x_1 y_1 + \ldots + x_n y_n = uv^t = uI_n v^t$.

(c) $\langle u,v\rangle = uAv^t = [1\ 0]\begin{bmatrix} 2 & 0 \\ 0 & 3 \end{bmatrix}\begin{bmatrix} 0 \\ 1 \end{bmatrix} = [2\ 0]\begin{bmatrix} 0 \\ 1 \end{bmatrix} = 0.$

$\|u\|^2 = uAu^t = [1\ 0]\begin{bmatrix} 2 & 0 \\ 0 & 3 \end{bmatrix}\begin{bmatrix} 1 \\ 0 \end{bmatrix} = [2\ 0]\begin{bmatrix} 1 \\ 0 \end{bmatrix} = 2$, so $\|u\| = \sqrt{2}$.

$\|v\|^2 = vAv^t = [0\ 1]\begin{bmatrix} 2 & 0 \\ 0 & 3 \end{bmatrix}\begin{bmatrix} 0 \\ 1 \end{bmatrix} = [0\ 3]\begin{bmatrix} 0 \\ 1 \end{bmatrix} = 3$, so $\|v\| = \sqrt{3}$.

$d(u,v)^2 = \|u-v\|^2 = [1\ -1]\begin{bmatrix} 2 & 0 \\ 0 & 3 \end{bmatrix}\begin{bmatrix} 1 \\ -1 \end{bmatrix} = [2\ -3]\begin{bmatrix} 1 \\ -1 \end{bmatrix} = 5$, so $d(u,v) = \sqrt{5}$.

$\langle u,v\rangle = uBv^t = [1\ 0]\begin{bmatrix} 1 & 2 \\ 2 & 3 \end{bmatrix}\begin{bmatrix} 0 \\ 1 \end{bmatrix} = [1\ 2]\begin{bmatrix} 0 \\ 1 \end{bmatrix} = 2.$

$\|u\|^2 = uBu^t = [1\ 0]\begin{bmatrix} 1 & 2 \\ 2 & 3 \end{bmatrix}\begin{bmatrix} 1 \\ 0 \end{bmatrix} = [1\ 2]\begin{bmatrix} 1 \\ 0 \end{bmatrix} = 1$, so $\|u\| = 1$.

$\|v\|^2 = vBv^t = [0\ 1]\begin{bmatrix} 1 & 2 \\ 2 & 3 \end{bmatrix}\begin{bmatrix} 0 \\ 1 \end{bmatrix} = [2\ 3]\begin{bmatrix} 0 \\ 1 \end{bmatrix} = 3$, so $\|v\| = \sqrt{3}$.

$d(u,v)^2 = \|u-v\|^2 = [1\ -1]\begin{bmatrix} 1 & 2 \\ 2 & 3 \end{bmatrix}\begin{bmatrix} 1 \\ -1 \end{bmatrix} = [-1\ -1]\begin{bmatrix} 1 \\ -1 \end{bmatrix} = 0$, so $d(u,v) = 0$.

$\langle u,v\rangle = uCv^t = [1\ 0]\begin{bmatrix} 2 & 3 \\ 3 & 5 \end{bmatrix}\begin{bmatrix} 0 \\ 1 \end{bmatrix} = [2\ 3]\begin{bmatrix} 0 \\ 1 \end{bmatrix} = 3.$

$\|u\|^2 = uCu^t = [1\ 0]\begin{bmatrix} 2 & 3 \\ 3 & 5 \end{bmatrix}\begin{bmatrix} 1 \\ 0 \end{bmatrix} = [2\ 3]\begin{bmatrix} 1 \\ 0 \end{bmatrix} = 2$, so $\|u\| = \sqrt{2}$.

$\|v\|^2 = vCv^t = [0\ 1]\begin{bmatrix} 2 & 3 \\ 3 & 5 \end{bmatrix}\begin{bmatrix} 0 \\ 1 \end{bmatrix} = [3\ 5]\begin{bmatrix} 0 \\ 1 \end{bmatrix} = 5$, so $\|v\| = \sqrt{5}$.

$d(u,v)^2 = \|u-v\|^2 = [1\ -1]\begin{bmatrix} 2 & 3 \\ 3 & 5 \end{bmatrix}\begin{bmatrix} 1 \\ -1 \end{bmatrix} = [-1\ -2]\begin{bmatrix} 1 \\ -1 \end{bmatrix} = 1$, so $d(u,v) = 1$.

(d) $u = (x_1, x_2)$ and $v = (y_1, y_2)$. $\langle u,v\rangle = x_1 y_1 + 4 x_2 y_2 = [x_1\ x_2]\begin{bmatrix} a & b \\ c & d \end{bmatrix}\begin{bmatrix} y_1 \\ y_2 \end{bmatrix}$

$= [ax_1 + cx_2\ \ bx_1 + dx_2]\begin{bmatrix} y_1 \\ y_2 \end{bmatrix} = (ax_1 + cx_2)y_1 + (bx_1 + dx_2)y_2$

$= ax_1y_1 + cx_2y_1 + bx_1y_2 + dx_2y_2$ and equating coefficients $a = 1$, $b = 0$, $c = 0$, $d = 4$.

27. Let w_1 and w_2 be elements of W and c a scalar. Then $<w_1,v>=0$ and $<w_2,v> = 0$. Thus $<w_1,v>+<w_2,v> = 0$ and $c<w_1,v>=0$. Properties of inner product give: $<w_1+w_2,v> = 0$ and $<cw_1,v>=0$. Thus w_1+w_2 and cw_1 are in W. W is closed under vector addition and under scalar multiplication. Thus it is a subspace of V.

Exercise Set 8.2

1. $d((x_1,x_2), (0,0))^2 = \|(x_1,x_2)\|^2 = x_1^2 + 4x_2^2$, so the equation of the circle with radius 1 and center at the origin is $x_1^2 + 4x_2^2 = 1$.

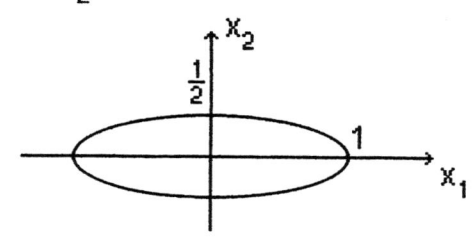

2. (a) $\|(1,0)\| = \sqrt{4 \times 1 \times 1 + 9 \times 0 \times 0} = 2$, $\|(0,1)\| = \sqrt{4 \times 0 \times 0 + 9 \times 1 \times 1} = 3$,

    $\|(1,1)\| = \sqrt{4 \times 1 \times 1 + 9 \times 1 \times 1} = \sqrt{13}$, $\|(2,3)\| = \sqrt{4 \times 2 \times 2 + 9 \times 3 \times 3} = \sqrt{97}$.

   (b) $<(2,1),(-9,8)> = 4 \times 2 \times -9 + 9 \times 1 \times 8 = 0$, so the vectors are orthogonal in this space.

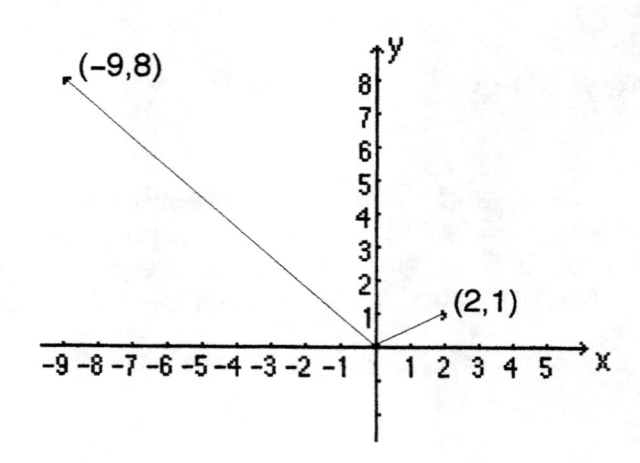

Section 8.2

(c) $d((1,0),(0,1)) = \|(1,-1)\| = \sqrt{4 \times 1 \times 1 + 9 \times -1 \times -1} = \sqrt{13}$.

(d) $d((x_1,x_2),(0,0))^2 = \|(x_1,x_2)\|^2 = 4x_1^2 + 9x_2^2$, so the equation of the circle with radius 1 and center at the origin is $4x_1^2 + 9x_2^2 = 1$.

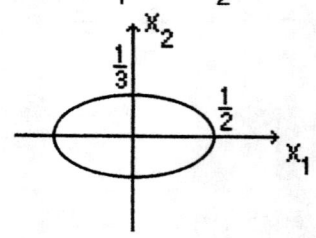

3. (a) $\|(1,0)\| = \sqrt{1 \times 1 + 16 \times 0 \times 0} = 1$, $\|(0,1)\| = \sqrt{0 \times 0 + 16 \times 1 \times 1} = 4$, and

$\|(1,1)\| = \sqrt{1 \times 1 + 16 \times 1 \times 1} = \sqrt{17}$.

(b) $<(1,1),(-16,1)> = 1 \times -16 + 16 \times 1 \times 1 = 0$, so the vectors are orthogonal.

(c) $d((5,0),(0,4)) = \|(5,-4)\| = \sqrt{5 \times 5 + 16 \times -4 \times -4} = \sqrt{281}$.
In Euclidean space the distance is $\sqrt{5^2 + 4^2} = \sqrt{41}$.

(d) $d((x_1,x_2),(0,0))^2 = \|(x_1,x_2)\|^2 = x_1^2 + 16x_2^2$, so the equation of the circle with radius 1 and center at the origin is $x_1^2 + 16x_2^2 = 1$.

296

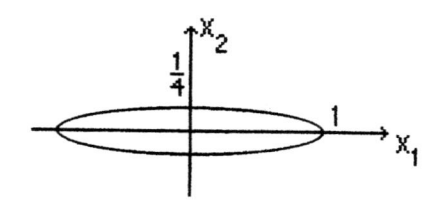

4. $\langle (x_1, x_2),(y_1, y_2)\rangle = \frac{1}{4}x_1y_1 + \frac{1}{25}x_2y_2$.

5. If $X = (1,0,0,1)$ then $\langle X,X\rangle = |-1^2-0^2-0^2+1^2| = 0$.

6. $d(R,Q)^2 = \|(4,0,0,-5)\|^2 = |-16 - 0 - 0 + 25| = 9$, so $d(R,Q) = 3$.

7. $d(M,P)^2 = \|(1,0,0,1)\|^2 = |-1 - 0 - 0 + 1| = 0$, so $d(M,P) = 0$.

8. $\langle(2,0,0,1),(1,0,0,2)\rangle$ $-2 - 0 - 0 + 2 = 0$, so the vectors are orthogonal.

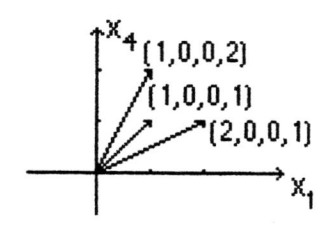

   $\langle(a,0,0,b),(b,0,0,a)\rangle = -ab - 0 - 0 + ba = 0$, so the vectors are orthogonal.

9. $d((x_1,0,0,x_4), (0,0,0,0))^2 = \|(x_1,0,0,x_4)\|^2 = |-x_1^2 + x_4^2|$, so the equations of the circles with radii $a = 1,2,3$ and center at the origin are $|-x_1^2 + x_4^2| = a^2$.

Section 8.2

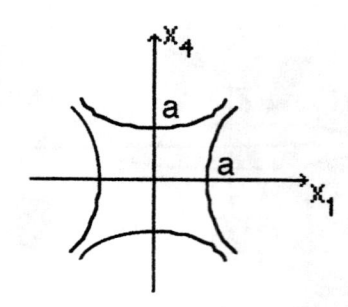

We use the following illustration for the remaining exercises in this section.

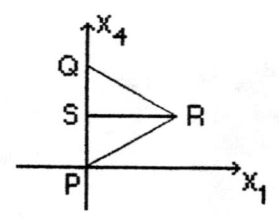

10. PQ = (0,0,0,20), so PS = (0,0,0,10) and PR = (8,0,0,10).

    ||PR|| = $\sqrt{-64-0-0+100}$ = 6, so the duration of the voyage for a person on the space ship is 2x6 = 12 years.

    Duration of voyage relative to Earth = $\frac{\text{distance in light years}}{\text{speed}}$, so speed = $\frac{16}{20}$ = .8 speed of light.

11. PQ = (0,0,0,120), so PS = (0,0,0,60) and PR = (45,0,0,60).

    ||PR|| = $\sqrt{-(45)^2-0-0+(60)^2}$ = $15\sqrt{7}$, so the duration of the voyage for a person on the spaceship is $30\sqrt{7}$ years.

12. Let PS = (0,0,0,t). Then PR = (410,0,0,t) and ||PR||² = |-(410)²-0-0+t²| = 400. We assume the speed of the traveler is less than the speed of light, so t > 410.

    -(410)² + t² = 400, so t² = 168,500 and t = 410.49 years. Therefore the duration of the voyage from Earth's point of view is 820.98 years. More than 8 centuries will have passed on Earth.

298

Section 8.3

13. Duration of voyage relative to Earth = $\dfrac{\text{distance in light years}}{\text{speed}} = \dfrac{1030}{.99999} = 1030.0103$ years. PR = (515,0,0,515.00515), so the duration of the voyage relative to the traveler is 2||PR|| = $2\sqrt{-(515)^2+(515.00515)^2} = 2\sqrt{5.3045} = 4.606$ years.

Exercise Set 8.3

1. The set $\left\{\dfrac{1}{\sqrt{2}}, \dfrac{\sqrt{3}}{\sqrt{2}} x\right\}$ is an orthonormal basis for $P_1[-1,1]$ (see the text).

   $\left\langle x^2, \dfrac{1}{\sqrt{2}} \right\rangle = \dfrac{1}{\sqrt{2}} \int_{-1}^{1} x^2\, dx = \left[\dfrac{1}{3\sqrt{2}} x^3\right]_{-1}^{1} = \dfrac{2}{3\sqrt{2}} = \dfrac{\sqrt{2}}{3}$ and

   $\left\langle x^2, \dfrac{\sqrt{3}}{\sqrt{2}} x \right\rangle = \dfrac{\sqrt{3}}{\sqrt{2}} \int_{-1}^{1} x^3\, dx = \left[\dfrac{\sqrt{3}}{\sqrt{2}} \dfrac{x^4}{4}\right]_{-1}^{1} = 0$, so that $\text{proj}_{P_1[-1,1]} x^2 = \dfrac{\sqrt{2}}{3} \dfrac{1}{\sqrt{2}} = \dfrac{1}{3}$

   is the least squares linear approximation to $f(x) = x^2$ over the interval $[-1,1]$.

2. The set $\left\{1, x-\dfrac{1}{2}\right\}$ is an orthogonal basis for $P_1[0,1]$, $||1|| = 1$, and

   $\left\|x-\dfrac{1}{2}\right\|^2 = \int_0^1 \left(x^2 - x + \dfrac{1}{4}\right) dx = \left[\dfrac{1}{3}x^3 - \dfrac{1}{2}x^2 + \dfrac{1}{4}x\right]_0^1 = \dfrac{1}{3} - \dfrac{1}{2} + \dfrac{1}{4} = \dfrac{1}{12}$, so the set

   $\left\{1, 2\sqrt{3}\left(x-\dfrac{1}{2}\right)\right\}$ is an orthonormal basis for $P_1[0,1]$.

   $\langle e^x, 1 \rangle = \int_0^1 e^x\, dx = [e^x]_0^1 = e - 1$, and $\langle e^x, 2\sqrt{3}x - \sqrt{3}\rangle = \int_0^1 (2\sqrt{3}\, xe^x - \sqrt{3}\, e^x)\, dx$

   $= [2\sqrt{3}(xe^x - e^x) - \sqrt{3}\, e^x]_0^1 = -\sqrt{3}\, e + 3\sqrt{3}$, so

   $\text{proj}_{P_1[0,1]} e^x = e - 1 + (-\sqrt{3}\, e + 3\sqrt{3})2\sqrt{3}\left(x - \dfrac{1}{2}\right) = 4e - 10 + (18 - 6e)x$ is the least squares linear approximation to e^x over the interval $[0,1]$.

3. The set $\left\{1, 2\sqrt{3}\left(x-\frac{1}{2}\right)\right\}$ is an orthonormal basis for $P_1[0,1]$ (see Exercise 2).

$$\langle\sqrt{x}, 1\rangle = \int_0^1 \sqrt{x}\, dx = \left[\frac{2}{3} x^{\frac{3}{2}}\right]_0^1 = \frac{2}{3}, \text{ and } \langle\sqrt{x}, 2\sqrt{3}\, x - \sqrt{3}\rangle = \int_0^1 (2\sqrt{3}\, x^{\frac{3}{2}} - \sqrt{3}\, x^{\frac{1}{2}})\, dx$$

$$= \left[\frac{4\sqrt{3}}{5} x^{\frac{5}{2}} - \frac{2\sqrt{3}}{3} x^{\frac{3}{2}}\right]_0^1 = \frac{4\sqrt{3}}{5} - \frac{2\sqrt{3}}{3} = \frac{2\sqrt{3}}{15}, \text{ so that }$$

$$\text{proj}_{P_1[0,1]} \sqrt{x} = \frac{2}{3} + \frac{2\sqrt{3}}{15}\, 2\sqrt{3}\left(x-\frac{1}{2}\right) = \frac{4}{5} x + \frac{4}{15}$$ is the least squares linear

approximation to $\sqrt{x}$ over the interval $[0,1]$.

4. The set $\left\{1, x - \frac{\pi}{2}\right\}$ is an orthogonal basis for $P_1[0,\pi]$, $\|1\| = \pi$, and

$$\left\|x - \frac{\pi}{2}\right\|^2 = \int_0^\pi \left(x^2 - \pi x + \frac{\pi^2}{4}\right) dx = \left[\frac{1}{3} x^3 - \frac{1}{2}\pi x^2 + \frac{\pi^2}{4} x\right]_0^\pi = \left(\frac{1}{3} - \frac{1}{2} + \frac{1}{4}\right)\pi^3 = \frac{\pi^3}{12}, \text{ so}$$

the set $\left\{\pi^{-\frac{1}{2}}, 2\sqrt{3}\, \pi^{-\frac{3}{2}}\left(x - \frac{\pi}{2}\right)\right\}$ is an orthonormal basis for $P_1[0,\pi]$.

$$\langle\cos x, \pi^{-\frac{1}{2}}\rangle = \pi^{-\frac{1}{2}} \int_0^\pi \cos x\, dx = \pi^{-\frac{1}{2}} [\sin x]_0^\pi = 0, \text{ and }$$

$$\left\langle\cos x, 2\sqrt{3}\, \pi^{-\frac{3}{2}}\left(x - \frac{\pi}{2}\right)\right\rangle = 2\sqrt{3}\, \pi^{-\frac{3}{2}} \int_0^\pi \left(x \cos x - \frac{\pi}{2}\cos x\right) dx$$

$$= 2\sqrt{3}\, \pi^{-\frac{3}{2}} \left[x \sin x + \cos x - \frac{\pi}{2} \sin x\right]_0^\pi = 2\sqrt{3}\, \pi^{-\frac{3}{2}} [-1-1] = -4\sqrt{3}\, \pi^{-\frac{3}{2}}, \text{ so that }$$

$$\text{proj}_{P_1[0,\pi]} \cos x = -4\sqrt{3}\, \pi^{-\frac{3}{2}} 2\sqrt{3}\, \pi^{-\frac{3}{2}}\left(x - \frac{\pi}{2}\right) = -24\pi^{-3} x + 12\pi^{-2}$$ is the least

squares linear approximation to $\cos x$ over the interval $[0,\pi]$.

Section 8.3

5. (a) The set $\left\{\frac{1}{\sqrt{2}}, \frac{\sqrt{3}}{\sqrt{2}} x, \frac{3\sqrt{5}}{2\sqrt{2}}\left(x^2 - \frac{1}{3}\right)\right\}$ is an orthonormal basis for $P_2[-1,1]$.

$\left\langle e^x, \frac{1}{\sqrt{2}} \right\rangle = \frac{1}{\sqrt{2}} \int_{-1}^{1} e^x \, dx = \left[\frac{1}{\sqrt{2}} e^x\right]_{-1}^{1} = \frac{1}{\sqrt{2}}(e - e^{-1})$, $\left\langle e^x, \frac{\sqrt{3}}{\sqrt{2}} x \right\rangle$

$= \frac{\sqrt{3}}{\sqrt{2}} \int_{-1}^{1} x e^x \, dx = \frac{\sqrt{3}}{\sqrt{2}} [xe^x - e^x]_{-1}^{1} = \frac{\sqrt{3}}{\sqrt{2}}(2e^{-1})$, and $\left\langle e^x, \frac{3\sqrt{5}}{2\sqrt{2}}\left(x^2 - \frac{1}{3}\right) \right\rangle$

$= \frac{3\sqrt{5}}{2\sqrt{2}} \int_{-1}^{1}\left(x^2 e^x - \frac{1}{3} e^x\right) dx = \frac{3\sqrt{5}}{2\sqrt{2}}\left[x^2 e^x - 2xe^x + 2e^x - \frac{1}{3} e^x\right]_{-1}^{1}$

$= \frac{3\sqrt{5}}{2\sqrt{2}}\left(\frac{2}{3} e - \frac{14}{3} e^{-1}\right)$, so that $\text{proj}_{P_2[-1,1]} e^x$

$= \frac{1}{\sqrt{2}}(e - e^{-1})\frac{1}{\sqrt{2}} + \frac{\sqrt{3}}{\sqrt{2}}(2e^{-1})\frac{\sqrt{3}}{\sqrt{2}} x + \frac{3\sqrt{5}}{2\sqrt{2}}\left(\frac{2}{3} e - \frac{14}{3} e^{-1}\right)\frac{3\sqrt{5}}{2\sqrt{2}}\left(x^2 - \frac{1}{3}\right)$

$= -\frac{3}{4} e + \frac{33}{4} e^{-1} + 3e^{-1}x + \frac{15}{4}(e - 7e^{-1})x^2$ is the least squares quadratic

approximation to $f(x) = e^x$ over the interval $[-1,1]$.

(b) The set $\left\{1, 2\sqrt{3}\left(x - \frac{1}{2}\right), 6\sqrt{5}\left(x^2 - x + \frac{1}{6}\right)\right\}$ is an orthonormal basis for

$P_2[0,1]$. From Exercise 2, $\langle e^x, 1\rangle = e - 1$ and $\langle e^x, 2\sqrt{3} x - \sqrt{3}\rangle = -\sqrt{3} e + 3\sqrt{3}$.

$\left\langle e^x, 6\sqrt{5}\left(x^2 - x + \frac{1}{6}\right) \right\rangle = \int_0^1 (6\sqrt{5} x^2 e^x - 6\sqrt{5} x e^x + \sqrt{5} e^x) \, dx$

$= [6\sqrt{5}(x^2 e^x - 2xe^x + 2e^x) - 6\sqrt{5}(xe^x - e^x) + \sqrt{5} e^x]_0^1 = 7\sqrt{5} e - 19\sqrt{5}$, so that

$\text{proj}_{P_2[0,1]} e^x = e - 1 + (-\sqrt{3} e + 3\sqrt{3})2\sqrt{3}\left(x - \frac{1}{2}\right) + (7\sqrt{5} e - 19\sqrt{5})6\sqrt{5}\left(x^2 - x + \frac{1}{6}\right)$

$= e - 1 + (-6e + 18)\left(x - \frac{1}{2}\right) + (210e - 570)\left(x^2 - x + \frac{1}{6}\right)$

301

$= 39e - 105 + (-216e + 588)x + (210e - 570)x^2$ is the least squares quadratic approximation to $f(x) = e^x$ over the interval $[0,1]$.

6. The set $\left\{1, 2\sqrt{3}\left(x-\frac{1}{2}\right), 6\sqrt{5}\left(x^2 - x + \frac{1}{6}\right)\right\}$ is an orthonormal basis for $P_2[0,1]$.

   $\langle \sqrt{x}, 1 \rangle = \frac{2}{3}$, $\langle \sqrt{x}, 2\sqrt{3}\,x - \sqrt{3} \rangle = \frac{2\sqrt{3}}{15}$, and $\left\langle \sqrt{x}, 6\sqrt{5}\left(x^2 - x + \frac{1}{6}\right) \right\rangle$

   $= 6\sqrt{5} \int_0^1 \left(x^{\frac{5}{2}} - x^{\frac{3}{2}} + \frac{1}{6} x^{\frac{1}{2}} \right) dx = 6\sqrt{5} \left[\frac{2}{7} x^{\frac{7}{2}} - \frac{2}{5} x^{\frac{5}{2}} + \frac{1}{9} x^{\frac{3}{2}} \right]_0^1 = 6\sqrt{5}\left(\frac{2}{7} - \frac{2}{5} + \frac{1}{9}\right)$

   $= -\frac{2\sqrt{5}}{105}$, so that $\operatorname{proj}_{P_2[0,1]} \sqrt{x} = \frac{2}{3} + \frac{2\sqrt{3}}{15} 2\sqrt{3}\left(x - \frac{1}{2}\right) - \frac{2\sqrt{5}}{105} 6\sqrt{5}\left(x^2 - x + \frac{1}{6}\right)$

   $= -\frac{4}{7} x^2 + \frac{48}{35} x + \frac{6}{35}$ is the least squares quadratic approximation to $f(x) = \sqrt{x}$ over the interval $[0,1]$.

7. $f(x) = x^2$ is in $P_2[0,1]$, so the least squares quadratic approximation to $f(x) = x^2$ is x^2.

8. The set $\left\{\pi^{-\frac{1}{2}}, 2\sqrt{3}\,\pi^{-\frac{3}{2}}\left(x - \frac{\pi}{2}\right), 6\sqrt{5}\,\pi^{-\frac{5}{2}}\left(x^2 - \pi x + \frac{\pi^2}{6}\right)\right\}$ is an orthonormal basis for $P_2[0,\pi]$. $\langle \sin x, \pi^{-\frac{1}{2}} \rangle = \pi^{-\frac{1}{2}} \int_0^\pi \sin x\, dx = \pi^{-\frac{1}{2}} [-\cos x]_0^\pi = 2\pi^{-\frac{1}{2}}$,

   $\left\langle \sin x, 2\sqrt{3}\,\pi^{-\frac{3}{2}}\left(x - \frac{\pi}{2}\right) \right\rangle = 2\sqrt{3}\,\pi^{-\frac{3}{2}} \int_0^\pi \left(x \sin x - \frac{\pi}{2} \sin x \right) dx$

   $= 2\sqrt{3}\,\pi^{-\frac{3}{2}} \left[-x \cos x + \sin x + \frac{\pi}{2} \cos x \right]_0^\pi = 0$, and

302

Section 8.3

$$\left\langle \sin x, 6\sqrt{5}\,\pi^{-\frac{5}{2}}\left(x^2 - \pi x + \frac{\pi^2}{6}\right)\right\rangle$$

$$= 6\sqrt{5}\,\pi^{-\frac{5}{2}}\int_0^\pi \left(x^2 \sin x - \pi x \sin x + \frac{\pi^2}{6}\sin x\right)dx$$

$$= 6\sqrt{5}\,\pi^{-\frac{5}{2}}\left[-x^2\cos x + 2x\sin x + 2\cos x + \pi x\cos x - \pi\sin x - \frac{\pi^2}{6}\cos x\right]_0^\pi$$

$$= 6\sqrt{5}\,\pi^{-\frac{5}{2}}\left(-4 + \frac{\pi^2}{3}\right), \text{ so}$$

$$\mathrm{proj}_{P_2[0,\pi]}\sin x = 2\pi^{-\frac{1}{2}}\pi^{-\frac{1}{2}} + 6\sqrt{5}\,\pi^{-\frac{5}{2}}\left(-4 + \frac{\pi^2}{3}\right)6\sqrt{5}\,\pi^{-\frac{5}{2}}\left(x^2 - \pi x + \frac{\pi^2}{6}\right)$$

$$= 180\pi^{-5}\left(-4 + \frac{\pi^2}{3}\right)x^2 - 180\pi^{-4}\left(-4 + \frac{\pi^2}{3}\right)x - 120\pi^{-3} + 12\pi^{-1}\text{ is the least squares}$$

quadratic approximation to $\cos x$ over the interval $[0,\pi]$.

9. The vectors that form an orthonormal basis over $[-\pi, \pi]$ also form an orthonormal basis over $[0,2\pi]$, so the Fourier approximations can be found in the same way, only with integration over $[0,2\pi]$.

$$a_0 = \frac{1}{2\pi}\int_0^{2\pi} x\,dx = \left[\frac{1}{4\pi}x^2\right]_0^{2\pi} = \pi,$$

$$a_k = \frac{1}{\pi}\int_0^{2\pi} x\cos kx\,dx = \frac{1}{\pi}\left[\frac{x}{k}\sin kx + \frac{1}{k^2}\cos kx\right]_0^{2\pi} = 0,\text{ and}$$

$$b_k = \frac{1}{\pi}\int_0^{2\pi} x\sin kx\,dx = \frac{1}{\pi}\left[-\frac{x}{k}\cos kx + \frac{1}{k^2}\sin kx\right]_0^{2\pi} = \frac{1}{\pi}\left(-\frac{2\pi}{k}\cos 2k\pi\right) = \frac{-2}{k},$$

so the fourth-order Fourier approximation to $f(x)$ over $[0,2\pi]$ is

$$g(x) = \pi + \sum_{k=1}^{4}\frac{-2}{k}\sin kx = \pi - 2\left(\sin x + \frac{1}{2}\sin 2x + \frac{1}{3}\sin 3x + \frac{1}{4}\sin 4x\right).$$

Section 8.3

10. The fourth-order Fourier approximation to $f(x) = 1 + x$ over the interval $[-\pi,\pi]$ is $1 + g(x)$, where $g(x)$ is the fourth-order Fourier approximation to $h(x) = x$, which was found in the text. $1 + g(x) = 1 + 2\left(\sin x - \frac{1}{2}\sin 2x + \frac{1}{3}\sin 3x - \frac{1}{4}\sin 4x\right)$.

11. $a_0 = \frac{1}{2\pi}\int_0^{2\pi} x^2\, dx = \left[\frac{1}{6\pi}x^3\right]_0^{2\pi} = \frac{4}{3}\pi^2$,

$a_k = \frac{1}{\pi}\int_0^{2\pi} x^2 \cos kx\, dx = \frac{1}{\pi}\left[\frac{x^2}{k}\sin kx - \frac{2}{k}\left(\frac{-x}{k}\cos kx + \frac{1}{k^2}\sin kx\right)\right]_0^{2\pi} = \frac{4}{k^2}$, and

$b_k = \frac{1}{\pi}\int_0^{2\pi} x^2 \sin kx\, dx = \frac{1}{\pi}\left[\frac{-x^2}{k}\cos kx - \frac{2}{k}\left(\frac{x}{k}\sin kx + \frac{1}{k^2}\cos kx\right)\right]_0^{2\pi} = -\frac{4\pi}{k}$, so

$g(x) = \frac{4}{3}\pi^2 + \sum_{k=1}^{4}\left(\frac{4}{k^2}\cos kx - \frac{4\pi}{k}\sin kx\right)$

$= \frac{4}{3}\pi^2 + 4\left(\cos x + \frac{1}{4}\cos 2x + \frac{1}{9}\cos 3x + \frac{1}{16}\cos 4x\right)$

$\quad - 4\pi\left(\sin x + \frac{1}{2}\sin 2x + \frac{1}{3}\sin 3x + \frac{1}{4}\sin 4x\right)$.

12. $\langle 1, \sin nx\rangle = \int_{-\pi}^{\pi}\sin nx\, dx = \left[-\frac{1}{n}\cos nx\right]_{-\pi}^{\pi} = 0$,

$\langle 1, \cos nx\rangle = \int_{-\pi}^{\pi}\cos nx\, dx = \left[\frac{1}{n}\sin nx\right]_{-\pi}^{\pi} = 0$,

$\langle \sin mx, \sin nx\rangle = \int_{-\pi}^{\pi}\sin mx \sin nx\, dx = \left[\frac{\sin(m-n)x}{2(m-n)} - \frac{\sin(m+n)x}{2(m+n)}\right]_{-\pi}^{\pi} = 0$,

$\langle \cos mx, \cos nx\rangle = \int_{-\pi}^{\pi}\cos mx \cos nx\, dx = \left[\frac{\sin(m-n)x}{2(m-n)} + \frac{\sin(m+n)x}{2(m+n)}\right]_{-\pi}^{\pi} = 0$, and

$\langle \sin mx, \cos nx\rangle = \int_{-\pi}^{\pi}\sin mx \cos nx\, dx = \left[\frac{\cos(m-n)x}{2(m-n)} - \frac{\cos(m+n)x}{2(m+n)}\right]_{-\pi}^{\pi} = 0$, so

the vectors are mutually orthogonal in $C[-\pi,\pi]$.

Section 8.4

13. (1,0,0,0,0,1,1), (0,1,0,0,1,0,1), (0,0,1,0,1,1,0), (0,0,0,1,1,1,1),
    (0,1,1,1,1,0,0), (1,0,1,1,0,1,0), (1,1,0,1,0,0,1), (1,1,1,0,0,0,0),
    (1,1,0,0,1,1,0), (1,0,1,0,1,0,1), (1,0,0,1,1,0,0), (1,1,1,1,1,1,1),
    (0,0,1,1,0,0,1), (0,1,0,1,0,1,0), (0,1,1,0,0,1,1), (0,0,0,0,0,0,0)

14. (a) center = (0,0,1,0,1,1,0)
       (1,0,1,0,1,1,0), (0,1,1,0,1,1,0), (0,0,0,0,1,1,0), (0,0,1,1,1,1,0), (0,0,1,0,0,1,0),
       (0,0,1,0,1,0,0), (0,0,1,0,1,1,1)

    (b) center = (1,1,0,1,0,0,1)
       (0,1,0,1,0,0,1), (1,0,0,1,0,0,1), (1,1,1,1,0,0,1), (1,1,0,0,0,0,1), (1,1,0,1,1,0,1),
       (1,1,0,1,0,1,1), (1,1,0,1,0,0,0)

    (c) center = (1,1,1,1,1,1,1)
       (0,1,1,1,1,1,1), (1,0,1,1,1,1,1), (1,1,0,1,1,1,1), (1,1,1,0,1,1,1), (1,1,1,1,0,1,1),
       (1,1,1,1,1,0,1), (1,1,1,1,1,1,0)

    (d) center = (0,0,0,0,0,0,0)
       (1,0,0,0,0,0,0), (0,1,0,0,0,0,0), (0,0,1,0,0,0,0), (0,0,0,1,0,0,0), (0,0,0,0,1,0,0),
       (0,0,0,0,0,1,0), (0,0,0,0,0,0,1)

15. (a) (0,0,0,0,0,0,0) (b) (0,0,1,1,0,0,1) (c) (0,0,1,0,1,1,0)
    (d) (1,0,1,1,0,1,0)

16. (a) Since each vector in V_{23} has 23 components and there are 2 possible values (zero and 1) for each component, there are 2^{23} vectors.

    (b) Each vector in $C_{23,12}$ is a linear combination of 12 basis vectors and each coefficient is either zero or 1. Thus there are $2^{12} = 4096$ vectors.

    (c) There are 23 ways to change exactly one component of the center vector, 23x22/2 = 253 ways to change exactly two components, and 23x22x21/6 = 1771 ways to change exactly three components. 1 + 23 + 253 + 1771 = 2048 vectors.

Exercise Set 8.4

1. $\text{Pinv} \begin{bmatrix} 1 & 3 \\ 0 & 1 \\ -1 & 2 \end{bmatrix} = \left(\begin{bmatrix} 1 & 0 & -1 \\ 3 & 1 & 2 \end{bmatrix} \begin{bmatrix} 1 & 3 \\ 0 & 1 \\ -1 & 2 \end{bmatrix} \right)^{-1} \begin{bmatrix} 1 & 0 & -1 \\ 3 & 1 & 2 \end{bmatrix} = \begin{bmatrix} 2 & 1 \\ 1 & 14 \end{bmatrix}^{-1} \begin{bmatrix} 1 & 0 & -1 \\ 3 & 1 & 2 \end{bmatrix}$

$$= \frac{1}{27} \begin{bmatrix} 14 & -1 \\ -1 & 2 \end{bmatrix} \begin{bmatrix} 1 & 0 & -1 \\ 3 & 1 & 2 \end{bmatrix} = \frac{1}{27} \begin{bmatrix} 11 & -1 & -16 \\ 5 & 2 & 5 \end{bmatrix}.$$

2. $\text{Pinv} \begin{bmatrix} 1 & 1 \\ 2 & 3 \\ 0 & 1 \end{bmatrix} = \left(\begin{bmatrix} 1 & 2 & 0 \\ 1 & 3 & 1 \end{bmatrix} \begin{bmatrix} 1 & 1 \\ 2 & 3 \\ 0 & 1 \end{bmatrix} \right)^{-1} \begin{bmatrix} 1 & 2 & 0 \\ 1 & 3 & 1 \end{bmatrix} = \frac{1}{6} \begin{bmatrix} 4 & 1 & -7 \\ -2 & 1 & 5 \end{bmatrix}.$

3. The pseudoinverse does not exist because $\left(\begin{bmatrix} 1 & 2 & 3 \\ 2 & 4 & 6 \end{bmatrix} \begin{bmatrix} 1 & 2 \\ 2 & 4 \\ 3 & 6 \end{bmatrix} \right)^{-1}$ does not exist.

4. $\text{Pinv} \begin{bmatrix} 1 & 2 \\ 3 & 4 \\ 5 & 6 \end{bmatrix} = \left(\begin{bmatrix} 1 & 3 & 5 \\ 2 & 4 & 6 \end{bmatrix} \begin{bmatrix} 1 & 2 \\ 3 & 4 \\ 5 & 6 \end{bmatrix} \right)^{-1} \begin{bmatrix} 1 & 3 & 5 \\ 2 & 4 & 6 \end{bmatrix} = \frac{1}{24} \begin{bmatrix} -32 & -8 & 16 \\ 26 & 8 & -10 \end{bmatrix}$

$$= \frac{1}{12} \begin{bmatrix} -16 & -4 & 8 \\ 13 & 4 & -5 \end{bmatrix}.$$

5. $\text{Pinv} \begin{bmatrix} 1 & 3 \\ 2 & 0 \\ 1 & -1 \end{bmatrix} = \left(\begin{bmatrix} 1 & 2 & 1 \\ 3 & 0 & -1 \end{bmatrix} \begin{bmatrix} 1 & 3 \\ 2 & 0 \\ 1 & -1 \end{bmatrix} \right)^{-1} \begin{bmatrix} 1 & 2 & 1 \\ 3 & 0 & -1 \end{bmatrix} = \frac{1}{56} \begin{bmatrix} 4 & 20 & 12 \\ 16 & -4 & -8 \end{bmatrix}$

$$= \frac{1}{14} \begin{bmatrix} 1 & 5 & 3 \\ 4 & -1 & -2 \end{bmatrix}.$$

6. There is no pseudoinverse. This is not the coefficient matrix of an overdetermined system of equations.

7. There is no pseudoinverse. This is not the coefficient matrix of an overdetermined system of equations.

8. $\text{Pinv} \begin{bmatrix} 1 & 2 \\ -3 & 5 \end{bmatrix} = \left(\begin{bmatrix} 1 & -3 \\ 2 & 5 \end{bmatrix} \begin{bmatrix} 1 & 2 \\ -3 & 5 \end{bmatrix} \right)^{-1} \begin{bmatrix} 1 & -3 \\ 2 & 5 \end{bmatrix} = \frac{1}{11} \begin{bmatrix} 5 & -2 \\ 3 & 1 \end{bmatrix}.$

9. The equation of the line will be $y = a + bx$. The data points give $0 = a + b$, $4 = a + 2b$, and

$7 = a + 3b$. Thus the system of equations is given by $A\begin{bmatrix} a \\ b \end{bmatrix} = \begin{bmatrix} 0 \\ 4 \\ 7 \end{bmatrix}$, where $A = \begin{bmatrix} 1 & 1 \\ 1 & 2 \\ 1 & 3 \end{bmatrix}$.

$\text{Pinv } A = \left(\begin{bmatrix} 1 & 1 & 1 \\ 1 & 2 & 3 \end{bmatrix} \begin{bmatrix} 1 & 1 \\ 1 & 2 \\ 1 & 3 \end{bmatrix} \right)^{-1} \begin{bmatrix} 1 & 1 & 1 \\ 1 & 2 & 3 \end{bmatrix} = \frac{1}{6} \begin{bmatrix} 8 & 2 & -4 \\ -3 & 0 & 3 \end{bmatrix}$.

$\text{Pinv } A \begin{bmatrix} 0 \\ 4 \\ 7 \end{bmatrix} = \frac{1}{6} \begin{bmatrix} -20 \\ 21 \end{bmatrix}$, so $a = \frac{-10}{3}$, $b = \frac{7}{2}$, and the least squares line is $y = \frac{-10}{3} + \frac{7}{2} x$.

10. The system of equations is given by $A \begin{bmatrix} a \\ b \end{bmatrix} = \begin{bmatrix} 1 \\ 3 \\ 7 \end{bmatrix}$, with A given in Exercise 9.

$\text{Pinv } A \begin{bmatrix} 1 \\ 3 \\ 7 \end{bmatrix} = \frac{1}{6} \begin{bmatrix} 8 & 2 & -4 \\ -3 & 0 & 3 \end{bmatrix} \begin{bmatrix} 1 \\ 3 \\ 7 \end{bmatrix} = \frac{1}{6} \begin{bmatrix} -14 \\ 18 \end{bmatrix}$, so

$a = -7/3$, $b = 3$, and the least squares line is $y = -7/3 + 3x$.

11. The system of equations is given by $A \begin{bmatrix} a \\ b \end{bmatrix} = \begin{bmatrix} 1 \\ 5 \\ 9 \end{bmatrix}$, with A given in Exercise 9.

$\text{Pinv } A \begin{bmatrix} 1 \\ 5 \\ 9 \end{bmatrix} = \frac{1}{6} \begin{bmatrix} 8 & 2 & -4 \\ -3 & 0 & 3 \end{bmatrix} \begin{bmatrix} 1 \\ 5 \\ 9 \end{bmatrix} = \frac{1}{6} \begin{bmatrix} -18 \\ 24 \end{bmatrix}$, so $a = -3$, $b = 4$, and the least

squares line is $y = -3 + 4x$.

12. The system of equations is given by $A \begin{bmatrix} a \\ b \end{bmatrix} = \begin{bmatrix} 1 \\ 6 \\ 9 \end{bmatrix}$, with A given in Exercise 9.

$\text{Pinv } A \begin{bmatrix} 1 \\ 6 \\ 9 \end{bmatrix} = \frac{1}{6} \begin{bmatrix} 8 & 2 & -4 \\ -3 & 0 & 3 \end{bmatrix} \begin{bmatrix} 1 \\ 6 \\ 9 \end{bmatrix} = \frac{1}{6} \begin{bmatrix} -16 \\ 24 \end{bmatrix}$, so

$a = -8/3$, $b = 4$, and the least squares line is $y = -8/3 + 4x$.

13. The system of equations is given by $A\begin{bmatrix} a \\ b \end{bmatrix} = \begin{bmatrix} 7 \\ 2 \\ 0 \end{bmatrix}$, with A given in Exercise 9.

$$\text{Pinv } A \begin{bmatrix} 7 \\ 2 \\ 0 \end{bmatrix} = \frac{1}{6}\begin{bmatrix} 8 & 2 & -4 \\ -3 & 0 & 3 \end{bmatrix}\begin{bmatrix} 7 \\ 2 \\ 0 \end{bmatrix} = \frac{1}{6}\begin{bmatrix} 60 \\ -21 \end{bmatrix}, \text{ so}$$

$a = 10$, $b = -7/2$, and the least squares line is $y = 10 - 7x/2$.

14. The system of equations is given by $A\begin{bmatrix} a \\ b \end{bmatrix} = \begin{bmatrix} 12 \\ 5 \\ 1 \end{bmatrix}$, with A given in Exercise 9.

$$\text{Pinv } A \begin{bmatrix} 12 \\ 5 \\ 1 \end{bmatrix} = \frac{1}{6}\begin{bmatrix} 8 & 2 & -4 \\ -3 & 0 & 3 \end{bmatrix}\begin{bmatrix} 12 \\ 5 \\ 1 \end{bmatrix} = \frac{1}{6}\begin{bmatrix} 102 \\ -33 \end{bmatrix}, \text{ so}$$

$a = 17$, $b = -11/2$, and the least squares line is $y = 17 - 11x/2$.

15. The equation of the line will be $y = a + bx$. The data points give $0 = a + b$, $3 = a + 2b$, $5 = a + 3b$, and $6 = a + 4b$. Thus the system of equations is given by $A\begin{bmatrix} a \\ b \end{bmatrix} = \begin{bmatrix} 0 \\ 3 \\ 5 \\ 6 \end{bmatrix}$,

where $A = \begin{bmatrix} 1 & 1 \\ 1 & 2 \\ 1 & 3 \\ 1 & 4 \end{bmatrix}$. Pinv $A = \left(\begin{bmatrix} 1 & 1 & 1 & 1 \\ 1 & 2 & 3 & 4 \end{bmatrix} \begin{bmatrix} 1 & 1 \\ 1 & 2 \\ 1 & 3 \\ 1 & 4 \end{bmatrix} \right)^{-1} \begin{bmatrix} 1 & 1 & 1 & 1 \\ 1 & 2 & 3 & 4 \end{bmatrix}$

$= \frac{1}{10}\begin{bmatrix} 10 & 5 & 0 & -5 \\ -3 & -1 & 1 & 3 \end{bmatrix}$. Pinv $A \begin{bmatrix} 0 \\ 3 \\ 5 \\ 6 \end{bmatrix} = \frac{1}{10}\begin{bmatrix} -15 \\ 20 \end{bmatrix}$, so

$a = -3/2$, $b = 2$, and the least squares line is $y = -3/2 + 2x$.

Section 8.4

16. The system of equations is given by $A\begin{bmatrix}a\\b\end{bmatrix} = \begin{bmatrix}1\\3\\5\\9\end{bmatrix}$, with A given in Exercise 15.

    Pinv $A\begin{bmatrix}1\\3\\5\\9\end{bmatrix} = \frac{1}{10}\begin{bmatrix}10 & 5 & 0 & -5\\-3 & -1 & 1 & 3\end{bmatrix}\begin{bmatrix}1\\3\\5\\9\end{bmatrix} = \frac{1}{10}\begin{bmatrix}-20\\26\end{bmatrix}$, so

    $a = -2$, $b = 13/5$, and the least squares line is $y = -2 + 13x/5$.

17. The system of equations is given by $A\begin{bmatrix}a\\b\end{bmatrix} = \begin{bmatrix}9\\7\\3\\2\end{bmatrix}$, with A given in Exercise 15.

    Pinv $A\begin{bmatrix}9\\7\\3\\2\end{bmatrix} = \frac{1}{10}\begin{bmatrix}10 & 5 & 0 & -5\\-3 & -1 & 1 & 3\end{bmatrix}\begin{bmatrix}9\\7\\3\\2\end{bmatrix} = \frac{1}{2}\begin{bmatrix}23\\-5\end{bmatrix}$, so

    $a = 23/2$, $b = -5/2$, and the least squares line is $y = 23/2 - 5x/2$.

18. The system of equations is given by $A\begin{bmatrix}a\\b\end{bmatrix} = \begin{bmatrix}21\\17\\12\\7\end{bmatrix}$, with A given in Exercise 15.

    Pinv $A\begin{bmatrix}21\\17\\12\\7\end{bmatrix} = \frac{1}{10}\begin{bmatrix}10 & 5 & 0 & -5\\-3 & -1 & 1 & 3\end{bmatrix}\begin{bmatrix}21\\17\\12\\7\end{bmatrix} = \frac{1}{10}\begin{bmatrix}260\\-47\end{bmatrix}$, so

    $a = 26$, $b = -47/10$, and the least squares line is $y = 26 - 47x/10$.

19. The equation of the line will be $y = a + bx$. The data points give $0 = a + b$, $1 = a + 2b$, $4 = a + 3b$, $7 = a + 4b$, and $9 = a + 5b$. Thus the system of equations is given by

$$A\begin{bmatrix}a\\b\end{bmatrix} = \begin{bmatrix}0\\1\\4\\7\\9\end{bmatrix}, \text{ where } A = \begin{bmatrix}1&1\\1&2\\1&3\\1&4\\1&5\end{bmatrix}.$$

$$\text{Pinv } A = \left(\begin{bmatrix}1&1&1&1&1\\1&2&3&4&5\end{bmatrix}\begin{bmatrix}1&1\\1&2\\1&3\\1&4\\1&5\end{bmatrix}\right)^{-1}\begin{bmatrix}1&1&1&1&1\\1&2&3&4&5\end{bmatrix}$$

$$= \frac{1}{10}\begin{bmatrix}8&5&2&-1&-4\\-2&-1&0&1&2\end{bmatrix}. \text{ Pinv } A\begin{bmatrix}0\\1\\4\\7\\9\end{bmatrix} = \frac{1}{10}\begin{bmatrix}-30\\24\end{bmatrix}, \text{ so}$$

$a = -3$, $b = 12/5$, and the least squares line is $y = -3 + 12x/5$.

20. The system of equations is given by $A\begin{bmatrix}a\\b\end{bmatrix} = \begin{bmatrix}1\\3\\4\\8\\11\end{bmatrix}$, with A given in Exercise 19.

$$\text{Pinv } A\begin{bmatrix}1\\3\\4\\8\\11\end{bmatrix} = \frac{1}{10}\begin{bmatrix}8&5&2&-1&-4\\-2&-1&0&1&2\end{bmatrix}\begin{bmatrix}1\\3\\4\\8\\11\end{bmatrix} = \frac{1}{10}\begin{bmatrix}-21\\25\end{bmatrix}, \text{ so}$$

$a = -21/10$, $b = 5/2$, and the least squares line is $y = -21/10 + 5x/2$.

21. The equation of the parabola will be $y = a + bx + cx^2$. The data points give $5 = a + b + c$, $2 = a + 2b + 4c$, $3 = a + 3b + 9c$, and $8 = a + 4b + 16c$. Thus the system of equations is

Section 8.4

given by $A\begin{bmatrix}a\\b\\c\end{bmatrix} = \begin{bmatrix}5\\2\\3\\8\end{bmatrix}$, where $A = \begin{bmatrix}1 & 1 & 1\\1 & 2 & 4\\1 & 3 & 9\\1 & 4 & 16\end{bmatrix}$.

$\text{Pinv } A = \left(\begin{bmatrix}1 & 1 & 1 & 1\\1 & 2 & 3 & 4\\1 & 4 & 9 & 16\end{bmatrix}\begin{bmatrix}1 & 1 & 1\\1 & 2 & 4\\1 & 3 & 9\\1 & 4 & 16\end{bmatrix}\right)^{-1}\begin{bmatrix}1 & 1 & 1 & 1\\1 & 2 & 3 & 4\\1 & 4 & 9 & 16\end{bmatrix}$

$= \begin{bmatrix}4 & 10 & 30\\10 & 30 & 100\\30 & 100 & 354\end{bmatrix}^{-1}\begin{bmatrix}1 & 1 & 1 & 1\\1 & 2 & 3 & 4\\1 & 4 & 9 & 16\end{bmatrix} = \frac{1}{80}\begin{bmatrix}620 & -540 & 100\\-540 & 516 & -100\\100 & -100 & 20\end{bmatrix}\begin{bmatrix}1 & 1 & 1 & 1\\1 & 2 & 3 & 4\\1 & 4 & 9 & 16\end{bmatrix}$

$= \frac{1}{20}\begin{bmatrix}45 & -15 & -25 & 15\\-31 & 23 & 27 & -19\\5 & -5 & -5 & 5\end{bmatrix}$. $\text{Pinv } A\begin{bmatrix}5\\2\\3\\8\end{bmatrix} = \begin{bmatrix}12\\-9\\2\end{bmatrix}$, so

$a = 12$, $b = -9$, $c = 2$, and the least squares parabola is $y = 12 - 9x + 2x^2$.

22. The system of equations is given by $A\begin{bmatrix}a\\b\\c\end{bmatrix} = \begin{bmatrix}4\\0\\3\\5\end{bmatrix}$ with A given in Exercise 21.

$\text{Pinv } A\begin{bmatrix}4\\0\\3\\5\end{bmatrix} = \frac{1}{20}\begin{bmatrix}45 & -15 & -25 & 15\\-31 & 23 & 27 & -19\\5 & -5 & -5 & 5\end{bmatrix}\begin{bmatrix}4\\0\\3\\5\end{bmatrix} = \frac{1}{20}\begin{bmatrix}180\\-138\\30\end{bmatrix}$, so

$a = 9$, $b = -69/10$, $c = 3/2$, and the least squares parabola is $y = 9 - 69x/10 + 3x^2/2$.

23. The system of equations is given by $A\begin{bmatrix}a\\b\\c\end{bmatrix} = \begin{bmatrix}2\\5\\7\\1\end{bmatrix}$ with A given in Exercise 21.

$\text{Pinv } A\begin{bmatrix}2\\5\\7\\1\end{bmatrix} = \frac{1}{20}\begin{bmatrix}45 & -15 & -25 & 15\\-31 & 23 & 27 & -19\\5 & -5 & -5 & 5\end{bmatrix}\begin{bmatrix}2\\5\\7\\1\end{bmatrix} = \frac{1}{20}\begin{bmatrix}-145\\223\\-45\end{bmatrix}$, so $a = \frac{-29}{4}$, $b = \frac{223}{20}$,

Section 8.4

$c = \frac{-9}{4}$, and the least squares parabola is $y = \frac{-29}{4} + \frac{223}{20}x - \frac{9}{4}x^2$.

24. The system of equations is given by $A\begin{bmatrix} a \\ b \\ c \end{bmatrix} = \begin{bmatrix} 23 \\ 4 \\ 7 \\ 17 \end{bmatrix}$ with A given in Exercise 21.

$\text{Pinv } A \begin{bmatrix} 23 \\ 4 \\ 7 \\ 17 \end{bmatrix} = \frac{1}{20}\begin{bmatrix} 45 & -15 & -25 & 15 \\ -31 & 23 & 27 & -19 \\ 5 & -5 & -5 & 5 \end{bmatrix}\begin{bmatrix} 23 \\ 4 \\ 7 \\ 17 \end{bmatrix} = \frac{1}{20}\begin{bmatrix} 1055 \\ -755 \\ 145 \end{bmatrix}$, so $a = \frac{211}{4}$,

$b = \frac{-151}{4}$, $c = \frac{29}{4}$, and the least squares parabola is $y = \frac{211}{4} - \frac{151}{4}x + \frac{29}{4}x^2$.

25. (a) The equation of the line will be $L = a + bF$. The data points give $5.9 = a + 2b$, $8 = a + 4b$, $10.3 = a + 6b$, and $12.2 = a + 8b$. Thus the system of equations is given

by $A\begin{bmatrix} a \\ b \end{bmatrix} = \begin{bmatrix} 5.9 \\ 8 \\ 10.3 \\ 12.2 \end{bmatrix}$, where $A = \begin{bmatrix} 1 & 2 \\ 1 & 4 \\ 1 & 6 \\ 1 & 8 \end{bmatrix}$. $\text{Pinv } A = \begin{bmatrix} 1 & .5 & 0 & -.5 \\ -.15 & -.05 & .05 & .15 \end{bmatrix}$,

as given in the text. $\text{Pinv } A \begin{bmatrix} 5.9 \\ 8 \\ 10.3 \\ 12.2 \end{bmatrix} = \begin{bmatrix} 3.8 \\ 1.06 \end{bmatrix}$, so $a = 3.8$, $b = 1.06$, and the least

squares equation is $L = 3.8 + 1.06F$. If the force is 15 ounces, then the predicted length of the spring is $L = 3.8 + 1.06 \times 15 = 19.7$ inches.

(b) The equation of the line will be $L = a + bF$. The data points give $7.4 = a + 2b$, $10.3 = a + 4b$, $13.7 = a + 6b$, and $16.7 = a + 8b$. Thus the system of equations is given by $A\begin{bmatrix} a \\ b \end{bmatrix} = \begin{bmatrix} 7.4 \\ 10.3 \\ 13.7 \\ 16.7 \end{bmatrix}$, where $A = \begin{bmatrix} 1 & 2 \\ 1 & 4 \\ 1 & 6 \\ 1 & 8 \end{bmatrix}$.

Pinv $A = \begin{bmatrix} 1 & .5 & 0 & -.5 \\ -.15 & -.05 & .05 & .15 \end{bmatrix}$, as given in the text.

Pinv $A \begin{bmatrix} 7.4 \\ 10.3 \\ 13.7 \\ 16.7 \end{bmatrix} = \begin{bmatrix} 4.2 \\ 1.565 \end{bmatrix}$, so a = 4.2, b = 1.565, and the least squares equation is

L = 4.2 + 1.565F. If the force is 15 ounces, then the predicted length of the spring is L = 4.2 + 1.565×15 = 27.675 inches.

26. The equation of the line will be G = a + bS. The data points give 18.25 = a + 30b, 20 = a + 40b, 16.32 = a + 50b, 15.77 = a + 60b, and 13.61 = a + 70b. Thus the system of

equations is given by $A\begin{bmatrix} a \\ b \end{bmatrix} = \begin{bmatrix} 18.25 \\ 20.00 \\ 16.32 \\ 15.77 \\ 13.61 \end{bmatrix}$, where $A = \begin{bmatrix} 1 & 30 \\ 1 & 40 \\ 1 & 50 \\ 1 & 60 \\ 1 & 70 \end{bmatrix}$. Pinv $A = (A^t A)^{-1} A^t$

$= \dfrac{1}{5000} \begin{bmatrix} 13500 & -250 \\ -250 & 5 \end{bmatrix} \begin{bmatrix} 1 & 1 & 1 & 1 & 1 \\ 30 & 40 & 50 & 60 & 70 \end{bmatrix} = \dfrac{1}{50} \begin{bmatrix} 60 & 35 & 10 & -15 & -40 \\ 1 & .5 & 0 & .5 & 1 \end{bmatrix}$.

Pinv $A \begin{bmatrix} 18.25 \\ 20.00 \\ 16.32 \\ 15.77 \\ 13.61 \end{bmatrix} = \dfrac{1}{50} \begin{bmatrix} 1177.25 \\ -6.755 \end{bmatrix}$, so a = 23.545, b = −.1351, and the least squares

line is G = 23.545 − .1351S. The predicted mileage per gallon of this car at 55 mph is G = 23.545 − .1351×55 = 16.1145 miles per gallon.

Section 8.4

27. The equation of the line will be $N = a + bX$. The data points give $101 = a + 20b$, $115 = a + 25b$, $92 = a + 30b$, $64 = a + 35b$, $60 = a + 40b$, $50 = a + 45b$, and $49 = a + 50b$.

Thus the system of equations is given by $A \begin{bmatrix} a \\ b \end{bmatrix} = \begin{bmatrix} 101 \\ 115 \\ 92 \\ 64 \\ 60 \\ 50 \\ 49 \end{bmatrix}$, where $A = \begin{bmatrix} 1 & 20 \\ 1 & 25 \\ 1 & 30 \\ 1 & 35 \\ 1 & 40 \\ 1 & 45 \\ 1 & 50 \end{bmatrix}$.

$\text{Pinv } A = (A^t A)^{-1} A^t = \dfrac{1}{4900} \begin{bmatrix} 9275 & -245 \\ -245 & 7 \end{bmatrix} \begin{bmatrix} 1 & 1 & 1 & 1 & 1 & 1 & 1 \\ 20 & 25 & 30 & 35 & 40 & 45 & 50 \end{bmatrix}$

$= \dfrac{1}{4900} \begin{bmatrix} 4375 & 3150 & 1925 & 700 & -525 & -1750 & -2975 \\ -105 & -70 & -35 & 0 & 35 & 70 & 105 \end{bmatrix}$.

$\text{Pinv } A \begin{bmatrix} 101 \\ 115 \\ 92 \\ 64 \\ 60 \\ 50 \\ 49 \end{bmatrix} = \dfrac{1}{4900} \begin{bmatrix} 761250 \\ -11130 \end{bmatrix}$, so $a = \dfrac{76125}{490}$, $b = \dfrac{-1113}{490}$, and the least

squares equation is $N = \dfrac{76125}{490} - \dfrac{1113}{490} X$. The predicted number of driver fatalities at age 22 is $\dfrac{76125 - 1113 \times 22}{490} = 105$.

28. The equation of the line will be $U = a + bD$. The data points give $520 = a + 1000b$, $540 = a + 1500b$, $582 = a + 2000b$, $600 = a + 2500b$, $610 = a + 3000b$, and $615 = a + 3500b$. Thus the system of equations is given by $A \begin{bmatrix} a \\ b \end{bmatrix} = \begin{bmatrix} 520 \\ 540 \\ 582 \\ 600 \\ 610 \\ 615 \end{bmatrix}$,

where $A = \begin{bmatrix} 1 & 1000 \\ 1 & 1500 \\ 1 & 2000 \\ 1 & 2500 \\ 1 & 3000 \\ 1 & 3500 \end{bmatrix}$. $\text{Pinv } A = (A^t A)^{-1} A^t$

$= \dfrac{1}{26250000} \begin{bmatrix} 34750000 & -13500 \\ -13500 & 6 \end{bmatrix} \begin{bmatrix} 1 & 1 & 1 & 1 & 1 & 1 \\ 1000 & 1500 & 2000 & 2500 & 3000 & 3500 \end{bmatrix}$

$= \dfrac{1}{262500} \begin{bmatrix} 212500 & 145000 & 77500 & 10000 & -57500 & -125000 \\ -75 & -45 & -15 & 15 & 45 & 75 \end{bmatrix}$.

$\text{Pinv } A \begin{bmatrix} 520 \\ 540 \\ 582 \\ 600 \\ 610 \\ 615 \end{bmatrix} = \dfrac{1}{262500} \begin{bmatrix} 127955000 \\ 10545 \end{bmatrix}$, so $a = \dfrac{127955000}{262500}$, $b = \dfrac{10545}{262500}$,

and the least squares line is $U = \dfrac{127955000 + 10545 D}{262500}$. Predicted sales when

$5000 is spent on advertising is $U = \dfrac{127955000 + 10545 \times 5000}{262500} = 688$ units. This is

probably an unrealistic prediction. The data do not appear to be linear.

29. The equation of the line will be $G = a + bS$. The system of equations is given by

$A \begin{bmatrix} a \\ b \end{bmatrix} = C$, where $C = \begin{bmatrix} 3.2 \\ 2.8 \\ 3.9 \\ 2.4 \\ 4.0 \\ 3.5 \\ 2.7 \\ 3.6 \\ 3.5 \\ 2.4 \end{bmatrix}$ and $A = \begin{bmatrix} 1 & 490 \\ 1 & 450 \\ 1 & 640 \\ 1 & 510 \\ 1 & 680 \\ 1 & 610 \\ 1 & 480 \\ 1 & 450 \\ 1 & 600 \\ 1 & 650 \end{bmatrix}$. $(\text{Pinv } A) C = (A^t A)^{-1} A^t C$

Section 8.4

$$= \frac{1}{70840}\begin{bmatrix} 316220 & -556 \\ -556 & 1 \end{bmatrix} A^t C = \frac{1}{70840}\begin{bmatrix} 122716 \\ 187 \end{bmatrix},$$ so the least squares line is

given by $G = \frac{122716 + 187S}{70840}$. $G = 1.732298137 + 0.0026397516s$.

Prediction of the graduating GPA of a student entering with an SAT of S = 670 is G = 3.5.

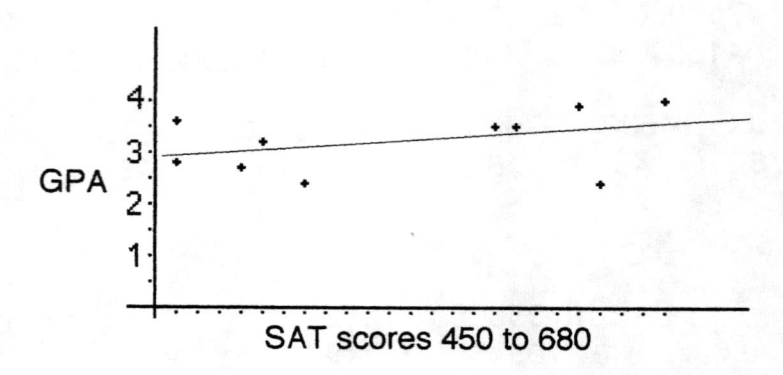

30. The equation of the least squares line will be % = a + bY. The system of equations is

given by $A\begin{bmatrix} a \\ b \end{bmatrix} = C$, where $C = \begin{bmatrix} .18 \\ .19 \\ .21 \\ .25 \\ .35 \\ .54 \\ .83 \end{bmatrix}$ and $A = \begin{bmatrix} 1 & 20 \\ 1 & 25 \\ 1 & 30 \\ 1 & 35 \\ 1 & 40 \\ 1 & 45 \\ 1 & 50 \end{bmatrix}$. Pinv $A = (A^t A)^{-1} A^t$

$$= \frac{1}{4900}\begin{bmatrix} 4375 & 3150 & 1925 & 700 & -525 & -1750 & -2975 \\ -105 & -70 & -35 & 0 & 35 & 70 & 105 \end{bmatrix},$$ as in Exercise 27.

(Pinv A) C $= \frac{1}{4900}\begin{bmatrix} -1632.75 \\ 97.65 \end{bmatrix}$, so the least squares line is given by

% $= \frac{-1632.75 + 97.65Y}{4900}$. The predicted percentage of deaths at 23 years of

age is $\frac{-1632.75 + 97.65 \times 23}{4900} = .125$.

It is clear from the drawing below that the least squares line does not fit the data well.

Section 8.4

The parabola fits better. The equation of the least squares parabola will be

$\% = a + bY + cY^2$. The system of equations is given by $A\begin{bmatrix} a \\ b \\ c \end{bmatrix} = C$, where

$A = \begin{bmatrix} 1 & 20 & 400 \\ 1 & 25 & 625 \\ 1 & 30 & 900 \\ 1 & 35 & 1225 \\ 1 & 40 & 1600 \\ 1 & 45 & 2025 \\ 1 & 50 & 2500 \end{bmatrix}$ and C is given above. $(\text{Pinv } A)C = \begin{bmatrix} .93643 \\ -.05907 \\ .00113 \end{bmatrix}$, so

the least squares parabola is $\% = .93643 - .05907Y + .00113Y^2$. The predicted percentage of deaths at 23 years of age is $.93643 - .05907 \times 23 + .00113 \times 529 = .17559$.

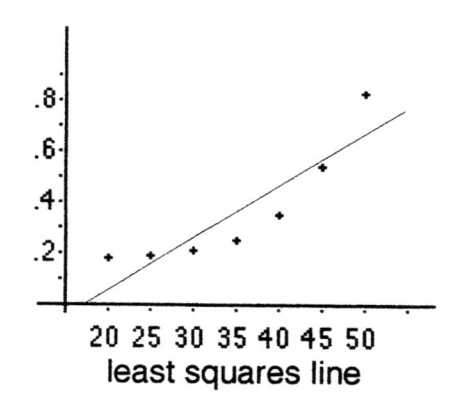

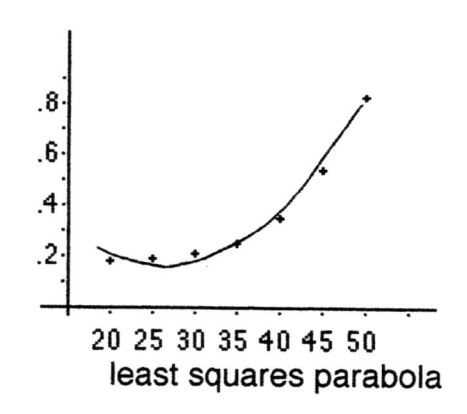

least squares line

least squares parabola

31. The cost-volume formula will be $C = a + bV$. The system of equations is given by

$A\begin{bmatrix} a \\ b \end{bmatrix} = D$, where $D = \begin{bmatrix} 6300 \\ 6500 \\ 6670 \\ 6450 \\ 6100 \\ 5950 \end{bmatrix}$ and $A = \begin{bmatrix} 1 & 120 \\ 1 & 140 \\ 1 & 155 \\ 1 & 135 \\ 1 & 110 \\ 1 & 105 \end{bmatrix}$. $(\text{Pinv } A)D = (A^tA)^{-1}A^tD$

$= \dfrac{1}{11025} \begin{bmatrix} 99375 & -765 \\ -765 & 6 \end{bmatrix} A^tD = \dfrac{1}{441} \begin{bmatrix} 2020440 \\ 6042 \end{bmatrix}$, so the least squares line is

given by $C = \dfrac{2020440 + 6042V}{441}$. For the next month, the predicted cost is

Section 8.4

$$C = \frac{2020440 + 6042 \times 165}{441} = \$6842.$$

32. Let the line be P = a + by. Let 1974 be year 0. Then 2000 is year 26. Use

$$A = \begin{bmatrix} 1 & 0 \\ 1 & 5 \\ 1 & 11 \\ 1 & 14 \\ 1 & 16 \\ 1 & 18 \\ 1 & 19 \end{bmatrix} \text{ and } y = \begin{bmatrix} 37.2 \\ 33.5 \\ 30.0 \\ 27.9 \\ 25.4 \\ 26.4 \\ 25.0 \end{bmatrix}.$$

Get Pin(A) = $\begin{bmatrix} .6133 & .4149 & .1769 & .0578 & -.0215 & -.1009 & .1405 \\ -.0397 & -.0229 & -.0029 & .0072 & .0139 & .0206 & .0239 \end{bmatrix}$

and Pinv(A)y = $\begin{bmatrix} 36.9111 \\ -.06383 \end{bmatrix}$. Thus P = 36.9111 - .06383y.

When y=26, P=20.3153. Prediction for the year 2000 is 20% smokers.

33. Let the line be E = a + by. Let 1940 be year 0. Then 2020 is year 80.

$$A = \begin{bmatrix} 1 & 0 \\ 1 & 10 \\ 1 & 20 \\ 1 & 30 \\ 1 & 40 \\ 1 & 50 \end{bmatrix} \text{ and } y = \begin{bmatrix} 62.9 \\ 68.2 \\ 69.7 \\ 70.8 \\ 73.7 \\ 75.4 \end{bmatrix}.$$

Get Pin(A) = $\begin{bmatrix} .5238 & .3810 & .2381 & .0952 & -.0476 & -.1905 \\ -.0143 & -.0086 & -.0029 & .0029 & .0086 & .0143 \end{bmatrix}$

and Pinv(A)y = $\begin{bmatrix} 64.3952 \\ .2289 \end{bmatrix}$. Thus E = 64.3952 + .2289y.

When y=80, E = 82.7072. Predicted life expectancy for the year 2020 is 82.7 years.

34. Let the lines be of the form y = a + bx with x measured from 1994. The statistics give

$$A = \begin{bmatrix} 1 & 0 \\ 1 & 1 \\ 1 & 2 \\ 1 & 3 \\ 1 & 4 \\ 1 & 5 \\ 1 & 6 \\ 1 & 7 \\ 1 & 8 \\ 1 & 9 \\ 1 & 10 \end{bmatrix} \text{ and } X = \begin{bmatrix} 6.831 & 8.174 \\ 6.781 & 8.165 \\ 6.752 & 8.186 \\ 6.765 & 8.234 \\ 6.811 & 8.300 \\ 6.892 & 8.412 \\ 6.970 & 8.492 \\ 7.041 & 8.566 \\ 7.116 & 8.622 \\ 7.161 & 8.641 \\ 7.216 & 8.676 \end{bmatrix}.$$

Get $\text{Pin}(A) = \begin{bmatrix} .3182 & .2727 & .2273 & .1818 & .1364 & .0909 & .0455 & 0 & .0455 & -.0909 & -.136 \\ -.0455 & -.0364 & -.0273 & -.0182 & -.0091 & 0 & .0091 & .0182 & .0273 & .0364 & .045 \end{bmatrix}$

and $\text{Pinv}(A)y = \begin{bmatrix} 6.7011 & 8.1072 \\ .0477 & .0598 \end{bmatrix}.$

Thus y=6.7011 + .0477x for men and y=8.1072+.0598x for women.

The slope of the women's line is .0598, greater than the slope .0477 for the men's line. The graphs will diverge. The gap between the number of women and men will get bigger.

35. The equations will be of the form $N = a + bY + cY^2$. Let 1950 be year 0. The system of equations for a, b, and c is $A \begin{bmatrix} a \\ b \\ c \end{bmatrix} = C$, where C is the appropriate column of

$$D = \begin{bmatrix} 554 & 84 & 122 & 166 & 2504 \\ 801 & 94 & 135 & 199 & 3014 \\ 873 & 99 & 143 & 214 & 3324 \\ 984 & 104 & 148 & 227 & 3683 \\ 1102 & 112 & 152 & 239 & 4076 \\ 1183 & 117 & 154 & 252 & 4453 \\ 1225 & 119 & 154 & 259 & 4685 \\ 1239 & 119 & 154 & 261 & 4763 \end{bmatrix}, \text{ and } A = \begin{bmatrix} 1 & 0 & 0 \\ 1 & 10 & 100 \\ 1 & 15 & 225 \\ 1 & 20 & 400 \\ 1 & 25 & 625 \\ 1 & 30 & 900 \\ 1 & 33 & 1089 \\ 1 & 34 & 1156 \end{bmatrix}.$$

Section 8.4

$$(\text{Pinv A})D = \begin{bmatrix} 555.61302 & 83.65595 & 121.31114 & 166.28642 & 2493.40194 \\ 24.11237 & 1.05255 & 1.78878 & 3.42106 & 48.06250 \\ -0.11360 & 0.00066 & -0.02385 & -0.01882 & 0.56248 \end{bmatrix},$$

so the least squares parabolas are

China $N = 555.61302 + 24.11237Y - 0.11360Y^2$
Japan $N = 83.65595 + 1.05255Y + 0.00066Y^2$
W. Europe $N = 121.31114 + 1.78878Y - 0.02385Y^2$
N. America $N = 166.28642 + 3.42106Y - 0.01882Y^2$
World $N = 2493.40194 + 48.06250Y + 0.56248Y^2$

The year 2000 is year 50. The least squares predictions are $N = 1477$ for China, 138 for Japan, 151 for W. Europe, 290 for N. America, and 6303 for the world.

36. (a) Let W be the plane given by the equation $x - y - z = 0$. For all points in this plane $x = y + z$, so $W = \{(y + z, y, z)\} = \{y(1,1,0) + z(1,0,1)\}$. Thus a basis for W is the set

$\{(1,1,0), (1,0,1)\}$ and $A = \begin{bmatrix} 1 & 1 \\ 1 & 0 \\ 0 & 1 \end{bmatrix}$. $(A^tA)^{-1} = \frac{1}{3}\begin{bmatrix} 2 & -1 \\ -1 & 2 \end{bmatrix}$, and the projection

matrix $P = A(A^tA)^{-1}A^t = \frac{1}{3}\begin{bmatrix} 2 & 1 & 1 \\ 1 & 2 & -1 \\ 1 & -1 & 2 \end{bmatrix}$. $P\begin{bmatrix} 1 \\ 2 \\ 0 \end{bmatrix} = \frac{1}{3}\begin{bmatrix} 4 \\ 5 \\ -1 \end{bmatrix}$, so the projection of

the vector $(1,2,0)$ onto the plane $x - y - z = 0$ is the vector $(4/3, 5/3, -1/3)$.

(b) Let W be the plane given by the equation $x - y + z = 0$. For all points in this plane $x = y - z$, so $W = \{(y - z, y, z)\} = \{y(1,1,0) + z(-1,0,1)\}$. Thus a basis for W is the set

$\{(1,1,0), (-1,0,1)\}$ and $A = \begin{bmatrix} 1 & -1 \\ 1 & 0 \\ 0 & 1 \end{bmatrix}$. $(A^tA)^{-1} = \frac{1}{3}\begin{bmatrix} 2 & 1 \\ 1 & 2 \end{bmatrix}$, and the projection

matrix $P = A(A^tA)^{-1}A^t = \frac{1}{3}\begin{bmatrix} 2 & 1 & -1 \\ 1 & 2 & 1 \\ -1 & 1 & 2 \end{bmatrix}$. $P\begin{bmatrix} 1 \\ 1 \\ 1 \end{bmatrix} = \frac{1}{3}\begin{bmatrix} 2 \\ 4 \\ 2 \end{bmatrix}$, so the projection of

the vector $(1,1,1)$ onto the plane $x - y + z = 0$ is the vector $(2/3, 4/3, 2/3)$.

(c) Let W be the plane given by the equation $x - 2y + z = 0$. For all points in this plane $x = 2y - z$, so $W = \{(2y - z, y, z)\} = \{y(2,1,0) + z(-1,0,1)\}$. Thus a basis for W is the set $\{(2,1,0), (-1,0,1)\}$ and $A = \begin{bmatrix} 2 & -1 \\ 1 & 0 \\ 0 & 1 \end{bmatrix}$. $(A^t A)^{-1} = \frac{1}{6} \begin{bmatrix} 2 & 2 \\ 2 & 5 \end{bmatrix}$, and the projection matrix $P = A(A^t A)^{-1} A^t = \frac{1}{6} \begin{bmatrix} 5 & 2 & -1 \\ 2 & 2 & 2 \\ -1 & 2 & 5 \end{bmatrix}$. $P \begin{bmatrix} 0 \\ 3 \\ 0 \end{bmatrix} = \frac{1}{6} \begin{bmatrix} 6 \\ 6 \\ 6 \end{bmatrix}$, so the projection of the vector $(0,3,0)$ onto the plane $x - 2y + z = 0$ is the vector $(1,1,1)$.

37. (a) $A = \begin{bmatrix} 1 & 1 \\ -1 & 1 \\ 1 & -1 \end{bmatrix}$ and $(A^t A)^{-1} = \frac{1}{8} \begin{bmatrix} 3 & 1 \\ 1 & 3 \end{bmatrix}$. Thus the projection matrix

$P = A(A^t A)^{-1} A^t = \frac{1}{8} \begin{bmatrix} 8 & 0 & 0 \\ 0 & 4 & -4 \\ 0 & -4 & 4 \end{bmatrix}$. $P \begin{bmatrix} -2 \\ 1 \\ 3 \end{bmatrix} = \frac{1}{8} \begin{bmatrix} -16 \\ -8 \\ 8 \end{bmatrix}$, so the projection of

the vector $(-2,1,3)$ onto W is the vector $(-2,-1,1)$.

(b) $A = \begin{bmatrix} 1 & 2 \\ 0 & 1 \\ 1 & 0 \end{bmatrix}$ and $(A^t A)^{-1} = \frac{1}{6} \begin{bmatrix} 5 & -2 \\ -2 & 2 \end{bmatrix}$. Thus the projection matrix

$P = A(A^t A)^{-1} A^t = \frac{1}{6} \begin{bmatrix} 5 & 2 & 1 \\ 2 & 2 & -2 \\ 1 & -2 & 5 \end{bmatrix}$. $P \begin{bmatrix} 6 \\ 0 \\ 12 \end{bmatrix} = \frac{1}{6} \begin{bmatrix} 42 \\ -12 \\ 66 \end{bmatrix}$, so the projection of

the vector $(6,0,12)$ onto W is the vector $(7,-2,11)$.

38. $A = \begin{bmatrix} 1 & 2 \\ 1 & 1 \\ 1 & 0 \end{bmatrix}$ and $(A^tA)^{-1} = \frac{1}{6}\begin{bmatrix} 5 & -3 \\ -3 & 3 \end{bmatrix}$. Thus the projection matrix

$P = A(A^tA)^{-1}A^t = \frac{1}{6}\begin{bmatrix} 5 & 2 & -1 \\ 2 & 2 & 2 \\ -1 & 2 & 5 \end{bmatrix}$. $P\begin{bmatrix} 3 \\ 0 \\ 1 \end{bmatrix} = \frac{1}{6}\begin{bmatrix} 14 \\ 8 \\ 2 \end{bmatrix}$, so the projection of the vector

(3,0,1) onto W is the vector (7/3,4/3,1/3). $P\begin{bmatrix} 6 \\ 3 \\ 0 \end{bmatrix} = \frac{1}{6}\begin{bmatrix} 36 \\ 18 \\ 0 \end{bmatrix} = \begin{bmatrix} 6 \\ 3 \\ 0 \end{bmatrix}$, so the projection

of the vector (6,3,0) onto W is the vector (6,3,0). This is because (6,3,0) is a multiple of a basis vector and is therefore in W.

39. $Ax = y$, so $A^tAx = A^ty$, and if $(A^tA)^{-1}$ exists then $x = (A^tA)^{-1}A^tAx = (A^tA)^{-1}A^ty$.

$x = (A^tA)^{-1}A^ty$ solves the system $A^tAx = A^ty$ but does not solve the original system unless A^t (and therefore A) is an invertible matrix. The first step of multiplying by A^t is not reversible unless A^t is invertible.

40. If A is invertible then A^{-1} and $(A^t)^{-1}$ exist and $(A^tA)^{-1} = A^{-1}(A^t)^{-1}$ so that

pinv(A) = $(A^tA)^{-1}A^t = A^{-1}(A^t)^{-1}A^t = A^{-1}$.

41. pinv(AB) = $((AB)^t(AB))^{-1}(AB)^t$. If A is mxn and B is nxk then pinv(A) is nxm and pinv(B) is kxn, AB is mxk, and pinv (AB) is kxm. In general pinv(A)pinv(B) does not exist and pinv(B)pinv(A) is kxm, so (a) is not possible, but (b) is. In fact (b) obviously holds if A and B are invertible. (b) also holds for some fairly simple examples where A and B are

not invertible, such as $A = \begin{bmatrix} 1 & 0 & 0 \\ 0 & 1 & 0 \\ 0 & 0 & 1 \\ 0 & 0 & 0 \end{bmatrix}$ and $B = \begin{bmatrix} 1 & 0 \\ 0 & 1 \\ 0 & 0 \end{bmatrix}$. However it is easy to find examples where (b) does not hold. If $A = \begin{bmatrix} 1 & 0 & 1 \\ 1 & 0 & 0 \\ 1 & 1 & 0 \\ 0 & 1 & 0 \end{bmatrix}$ and $B = \begin{bmatrix} 1 & 1 \\ 0 & 1 \\ 1 & 0 \end{bmatrix}$ then

$$\text{pinv}(B)\text{pinv}(A) = \frac{1}{3}\begin{bmatrix} 1 & -1 & 2 \\ 1 & 2 & -1 \end{bmatrix} \frac{1}{3}\begin{bmatrix} 0 & 2 & 1 & -1 \\ 0 & -1 & 1 & 2 \\ 3 & -2 & -1 & 1 \end{bmatrix} = \frac{1}{9}\begin{bmatrix} 6 & -1 & -2 & -1 \\ -3 & 2 & 4 & 2 \end{bmatrix}$$ and

$$\text{pinv}(AB) = \frac{1}{17}\begin{bmatrix} 9 & 2 & -3 & -5 \\ -4 & 1 & 7 & 6 \end{bmatrix}.$$

42. (a) $\text{Pinv}(cA) = ((cA)^t(cA))^{-1}(cA)^t = (c^2 A^t A)^{-1} cA^t = \frac{1}{c^2}(A^t A)^{-1} cA^t = \frac{1}{c}(A^t A)^{-1} A^t$.

    (b) Pinv(A) exists if and only if $(A^t A)^{-1}$ exists, and $(A^t A)^{-1}$ exists if and only if $|A^t A| \neq 0$. Since A (and therefore A^t) is square, $|A^t A| = |A^t||A|$, so $|A^t A| \neq 0$ if and only if $|A| = |A^t| \neq 0$. Thus pinv(A) exists if and only if A is invertible, and therefore pinv(A) = A^{-1} (see Exercise 40). Thus pinv(pinv(A)) = pinv(A^{-1}) = $(A^{-1})^{-1}$ = A.

    (c) As in (b), pinv(A) = A^{-1} and pinv(A^t) = $(A^t)^{-1}$ = $(A^{-1})^t$ = (pinv(A))^t.

43. If A is m×n, then A^t is n×m and $A^t A$ is n×n. Pinv(A) exists if and only if $(A^t A)^{-1}$ exists, and $(A^t A)^{-1}$ exists if and only if $A^t A$ has rank n. We must show that $A^t A$ has rank n if and only if A has rank n. A is the matrix of a linear transformation T from $\mathbf{R}^n$ to $\mathbf{R}^m$, A^t is the matrix of a linear transformation S from $\mathbf{R}^m$ to $\mathbf{R}^n$, and $A^t A$ is the matrix of the linear transformation ST from $\mathbf{R}^n$ to $\mathbf{R}^n$. Thus dim range(T) + dim ker(T) = dim domain(T) = dim($\mathbf{R}^n$) = n and dim range(ST) + dim ker(ST) = dim domain(ST) = dim($\mathbf{R}^n$) = n. Dim range(T) = rank(A) and dim range(ST) = rank($A^t A$), so rank(A) + dim ker(T) = n

Chapter 8 Review Exercises

$= \text{rank}(A^tA) + \dim \ker(ST)$. We show that $\ker(T) = \ker(ST)$.

If $\mathbf{x}$ is in $\ker(T)$ then $A\mathbf{x} = \mathbf{0}$ and $A^tA\mathbf{x} = \mathbf{0}$ so $\mathbf{x}$ is in $\ker(ST)$. If $\mathbf{x}$ is in $\ker(ST)$ then $A^tA\mathbf{x} = \mathbf{0}$ and $\mathbf{x}^tA^tA\mathbf{x} = 0$. But $\mathbf{x}^tA^tA\mathbf{x} = (A\mathbf{x})^tA\mathbf{x} = \|A\mathbf{x}\|^2$ and if $\|A\mathbf{x}\|^2 = 0$ then $\|A\mathbf{x}\| = 0$ and $A\mathbf{x} = \mathbf{0}$. Thus $\mathbf{x}$ is in $\ker(T)$. So $\dim \ker(T) = 0$ if and only if $\dim \ker(ST) = 0$ and $\text{rank}(A) = n$ if and only if $\text{rank}(A^tA) = n$.

If $m < n$, then $\text{rank}(A) \leq m < n$, so $\text{pinv}(A)$ does not exist.

44. $P = A(A^tA)^{-1}A^t$, where the columns of A are a basis for the vector space.

   (a) $P^t = (A(A^tA)^{-1}A^t)^t = (A^t)^t((A^tA)^{-1})^t A^t = A((A^tA)^t)^{-1}A^t = A(A^tA)^{-1}A^t = P$.

   (b) $P^2 = (A(A^tA)^{-1}A^t)(A(A^tA)^{-1}A^t) = A(A^tA)^{-1}(A^tA)(A^tA)^{-1}A^t = A(A^tA)^{-1}A^t = P$.

Chapter 8 Review Exercises

1. $\mathbf{u} = (x_1, x_2)$, $\mathbf{v} = (y_1, y_2)$ $\mathbf{w} = (z_1, z_2)$, and c is a scalar.

   $\langle \mathbf{u}, \mathbf{v} \rangle = 2x_1 y_1 + 3x_2 y_2 = 2y_1 x_1 + 3y_2 x_2 = \langle \mathbf{v}, \mathbf{u} \rangle$,

   $\langle \mathbf{u}+\mathbf{v}, \mathbf{w} \rangle = 2(x_1 + y_1)z_1 + 3(x_2 + y_2)z_2 = 2x_1 z_1 + 2y_1 z_1 + 3x_2 z_2 + 3y_2 z_2$

   $= 2x_1 z_1 + 3x_2 z_2 + 2y_1 z_1 + 3y_2 z_2 = \langle \mathbf{u}, \mathbf{w} \rangle + \langle \mathbf{v}, \mathbf{w} \rangle$,

   $\langle c\mathbf{u}, \mathbf{v} \rangle = 2cx_1 y_1 + 3cx_2 y_2 = c(2x_1 y_1 + 3x_2 y_2) = c\langle \mathbf{u}, \mathbf{v} \rangle$, and

   $\langle \mathbf{u}, \mathbf{u} \rangle = 2x_1 x_1 + 3x_2 x_2 = 2x_1^2 + 3x_2^2 \geq 0$ and equality holds if and only if $2x_1^2 = 0$ and $3x_2^2 = 0$, i.e. if and only if $x_1 = x_2 = 0$. Thus the given function is an inner product on $\mathbf{R}^2$.

2. $\langle f, g \rangle = \int_0^1 (3x-1)(5x+3)\,dx = \int_0^1 (15x^2 + 4x - 3)\,dx = [5x^3 + 2x^2 - 3x]_0^1 = 4$.

Chapter 8 Review Exercises

$\|f\|^2 = \int_0^1 (3x-1)^2 \, dx = \int_0^1 (9x^2-6x+1) \, dx = [3x^3-3x^2+x]_0^1 = 1$, so $\|f\| = 1$.

$\|g\|^2 = \int_0^1 (5x+3)^2 \, dx = \int_0^1 (25x^2+30x+9) \, dx = \left[\frac{25}{3}x^3 + 15x^2 + 9x\right]_0^1 = \frac{97}{3}$, so $\|g\| = \frac{\sqrt{97}}{\sqrt{3}}$.

$d(f,g)^2 = \|f-g\|^2 = \|-2x-4\|^2 = \int_0^1 (-2x-4)^2 \, dx = 4\int_0^1 (x^2+4x+4) \, dx$

$= 4\left[\frac{x^3}{3} + 2x^2 + 4x\right]_0^1 = 4\left(\frac{19}{3}\right)$, so $d(f,g) = 2\frac{\sqrt{19}}{\sqrt{3}}$.

3. $\|(1,-2)\| = \sqrt{2(1)+3(4)} = \sqrt{14}$, $\|(3,2)\| = \sqrt{2(9)+3(4)} = \sqrt{30}$, and

   $d((1,-2),(3,2)) = \|(-2,-4)\| = \sqrt{2(4)+3(16)} = \sqrt{56} = 2\sqrt{14}$.

4. $d((x_1,x_2),(0,0))^2 = \|(x_1,x_2)\|^2 = 2x_1^2 + 3x_2^2$, so the equation of the circle with radius 1 and center at the origin is $2x_1^2 + 3x_2^2 = 1$. In the sketch below, $a = 1/\sqrt{2}$ and $b = 1/\sqrt{3}$.

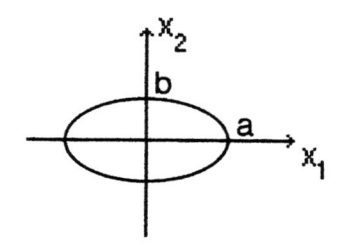

5. The set $\left\{\frac{1}{\sqrt{2}}, \frac{\sqrt{3}}{\sqrt{2}}x\right\}$ is an orthonormal basis for $P_1[-1,1]$ (see the text).

   $\left\langle x^2 + 2x - 1, \frac{1}{\sqrt{2}}\right\rangle = \frac{1}{\sqrt{2}}\int_{-1}^1 (x^2 + 2x - 1) \, dx = \frac{1}{\sqrt{2}}\left[\frac{x^3}{3} + x^2 - x\right]_{-1}^1 = -\frac{4}{3\sqrt{2}}$ and

   $\left\langle x^2 + 2x - 1, \frac{\sqrt{3}}{\sqrt{2}}x\right\rangle = \frac{\sqrt{3}}{\sqrt{2}}\int_{-1}^1 (x^3 + 2x^2 - x) \, dx = \frac{\sqrt{3}}{\sqrt{2}}\left[\frac{x^4}{4} + 2\frac{x^3}{3} - \frac{x^2}{2}\right]_{-1}^1$

   $= \frac{\sqrt{3}}{\sqrt{2}}\frac{4}{3}$, so that $\text{proj}_{P_1[-1,1]} x^2 + 2x - 1 = -\frac{4}{3\sqrt{2}}\frac{1}{\sqrt{2}} + \frac{\sqrt{3}}{\sqrt{2}}\frac{4}{3}\frac{\sqrt{3}}{\sqrt{2}}x = 2x - \frac{2}{3}$

   is the least squares linear approximation to $f(x) = x^2 + 2x - 1$ over the interval $[-1,1]$.

6. $a_0 = \dfrac{1}{2\pi} \displaystyle\int_0^{2\pi} (2x-1)\, dx = \dfrac{1}{2\pi} [x^2 - x]_0^{2\pi} = 2\pi - 1,$

$a_k = \dfrac{1}{\pi} \displaystyle\int_0^{2\pi} (2x-1)\cos kx\, dx = \dfrac{1}{\pi}\left[\dfrac{2x}{k}\sin kx + \dfrac{2}{k^2}\cos kx - \dfrac{1}{k}\sin kx\right]_0^{2\pi} = 0,$ and

$b_k = \dfrac{1}{\pi} \displaystyle\int_0^{2\pi} (2x-1)\sin kx\, dx = \dfrac{1}{\pi}\left[-\dfrac{2x}{k}\cos kx + \dfrac{2}{k^2}\sin kx + \dfrac{1}{k}\cos kx\right]_0^{2\pi} = \dfrac{-4}{k},$ so the

fourth-order Fourier approximation to $f(x) = 2x-1$ over $[0, 2\pi]$ is

$g(x) = 2\pi - 1 + \displaystyle\sum_{k=1}^{4} \dfrac{-4}{k}\sin kx = 2\pi - 1 - 4\left(\sin x + \dfrac{1}{2}\sin 2x + \dfrac{1}{3}\sin 3x + \dfrac{1}{4}\sin 4x\right).$

7. center = (0,1,1,0,0,1,1)
(1,1,1,0,0,1,1), (0,0,1,0,0,1,1), (0,1,0,0,0,1,1), (0,1,1,1,0,1,1), (0,1,1,0,1,1,1),
(0,1,1,0,0,0,1), (0,1,1,0,0,1,0)

8. Pinv $\begin{bmatrix} 1 & 2 \\ 3 & 1 \\ 4 & 2 \end{bmatrix} = \left(\begin{bmatrix} 1 & 3 & 4 \\ 2 & 1 & 2 \end{bmatrix}\begin{bmatrix} 1 & 2 \\ 3 & 1 \\ 4 & 2 \end{bmatrix}\right)^{-1}\begin{bmatrix} 1 & 3 & 4 \\ 2 & 1 & 2 \end{bmatrix} = \dfrac{1}{65}\begin{bmatrix} -17 & 14 & 10 \\ 39 & -13 & 0 \end{bmatrix}.$

9. The equation of the parabola will be $y = a + bx + cx^2$. The data points give $6 = a + b + c$, $2 = a + 2b + 4c$, $5 = a + 3b + 9c$, and $9 = a + 4b + 16c$. Thus the system of equations is

given by $A\begin{bmatrix} a \\ b \\ c \end{bmatrix} = \begin{bmatrix} 6 \\ 2 \\ 5 \\ 9 \end{bmatrix}$, where $A = \begin{bmatrix} 1 & 1 & 1 \\ 1 & 2 & 4 \\ 1 & 3 & 9 \\ 1 & 4 & 16 \end{bmatrix}.$

Pinv $A = \left(\begin{bmatrix} 1 & 1 & 1 & 1 \\ 1 & 2 & 3 & 4 \\ 1 & 4 & 9 & 16 \end{bmatrix}\begin{bmatrix} 1 & 1 & 1 \\ 1 & 2 & 4 \\ 1 & 3 & 9 \\ 1 & 4 & 16 \end{bmatrix}\right)^{-1}\begin{bmatrix} 1 & 1 & 1 & 1 \\ 1 & 2 & 3 & 4 \\ 1 & 4 & 9 & 16 \end{bmatrix}$

$$= \begin{bmatrix} 4 & 10 & 30 \\ 10 & 30 & 100 \\ 30 & 100 & 354 \end{bmatrix}^{-1} \begin{bmatrix} 1 & 1 & 1 & 1 \\ 1 & 2 & 3 & 4 \\ 1 & 4 & 9 & 16 \end{bmatrix} = \frac{1}{80} \begin{bmatrix} 620 & -540 & 100 \\ -540 & 516 & -100 \\ 100 & -100 & 20 \end{bmatrix} \begin{bmatrix} 1 & 1 & 1 & 1 \\ 1 & 2 & 3 & 4 \\ 1 & 4 & 9 & 16 \end{bmatrix}$$

$$= \frac{1}{20} \begin{bmatrix} 45 & -15 & -25 & 15 \\ -31 & 23 & 27 & -19 \\ 5 & -5 & -5 & 5 \end{bmatrix}. \quad \text{Pinv } A \begin{bmatrix} 6 \\ 2 \\ 5 \\ 9 \end{bmatrix} = \begin{bmatrix} 12.5 \\ -8.8 \\ 2 \end{bmatrix}, \text{ so } a = 12.5, b = -8.8, c = 2,$$

and the least squares parabola is $y = 12.5 - 8.8x + 2x^2$.

10. $B = \text{Pinv } A = (A^t A)^{-1} A^t$. If A is $n \times m$, then B is $m \times n$.

   (a) $ABA = A(A^t A)^{-1} A^t A = AI_m = A$.   (b) $BAB = (A^t A)^{-1} A^t AB = I_m B = B$.

   (c) AB is the projection matrix for the subspace spanned by the columns of A. This matrix was shown to be symmetric in Exercise 44(a) of Section 7.5.

   (d) $BA = (A^t A)^{-1} A^t A = I_m$ and is therefore symmetric.

Chapter 9

Exercise Set 9.1

1. (a) Yes. (b) Yes. (c) Yes.

   (d) No. The leading 1 of row three is not to the right of the leading 1 of row two.

2. (a) Yes. (b) No. The leading 1 in row 3 is not to the right of the leading 1 in row 2.

   (c) No. The leading nonzero number in row 2 is not a 1.

3. $\begin{bmatrix} 1 & 1 & 1 & 6 \\ 1 & -1 & 1 & 2 \\ 1 & 2 & 3 & 14 \end{bmatrix} \underset{R3+(-1)R1}{\overset{R2+(-1)R1}{\approx}} \begin{bmatrix} 1 & 1 & 1 & 6 \\ 0 & -2 & 0 & -4 \\ 0 & 1 & 2 & 8 \end{bmatrix} \overset{(-1/2)R2}{\approx} \begin{bmatrix} 1 & 1 & 1 & 6 \\ 0 & 1 & 0 & 2 \\ 0 & 1 & 2 & 8 \end{bmatrix}$

   $\overset{R3+(-1)R2}{\approx} \begin{bmatrix} 1 & 1 & 1 & 6 \\ 0 & 1 & 0 & 2 \\ 0 & 0 & 2 & 6 \end{bmatrix} \overset{(1/2)R3}{\approx} \begin{bmatrix} 1 & 1 & 1 & 6 \\ 0 & 1 & 0 & 2 \\ 0 & 0 & 1 & 3 \end{bmatrix}.$

   (a) $x_1 + x_2 + x_3 = 6$, $x_2 = 2$, and $x_3 = 3$. Back substituting $x_2 = 2$ and $x_3 = 3$ into the first equation, $x_1 = 1$.

   (b) $\begin{bmatrix} 1 & 1 & 1 & 6 \\ 0 & 1 & 0 & 2 \\ 0 & 0 & 1 & 3 \end{bmatrix} \overset{R1+(-1)R3}{\approx} \begin{bmatrix} 1 & 1 & 0 & 3 \\ 0 & 1 & 0 & 2 \\ 0 & 0 & 1 & 3 \end{bmatrix} \overset{R1+(-1)R2}{\approx} \begin{bmatrix} 1 & 0 & 0 & 1 \\ 0 & 1 & 0 & 2 \\ 0 & 0 & 1 & 3 \end{bmatrix}$, so again
   $x_1 = 1, x_2 = 2, x_3 = 3.$

4. $\begin{bmatrix} 1 & -1 & -1 & 2 \\ 1 & -1 & 1 & 2 \\ 3 & -2 & 1 & 5 \end{bmatrix} \underset{R3+(-3)R1}{\overset{R2+(-1)R1}{\approx}} \begin{bmatrix} 1 & -1 & -1 & 2 \\ 0 & 0 & -2 & 0 \\ 0 & 1 & 4 & -1 \end{bmatrix} \overset{R2 \leftrightarrow R3}{\approx} \begin{bmatrix} 1 & -1 & -1 & 2 \\ 0 & 1 & 4 & -1 \\ 0 & 0 & -2 & 0 \end{bmatrix}$

   $\overset{(-1/2)R3}{\approx} \begin{bmatrix} 1 & -1 & -1 & 2 \\ 0 & 1 & 4 & -1 \\ 0 & 0 & 1 & 0 \end{bmatrix}.$

   (a) $x_1 - x_2 - x_3 = 2$, $x_2 + 4x_3 = -1$, and $x_3 = 0$. Back substituting $x_3 = 0$ into the

328

second equation, $x_2 = -1$, and substituting for x_2 and x_3 in the first equation, $x_1 = 1$.

(b) $\begin{bmatrix} 1 & -1 & -1 & 2 \\ 0 & 1 & 4 & -1 \\ 0 & 0 & 1 & 0 \end{bmatrix} \underset{R2+(-4)R3}{\overset{R1+(-1)R3}{\approx}} \begin{bmatrix} 1 & -1 & 0 & 2 \\ 0 & 1 & 0 & -1 \\ 0 & 0 & 1 & 0 \end{bmatrix} \underset{R1+R2}{\approx} \begin{bmatrix} 1 & 0 & 0 & 1 \\ 0 & 1 & 0 & -1 \\ 0 & 0 & 1 & 0 \end{bmatrix}$, so again

$x_1 = 1, x_2 = -1, x_3 = 0$.

5. $\begin{bmatrix} 1 & -1 & 2 & 3 \\ 2 & -2 & 5 & 4 \\ 1 & 2 & -1 & -3 \\ 0 & 2 & 2 & 1 \end{bmatrix} \underset{R3+(-1)R1}{\overset{R2+(-2)R1}{\approx}} \begin{bmatrix} 1 & -1 & 2 & 3 \\ 0 & 0 & 1 & -2 \\ 0 & 3 & -3 & -6 \\ 0 & 2 & 2 & 1 \end{bmatrix} \underset{R2 \leftrightarrow R3}{\approx} \begin{bmatrix} 1 & -1 & 2 & 3 \\ 0 & 3 & -3 & -6 \\ 0 & 0 & 1 & -2 \\ 0 & 2 & 2 & 1 \end{bmatrix}$

$\underset{(1/3)R2}{\approx} \begin{bmatrix} 1 & -1 & 2 & 3 \\ 0 & 1 & -1 & -2 \\ 0 & 0 & 1 & -2 \\ 0 & 2 & 2 & 1 \end{bmatrix} \underset{R4+(-2)R2}{\approx} \begin{bmatrix} 1 & -1 & 2 & 3 \\ 0 & 1 & -1 & -2 \\ 0 & 0 & 1 & -2 \\ 0 & 0 & 4 & 5 \end{bmatrix}$, and there is no reason to

continue. The last two rows give inconsistent equations, so there is no solution.

6. $\begin{bmatrix} 1 & -1 & 1 & 2 & -2 & 1 \\ 2 & -1 & -1 & 3 & -1 & 3 \\ -1 & -1 & 5 & 0 & -4 & -3 \end{bmatrix} \underset{R3+R1}{\overset{R2+(-2)R1}{\approx}} \begin{bmatrix} 1 & -1 & 1 & 2 & -2 & 1 \\ 0 & 1 & -3 & -1 & 3 & 1 \\ 0 & -2 & 6 & 2 & -6 & -2 \end{bmatrix}$

$\underset{R3+2R2}{\approx} \begin{bmatrix} 1 & -1 & 1 & 2 & -2 & 1 \\ 0 & 1 & -3 & -1 & 3 & 1 \\ 0 & 0 & 0 & 0 & 0 & 0 \end{bmatrix}$.

(a) $x_1 - x_2 + x_3 + 2x_4 - 2x_5 = 1$ and $x_2 - 3x_3 - x_4 + 3x_5 = 1$.
Back substituting $x_3 = r$, $x_4 = s$, $x_5 = t$ into the second equation, $x_2 = 1 + 3r + s - 3t$, and substituting for x_2, x_3, x_4, and x_5 in the first equation, $x_1 = 2 + 2r - s - t$.

(b) $\begin{bmatrix} 1 & -1 & 1 & 2 & -2 & 1 \\ 0 & 1 & -3 & -1 & 3 & 1 \\ 0 & 0 & 0 & 0 & 0 & 0 \end{bmatrix} \underset{R1+R2}{\approx} \begin{bmatrix} 1 & 0 & -2 & 1 & 1 & 2 \\ 0 & 1 & -3 & -1 & 3 & 1 \\ 0 & 0 & 0 & 0 & 0 & 0 \end{bmatrix}$, and again the general

solution is $x_1 = 2 + 2r - s - t$, $x_2 = 1 + 3r + s - 3t$, $x_3 = r$, $x_4 = s$, $x_5 = t$.

7 and 8. The augmented matrix is 4x(n+1), where n is the number of variables.

Section 9.1

Gauss-Jordan elimination:

$$\begin{bmatrix} * & * & * & * & * & \cdots & * \\ * & * & * & * & * & \cdots & * \\ * & * & * & * & * & \cdots & * \\ * & * & * & * & * & \cdots & * \end{bmatrix} \underset{n \text{ mults}}{\approx} \begin{bmatrix} 1 & * & * & * & * & \cdots & * \\ * & * & * & * & * & \cdots & * \\ * & * & * & * & * & \cdots & * \\ * & * & * & * & * & \cdots & * \end{bmatrix} \underset{\substack{3n \text{ mults} \\ 3n \text{ adds}}}{\approx} \begin{bmatrix} 1 & * & * & * & * & \cdots & * \\ 0 & * & * & * & * & \cdots & * \\ 0 & * & * & * & * & \cdots & * \\ 0 & * & * & * & * & \cdots & * \end{bmatrix}$$

$$\underset{n-1 \text{ mults}}{\approx} \begin{bmatrix} 1 & * & * & * & \cdots & * \\ 0 & 1 & * & * & * & \cdots & * \\ 0 & * & * & * & * & \cdots & * \\ 0 & * & * & * & * & \cdots & * \end{bmatrix} \underset{\substack{3(n-1) \text{ mults} \\ 3(n-1) \text{ adds}}}{\approx} \begin{bmatrix} 1 & 0 & * & * & \cdots & * \\ 0 & 1 & * & * & * & \cdots & * \\ 0 & 0 & * & * & * & \cdots & * \\ 0 & 0 & * & * & * & \cdots & * \end{bmatrix}$$

$$\underset{n-2 \text{ mults}}{\approx} \begin{bmatrix} 1 & 0 & * & * & * & \cdots & * \\ 0 & 1 & * & * & * & \cdots & * \\ 0 & 0 & 1 & * & * & \cdots & * \\ 0 & 0 & * & * & * & \cdots & * \end{bmatrix} \underset{\substack{3(n-2) \text{ mults} \\ 3(n-2) \text{ adds}}}{\approx} \begin{bmatrix} 1 & 0 & 0 & * & * & \cdots & * \\ 0 & 1 & 0 & * & * & \cdots & * \\ 0 & 0 & 1 & * & * & \cdots & * \\ 0 & 0 & 0 & * & * & \cdots & * \end{bmatrix}$$

$$\underset{n-3 \text{ mults}}{\approx} \begin{bmatrix} 1 & 0 & 0 & * & * & \cdots & * \\ 0 & 1 & 0 & * & * & \cdots & * \\ 0 & 0 & 1 & * & * & \cdots & * \\ 0 & 0 & 0 & 1 & * & \cdots & * \end{bmatrix} \underset{\substack{3(n-3) \text{ mults} \\ 3(n-3) \text{ adds}}}{\approx} \begin{bmatrix} 1 & 0 & 0 & 0 & * & \cdots & * \\ 0 & 1 & 0 & 0 & * & \cdots & * \\ 0 & 0 & 1 & 0 & * & \cdots & * \\ 0 & 0 & 0 & 1 & * & \cdots & * \end{bmatrix}.$$

number of mults: $n + (n-1) + (n-2) + (n-3) + 3[n + (n-1) + (n-2) + (n-3)]$
number of adds: $3[n + (n-1) + (n-2) + (n-3)]$
total number of operations: $7[n + (n-1) + (n-2) + (n-3)] = 7(4n-6)$

Gaussian elimination:

$$\begin{bmatrix} * & * & * & * & * & \cdots & * \\ * & * & * & * & * & \cdots & * \\ * & * & * & * & * & \cdots & * \\ * & * & * & * & * & \cdots & * \end{bmatrix} \underset{n \text{ mults}}{\approx} \begin{bmatrix} 1 & * & * & * & * & \cdots & * \\ * & * & * & * & * & \cdots & * \\ * & * & * & * & * & \cdots & * \\ * & * & * & * & * & \cdots & * \end{bmatrix} \underset{\substack{3n \text{ mults} \\ 3n \text{ adds}}}{\approx} \begin{bmatrix} 1 & * & * & * & * & \cdots & * \\ 0 & * & * & * & * & \cdots & * \\ 0 & * & * & * & * & \cdots & * \\ 0 & * & * & * & * & \cdots & * \end{bmatrix}$$

$$\underset{n-1 \text{ mults}}{\approx} \begin{bmatrix} 1 & * & * & * & * & \cdots & * \\ 0 & 1 & * & * & * & \cdots & * \\ 0 & * & * & * & * & \cdots & * \\ 0 & * & * & * & * & \cdots & * \end{bmatrix} \underset{\substack{2(n-1) \text{ mults} \\ 2(n-1) \text{ adds}}}{\approx} \begin{bmatrix} 1 & * & * & * & * & \cdots & * \\ 0 & 1 & * & * & * & \cdots & * \\ 0 & 0 & * & * & * & \cdots & * \\ 0 & 0 & * & * & * & \cdots & * \end{bmatrix}$$

$$n-2 \overset{\approx}{\text{mults}} \begin{bmatrix} 1 & * & * & * & * & \cdots & * \\ 0 & 1 & * & * & * & \cdots & * \\ 0 & 0 & 1 & * & * & \cdots & * \\ 0 & 0 & * & * & * & \cdots & * \end{bmatrix} \quad \begin{matrix} n-2 \overset{\approx}{\text{mults}} \\ n-2 \text{ adds} \end{matrix} \begin{bmatrix} 1 & * & * & * & * & \cdots & * \\ 0 & 1 & * & * & * & \cdots & * \\ 0 & 0 & 1 & * & * & \cdots & * \\ 0 & 0 & 0 & * & * & \cdots & * \end{bmatrix}$$

$$n-3 \overset{\approx}{\text{mults}} \begin{bmatrix} 1 & * & * & * & * & \cdots & * \\ 0 & 1 & * & * & * & \cdots & * \\ 0 & 0 & 1 & * & * & \cdots & * \\ 0 & 0 & 0 & 1 & * & \cdots & * \end{bmatrix}.$$

number of mults: $n + (n-1) + (n-2) + (n-3) + 3n + 2(n-1) + (n-2)$
number of adds: $3n + 2(n-1) + (n-2)$
total number of operations: $16n - 14$

Back substitution:

$$\begin{bmatrix} 1 & * & * & * & * & \cdots & * \\ 0 & 1 & * & * & * & \cdots & * \\ 0 & 0 & 1 & * & * & \cdots & * \\ 0 & 0 & 0 & 1 & * & \cdots & * \end{bmatrix} \quad \begin{matrix} 3(n-3) \overset{\approx}{\text{mults}} \\ 3(n-3) \text{ adds} \end{matrix} \begin{bmatrix} 1 & * & * & 0 & * & \cdots & * \\ 0 & 1 & * & 0 & * & \cdots & * \\ 0 & 0 & 1 & 0 & * & \cdots & * \\ 0 & 0 & 0 & 1 & * & \cdots & * \end{bmatrix}$$

$$\begin{matrix} 2(n-3) \overset{\approx}{\text{mults}} \\ 2(n-3) \text{ adds} \end{matrix} \begin{bmatrix} 1 & * & 0 & 0 & * & \cdots & * \\ 0 & 1 & 0 & 0 & * & \cdots & * \\ 0 & 0 & 1 & 0 & * & \cdots & * \\ 0 & 0 & 0 & 1 & * & \cdots & * \end{bmatrix} \quad \begin{matrix} (n-3) \overset{\approx}{\text{mults}} \\ (n-3) \text{ adds} \end{matrix} \begin{bmatrix} 1 & 0 & 0 & 0 & * & \cdots & * \\ 0 & 1 & 0 & 0 & * & \cdots & * \\ 0 & 0 & 1 & 0 & * & \cdots & * \\ 0 & 0 & 0 & 1 & * & \cdots & * \end{bmatrix}.$$

number of mults: $3(n-3) + 2(n-3) + (n-3)$ number of adds: $3(n-3) + 2(n-3) + (n-3)$
total number of operations: $12(n-3)$

In both exercises, eight operations are saved using Gaussian elimination. One addition and one multiplication are saved in getting zeros in each of the positions above the diagonal in column 3 and two additions and two multiplications are saved in getting the zero above the diagonal in column 2. The basic reason for the savings is that in back substituting, one is working from right to left. This provides the same savings in operations above the diagonal that one has below the diagonal working from left to right.

Let $n = 5$ (Exercise 7).
For Gauss-Jordan elimination the number of operations is $7(4n-6) = 98$.
For Gaussian elimination the total number of operations is $16n - 14 + 12(n-3) = 90$.

Let $n = 6$ (Exercise 8).
For Gauss-Jordan elimination the number of operations is $7(4n-6) = 126$.
For Gaussian elimination the total number of operations is $16n - 14 + 12(n-3) = 118$.

9. Yes. Consider a 2x3 matrix with no zeros in the first column. Find the echelon form leaving the rows in the original order, then do it switching the order of the rows first.

10. $\begin{bmatrix} * & * & \cdots & * \\ * & * & \cdots & * \\ \vdots & \vdots & & \vdots \\ * & * & \cdots & * \end{bmatrix}$ n mults $\approx$ $\begin{bmatrix} 1 & * & \cdots & * \\ * & * & \cdots & * \\ \vdots & \vdots & & \vdots \\ * & * & \cdots & * \end{bmatrix}$ $\begin{array}{c}(n-1)n \text{ mults} \\ (n-1)n \text{ adds}\end{array}$ $\approx$ $\begin{bmatrix} 1 & * & \cdots & * \\ 0 & * & \cdots & * \\ \vdots & \vdots & & \vdots \\ 0 & * & \cdots & * \end{bmatrix}$

$\approx$ n−1 mults $\begin{bmatrix} 1 & * & \cdots & * \\ 0 & 1 & \cdots & * \\ \vdots & \vdots & & \vdots \\ 0 & * & \cdots & * \end{bmatrix}$ $\begin{array}{c}(n-2)(n-1) \text{ mults} \\ (n-2)(n-1) \text{ adds}\end{array}$ $\approx$ $\begin{bmatrix} 1 & * & \cdots & * \\ 0 & 1 & \cdots & * \\ \vdots & \vdots & & \vdots \\ 0 & 0 & \cdots & * \end{bmatrix}$ $\approx$ n−2 mults $\cdots$

Back substitution

$\approx$ 1 mult $\begin{bmatrix} 1 & * & \cdots & * & * \\ 0 & 1 & \cdots & * & * \\ \vdots & \vdots & & \vdots & \vdots \\ 0 & 0 & \cdots & 1 & * \end{bmatrix}$ $\begin{array}{c}\text{n−1 mults} \\ \text{n−1 adds}\end{array}$ $\approx$ $\begin{bmatrix} 1 & * & \cdots & 0 & * \\ 0 & 1 & \cdots & 0 & * \\ \vdots & \vdots & & \vdots & \vdots \\ 0 & 0 & \cdots & 1 & * \end{bmatrix}$ $\approx$ $\begin{array}{c}\text{n−2 mults} \\ \text{n−2 adds}\end{array}$ $\cdots$

$\approx$ 1 mult 1 add $\begin{bmatrix} 1 & 0 & \cdots & 0 & * \\ 0 & 1 & \cdots & 0 & * \\ \vdots & \vdots & & \vdots & \vdots \\ 0 & 0 & \cdots & 1 & * \end{bmatrix}$.

So the number of multiplications in Gaussian elimination is

$n + (n-1)n + (n-2) + (n-2)(n-1) + \cdots + 2 + (1)2 + 1 = n^2 + (n-1)^2 + \cdots + 1$

$= n(n+1)(2n+1)/6 = n^3/3 + n^2/2 + n/6$. The number of additions is

$(n-1)n + (n-2)(n-1) + \cdots + 2(3) + 1(2) = n^2 - n + (n-1)^2 - (n-1) + \cdots + 2^2 - 2 + (1-1)$

$= n^2 + (n-1)^2 + \cdots + 2^2 + 1 - [n + (n-1) + \cdots + 2 + 1] = n(n+1)(2n+1)/6 - n(n+1)/2$

$= n^3/3 + n^2/2 + n/6 - n^2/2 - n/2 = n^3/3 - n/3$.

Back substitution requires equal numbers of multiplications and additions:

332

$(n-1) + (n-2) + 2 + 1 = (n-1)n/2 = n^2/2 - n/2$. The total number of multiplications is therefore $n^3/3 + n^2/2 + n/6 + n^2/2 - n/2 = n^3/3 + n^2 - n/3$ and the total number of additions is $n^3/3 - n/3 + n^2/2 - n/2 = n^3/3 + n^2/2 - 5n/6$.

11.

	Gauss-Jordan		Gaussian	
	mult	add	mult	add
n	$n^3/2 + n^2/2$	$n^3/2 - n/2$	$n^3/3 + n^2 - n/3$	$n^3/3 + n^2/2 - 5n/6$
2	6	3	6	3
3	18	12	17	7
4	40	30	36	26
5	75	60	65	50
6	126	105	106	85
7	196	168	161	133
8	288	252	232	196
9	405	360	321	276
10	550	495	430	375

Exercise Set 9.2

1. From the first equation $x_1 = 1$, and from the second equation $2 - x_2 = -2$, so $x_2 = 4$. From the third equation $3 + 4 - x_3 = 8$, so $x_3 = -1$.

2. From the first equation $x_1 = 2$, and from the second equation $-2 + x_2 = 1$, so $x_2 = 3$. From the third equation $2 + 3 + x_3 = 5$, so $x_3 = 0$.

3. From the first equation $x_1 = -2$, and from the second equation $-6 + x_2 = -5$, so $x_2 = 1$. From the third equation $-2 + 4 + 2x_3 = 4$, so $x_3 = 1$.

Section 9.2

4. From the third equation $x_3 = 2$, and from the second equation $2x_2 + 2 = 3$, so $x_2 = 1/2$.

   From the first equation $x_1 + 2 + 1/2 = 3$, so $x_1 = 1/2$.

5. From the third equation $x_3 = 3$, and from the second equation $x_2 - 9 = -7$, so $x_2 = 2$.

   From the first equation $2x_1 - 2 + 3 = 3$, so $x_1 = 1$.

6. From the third equation $x_3 = 7$, and from the second equation $2x_2 + 28 = 26$, so $x_2 = -1$.

   From the first equation $3x_1 - 2 - 7 = -6$, so $x_1 = 1$.

7. The "rule" is Ri – aRj causes $a_{ij} = a$, where a_{ij} is the (i,j)th position in L.

   (a) $L = \begin{bmatrix} 1 & 0 & 0 \\ 1 & 1 & 0 \\ -1 & -1 & 1 \end{bmatrix}$.
   (b) $L = \begin{bmatrix} 1 & 0 & 0 \\ -5 & 1 & 0 \\ 2 & -7 & 1 \end{bmatrix}$.

   (c) $L = \begin{bmatrix} 1 & 0 & 0 \\ -1/2 & 1 & 0 \\ -1/5 & 1 & 1 \end{bmatrix}$.
   (d) $L = \begin{bmatrix} 1 & 0 & 0 & 0 \\ -4 & 1 & 0 & 0 \\ 2 & 2 & 1 & 0 \\ 8 & -2 & 6 & 1 \end{bmatrix}$.

8. (a) R2 – 2R1, R3 + 3R1, R3 – 5R2

   (b) R2 + R1, R3 – 4R1, R3 – 2R2

   (c) R3 – $\frac{1}{7}$R1, R3 + $\frac{3}{4}$R2

   (d) R2 + 2R1, R3 – 3R1, R4 – 6R1, R3 – 5R2, R4 + 3R3

Section 9.2

9. The matrix equation $A\mathbf{x} = \mathbf{b}$ is $\begin{bmatrix} 1 & 2 & -1 \\ -2 & -1 & 3 \\ 1 & -1 & -4 \end{bmatrix} \begin{bmatrix} x_1 \\ x_2 \\ x_3 \end{bmatrix} = \begin{bmatrix} 2 \\ 3 \\ -7 \end{bmatrix}$.

$\begin{bmatrix} 1 & 2 & -1 \\ -2 & -1 & 3 \\ 1 & -1 & -4 \end{bmatrix} \begin{array}{c} \approx \\ \text{R2+2R1} \\ \text{R3+(-1)R1} \end{array} \begin{bmatrix} 1 & 2 & -1 \\ 0 & 3 & 1 \\ 0 & -3 & -3 \end{bmatrix} \begin{array}{c} \approx \\ \text{R3+R2} \end{array} \begin{bmatrix} 1 & 2 & -1 \\ 0 & 3 & 1 \\ 0 & 0 & -2 \end{bmatrix} = U$, and

$L = \begin{bmatrix} 1 & 0 & 0 \\ -2 & 1 & 0 \\ 1 & -1 & 1 \end{bmatrix}$. $L\mathbf{y} = \begin{bmatrix} 2 \\ 3 \\ -7 \end{bmatrix}$ gives $\mathbf{y} = \begin{bmatrix} 2 \\ 7 \\ -2 \end{bmatrix}$, and $U\mathbf{x} = \mathbf{y}$ gives $\mathbf{x} = \begin{bmatrix} -1 \\ 2 \\ 1 \end{bmatrix}$.

10. The matrix equation $A\mathbf{x} = \mathbf{b}$ is $\begin{bmatrix} 2 & 1 & 0 \\ 6 & 4 & -1 \\ 2 & 0 & 4 \end{bmatrix} \begin{bmatrix} x_1 \\ x_2 \\ x_3 \end{bmatrix} = \begin{bmatrix} 9 \\ 25 \\ 20 \end{bmatrix}$.

$\begin{bmatrix} 2 & 1 & 0 \\ 6 & 4 & -1 \\ 2 & 0 & 4 \end{bmatrix} \begin{array}{c} \approx \\ \text{R2+(-3)R1} \\ \text{R3+(-1)R1} \end{array} \begin{bmatrix} 2 & 1 & 0 \\ 0 & 1 & -1 \\ 0 & -1 & 4 \end{bmatrix} \begin{array}{c} \approx \\ \text{R3+R2} \end{array} \begin{bmatrix} 2 & 1 & 0 \\ 0 & 1 & -1 \\ 0 & 0 & 3 \end{bmatrix} = U$, and

$L = \begin{bmatrix} 1 & 0 & 0 \\ 3 & 1 & 0 \\ 1 & -1 & 1 \end{bmatrix}$. $L\mathbf{y} = \begin{bmatrix} 9 \\ 25 \\ 20 \end{bmatrix}$ gives $\mathbf{y} = \begin{bmatrix} 9 \\ -2 \\ 9 \end{bmatrix}$, and $U\mathbf{x} = \mathbf{y}$ gives $\mathbf{x} = \begin{bmatrix} 4 \\ 1 \\ 3 \end{bmatrix}$.

11. The matrix equation $A\mathbf{x} = \mathbf{b}$ is $\begin{bmatrix} 3 & -1 & 1 \\ -3 & 2 & 1 \\ 9 & 5 & -3 \end{bmatrix} \begin{bmatrix} x_1 \\ x_2 \\ x_3 \end{bmatrix} = \begin{bmatrix} 10 \\ -8 \\ 24 \end{bmatrix}$.

$\begin{bmatrix} 3 & -1 & 1 \\ -3 & 2 & 1 \\ 9 & 5 & -3 \end{bmatrix} \begin{array}{c} \approx \\ \text{R2+R1} \\ \text{R3+(-3)R1} \end{array} \begin{bmatrix} 3 & -1 & 1 \\ 0 & 1 & 2 \\ 0 & 8 & -6 \end{bmatrix} \begin{array}{c} \approx \\ \text{R3+(-8)R2} \end{array} \begin{bmatrix} 3 & -1 & 1 \\ 0 & 1 & 2 \\ 0 & 0 & -22 \end{bmatrix} = U$, and

$L = \begin{bmatrix} 1 & 0 & 0 \\ -1 & 1 & 0 \\ 3 & 8 & 1 \end{bmatrix}$. $L\mathbf{y} = \begin{bmatrix} 10 \\ -8 \\ 24 \end{bmatrix}$ gives $\mathbf{y} = \begin{bmatrix} 10 \\ 2 \\ -22 \end{bmatrix}$, and $U\mathbf{x} = \mathbf{y}$ gives $\mathbf{x} = \begin{bmatrix} 3 \\ 0 \\ 1 \end{bmatrix}$.

Section 9.2

12. The matrix equation $A\mathbf{x} = \mathbf{b}$ is $\begin{bmatrix} 1 & 2 & 3 \\ 2 & 8 & 7 \\ -1 & 14 & -1 \end{bmatrix} \begin{bmatrix} x_1 \\ x_2 \\ x_3 \end{bmatrix} = \begin{bmatrix} -5 \\ -9 \\ 15 \end{bmatrix}$.

$\begin{bmatrix} 1 & 2 & 3 \\ 2 & 8 & 7 \\ -1 & 14 & -1 \end{bmatrix} \underset{R3+R1}{\overset{R2+(-2)R1}{\approx}} \begin{bmatrix} 1 & 2 & 3 \\ 0 & 4 & 1 \\ 0 & 16 & 2 \end{bmatrix} \overset{R3+(-4)R2}{\approx} \begin{bmatrix} 1 & 2 & 3 \\ 0 & 4 & 1 \\ 0 & 0 & -2 \end{bmatrix} = U$, and

$L = \begin{bmatrix} 1 & 0 & 0 \\ 2 & 1 & 0 \\ -1 & 4 & 1 \end{bmatrix}$. $L\mathbf{y} = \begin{bmatrix} -5 \\ -9 \\ 15 \end{bmatrix}$ gives $\mathbf{y} = \begin{bmatrix} -5 \\ 1 \\ 6 \end{bmatrix}$, and $U\mathbf{x} = \mathbf{y}$ gives $\mathbf{x} = \begin{bmatrix} 2 \\ 1 \\ -3 \end{bmatrix}$.

13. The matrix equation $A\mathbf{x} = \mathbf{b}$ is $\begin{bmatrix} 4 & 1 & 2 \\ -12 & -4 & -3 \\ 0 & -5 & 16 \end{bmatrix} \begin{bmatrix} x_1 \\ x_2 \\ x_3 \end{bmatrix} = \begin{bmatrix} 18 \\ -56 \\ -10 \end{bmatrix}$.

$\begin{bmatrix} 4 & 1 & 2 \\ -12 & -4 & -3 \\ 0 & -5 & 16 \end{bmatrix} \overset{R2+(3)R1}{\approx} \begin{bmatrix} 4 & 1 & 2 \\ 0 & -1 & 3 \\ 0 & -5 & 16 \end{bmatrix} \overset{R3+(-5)R2}{\approx} \begin{bmatrix} 4 & 1 & 2 \\ 0 & -1 & 3 \\ 0 & 0 & 1 \end{bmatrix} = U$, and

$L = \begin{bmatrix} 1 & 0 & 0 \\ -3 & 1 & 0 \\ 0 & 5 & 1 \end{bmatrix}$. $L\mathbf{y} = \begin{bmatrix} 18 \\ -56 \\ -10 \end{bmatrix}$ gives $\mathbf{y} = \begin{bmatrix} 18 \\ -2 \\ 0 \end{bmatrix}$, and $U\mathbf{x} = \mathbf{y}$ gives $\mathbf{x} = \begin{bmatrix} 4 \\ 2 \\ 0 \end{bmatrix}$.

14. The matrix equation $A\mathbf{x} = \mathbf{b}$ is $\begin{bmatrix} -2 & 0 & 3 \\ -4 & 3 & 5 \\ 8 & 9 & -11 \end{bmatrix} \begin{bmatrix} x_1 \\ x_2 \\ x_3 \end{bmatrix} = \begin{bmatrix} 3 \\ 11 \\ 7 \end{bmatrix}$.

$\begin{bmatrix} -2 & 0 & 3 \\ -4 & 3 & 5 \\ 8 & 9 & -11 \end{bmatrix} \underset{R3+4R1}{\overset{R2+(-2)R1}{\approx}} \begin{bmatrix} -2 & 0 & 3 \\ 0 & 3 & -1 \\ 0 & 9 & 1 \end{bmatrix} \overset{R3+(-3)R2}{\approx} \begin{bmatrix} -2 & 0 & 3 \\ 0 & 3 & -1 \\ 0 & 0 & 4 \end{bmatrix} = U$, and

$L = \begin{bmatrix} 1 & 0 & 0 \\ 2 & 1 & 0 \\ -4 & 3 & 1 \end{bmatrix}$. $L\mathbf{y} = \begin{bmatrix} 3 \\ 11 \\ 7 \end{bmatrix}$ gives $\mathbf{y} = \begin{bmatrix} 3 \\ 5 \\ 4 \end{bmatrix}$, and $U\mathbf{x} = \mathbf{y}$ gives $\mathbf{x} = \begin{bmatrix} 0 \\ 2 \\ 1 \end{bmatrix}$.

Section 9.2

15. The matrix equation $A\mathbf{x} = \mathbf{b}$ is $\begin{bmatrix} -2 & 0 & 3 \\ -4 & 3 & 5 \\ 8 & 9 & -11 \end{bmatrix} \begin{bmatrix} x_1 \\ x_2 \\ x_3 \end{bmatrix} = \begin{bmatrix} 14 \\ 30 \\ -34 \end{bmatrix}$. Since the matrix A is the same as in Exercise 14, so are L and U. $L\mathbf{y} = \begin{bmatrix} 14 \\ 30 \\ -34 \end{bmatrix}$ gives $\mathbf{y} = \begin{bmatrix} 14 \\ 2 \\ 16 \end{bmatrix}$, and $U\mathbf{x} = \mathbf{y}$ gives $\mathbf{x} = \begin{bmatrix} -1 \\ 2 \\ 4 \end{bmatrix}$.

16. The matrix equation $A\mathbf{x} = \mathbf{b}$ is $\begin{bmatrix} 2 & -3 & 1 \\ 4 & -5 & 6 \\ -10 & 19 & 9 \end{bmatrix} \begin{bmatrix} x_1 \\ x_2 \\ x_3 \end{bmatrix} = \begin{bmatrix} -5 \\ 2 \\ 55 \end{bmatrix}$.

$\begin{bmatrix} 2 & -3 & 1 \\ 4 & -5 & 6 \\ -10 & 19 & 9 \end{bmatrix} \underset{\substack{R2+(-2)R1 \\ R3+5R1}}{\approx} \begin{bmatrix} 2 & -3 & 1 \\ 0 & 1 & 4 \\ 0 & 4 & 14 \end{bmatrix} \underset{R3+(-4)R2}{\approx} \begin{bmatrix} 2 & -3 & 1 \\ 0 & 1 & 4 \\ 0 & 0 & -2 \end{bmatrix} = U$, and

$L = \begin{bmatrix} 1 & 0 & 0 \\ 2 & 1 & 0 \\ -5 & 4 & 1 \end{bmatrix}$. $L\mathbf{y} = \begin{bmatrix} -5 \\ 2 \\ 55 \end{bmatrix}$ gives $\mathbf{y} = \begin{bmatrix} -5 \\ 12 \\ -18 \end{bmatrix}$, and $U\mathbf{x} = \mathbf{y}$ gives $\mathbf{x} = \begin{bmatrix} -43 \\ -24 \\ 9 \end{bmatrix}$.

17. The matrix equation $A\mathbf{x} = \mathbf{b}$ is $\begin{bmatrix} 4 & 1 & 3 \\ 12 & 1 & 10 \\ -8 & -16 & 6 \end{bmatrix} \begin{bmatrix} x_1 \\ x_2 \\ x_3 \end{bmatrix} = \begin{bmatrix} 11 \\ 28 \\ -62 \end{bmatrix}$.

$\begin{bmatrix} 4 & 1 & 3 \\ 12 & 1 & 10 \\ -8 & -16 & 6 \end{bmatrix} \underset{\substack{R2+(-3)R1 \\ R3+2R1}}{\approx} \begin{bmatrix} 4 & 1 & 3 \\ 0 & -2 & 1 \\ 0 & -14 & 12 \end{bmatrix} \underset{R3+(-7)R2}{\approx} \begin{bmatrix} 4 & 1 & 3 \\ 0 & -2 & 1 \\ 0 & 0 & 5 \end{bmatrix} = U$, and

$L = \begin{bmatrix} 1 & 0 & 0 \\ 3 & 1 & 0 \\ -2 & 7 & 1 \end{bmatrix}$. $L\mathbf{y} = \begin{bmatrix} 11 \\ 28 \\ -62 \end{bmatrix}$ gives $\mathbf{y} = \begin{bmatrix} 11 \\ -5 \\ -5 \end{bmatrix}$, and $U\mathbf{x} = \mathbf{y}$ gives $\mathbf{x} = \begin{bmatrix} 3 \\ 2 \\ -1 \end{bmatrix}$.

Section 9.2

18. The matrix equation $A\mathbf{x} = \mathbf{b}$ is $\begin{bmatrix} 1 & 2 & -1 \\ 2 & 5 & 1 \\ -1 & -1 & 4 \end{bmatrix} \begin{bmatrix} x_1 \\ x_2 \\ x_3 \end{bmatrix} = \begin{bmatrix} 2 \\ 3 \\ -3 \end{bmatrix}$.

$\begin{bmatrix} 1 & 2 & -1 \\ 2 & 5 & 1 \\ -1 & -1 & 4 \end{bmatrix} \underset{R3+R1}{\overset{R2+(-2)R1}{\approx}} \begin{bmatrix} 1 & 2 & -1 \\ 0 & 1 & 3 \\ 0 & 1 & 3 \end{bmatrix} \overset{R3+(-1)R2}{\approx} \begin{bmatrix} 1 & 2 & -1 \\ 0 & 1 & 3 \\ 0 & 0 & 0 \end{bmatrix} = U$, and

$L = \begin{bmatrix} 1 & 0 & 0 \\ 2 & 1 & 0 \\ -1 & 1 & 1 \end{bmatrix}$. $L\mathbf{y} = \begin{bmatrix} 2 \\ 3 \\ -3 \end{bmatrix}$ gives $\mathbf{y} = \begin{bmatrix} 2 \\ -1 \\ 0 \end{bmatrix}$, and $U\mathbf{x} = \mathbf{y}$ gives $\mathbf{x} = \begin{bmatrix} 4+7r \\ -1-3r \\ r \end{bmatrix}$.

19. The matrix equation $A\mathbf{x} = \mathbf{b}$ is $\begin{bmatrix} 4 & 1 & -2 \\ -4 & 2 & 3 \\ 8 & -7 & -7 \end{bmatrix} \begin{bmatrix} x_1 \\ x_2 \\ x_3 \end{bmatrix} = \begin{bmatrix} 3 \\ 1 \\ -2 \end{bmatrix}$.

$\begin{bmatrix} 4 & 1 & -2 \\ -4 & 2 & 3 \\ 8 & -7 & -7 \end{bmatrix} \underset{R3+(-2)R1}{\overset{R2+R1}{\approx}} \begin{bmatrix} 4 & 1 & -2 \\ 0 & 3 & 1 \\ 0 & -9 & -3 \end{bmatrix} \overset{R3+3R2}{\approx} \begin{bmatrix} 4 & 1 & -2 \\ 0 & 3 & 1 \\ 0 & 0 & 0 \end{bmatrix} = U$, and

$L = \begin{bmatrix} 1 & 0 & 0 \\ -1 & 1 & 0 \\ 2 & -3 & 1 \end{bmatrix}$. $L\mathbf{y} = \begin{bmatrix} 3 \\ 1 \\ -2 \end{bmatrix}$ gives $\mathbf{y} = \begin{bmatrix} 3 \\ 4 \\ 4 \end{bmatrix}$, and $U\mathbf{x} = \mathbf{y}$ is a system with no solution.

20. The matrix equation $A\mathbf{x} = \mathbf{b}$ is $\begin{bmatrix} 1 & 1 & -1 & 2 \\ 1 & 3 & 2 & 2 \\ -1 & -3 & -4 & 6 \\ 0 & 4 & 7 & -2 \end{bmatrix} \begin{bmatrix} x_1 \\ x_2 \\ x_3 \\ x_4 \end{bmatrix} = \begin{bmatrix} 7 \\ 6 \\ 12 \\ -7 \end{bmatrix}$.

$\begin{bmatrix} 1 & 1 & -1 & 2 \\ 1 & 3 & 2 & 2 \\ -1 & -3 & -4 & 6 \\ 0 & 4 & 7 & -2 \end{bmatrix} \underset{R3+R1}{\overset{R2+(-1)R1}{\approx}} \begin{bmatrix} 1 & 1 & -1 & 2 \\ 0 & 2 & 3 & 0 \\ 0 & -2 & -5 & 8 \\ 0 & 4 & 7 & -2 \end{bmatrix} \underset{R4+(-2)R2}{\overset{R3+R2}{\approx}} \begin{bmatrix} 1 & 1 & -1 & 2 \\ 0 & 2 & 3 & 0 \\ 0 & 0 & -2 & 8 \\ 0 & 0 & 1 & -2 \end{bmatrix}$

$\overset{R4+(1/2)R3}{\approx} \begin{bmatrix} 1 & 1 & -1 & 2 \\ 0 & 2 & 3 & 0 \\ 0 & 0 & -2 & 8 \\ 0 & 0 & 0 & 2 \end{bmatrix} = U$, and $L = \begin{bmatrix} 1 & 0 & 0 & 0 \\ 1 & 1 & 0 & 0 \\ -1 & -1 & 1 & 0 \\ 0 & 2 & -1/2 & 1 \end{bmatrix}$.

$$L\mathbf{y} = \begin{bmatrix} 7 \\ 6 \\ 12 \\ -7 \end{bmatrix} \text{ gives } \mathbf{y} = \begin{bmatrix} 7 \\ -1 \\ 18 \\ 4 \end{bmatrix}, \text{ and } U\mathbf{x} = \mathbf{y} \text{ gives } \mathbf{x} = \begin{bmatrix} 1 \\ 1 \\ -1 \\ 2 \end{bmatrix}.$$

21. The matrix equation $A\mathbf{x} = \mathbf{b}$ is
$$\begin{bmatrix} 2 & 0 & 1 & -1 \\ 6 & 3 & 2 & -1 \\ 4 & 3 & -2 & 3 \\ -2 & -6 & 2 & -14 \end{bmatrix} \begin{bmatrix} x_1 \\ x_2 \\ x_3 \\ x_4 \end{bmatrix} = \begin{bmatrix} 6 \\ 15 \\ 3 \\ 12 \end{bmatrix}.$$

$$\begin{bmatrix} 2 & 0 & 1 & -1 \\ 6 & 3 & 2 & -1 \\ 4 & 3 & -2 & 3 \\ -2 & -6 & 2 & -14 \end{bmatrix} \underset{\substack{R2+(-3)R2 \\ R3+(-2)R1 \\ R4+R1}}{\approx} \begin{bmatrix} 2 & 0 & 1 & -1 \\ 0 & 3 & -1 & 2 \\ 0 & 3 & -4 & 5 \\ 0 & -6 & 3 & -15 \end{bmatrix} \underset{\substack{R3+(-1)R2 \\ R4+2R2}}{\approx} \begin{bmatrix} 2 & 0 & 1 & -1 \\ 0 & 3 & -1 & 2 \\ 0 & 0 & -3 & 3 \\ 0 & 0 & 1 & -11 \end{bmatrix}$$

$$\underset{R4+(1/3)R3}{\approx} \begin{bmatrix} 2 & 0 & 1 & -1 \\ 0 & 3 & -1 & 2 \\ 0 & 0 & -3 & 3 \\ 0 & 0 & 0 & -10 \end{bmatrix} = U, \text{ and } L = \begin{bmatrix} 1 & 0 & 0 & 0 \\ 3 & 1 & 0 & 0 \\ 2 & 1 & 1 & 0 \\ -1 & -2 & -1/3 & 1 \end{bmatrix}.$$

$$L\mathbf{y} = \begin{bmatrix} 6 \\ 15 \\ 3 \\ 12 \end{bmatrix} \text{ gives } \mathbf{y} = \begin{bmatrix} 6 \\ -3 \\ -6 \\ 10 \end{bmatrix}, \text{ and } U\mathbf{x} = \mathbf{y} \text{ gives } \mathbf{x} = \begin{bmatrix} 2 \\ 0 \\ 1 \\ -1 \end{bmatrix}.$$

22. Let $B = A^{-1}$. The (i,j)th element of the product $AB = I_n$ is $a_{i1}b_{1j} + a_{i2}b_{2j} + \ldots + a_{in}b_{nj}$. For $i = 1$, $a_{11}b_{1j} + a_{12}b_{2j} + \ldots + a_{1n}b_{nj} = a_{11}b_{1j}$, because $a_{12} = a_{13} = \ldots = a_{1n} = 0$. This means that $a_{11}b_{1j} = 0$ for $j > 1$. But $a_{11} \neq 0$ (because $|A| \neq 0$), so $b_{1j} = 0$ for $j > 1$. For $i = 2$, $a_{21}b_{1j} + a_{22}b_{2j} + \ldots + a_{2n-1}b_{nj} = a_{22}b_{2j}$, for $j > 1$, because $b_{1j} = a_{23} = a_{24} = \ldots = a_{2n} = 0$. This means that $a_{22}b_{2j} = 0$ for $j > 2$. But $a_{22} \neq 0$, so $b_{2j} = 0$ for $j > 2$.

In the same manner each row of B is shown to have zero in all positions for which the column number is greater than the row number. Thus B is lower triangular.

23. Let A and B be two $n \times n$ lower triangular matrices. The (i,j)th element of the product AB is $a_{i1}b_{1j} + a_{i2}b_{2j} + \ldots + a_{in}b_{nj}$. If $1 \leq i < j \leq n$, the elements $a_{ii+1}, a_{ii+2}, \ldots, a_{in}$ of A and

339

Section 9.2

$b_{1j}, b_{2j}, \ldots, b_{j-1,j}$ of B are all zero. Thus the first $j-1$ terms of $a_{i1}b_{1j} + a_{i2}b_{2j} + \ldots + a_{in}b_{nj}$ are zero because $b_{1j} = b_{2j} = \ldots = b_{j-1,j} = 0$ and the remainder of the terms are zero because $j > i$ and therefore $a_{ij} = a_{ij+1} = \ldots = a_{in} = 0$. Thus if $1 \le i < j \le n$, the (i,j)th element of the product is zero, so the product is a lower triangular matrix.

24. $\begin{bmatrix} 6 & -2 \\ 12 & 8 \end{bmatrix} = \begin{bmatrix} 1 & 0 \\ 2 & 1 \end{bmatrix} \begin{bmatrix} 6 & -2 \\ 0 & 12 \end{bmatrix} = \begin{bmatrix} 2 & 0 \\ 4 & 2 \end{bmatrix} \begin{bmatrix} 3 & -1 \\ 0 & 6 \end{bmatrix}$.

25. If $\begin{bmatrix} 1 & 0 & 0 \\ 0 & 0 & 1 \\ 0 & 1 & 0 \end{bmatrix} = LU = \begin{bmatrix} a & 0 & 0 \\ b & c & 0 \\ d & e & f \end{bmatrix} \begin{bmatrix} p & q & r \\ 0 & s & t \\ 0 & 0 & u \end{bmatrix}$, then $ap = 1$, so $a \ne 0$ and $p \ne 0$.

$aq = 0$ and $ar = 0$, so $q = r = 0$. Also $bp = 0$ and $dp = 0$, so $b = d = 0$. Thus

$\begin{bmatrix} 1 & 0 & 0 \\ 0 & 0 & 1 \\ 0 & 1 & 0 \end{bmatrix} = \begin{bmatrix} a & 0 & 0 \\ 0 & c & 0 \\ 0 & e & f \end{bmatrix} \begin{bmatrix} p & 0 & 0 \\ 0 & s & t \\ 0 & 0 & u \end{bmatrix} = \begin{bmatrix} 1 & 0 & 0 \\ 0 & cs & ct \\ 0 & es & et+fu \end{bmatrix}$. $cs = 0$, so one of c and s must be

zero, but $ct = es = 1$, so neither c nor s can be zero. Since no number can be both zero and nonzero, we must conclude that the matrices L and U do not exist.

26. The first pair of row operations below require three multiplications and two additions each. The last row operation requires two multiplications and one addition. Thus a total of thirteen arithmetic operations are required to obtain U.

$\begin{bmatrix} a & b & c \\ d & e & f \\ g & h & i \end{bmatrix} \underset{R3-(g/a)R1}{\overset{R2-(d/a)R1}{\approx}} \begin{bmatrix} a & b & c \\ 0 & j & k \\ 0 & m & n \end{bmatrix} \overset{R3-(m/j)R2}{\approx} \begin{bmatrix} a & b & c \\ 0 & j & k \\ 0 & 0 & s \end{bmatrix} = U$. No arithmetic operations

are required for L since it can be written down with no additional calculation.

$L = \begin{bmatrix} 1 & 0 & 0 \\ -p & 1 & 0 \\ -q & -r & 0 \end{bmatrix}$, where $p = \dfrac{d}{a}$, $q = \dfrac{g}{a}$, and $r = \dfrac{m}{j}$.

To solve $L\mathbf{y} = \mathbf{b}$ requires six operations as follows: $y_1 = b_1$ requires no operations, $y_2 = b_2 + pb_1$ requires one multiplication and one addition, and $y_3 = b_3 + qy_1 + ry_2$ requires two multiplications and two additions.

To solve $U\mathbf{x} = \mathbf{y}$ requires nine operations because there is an additional multiplication at each stage: $x_3 = y_3/s$ requires one multiplication, $x_2 = (y_2 - kx_3)/j$ requires two

Section 9.3

multiplications and one addition, and $x_1 = (y_1 - bx_2 - cx_3)/a$ requires three multiplications and two additions.

Exercise Set 9.3

1. (a) max{1+3, 2+4} = max{4,6} = 6

    (b) max{5+1, 2+3} = max{6,5} = 6

    (c) max{1+3+6, 2+5+1, 0+4+2} = max{10,8,6} = 10

    (d) max{1+6+5, 24+247+7, 3+56+219} = max{12,278,278} = 278

2. (a) $A = \begin{bmatrix} 1 & 2 \\ 3 & 4 \end{bmatrix}$ and $A^{-1} = \begin{bmatrix} -2 & 1 \\ 3/2 & -1/2 \end{bmatrix}$, so $\|A\| = \max\{4,6\} = 6$ and

    $\|A^{-1}\| = \max\{7/2, 3/2\} = 7/2$. Thus $c(A) = 21$.

    (b) $A = \begin{bmatrix} 2 & -2 \\ 3 & 1 \end{bmatrix}$ and $A^{-1} = \begin{bmatrix} 1/8 & 1/4 \\ -3/8 & 1/4 \end{bmatrix}$, so $\|A\| = \max\{5,3\} = 5$ and

    $\|A^{-1}\| = \max\{1/2, 1/2\} = 1/2$. Thus $c(A) = 5/2$.

    (c) $A = \begin{bmatrix} 5 & 2 \\ 8 & 3 \end{bmatrix}$ and $A^{-1} = \begin{bmatrix} -3 & 2 \\ 8 & -5 \end{bmatrix}$, so $\|A\| = \max\{13,5\} = 13$ and

    $\|A^{-1}\| = \max\{11,7\} = 11$. Thus $c(A) = 143$.

    (d) $A = \begin{bmatrix} 24 & 21 \\ 6 & 5 \end{bmatrix}$ and $A^{-1} = \begin{bmatrix} -5/6 & 7/2 \\ 1 & -4 \end{bmatrix}$, so $\|A\| = \max\{30,26\} = 30$ and

    $\|A^{-1}\| = \max\{11/6, 15/2\} = 15/2$. Thus $c(A) = 225$.

    (e) $A = \begin{bmatrix} 5.2 & 3.7 \\ 3.8 & 2.6 \end{bmatrix}$ and $A^{-1} = \begin{bmatrix} -2.6/.54 & 3.7/.54 \\ 3.8/.54 & -5.2/.54 \end{bmatrix}$, so $\|A\| = \max\{9,6.3\} = 9$ and

    $\|A^{-1}\| = \max\{6.4/.54, 8.9/.54\} = 8.9/.54$. Thus $c(A) = 148\frac{1}{3}$.

3. (a) $A = \begin{bmatrix} 9 & 2 \\ 4 & 7 \end{bmatrix}$ and $A^{-1} = \begin{bmatrix} 7/55 & -2/55 \\ -4/55 & 9/55 \end{bmatrix}$, so $\|A\| = \max\{13,9\} = 13$ and

$\|A^{-1}\| = \max\{1/5, 1/5\} = 1/5$. Thus $c(A) = 13/5 = 2.6 = .26 \times 10^1$.

The solution of a system of equations $AX = B$ can have one fewer significant digit of accuracy than the elements of A.

(b) $A = \begin{bmatrix} 51 & 3 \\ -1 & 27 \end{bmatrix}$ and $A^{-1} = \begin{bmatrix} 27/1380 & -3/1380 \\ 1/1380 & 51/1380 \end{bmatrix}$, so $\|A\| = \max\{52,30\} = 52$ and

$\|A^{-1}\| = \max\{28/1380, 54/1380\} = 54/1380$. Thus $c(A) = 2.035 = .2035 \times 10^1$.

The solution of a system of equations $AX = B$ can have one fewer significant digit of accuracy than the elements of A.

(c) $A = \begin{bmatrix} 300 & 1001 \\ 75 & 250 \end{bmatrix}$ and $A^{-1} = \begin{bmatrix} -10/3 & 1001/75 \\ 1 & -4 \end{bmatrix}$, so $\|A\| = \max\{375, 1251\}$

$= 1251$ and $\|A^{-1}\| = \max\{13/3, 1301/75\} = 1301/75$. Thus $c(A) = 21700.7 = .217 \times 10^5$.

The solution of a system of equations $AX = B$ can have five fewer significant digits of accuracy than the elements of A.

(d) $A = \begin{bmatrix} 3 & 4 \\ 1 & 2 \end{bmatrix}$ and $A^{-1} = \begin{bmatrix} 1 & -2 \\ -1/2 & 3/2 \end{bmatrix}$, so $\|A\| = \max\{4,6\} = 6$ and

$\|A^{-1}\| = \max\{3/2, 7/2\} = 7/2$. Thus $c(A) = 21 = .21 \times 10^2$.

The solution of a system of equations $AX = B$ can have two fewer significant digits of accuracy than the elements of A.

(e) $A = \begin{bmatrix} 5 & 3 \\ 4 & 2 \end{bmatrix}$ and $A^{-1} = \begin{bmatrix} -1 & 3/2 \\ 2 & -5/2 \end{bmatrix}$, so $\|A\| = \max\{9,5\} = 9$ and

$\|A^{-1}\| = \max\{3,4\} = 4$. Thus $c(A) = 36 = .36 \times 10^2$.

Section 9.3

The solution of a system of equations AX = B can have two fewer significant digits of accuracy than the elements of A.

4. (a) $A = \begin{bmatrix} 1 & 1 & -1 \\ 1 & 0 & 2 \\ 1 & -2 & 0 \end{bmatrix}$ and $A^{-1} = \begin{bmatrix} 1/2 & 1/4 & 1/4 \\ 1/4 & 1/8 & -3/8 \\ -1/4 & 3/8 & -1/8 \end{bmatrix}$, so $\|A\|$ = max{3,3,3} = 3 and

$\|A^{-1}\|$ = max{1, 3/4, 3/4} = 1. Thus c(A) = 3 = .3×10^1.

The solution of a system of equations AX = B can have one fewer significant digit of accuracy than the elements of A.

(b) $C = \begin{bmatrix} 4/5 & -1/5 & -3/10 \\ -1/2 & 1/2 & 0 \\ 0 & 0 & 4/5 \end{bmatrix}$ = I − A from Example 2 in Section 2.5, and

$C^{-1} = \begin{bmatrix} 5/3 & 2/3 & 5/8 \\ 5/3 & 8/3 & 5/8 \\ 0 & 0 & 5/4 \end{bmatrix}$, so $\|C\|$ = max{13/10, 7/10, 11/10} = 13/10 and

$\|C^{-1}\|$ = max{10/3, 10/3, 5/2} = 10/3. Thus c(C) = 13/3 = 4.33 = .433×10^1.

The solution of a system of equations CX = B can have one fewer significant digit of accuracy than the elements of C.

5. $A = \begin{bmatrix} 1/3 & 1/4 & 1/5 \\ 1/4 & 1/5 & 1/6 \\ 1/5 & 1/6 & 1/7 \end{bmatrix}$ and $A^{-1} = \begin{bmatrix} 300 & -900 & 630 \\ -900 & 2880 & -2100 \\ 630 & -2100 & 1575 \end{bmatrix}$, so $\|A\|$ = 47/60 and

$\|A^{-1}\|$ = 5880. Thus c(A) = 4606.

6. $A = \begin{bmatrix} 1 & k \\ 1 & 1 \end{bmatrix}$, k ≠ 1, and $A^{-1} = \begin{bmatrix} 1/(1-k) & -k/(1-k) \\ -1/(1-k) & 1/(1-k) \end{bmatrix}$, so $\|A\|$ = max{2, 1+|k|} and

$\|A^{-1}\|$ = max{2/|1−k|, (1+|k|)/|1−k|}. Thus if k > 1, c(A) = (1+k)²/k−1. Solutions to c(A) = 100 are k = 1.0417 and k = 96.9583. c(A) > 100 if k is in the interval

343

(1, 1.0417) or $k > 96.9583$. If $-1 \le k < 1$, $c(A) = 4/(1-k)$, which equals 100 if $k = .96$. $c(A) > 100$ if $.96 < k < 1$. If $k < -1$, $c(A) = 1 - k$, which is greater than 100 if $k < -99$.

7. (a) Since $\|A\|$ is the sum of nonnegative numbers, $\|A\|$ is nonnegative.

   (b) For any j, $0 \le j \le n$, $|a_{1j}| + |a_{2j}| + \ldots + |a_{nj}| = 0$ if and only if $a_{ij} = 0$ for $i = 1, 2, \ldots, n$.

   $\|A\| = 0$ if and only if $|a_{1j}| + |a_{2j}| + \ldots + |a_{nj}| = 0$ for $j = 1, 2, \ldots, n$. Thus $\|A\| = 0$ if and only if $a_{ij} = 0$ for $i = 1, 2, \ldots, n$ and $j = 1, 2, \ldots, n$, so $\|A\| = 0$ if and only if A is the zero matrix.

   (c) $\|cA\| = \max\{|ca_{1j}| + |ca_{2j}| + \ldots + |ca_{nj}|\} = \max\{|c|(|a_{1j}| + |a_{2j}| + \ldots + |a_{nj}|)\}$

   $= |c| \max\{|a_{1j}| + |a_{2j}| + \ldots + |a_{nj}|\} = |c| \|A\|$.

   (d) $|a_{1j}+b_{1j}|+|a_{2j}+b_{2j}|+\ldots+|a_{nj}+b_{nj}| \le |a_{1j}|+|b_{1j}|+|a_{2j}|+|b_{2j}|+\ldots+|a_{nj}|+|b_{nj}|$

   $= |a_{1j}| + |a_{2j}| + \ldots + |a_{nj}|+|b_{1j}| + |b_{2j}| + \ldots + |b_{nj}|$, for $j = 1, 2, \ldots, n$, so

   $\|A+B\| = \max\{|a_{1j}+b_{1j}|+|a_{2j}+b_{2j}|+\ldots+|a_{nj}+b_{nj}|\}$

   $\le \max\{|a_{1j}| + |a_{2j}| + \ldots + |a_{nj}|+|b_{1j}| + |b_{2j}| + \ldots + |b_{nj}|\}$

   $\le \max\{|a_{1j}| + |a_{2j}| + \ldots + |a_{nj}|\} + \max\{|b_{1j}| + |b_{2j}| + \ldots + |b_{nj}|\} = \|A\| + \|B\|$.

The sum of the absolute values of the elements of a column of AB is

$|a_{11}b_{1j}+a_{12}b_{2j}+\ldots+a_{1n}b_{nj}| + |a_{21}b_{1j}+a_{22}b_{2j}+\ldots+a_{2n}b_{nj}|+\ldots+|a_{n1}b_{1j}+a_{n2}b_{2j}+\ldots+a_{nn}b_{nj}|$

$\le |a_{11}b_{1j}|+|a_{12}b_{2j}|+\ldots+|a_{1n}b_{nj}|+|a_{21}b_{1j}|+|a_{22}b_{2j}|+\ldots+|a_{2n}b_{nj}|+\ldots+|a_{n1}b_{1j}|+|a_{n2}b_{2j}|$

$+\ldots+|a_{nn}b_{nj}| = (|a_{11}|+|a_{21}|+\ldots+|a_{n1}|)|b_{1j}|+(|a_{12}|+|a_{22}|+\ldots+|a_{n2}|)|b_{2j}| +\ldots$

$+(|a_{1n}|+|a_{2n}|+\ldots+|a_{nn}|)|b_{nj}| \le (\max\{|a_{1j}| + |a_{2j}| + \ldots + |a_{nj}|\})(|b_{1j}|+|b_{2j}|+\ldots+|b_{nj}|)$

$= \|A\|(|b_{1j}|+|b_{2j}|+\ldots+|b_{nj}|)$, so $\|AB\| \le \max\{\|A\|(|b_{1j}|+|b_{2j}|+\ldots+|b_{nj}|)\}$

$= \|A\| \max\{|b_{1j}|+|b_{2j}|+\ldots+|b_{nj}|\} = \|A\| \|B\|$.

Section 9.3

8. (a) $I^{-1} = I$, so both have 1-norm of 1 and $C(I) = 1$.

   (b) $1 = \|AA^{-1}\| \leq \|A\| \|A^{-1}\| = c(A)$.

9. $c(kA) = \|kA\|\|(kA)^{-1}\| = \|kA\| \|\frac{1}{k} A^{-1}\| = k\|A\| \frac{1}{k} \|A^{-1}\| = \|A\| \|A^{-1}\| = c(A)$. Thus we see that multiplying the system by a constant does not alter the expected accuracy.

10. If A is a diagonal matrix with diagonal elements a_{ii}, then A^{-1} is a diagonal matrix with diagonal elements $1/a_{ii}$. $\|A\| = \max\{a_{ii}\}$ and $\|A^{-1}\| = \max\{1/a_{ii}\} = 1/\min\{a_{ii}\}$, so $c(A) = (\max\{a_{ii}\})(1/\min\{a_{ii}\})$.

11. $\begin{bmatrix} 0 & -1 & 2 \\ 1 & 2 & -1 \\ 1 & 2 & 2 \end{bmatrix} \begin{bmatrix} x_1 \\ x_2 \\ x_3 \end{bmatrix} = \begin{bmatrix} 4 \\ 1 \\ 4 \end{bmatrix}$. $\begin{bmatrix} 0 & -1 & 2 & 4 \\ 1 & 2 & -1 & 1 \\ 1 & 2 & 2 & 4 \end{bmatrix} \underset{R1 \leftrightarrow R2}{\approx} \begin{bmatrix} 1 & 2 & -1 & 1 \\ 0 & -1 & 2 & 4 \\ 1 & 2 & 2 & 4 \end{bmatrix}$

$\underset{R3+(-1)R1}{\approx} \begin{bmatrix} 1 & 2 & -1 & 1 \\ 0 & -1 & 2 & 4 \\ 0 & 0 & 3 & 3 \end{bmatrix} \underset{(1/3)R3}{\overset{(-1)R2}{\approx}} \begin{bmatrix} 1 & 2 & -1 & 1 \\ 0 & 1 & -2 & -4 \\ 0 & 0 & 1 & 1 \end{bmatrix} \underset{R1+(-2)R2}{\approx} \begin{bmatrix} 1 & 0 & 3 & 9 \\ 0 & 1 & -2 & -4 \\ 0 & 0 & 1 & 1 \end{bmatrix}$

$\underset{R2+2R3}{\overset{R1+(-3)R3}{\approx}} \begin{bmatrix} 1 & 0 & 0 & 6 \\ 0 & 1 & 0 & -2 \\ 0 & 0 & 1 & 1 \end{bmatrix}$, so the exact solution is $\begin{bmatrix} x_1 \\ x_2 \\ x_3 \end{bmatrix} = \begin{bmatrix} 6 \\ -2 \\ 1 \end{bmatrix}$.

12. $\begin{bmatrix} -1 & 1 & 2 \\ 2 & 4 & -1 \\ 1 & 2 & 2 \end{bmatrix} \begin{bmatrix} x_1 \\ x_2 \\ x_3 \end{bmatrix} = \begin{bmatrix} 8 \\ 10 \\ 2 \end{bmatrix}$. $\begin{bmatrix} -1 & 1 & 2 & 8 \\ 2 & 4 & -1 & 10 \\ 1 & 2 & 2 & 2 \end{bmatrix} \underset{R1 \leftrightarrow R2}{\approx} \begin{bmatrix} 2 & 4 & -1 & 10 \\ -1 & 1 & 2 & 8 \\ 1 & 2 & 2 & 2 \end{bmatrix}$

$\underset{(1/2)R1}{\approx} \begin{bmatrix} 1 & 2 & -1/2 & 5 \\ -1 & 1 & 2 & 8 \\ 1 & 2 & 2 & 2 \end{bmatrix} \underset{R3+(-1)R1}{\overset{R2+R1}{\approx}} \begin{bmatrix} 1 & 2 & -1/2 & 5 \\ 0 & 3 & 3/2 & 13 \\ 0 & 0 & 5/2 & -3 \end{bmatrix}$

$\underset{(2/5)R3}{\overset{(1/3)R2}{\approx}} \begin{bmatrix} 1 & 2 & -1/2 & 5 \\ 0 & 1 & 1/2 & 4.33 \\ 0 & 0 & 1 & -6/5 \end{bmatrix} \underset{R1+(-2)R2}{\approx} \begin{bmatrix} 1 & 0 & -3/2 & -3.66 \\ 0 & 1 & 1/2 & 4.33 \\ 0 & 0 & 1 & -1.2 \end{bmatrix}$

(Here is the first round-off error.)   (The round-off error gets multiplied here.)

345

Section 9.3

$$\underset{\substack{R1+(1.5)R3\\R2+(-.5)R3}}{\approx} \begin{bmatrix} 1 & 0 & 0 & -5.46 \\ 0 & 1 & 0 & 4.93 \\ 0 & 0 & 1 & -1.2 \end{bmatrix}, \text{ so } \begin{bmatrix} x_1 \\ x_2 \\ x_3 \end{bmatrix} = \begin{bmatrix} -5.46 \\ 4.93 \\ -1.2 \end{bmatrix}. \text{ Substituting into the original}$$

equations $5.46 + 4.93 - 2.4 = 7.99$, $-10.92 + 19.72 + 1.2 = 10$, and $-5.46 + 9.86 - 2.4 = 2$. The exact solution is $x_1 = -82/15$, $x_2 = 74/15$, $x_3 = -6/5$.

13. Here one should substitute $.001x_2 = y_2$, $x_1 = y_1$, $x_3 = y_3$, which gives the matrix equation

$$\begin{bmatrix} -1 & 2 & 0 \\ 1 & 0 & 2 \\ 0 & 1 & 1 \end{bmatrix} \begin{bmatrix} y_1 \\ y_2 \\ y_3 \end{bmatrix} = \begin{bmatrix} 0 \\ -2 \\ 1 \end{bmatrix}.$$ The method from here on is straightforward and will

result in inconsistent equations. It is easy to see that the equations are inconsistent. The first equation gives $y_1 = 2y_2$ and the third equation gives $y_3 = 1 - y_2$. Substituting in the second equation $2y_2 + 2(1 - y_2) = 2$, not -2.

14. First multiply the second equation by $1/.0001$. Then the matrix equation is

$$\begin{bmatrix} 2 & 0 & 1 \\ 1 & 2 & 4 \\ 1 & -2 & -3 \end{bmatrix} \begin{bmatrix} x_1 \\ x_2 \\ x_3 \end{bmatrix} = \begin{bmatrix} 1 \\ 4 \\ -3 \end{bmatrix}. \quad \begin{bmatrix} 2 & 0 & 1 & 1 \\ 1 & 2 & 4 & 4 \\ 1 & -2 & -3 & -3 \end{bmatrix} \underset{(1/2)R1}{\approx} \begin{bmatrix} 1 & 0 & 1/2 & 1/2 \\ 1 & 2 & 4 & 4 \\ 1 & -2 & -3 & -3 \end{bmatrix}$$

$$\underset{\substack{R2+(-1)R1\\R3+(-1)R1}}{\approx} \begin{bmatrix} 1 & 0 & 1/2 & 1/2 \\ 0 & 2 & 7/2 & 7/2 \\ 0 & -2 & -7/2 & -7/2 \end{bmatrix} \underset{\substack{(1/2)R2\\R3+2R2}}{\approx} \begin{bmatrix} 1 & 0 & 1/2 & 1/2 \\ 0 & 1 & 7/4 & 7/4 \\ 0 & 0 & 0 & 0 \end{bmatrix}, \text{ so that there are many}$$

exact solutions: $x_1 = 1/2 - r/2$, $x_2 = 7/4 - 7r/4$, and $x_3 = r$.

15. First we multiply equation 2 by $1/.01$, then let $.1x_3 = y_3$, $x_2 = y_2$, $x_1 = y_1$. The matrix

equation becomes $\begin{bmatrix} 0 & 1 & -1 \\ 2 & 1 & 0 \\ 1 & -4 & 1 \end{bmatrix} \begin{bmatrix} y_1 \\ y_2 \\ y_3 \end{bmatrix} = \begin{bmatrix} 2 \\ 1 \\ 2 \end{bmatrix}.$

346

$$\begin{bmatrix} 0 & 1 & -1 & 2 \\ 2 & 1 & 0 & 1 \\ 1 & -4 & 1 & 2 \end{bmatrix} \underset{R1 \leftrightarrow R2}{\approx} \begin{bmatrix} 2 & 1 & 0 & 1 \\ 0 & 1 & -1 & 2 \\ 1 & -4 & 1 & 2 \end{bmatrix} \underset{(1/2)R1}{\approx} \begin{bmatrix} 1 & 1/2 & 0 & 1/2 \\ 0 & 1 & -1 & 2 \\ 1 & -4 & 1 & 2 \end{bmatrix}$$

$$\underset{R3+(-1)R1}{\approx} \begin{bmatrix} 1 & 1/2 & 0 & 1/2 \\ 0 & 1 & -1 & 2 \\ 0 & -9/2 & 1 & 3/2 \end{bmatrix} \underset{R2 \leftrightarrow R3}{\approx} \begin{bmatrix} 1 & 1/2 & 0 & 1/2 \\ 0 & -9/2 & 1 & 3/2 \\ 0 & 1 & -1 & 2 \end{bmatrix}$$

$$\underset{(-2/9)R2}{\approx} \begin{bmatrix} 1 & 1/2 & 0 & 1/2 \\ 0 & 1 & -.2222 & -.3333 \\ 0 & 1 & -1 & 2 \end{bmatrix} \underset{\substack{R1+(-1/2)R2 \\ R3+(-1)R2}}{\approx} \begin{bmatrix} 1 & 0 & .1111 & .6667 \\ 0 & 1 & -.2222 & -.3333 \\ 0 & 0 & -.7778 & 2.3333 \end{bmatrix}$$

(Four decimal places for **y** will result in three decimal places for **x**.)

$$\underset{(-1/.7778)R3}{\approx} \begin{bmatrix} 1 & 0 & .1111 & .6667 \\ 0 & 1 & -.2222 & -.3333 \\ 0 & 0 & 1 & -2.9999 \end{bmatrix} \underset{\substack{R1+(-.1111)R3 \\ R2+(.2222)R3}}{\approx} \begin{bmatrix} 1 & 0 & 0 & 1.0000 \\ 0 & 1 & 0 & -.9999 \\ 0 & 0 & 1 & -2.9999 \end{bmatrix}, \text{ so}$$

$$\begin{bmatrix} y_1 \\ y_2 \\ y_3 \end{bmatrix} = \begin{bmatrix} 1.0000 \\ -.9999 \\ -2.9999 \end{bmatrix}. \text{ Therefore } x_1 = 1.000, x_2 = -1.000, \text{ and } x_3 = -29.999.$$

The exact solution is $x_1 = 1$, $x_2 = -1$, and $x_3 = -30$.

16. First multiply equations 2 and 3 by 10, then let $.01x_2 = y_2$, $x_1 = y_1$, $x_3 = y_3$. The matrix equation becomes $\begin{bmatrix} -1 & 0 & 2 \\ 1 & 1 & 0 \\ 1 & 2 & 0 \end{bmatrix} \begin{bmatrix} y_1 \\ y_2 \\ y_3 \end{bmatrix} = \begin{bmatrix} 1 \\ 1 \\ 2 \end{bmatrix}.$

$$\begin{bmatrix} -1 & 0 & 2 & 1 \\ 1 & 1 & 0 & 1 \\ 1 & 2 & 0 & 2 \end{bmatrix} \underset{(-1)R1}{\approx} \begin{bmatrix} 1 & 0 & -2 & -1 \\ 1 & 1 & 0 & 1 \\ 1 & 2 & 0 & 2 \end{bmatrix} \underset{\substack{R2+(-1)R1 \\ R3+(-1)R1}}{\approx} \begin{bmatrix} 1 & 0 & -2 & -1 \\ 0 & 1 & 2 & 2 \\ 0 & 2 & 2 & 3 \end{bmatrix}$$

$$\underset{R2 \leftrightarrow R3}{\approx} \begin{bmatrix} 1 & 0 & -2 & -1 \\ 0 & 2 & 2 & 3 \\ 0 & 1 & 2 & 2 \end{bmatrix} \underset{(1/2)R2}{\approx} \begin{bmatrix} 1 & 0 & -2 & -1 \\ 0 & 1 & 1 & 3/2 \\ 0 & 1 & 2 & 2 \end{bmatrix} \underset{R3+(-1)R2}{\approx} \begin{bmatrix} 1 & 0 & -2 & -1 \\ 0 & 1 & 1 & 3/2 \\ 0 & 0 & 1 & 1/2 \end{bmatrix}$$

Section 9.4

$$\underset{R2+(-1)R3}{\overset{R1+2R3}{\approx}} \begin{bmatrix} 1 & 0 & 0 & 0 \\ 0 & 1 & 0 & 1 \\ 0 & 0 & 1 & 1/2 \end{bmatrix}, \text{ so } \begin{bmatrix} y_1 \\ y_2 \\ y_3 \end{bmatrix} = \begin{bmatrix} 0 \\ 1 \\ 1/2 \end{bmatrix}. \text{ Thus } x_1 = 0, x_2 = 100, \text{ and } x_3 = \frac{1}{2}.$$

17. $\begin{bmatrix} 1 & -2 & -1 \\ 1 & -.001 & -1 \\ 1 & 3 & -.002 \end{bmatrix} \begin{bmatrix} x_1 \\ x_2 \\ x_3 \end{bmatrix} = \begin{bmatrix} 1 \\ 2 \\ -1 \end{bmatrix}.$

$$\begin{bmatrix} 1 & -2 & -1 & 1 \\ 1 & -.001 & -1 & 2 \\ 1 & 3 & -.002 & -1 \end{bmatrix} \underset{R3+(-1)R1}{\overset{R2+(-1)R1}{\approx}} \begin{bmatrix} 1 & -2 & -1 & 1 \\ 0 & 1.999 & 0 & 1 \\ 0 & 5 & .998 & -2 \end{bmatrix}$$

$$\underset{R2 \leftrightarrow R3}{\approx} \begin{bmatrix} 1 & -2 & -1 & 1 \\ 0 & 5 & .998 & -2 \\ 0 & 1.999 & 0 & 1 \end{bmatrix} \underset{(1/5)R2}{\approx} \begin{bmatrix} 1 & -2 & -1 & 1 \\ 0 & 1 & .200 & -.4 \\ 0 & 1.999 & 0 & 1 \end{bmatrix}$$

(The first round-off error is here.)

$$\underset{R3+(-1.999)R2}{\overset{R1+2R2}{\approx}} \begin{bmatrix} 1 & 0 & -.600 & .2 \\ 0 & 1 & .200 & -.4 \\ 0 & 0 & -.400 & 1.800 \end{bmatrix} \underset{(-1/.400)R3}{\approx} \begin{bmatrix} 1 & 0 & -.600 & .2 \\ 0 & 1 & .200 & -.4 \\ 0 & 0 & 1 & -4.500 \end{bmatrix}$$

$$\underset{R2+(-.200)R3}{\overset{R1+.600R3}{\approx}} \begin{bmatrix} 1 & 0 & 0 & -2.500 \\ 0 & 1 & 0 & .500 \\ 0 & 0 & 1 & -4.500 \end{bmatrix}, \text{ so } \begin{bmatrix} x_1 \\ x_2 \\ x_3 \end{bmatrix} = \begin{bmatrix} -2.500 \\ .500 \\ -4.500 \end{bmatrix}. \text{ Substituting in the original}$$

equations $-2.500 - 1.000 + 4.500 = 1$, $-2.500 - .001 + 4.500 = 2 - .001 = 1.999$, and $-2.500 + 1.500 + .009 = -1 + .009 = -.991$.

Exercise Set 9.4

For Exercises 1–6 we give iterations found using a computer program. We stop when for all variables two successive iterations have given the same value to two decimal places.

1.
x	y	z
1	2	3
2.25	0	1.9375
2.484375	0.425	1.98515625
2.390039063	0.4059375	2.003974609
2.399509277	0.3984101562	1.99972522
2.400328766	0.4001099121	1.999945287

Thus to two decimal places x = 2.40, y = 0.40, and z = 2.00. In fact this is the exact solution.

2.

x	y	z
1	2	3
2	0.75	1.75
1.6875	0.59375	1.78125
1.6484375	0.62109375	1.79453125
1.655273438	0.6209960937	1.793144531
1.655249023	0.6206616211	1.79308252

Thus to two decimal places x = 1.66, y = 0.62, and z = 1.79. These values are the two-decimal-place approximations to the exact solution x = 144/87, y = 54/87, z = 156/87.

3.

x	y	z
0	0	0
4	5.5	2.875
4.525	5.2375	2.678125
4.511875	5.2440625	2.683046875
4.512203125	5.243898438	2.682923828

Thus to two decimal places x = 4.51, y = 5.24, and z = 2.68. These values are the two-decimal-place approximations to the exact solution x = 185/41, y = 215/41, z = 110/41.

4.

x	y	z
5	6	7
4.166666667	1.270833333	3.29375
5.125347222	2.317230903	3.206653646
4.762031973	2.293590585	3.276952664
4.781628582	2.278465407	3.271520824
4.785765002	2.280340419	3.270881042
4.785033367	2.280408911	3.271034218

Thus to two decimal places x = 4.79, y = 2.28, and z = 3.27. These values are the two-decimal-place approximations to the exact solution x = 512/107, y = 244/107, z = 350/107.

Section 9.4

5.

x	y	z
20	30	−40
30	−22.5	11.3
−1.02	5.835	−0.571
9.3954	−2.34045	3.14717
6.273042	0.1502715	2.0245541
7.22023266	−0.603977805	2.364842093
6.933267602	−0.3754232776	2.261738176
7.020220074	−0.4446754931	2.292979113
6.993873256	−0.4236918496	2.283513021
7.001856422	−0.4300499555	2.286381275
6.999437499	−0.4281234305	2.285512186

Thus to two decimal places x = 7.00, y = −0.43, and z = 2.29. These values are the two-decimal-place approximations to the exact solution x = 7, y = −3/7, z = 16/7.

6.

x	y	z	w
0	0	0	0
3.333333333	3.333333333	6.666666667	2.333333333
2.388888888	1.784722222	5.989583333	2.319097222
2.246006944	1.971853299	5.939326534	2.332596571
2.283321699	1.979753154	5.939729234	2.32525443
2.285794915	1.979343327	5.942657121	2.325638153
2.285174675	1.978688885	5.942534914	2.325725539

Thus to two decimal places x = 2.29, y = 1.98, z = 5.94, and w = 2.33. These values are the two-decimal-place approximations to the exact solution x = 7310/3199, y = 6330/3199, z = 19010/3199, w = 7440/3199.

7.

x	y	z
0	0	0
1.6	−2.4	0.1
3.54	−0.36	−2.86
1.316	−5.544	1.956
6.4264	4.3824	−8.1176
−3.52944	−15.64704	11.85296
16.488224	24.341184	−28.158816
−23.5047104	−55.6635264	51.8364736
56.49811584	104.3345894	−108.1654106
−103.5007537	−215.6661642	211.8338358
216.4996985	424.3335343	−428.1664657

The method doesn't work because the system doesn't satisfy the conditions needed for the method to work. The solution is x = 19/6, y = −7/3, z = −3/2.

Exercise Set 9.5

In Exercises 1–4 we choose $\mathbf{x} = \begin{bmatrix} 1 \\ 2 \\ 1 \end{bmatrix}$. Results are recorded to five decimal places. The process is stopped when successive results agree to three decimal places.

1.

$\dfrac{A\mathbf{x}\cdot\mathbf{x}}{\mathbf{x}\cdot\mathbf{x}}$	Scaled $A\mathbf{x}$
2	$\begin{bmatrix} 1 \\ 0.5 \\ -0.5 \end{bmatrix}$
8	$\begin{bmatrix} 1 \\ 0.125 \\ -0.875 \end{bmatrix}$
8.31579	$\begin{bmatrix} 1 \\ 0.03125 \\ -0.96875 \end{bmatrix}$
8.09063	$\begin{bmatrix} 1 \\ 0.00781 \\ -0.99219 \end{bmatrix}$
8.02325	$\begin{bmatrix} 1 \\ 0.00195 \\ -0.99805 \end{bmatrix}$
8.00585	$\begin{bmatrix} 1 \\ 0.00049 \\ -0.99951 \end{bmatrix}$
8.00146	$\begin{bmatrix} 1 \\ 0.00012 \\ -0.99988 \end{bmatrix}$
8.00037	$\begin{bmatrix} 1 \\ 0.00003 \\ -0.99997 \end{bmatrix}$

Hence $\lambda = 8$ and the dominant eigenvector is $r\begin{bmatrix} 1 \\ 0 \\ -1 \end{bmatrix}$.

2.

$\dfrac{A\mathbf{x}\cdot\mathbf{x}}{\mathbf{x}\cdot\mathbf{x}}$	Scaled $A\mathbf{x}$
5	$\begin{bmatrix} 1 \\ 0.15385 \\ 1 \end{bmatrix}$
5.56140	$\begin{bmatrix} 1 \\ 0.02740 \\ 1 \end{bmatrix}$
5.10805	$\begin{bmatrix} 1 \\ 0.00536 \\ 1 \end{bmatrix}$
5.02139	$\begin{bmatrix} 1 \\ 0.00107 \\ 1 \end{bmatrix}$
5.00427	$\begin{bmatrix} 1 \\ 0.00021 \\ 1 \end{bmatrix}$
5.00085	$\begin{bmatrix} 1 \\ 0.00004 \\ 1 \end{bmatrix}$

Hence $\lambda = 5$ and the dominant eigenvector is $r\begin{bmatrix} 1 \\ 0 \\ 1 \end{bmatrix}$.

Section 9.5

3.

$\dfrac{A\mathbf{x}\cdot\mathbf{x}}{\mathbf{x}\cdot\mathbf{x}}$	Scaled A**x**
6	$\begin{bmatrix} 1 \\ -0.1 \\ 1 \end{bmatrix}$
5.26866	$\begin{bmatrix} 1 \\ 0.01887 \\ 1 \end{bmatrix}$
6.13081	$\begin{bmatrix} 1 \\ -0.00308 \\ 1 \end{bmatrix}$
5.97843	$\begin{bmatrix} 1 \\ 0.00051 \\ 1 \end{bmatrix}$
6.00360	$\begin{bmatrix} 1 \\ -0.00008 \\ 1 \end{bmatrix}$
5.99940	$\begin{bmatrix} 1 \\ 0.00001 \\ 1 \end{bmatrix}$

Hence $\lambda = 6$ and the dominant eigenvector is $r\begin{bmatrix} 1 \\ 0 \\ 1 \end{bmatrix}$.

4.

$\dfrac{A\mathbf{x}\cdot\mathbf{x}}{\mathbf{x}\cdot\mathbf{x}}$	Scaled A**x**
9	$\begin{bmatrix} 1 \\ 0.08 \\ 1 \end{bmatrix}$
9.61244	$\begin{bmatrix} 1 \\ 0.00829 \\ 1 \end{bmatrix}$
9.06611	$\begin{bmatrix} 1 \\ 0.00092 \\ 1 \end{bmatrix}$
9.00732	$\begin{bmatrix} 1 \\ 0.00010 \\ 1 \end{bmatrix}$
9.00081	$\begin{bmatrix} 1 \\ 0.00001 \\ 1 \end{bmatrix}$
9.00009	$\begin{bmatrix} 1 \\ 0.00000 \\ 1 \end{bmatrix}$

Hence $\lambda = 9$ and the dominant eigenvector is $r\begin{bmatrix} 1 \\ 0 \\ 1 \end{bmatrix}$.

In Exercises 5–8 we choose $\mathbf{x} = \begin{bmatrix} 1 \\ 2 \\ 1 \\ 2 \end{bmatrix}$. Results are recorded to five decimal places. The process is stopped when successive results agree to three decimal places.

Section 9.5

5.

$\dfrac{A\mathbf{x}\cdot\mathbf{x}}{\mathbf{x}\cdot\mathbf{x}}$	Scaled A**x**
4.8	$\begin{bmatrix} 0.28571 \\ 0.14286 \\ 0.85714 \\ 1 \end{bmatrix}$
5.46667	$\begin{bmatrix} 0.10526 \\ 0.05263 \\ 0.94737 \\ 1 \end{bmatrix}$
5.86087	$\begin{bmatrix} 0.03636 \\ 0.01818 \\ 0.98182 \\ 1 \end{bmatrix}$
5.95964	$\begin{bmatrix} 0.01227 \\ 0.00613 \\ 0.99387 \\ 1 \end{bmatrix}$
5.98728	$\begin{bmatrix} 0.00411 \\ 0.00205 \\ 0.99795 \\ 1 \end{bmatrix}$
5.99584	$\begin{bmatrix} 0.00137 \\ 0.00069 \\ 0.99931 \\ 1 \end{bmatrix}$
5.99862	$\begin{bmatrix} 0.00046 \\ 0.00023 \\ 0.99977 \\ 1 \end{bmatrix}$
5.99954	$\begin{bmatrix} 0.00015 \\ 0.00008 \\ 0.99992 \\ 1 \end{bmatrix}$

Hence $\lambda = 6$ and the dominant eigenvector is $r\begin{bmatrix} 0 \\ 0 \\ 1 \\ 1 \end{bmatrix}$.

6.

$\dfrac{A\mathbf{x}\cdot\mathbf{x}}{\mathbf{x}\cdot\mathbf{x}}$	Scaled A**x**
6.4	$\begin{bmatrix} 0 \\ 0.09091 \\ 0.72727 \\ 1 \end{bmatrix}$
9.56989	$\begin{bmatrix} -0.04255 \\ 0.10638 \\ 0.97872 \\ 1 \end{bmatrix}$
10.42444	$\begin{bmatrix} -0.02449 \\ 0.02857 \\ 0.98776 \\ 1 \end{bmatrix}$
10.13543	$\begin{bmatrix} -0.01134 \\ 0.01377 \\ 0.99919 \\ 1 \end{bmatrix}$
10.07516	$\begin{bmatrix} -0.00482 \\ 0.00498 \\ 0.99952 \\ 1 \end{bmatrix}$
10.02848	$\begin{bmatrix} -0.00199 \\ 0.00208 \\ 0.99997 \\ 1 \end{bmatrix}$
10.01222	$\begin{bmatrix} -0.00081 \\ 0.00081 \\ 0.99998 \\ 1 \end{bmatrix}$
10.00483	$\begin{bmatrix} -0.00032 \\ 0.00033 \\ 1.00000 \\ 1 \end{bmatrix}$

Hence $\lambda = 10$ and the dominant eigenvector is $r\begin{bmatrix} 0 \\ 0 \\ 1 \\ 1 \end{bmatrix}$.

Section 9.5

7.

$\dfrac{A\mathbf{x}\cdot\mathbf{x}}{\mathbf{x}\cdot\mathbf{x}}$	Scaled A$\mathbf{x}$
14.2	$\begin{bmatrix} 1 \\ 0 \\ 0.03030 \\ 0.20202 \end{bmatrix}$
82.22351	$\begin{bmatrix} 1 \\ 0.00976 \\ -0.00940 \\ 0.04785 \end{bmatrix}$
84.30649	$\begin{bmatrix} 1 \\ 0.01118 \\ -0.01117 \\ 0.04011 \end{bmatrix}$
84.31129	$\begin{bmatrix} 1 \\ 0.01125 \\ -0.01125 \\ 0.03976 \end{bmatrix}$
84.31129	$\begin{bmatrix} 1 \\ 0.01126 \\ -0.01126 \\ 0.03975 \end{bmatrix}$

Hence $\lambda = 84.311$ and the dominant eigenvector is $r\begin{bmatrix} 1 \\ 0.011 \\ -0.011 \\ 0.040 \end{bmatrix}$.

8.

$\dfrac{A\mathbf{x}\cdot\mathbf{x}}{\mathbf{x}\cdot\mathbf{x}}$	Scaled A$\mathbf{x}$
6.8	$\begin{bmatrix} 1 \\ 0.04762 \\ 0.14286 \\ 1 \end{bmatrix}$
9.33184	$\begin{bmatrix} 1 \\ 0.02513 \\ 0.00503 \\ 0.98995 \end{bmatrix}$
10.08787	$\begin{bmatrix} 1 \\ 0.00350 \\ 0.00150 \\ 0.99700 \end{bmatrix}$
10.00593	$\begin{bmatrix} 1 \\ 0.00085 \\ 0.00005 \\ 0.99930 \end{bmatrix}$
10.00330	$\begin{bmatrix} 1 \\ 0.00015 \\ 0.00001 \\ 0.99985 \end{bmatrix}$
10.00054	$\begin{bmatrix} 1 \\ 0.00003 \\ 0.00000 \\ 0.99997 \end{bmatrix}$

Hence $\lambda = 10$ and the dominant eigenvector is $r\begin{bmatrix} 1 \\ 0 \\ 0 \\ 1 \end{bmatrix}$.

In exercises 9–11 we choose $\mathbf{x} = \begin{bmatrix} 1 \\ 2 \\ 1 \end{bmatrix}$. Results are recorded to five decimal places. The process is stopped when successive results agree to three decimal places.

354

9.

$\dfrac{Ax \cdot x}{x \cdot x}$	9.16667	9.99083	9.99991	10.00000	10.00000
Scaled Ax	$\begin{bmatrix} 0.9375 \\ 1 \\ 0.5 \end{bmatrix}$	$\begin{bmatrix} 0.99359 \\ 1 \\ 0.5 \end{bmatrix}$	$\begin{bmatrix} 0.99936 \\ 1 \\ 0.5 \end{bmatrix}$	$\begin{bmatrix} 0.99994 \\ 1 \\ 0.5 \end{bmatrix}$	$\begin{bmatrix} 0.99999 \\ 1 \\ 0.5 \end{bmatrix}$

Thus the dominant eigenvalue is $\lambda = 10$ and the dominant eigenvector is $r\begin{bmatrix} 1 \\ 1 \\ 0.5 \end{bmatrix}$.

The unit eigenvector is $\mathbf{y} = \begin{bmatrix} 2/3 \\ 2/3 \\ 1/3 \end{bmatrix}$, and $B = A - \lambda \mathbf{y}\mathbf{y}^t = \dfrac{1}{9}\begin{bmatrix} 5 & -4 & -2 \\ -4 & 5 & -2 \\ -2 & -2 & 8 \end{bmatrix}$.

Let $C = 9B$.

$\dfrac{Cx \cdot x}{x \cdot x}$	0.83333	9	9
Cx	$\begin{bmatrix} -1 \\ 0.8 \\ 0.4 \end{bmatrix}$	$\begin{bmatrix} -1 \\ 0.8 \\ 0.4 \end{bmatrix}$	$\begin{bmatrix} -1 \\ 0.8 \\ 0.4 \end{bmatrix}$

Thus the dominant eigenvalue of C is $\lambda = 9$ with eigenvector $\begin{bmatrix} -1 \\ 0.8 \\ 0.4 \end{bmatrix}$, so the dominant eigenvalue of B (and therefore an eigenvalue of A) is 1 with eigenvector $\begin{bmatrix} -1 \\ 0.8 \\ 0.4 \end{bmatrix}$.

The eigenspace of $\lambda = 1$ has dimension 2 with basis $\begin{bmatrix} -1 \\ 1 \\ 0 \end{bmatrix}$ and $\begin{bmatrix} 0 \\ -1 \\ 2 \end{bmatrix}$, so there are no more eigenvalues.

10. Using $\mathbf{x} = \begin{bmatrix} 1 \\ 2 \\ 1 \end{bmatrix}$, it takes far too many iterations to find dominant eigenvalue $\lambda = 6$ and

dominant eigenvector $r\begin{bmatrix} 1 \\ 0 \\ 1 \end{bmatrix}$. The unit eigenvector is $\mathbf{y} = \begin{bmatrix} 1/\sqrt{2} \\ 0 \\ 1/\sqrt{2} \end{bmatrix}$, and

$$B = A - \lambda \mathbf{y}\mathbf{y}^t = \begin{bmatrix} -1 & 0 & 1 \\ 0 & 4 & 0 \\ 1 & 0 & -1 \end{bmatrix}.$$

$\dfrac{B\mathbf{x} \cdot \mathbf{x}}{\mathbf{x} \cdot \mathbf{x}}$	2.66667	4	4
$B\mathbf{x}$	$\begin{bmatrix} 0 \\ 1 \\ 0 \end{bmatrix}$	$\begin{bmatrix} 0 \\ 1 \\ 0 \end{bmatrix}$	$\begin{bmatrix} 0 \\ 1 \\ 0 \end{bmatrix}$

Thus $\lambda = 4$ with unit eigenvector $\mathbf{z} = \begin{bmatrix} 0 \\ 1 \\ 0 \end{bmatrix}$. $C = B - \lambda \mathbf{z}\mathbf{z}^t = \begin{bmatrix} -1 & 0 & 1 \\ 0 & 0 & 0 \\ 1 & 0 & -1 \end{bmatrix}$. It is clear that

$\begin{bmatrix} 1 \\ 2 \\ 1 \end{bmatrix}$ is not a good choice for $\mathbf{x}$, since $C\begin{bmatrix} 1 \\ 2 \\ 1 \end{bmatrix} = \mathbf{0}$, so we choose $\mathbf{x} = \begin{bmatrix} 2 \\ 1 \\ 1 \end{bmatrix}$.

Three iterations produce $\lambda = 2$ with eigenvector $r\begin{bmatrix} 1 \\ 0 \\ -1 \end{bmatrix}$.

11.

$\dfrac{A\mathbf{x} \cdot \mathbf{x}}{\mathbf{x} \cdot \mathbf{x}}$	5.33333	11.47368	11.98459	11.99957	11.99999	12.00000
Scaled $A\mathbf{x}$	$\begin{bmatrix} 1 \\ 0.33333 \\ 1 \end{bmatrix}$	$\begin{bmatrix} 1 \\ 0.05556 \\ 1 \end{bmatrix}$	$\begin{bmatrix} 1 \\ 0.00926 \\ 1 \end{bmatrix}$	$\begin{bmatrix} 1 \\ 0.00154 \\ 1 \end{bmatrix}$	$\begin{bmatrix} 1 \\ 0.00026 \\ 1 \end{bmatrix}$	$\begin{bmatrix} 1 \\ 0.00004 \\ 1 \end{bmatrix}$

Thus $\lambda = 12$ with dominant eigenvector $r\begin{bmatrix} 1 \\ 0 \\ 1 \end{bmatrix}$. The unit eigenvector is $\mathbf{y} = \begin{bmatrix} 1/\sqrt{2} \\ 0 \\ 1/\sqrt{2} \end{bmatrix}$, and

$B = A - \lambda yy^t = \begin{bmatrix} 1 & 0 & -1 \\ 0 & 2 & 0 \\ -1 & 0 & 1 \end{bmatrix}$. Three iterations of the process confirms that $\lambda = 2$ with

dominant eigenvector $\begin{bmatrix} 0 \\ 1 \\ 0 \end{bmatrix}$. The eigenspace of $\lambda = 2$ has dimension 2 with basis

vectors $\begin{bmatrix} 0 \\ 1 \\ 0 \end{bmatrix}$ and $\begin{bmatrix} 1 \\ 0 \\ -1 \end{bmatrix}$, so there are no additional eigenvalues of A.

In Exercises 12 and 13 we choose $x = \begin{bmatrix} 1 \\ 2 \\ 1 \\ 2 \end{bmatrix}$. We record to five decimal places successive

values of $\dfrac{Ax \cdot x}{x \cdot x}$.

12. $\dfrac{Ax \cdot x}{x \cdot x}$: 3.8, 9.36066, 9.92232, 9.97852, 9.99251, 9.99731, 9.99903, 9.99965.

$\lambda = 10$ and the dominant eigenvector is $r\begin{bmatrix} 1 \\ 0 \\ 0 \\ 1 \end{bmatrix}$. The unit eigenvector is $y = \begin{bmatrix} 1/\sqrt{2} \\ 0 \\ 0 \\ 1/\sqrt{2} \end{bmatrix}$

and $B = A - \lambda yy^t = \begin{bmatrix} -1 & 0 & 0 & 1 \\ 0 & 2 & -4 & 0 \\ 0 & -4 & 2 & 0 \\ 1 & 0 & 0 & -1 \end{bmatrix}$.

$\dfrac{Bx \cdot x}{x \cdot x}$: −0.7, 1.78947, 5.12088, 5.89175, 5.98783, 5.99865, 5.99985, 5.99998.

$\lambda = 6$ and the dominant eigenvector is $r\begin{bmatrix} 0 \\ 1 \\ -1 \\ 0 \end{bmatrix}$. The unit eigenvector is $z = \begin{bmatrix} 0 \\ 1/\sqrt{2} \\ -1/\sqrt{2} \\ 0 \end{bmatrix}$

and $C = B - \lambda zz^t = \begin{bmatrix} -1 & 0 & 0 & 1 \\ 0 & -1 & -1 & 0 \\ 0 & -1 & -1 & 0 \\ 1 & 0 & 0 & -1 \end{bmatrix}$. $\dfrac{Cx \cdot x}{x \cdot x}$: $-1, -2, -2$.

$\lambda = -2$ with dominant eigenvector $r \begin{bmatrix} -1/3 \\ 1 \\ 1 \\ 1/3 \end{bmatrix}$. The eigenspace of $\lambda = -2$ is two-

dimensional with basis vectors $\begin{bmatrix} -1 \\ 0 \\ 0 \\ 1 \end{bmatrix}$ and $\begin{bmatrix} 0 \\ 1 \\ 1 \\ 0 \end{bmatrix}$, so there are no more eigenvalues.

13. $\dfrac{Ax \cdot x}{x \cdot x}$: $8.7, 9.84072, 9.96468, 9.99130, 9.99783, 9.99946, 9.99986$.

$\lambda = 10$ and the dominant eigenvector is $r \begin{bmatrix} 1 \\ 1 \\ 1/2 \\ 1/2 \end{bmatrix}$. The unit eigenvector is $y = \begin{bmatrix} 2/\sqrt{10} \\ 2/\sqrt{10} \\ 1/\sqrt{10} \\ 1/\sqrt{10} \end{bmatrix}$

and $B = A - \lambda yy^t = \begin{bmatrix} 1 & 0 & -1 & -1 \\ 0 & 1 & -1 & -1 \\ -1 & -1 & 3 & 1 \\ -1 & -1 & 1 & 3 \end{bmatrix}$.

$\dfrac{Bx \cdot x}{x \cdot x}$: $.6, 4.68, 4.95447, 4.99305, 4.99890, 4.99983, 4.99997$.

$\lambda = 5$ and the dominant eigenvector is $r \begin{bmatrix} -1/2 \\ -1/2 \\ 1 \\ 1 \end{bmatrix}$. The unit eigenvector is $z = \begin{bmatrix} -1/\sqrt{10} \\ -1/\sqrt{10} \\ 2/\sqrt{10} \\ 2/\sqrt{10} \end{bmatrix}$

and $C = B - \lambda zz^t = \begin{bmatrix} .5 & -.5 & 0 & 0 \\ -.5 & .5 & 0 & 0 \\ 0 & 0 & 1 & -1 \\ 0 & 0 & -1 & 1 \end{bmatrix}$.

Section 9.5

$\dfrac{C\mathbf{x} \cdot \mathbf{x}}{\mathbf{x} \cdot \mathbf{x}}$: .15, 1.8, 1.94118, 1.98462, 1.99611, 1.99902, 1.99976.

$\lambda = 2$ with dominant eigenvector $\mathbf{r}\begin{bmatrix} 0 \\ 0 \\ -1 \\ 1 \end{bmatrix}$. The unit eigenvector is $\mathbf{u} = \begin{bmatrix} 0 \\ 0 \\ -1/\sqrt{2} \\ 1/\sqrt{2} \end{bmatrix}$

and $D = C - \lambda \mathbf{u}\mathbf{u}^t = \begin{bmatrix} .5 & -.5 & 0 & 0 \\ -.5 & .5 & 0 & 0 \\ 0 & 0 & 0 & 0 \\ 0 & 0 & 0 & 0 \end{bmatrix}$. $\dfrac{D\mathbf{x} \cdot \mathbf{x}}{\mathbf{x} \cdot \mathbf{x}}$: 0.05, 1, 1.

$\lambda = 1$ with dominant eigenvector $\mathbf{r}\begin{bmatrix} -1 \\ 1 \\ 0 \\ 0 \end{bmatrix}$.

14. (a) $B = A + I = \begin{bmatrix} 1 & 1 & 0 & 0 \\ 1 & 1 & 1 & 0 \\ 0 & 1 & 1 & 1 \\ 0 & 0 & 1 & 1 \end{bmatrix}$ and $\mathbf{x} = \begin{bmatrix} 1 \\ 2 \\ 2 \\ 1 \end{bmatrix}$.

$\dfrac{B\mathbf{x} \cdot \mathbf{x}}{\mathbf{x} \cdot \mathbf{x}}$	2.6	2.6176	2.6180
Scaled $B\mathbf{x}$	$\begin{bmatrix} 0.6 \\ 1 \\ 1 \\ 0.6 \end{bmatrix}$	$\begin{bmatrix} 0.615 \\ 1 \\ 1 \\ 0.615 \end{bmatrix}$	$\begin{bmatrix} 0.618 \\ 1 \\ 1 \\ 0.618 \end{bmatrix}$

The sum of the entries is 3.236, so the Gould accessibility indices are .618/3.236 = .191, 1/3.236 = .309, 1/3.236 = .309, and .618/3.236 = .191. The given initial vector yields three decimal place accuracy after only three iterations.

Section 9.5

(b) $B = A + I = \begin{bmatrix} 1 & 0 & 1 & 0 & 0 & 0 \\ 0 & 1 & 1 & 0 & 0 & 0 \\ 1 & 1 & 1 & 1 & 0 & 0 \\ 0 & 0 & 1 & 1 & 1 & 1 \\ 0 & 0 & 0 & 1 & 1 & 0 \\ 0 & 0 & 0 & 1 & 0 & 1 \end{bmatrix}$ and $x = \begin{bmatrix} 1 \\ 1 \\ 3 \\ 3 \\ 1 \\ 1 \end{bmatrix}$.

$\dfrac{Bx \cdot x}{x \cdot x}$	2.90909	3	3
Scaled Bx	$\begin{bmatrix} 0.5 \\ 0.5 \\ 1 \\ 1 \\ 0.5 \\ 0.5 \end{bmatrix}$	$\begin{bmatrix} 0.5 \\ 0.5 \\ 1 \\ 1 \\ 0.5 \\ 0.5 \end{bmatrix}$	$\begin{bmatrix} 0.5 \\ 0.5 \\ 1 \\ 1 \\ 0.5 \\ 0.5 \end{bmatrix}$

The sum of the entries is 4, so the Gould accessibility indices are 1/8, 1/8, 1/4, 1/4, 1/8, and 1/8. The given initial vector yields an eigenvector for the largest positive eigenvalue immediately.

(c) $B = A + I = \begin{bmatrix} 1 & 1 & 1 & 1 & 1 \\ 1 & 1 & 0 & 0 & 0 \\ 1 & 0 & 1 & 0 & 0 \\ 1 & 0 & 0 & 1 & 0 \\ 1 & 0 & 0 & 0 & 1 \end{bmatrix}$ and $x = \begin{bmatrix} 2 \\ 1 \\ 1 \\ 1 \\ 1 \end{bmatrix}$.

$\dfrac{Bx \cdot x}{x \cdot x}$	3	3	3
Scaled Bx	$\begin{bmatrix} 1 \\ 0.5 \\ 0.5 \\ 0.5 \\ 0.5 \end{bmatrix}$	$\begin{bmatrix} 1 \\ 0.5 \\ 0.5 \\ 0.5 \\ 0.5 \end{bmatrix}$	$\begin{bmatrix} 1 \\ 0.5 \\ 0.5 \\ 0.5 \\ 0.5 \end{bmatrix}$

The sum of the entries is 3, so the Gould accessibility indices are 1/3, 1/6, 1/6, 1/6, and 1/6. In this case the initial vector is an eigenvector for the largest positive eigenvalue.

Chapter 9 Review Exercises

1. From the first equation $x_1 = 2$, and from the second equation $6 - x_2 = 7$, so $x_2 = -1$.

   From the third equation $2 - 3 + 2x_3 = 1$, so $x_3 = 1$.

2. The "rule" is $R_i - aR_j$ causes the (i,j)th position in L to be a. $L = \begin{bmatrix} 1 & 0 & 0 \\ 3/2 & 1 & 0 \\ -2/7 & 4 & 1 \end{bmatrix}$.

3. The matrix equation $A\mathbf{x} = \mathbf{b}$ is $\begin{bmatrix} 6 & 1 & -1 \\ -6 & 1 & 1 \\ 12 & 12 & 1 \end{bmatrix} \begin{bmatrix} x_1 \\ x_2 \\ x_3 \end{bmatrix} = \begin{bmatrix} 5 \\ 1 \\ 52 \end{bmatrix}$.

$\begin{bmatrix} 6 & 1 & -1 \\ -6 & 1 & 1 \\ 12 & 12 & 1 \end{bmatrix} \underset{R3+(-2)R1}{\overset{R2+R1}{\approx}} \begin{bmatrix} 6 & 1 & -1 \\ 0 & 2 & 0 \\ 0 & 10 & 3 \end{bmatrix} \overset{R3+(-5)R2}{\approx} \begin{bmatrix} 6 & 1 & -1 \\ 0 & 2 & 0 \\ 0 & 0 & 3 \end{bmatrix} = U$, and

$L = \begin{bmatrix} 1 & 0 & 0 \\ -1 & 1 & 0 \\ 2 & 5 & 1 \end{bmatrix}$. $L\mathbf{y} = \begin{bmatrix} 5 \\ 1 \\ 52 \end{bmatrix}$ gives $\mathbf{y} = \begin{bmatrix} 5 \\ 6 \\ 12 \end{bmatrix}$, and $U\mathbf{x} = \mathbf{y}$ gives $\mathbf{x} = \begin{bmatrix} 1 \\ 3 \\ 4 \end{bmatrix}$.

4. The arrays below show the numbers of multiplications and additions needed to compute each entry in the matrix $LU = A$.

```
1 1 1 ... 1   1           0 0 0 ... 0   0
1 2 2 ... 2   2           0 1 1 ... 1   1
1 2 3 ... 3   3           0 1 2 ... 2   2
: : :     :   :           : : :     :   :
1 2 3 ... n-1 n-1         0 1 2 ... n-2 n-2
1 2 3 ... n-1 n           0 1 2 ... n-2 n-1
   multiplications              additions
```

In the multiplication array, for any, i the sum of the numbers in the ith column down to a_{ii} and the numbers in the ith row to and including a_{ii} is i^2:

Chapter 9 Review Exercises

$a_{i1} + a_{1i} + a_{i2} + a_{2i} + \ldots + a_{i\,i-1} + a_{i-1\,i} + a_{ii} = 2\times 1 + 2\times 2 + \ldots + 2(i-1) + i$

$= 2(1+2+\ldots+(i-1)) + i = 2\frac{(i-1)i}{2} + i = i^2$. Thus the number of multiplications needed to calculate LU is the sum of the squares of the numbers from 1 to n:

$1^2 + 2^2 + \ldots + (n-1)^2 + n^2 = \frac{n(n+1)(2n+1)}{6} = \frac{2n^3+3n^2+n}{6}$. Likewise, the number of additions is the sum of the squares of the numbers from 1 to n–1:

$1^2 + 2^2 + \ldots + (n-2)^2 + (n-1)^2 = \frac{(n-1)n(2n-1)}{6} = \frac{2n^3-3n^2+n}{6}$. Thus the total number of operations is $\frac{2n^3+3n^2+n}{6} + \frac{2n^3-3n^2+n}{6} = \frac{2n^3+n}{3}$.

If one assumes the elements on the diagonal of L are all 1, the number of multiplications for each term on or above the diagonal is decreased by 1. There are

$n + (n-1) + \ldots + 2 + 1 = \frac{n(n+1)}{2}$ such terms, so the number of multiplications in this case is $\frac{2n^3+3n^2+n}{6} - \frac{n(n+1)}{2} = \frac{n^3-n}{3}$ and the total number of operations is

$\frac{2n^3-3n^2+n}{6} + \frac{n^3-n}{3} = \frac{4n^3-3n^2-n}{6}$.

5. $A = \begin{bmatrix} 250 & 401 \\ 125 & 201 \end{bmatrix}$ and $A^{-1} = \begin{bmatrix} 201/125 & -401/125 \\ -1 & 2 \end{bmatrix}$, so $\|A\| = \max\{375, 602\} = 602$ and

$\|A^{-1}\| = \max\{326/125, 651/125\} = 651/125$. Thus $c(A) = 3135.216 = .3135216 \times 10^4$.

The solution of a system of equations AX = B can have four fewer significant digits of accuracy than the elements of A or B.

6. $A = \begin{bmatrix} 1 & 1 & -1 \\ 1 & 0 & 2 \\ 0 & 3 & 2 \end{bmatrix}$ and $A^{-1} = \begin{bmatrix} 6/11 & 5/11 & -2/11 \\ 2/11 & -2/11 & 3/11 \\ -3/11 & 3/11 & 1/11 \end{bmatrix}$, so $\|A\| = \max\{2,4,5\} = 5$ and

Chapter 9 Review Exercises

$\|A^{-1}\| = \max\{1, 10/11, 6/11\} = 1$. Thus $c(A) = 5 = .5 \times 10^1$. The solution of a system of equations $AX = B$ can have one fewer significant digit of accuracy than the elements of A.

$$A = \begin{bmatrix} 1 & -1 & -1 & 0 & 0 \\ 0 & 0 & 1 & -1 & -1 \\ 1 & 1 & 0 & 0 & 0 \\ 1 & 0 & 0 & 2 & 0 \\ 0 & 0 & 0 & 2 & 2 \end{bmatrix} \text{ and } A^{-1} = \begin{bmatrix} 1/2 & 1/2 & 1/2 & 0 & 1/4 \\ -1/2 & -1/2 & 1/2 & 0 & -1/4 \\ 0 & 1 & 0 & 0 & 1/2 \\ -1/4 & -1/4 & -1/4 & 1/2 & -1/8 \\ 1/4 & 1/4 & 1/4 & -1/2 & 5/8 \end{bmatrix},$$

so $\|A\| = \max\{3,2,2,5,3\} = 5$ and $\|A^{-1}\| = \max\{3/2, 5/2, 3/2, 1, 7/4\} = 5/2$. Thus $c(A) = 25/2 = 12.5 = .125 \times 10^2$. The solution of a system of equations $AX = B$ can have two fewer significant digits of accuracy than the elements of A.

7. $c(A) = \|A\| \|A^{-1}\| = \|A^{-1}\| \|A\| = \|A^{-1}\| \|(A^{-1})^{-1}\| = c(A^{-1})$.

8. $c(A)$ is defined only for nonsingular matrices, so it is not a mapping of M_{nn}.

9. (a) Multiply the first equation by 10, then let $100x_1 = y_1$, $x_2 = y_2$, $.01x_3 = y_3$. The matrix equation becomes $\begin{bmatrix} 0 & -4 & 2 \\ 1 & 1 & -1 \\ 2 & 2 & -3 \end{bmatrix} \begin{bmatrix} y_1 \\ y_2 \\ y_3 \end{bmatrix} = \begin{bmatrix} 2 \\ 1 \\ 1 \end{bmatrix}$.

$$\begin{bmatrix} 0 & -4 & 2 & 2 \\ 1 & 1 & -1 & 1 \\ 2 & 2 & -3 & 1 \end{bmatrix} \underset{R1 \leftrightarrow R3}{\approx} \begin{bmatrix} 2 & 2 & -3 & 1 \\ 1 & 1 & -1 & 1 \\ 0 & -4 & 2 & 2 \end{bmatrix} \underset{(1/2)R1}{\approx} \begin{bmatrix} 1 & 1 & -3/2 & 1/2 \\ 1 & 1 & -1 & 1 \\ 0 & -4 & 2 & 2 \end{bmatrix}$$

$$\underset{R2+(-1)R1}{\approx} \begin{bmatrix} 1 & 1 & -3/2 & 1/2 \\ 0 & 0 & 1/2 & 1/2 \\ 0 & -4 & 2 & 2 \end{bmatrix} \underset{R2 \leftrightarrow R3}{\approx} \begin{bmatrix} 1 & 1 & -3/2 & 1/2 \\ 0 & -4 & 2 & 2 \\ 0 & 0 & 1/2 & 1/2 \end{bmatrix}$$

$$\underset{(-1/4)R2}{\approx} \begin{bmatrix} 1 & 1 & -3/2 & 1/2 \\ 0 & 1 & -1/2 & -1/2 \\ 0 & 0 & 1/2 & 1/2 \end{bmatrix} \underset{R1+(-1)R2}{\approx} \begin{bmatrix} 1 & 0 & -1 & 1 \\ 0 & 1 & -1/2 & -1/2 \\ 0 & 0 & 1/2 & 1/2 \end{bmatrix}$$

$$\underset{(2)R3}{\approx} \begin{bmatrix} 1 & 0 & -1 & 1 \\ 0 & 1 & -1/2 & -1/2 \\ 0 & 0 & 1 & 1 \end{bmatrix} \underset{R2+(1/2)R3}{\overset{R1+R3}{\approx}} \begin{bmatrix} 1 & 0 & 0 & 2 \\ 0 & 1 & 0 & 0 \\ 0 & 0 & 1 & 1 \end{bmatrix}, \text{ so } \begin{bmatrix} y_1 \\ y_2 \\ y_3 \end{bmatrix} = \begin{bmatrix} 2 \\ 0 \\ 1 \end{bmatrix}.$$

Thus $x_1 = 1/50$, $x_2 = 0$, and $x_3 = 100$.

(b) Multiply the second equation by 10 and the third equation by 100, then let $.001x_3 = y_3$, $x_1 = y_1$, $x_2 = y_2$. The matrix equation becomes

$$\begin{bmatrix} 0 & -1 & 1 \\ 1 & 0 & 2 \\ 1 & 1 & 3 \end{bmatrix} \begin{bmatrix} y_1 \\ y_2 \\ y_3 \end{bmatrix} = \begin{bmatrix} 6 \\ -2 \\ 2 \end{bmatrix}. \quad \begin{bmatrix} 0 & -1 & 1 & 6 \\ 1 & 0 & 2 & -2 \\ 1 & 1 & 3 & 2 \end{bmatrix} \underset{R1 \leftrightarrow R2}{\approx} \begin{bmatrix} 1 & 0 & 2 & -2 \\ 0 & -1 & 1 & 6 \\ 1 & 1 & 3 & 2 \end{bmatrix}$$

$$\underset{R3+(-1)R1}{\approx} \begin{bmatrix} 1 & 0 & 2 & -2 \\ 0 & -1 & 1 & 6 \\ 0 & 1 & 1 & 4 \end{bmatrix} \underset{(-1)R2}{\approx} \begin{bmatrix} 1 & 0 & 2 & -2 \\ 0 & 1 & -1 & -6 \\ 0 & 1 & 1 & 4 \end{bmatrix} \underset{R3+(-1)R2}{\approx} \begin{bmatrix} 1 & 0 & 2 & -2 \\ 0 & 1 & -1 & -6 \\ 0 & 0 & 2 & 10 \end{bmatrix}$$

$$\underset{(1/2)R3}{\approx} \begin{bmatrix} 1 & 0 & 2 & -2 \\ 0 & 1 & -1 & -6 \\ 0 & 0 & 1 & 5 \end{bmatrix} \underset{R2+R3}{\overset{R1+(-2)R3}{\approx}} \begin{bmatrix} 1 & 0 & 0 & -12 \\ 0 & 1 & 0 & -1 \\ 0 & 0 & 1 & 5 \end{bmatrix}, \text{ so } \begin{bmatrix} y_1 \\ y_2 \\ y_3 \end{bmatrix} = \begin{bmatrix} -12 \\ -1 \\ 5 \end{bmatrix}.$$

Thus $x_1 = -12$, $x_2 = -1$, and $x_3 = 5000$.

10. These iterations were found using a computer program. We stop when for all variables two successive iterations have given the same value to two decimal places.

x	y	z
1	2	-3
3.666666667	-1.047619048	3.845238095
5.31547619	1.414965986	2.817389456
4.733737245	1.140670554	3.03139805
4.815121249	1.20913595	2.9939357
4.797466625	1.198124836	3.001102135
4.800496217	1.200330567	2.999793304

Thus to two decimal places $x = 4.80$, $y = 1.20$, and $z = 3.00$. Substitution of these values into the given equations will show that, in fact, this is the exact solution.

Chapter 9 Review Exercises

11. We choose $\mathbf{x} = \begin{bmatrix} 1 \\ 2 \\ 1 \end{bmatrix}$. Results are recorded to 5 decimal places. The process is stopped when successive results agree to three significant digits.

$\dfrac{A\mathbf{x}\cdot\mathbf{x}}{\mathbf{x}\cdot\mathbf{x}}$	Scaled $A\mathbf{x}$
9.83333	$\begin{bmatrix} 0.61111 \\ 0.83333 \\ 1 \end{bmatrix}$
11.71045	$\begin{bmatrix} 0.64 \\ 1 \\ 0.955 \end{bmatrix}$
11.65458	$\begin{bmatrix} 0.64954 \\ 0.98520 \\ 1 \end{bmatrix}$
11.72116	$\begin{bmatrix} 0.65327 \\ 1 \\ 0.99933 \end{bmatrix}$
11.71197	$\begin{bmatrix} 0.65228 \\ 0.99646 \\ 1 \end{bmatrix}$
11.71629	$\begin{bmatrix} 0.65256 \\ 0.99752 \\ 1 \end{bmatrix}$
11.71567	$\begin{bmatrix} 0.65246 \\ 0.99723 \\ 1 \end{bmatrix}$
11.71594	$\begin{bmatrix} 0.65248 \\ 0.99731 \\ 1 \end{bmatrix}$

Hence $\lambda = 11.716$ and the dominant eigenvector is $r \begin{bmatrix} 0.652 \\ 0.997 \\ 1 \end{bmatrix}$.

12. We choose $\mathbf{x} = \begin{bmatrix} 1 \\ 2 \\ 1 \end{bmatrix}$. Values of $\dfrac{A\mathbf{x} \cdot \mathbf{x}}{\mathbf{x} \cdot \mathbf{x}}$ are 4, 5.5, 5.74194, 5.87645, 5.94314, 5.97432, 5.98851, 5.99487, 5.99772, 5.99899. Thus $\lambda = 6$. The dominant eigenvector is $r\begin{bmatrix} 1 \\ 2 \\ -1 \end{bmatrix}$ and the unit eigenvector is $\mathbf{y} = \begin{bmatrix} 1/\sqrt{6} \\ 2/\sqrt{6} \\ -1/\sqrt{6} \end{bmatrix}$.

$B = A - \lambda \mathbf{y}\mathbf{y}^t = \begin{bmatrix} 2 & 0 & 2 \\ 0 & 0 & 0 \\ 2 & 0 & 2 \end{bmatrix}$. Three iterations of the process confirm that $\lambda = 4$ with dominant eigenvector $s\begin{bmatrix} 1 \\ 0 \\ 1 \end{bmatrix}$. The unit eigenvector is $\mathbf{z} = \begin{bmatrix} 1/\sqrt{2} \\ 0 \\ 1/\sqrt{2} \end{bmatrix}$, and $C = A - \lambda \mathbf{z}\mathbf{z}^t$ is the zero matrix. Thus the remaining eigenvalue is zero. The corresponding eigenvector is $t\begin{bmatrix} -1 \\ 1 \\ 1 \end{bmatrix}$.

Chapter 10

Exercise Set 10.1

For Exercises 1 - 14 in this section, we refer to the drawing at the right. The points A, B, C, E, and F will be identified, and the objective function will be evaluated at the three nonzero vertices A, B, and C of the feasible region.

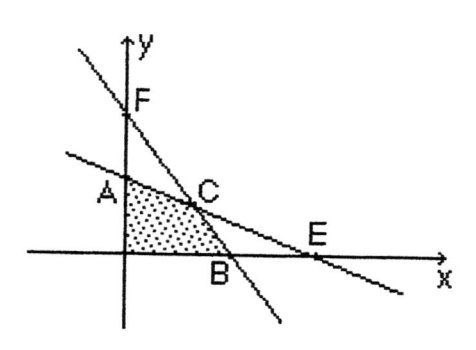

1. $4x + y = 36$ $F = (0,36)$ $B = (9,0)$

    $4x + 3y = 60$ $A = (0,20)$ $E = (15,0)$ $C = (6,12)$

    At A, $f = 2(0) + 20 = 20$; at B, $f = 2(9) + 0 = 18$; and at C, $f = 2(6) + 12 = 24$, so the maximum value of f is 24 and it occurs at the point C.

2. $x + 2y = 4$ $F = (0,2)$ $B = (4,0)$

    $x + 6y = 8$ $A = (0,4/3)$ $E = (8,0)$ $C = (2,1)$

    At A, $f = 0 - 4(4/3) = -16/3$; at B, $f = 4 - 4(0) = 4$; and at C, $f = 2 - 4(1) = -2$, so the maximum value of f is 4 and it occurs at the point B.

3. $x + 3y = 15$ $A = (0,5)$ $E = (15,0)$

    $2x + y = 10$ $F = (0,10)$ $B = (5,0)$ $C = (3,4)$

    At A, $f = 4(0) + 2(5) = 10$; at B, $f = 4(5) + 2(0) = 20$; and at C, $f = 4(3) + 2(4) = 20$, so the maximum value of f is 20 and it occurs at every point on the line segment joining B and C.

Section 10.1

4. $4x + y = 16$ $F = (0,16)$ $B = (4,0)$

     $x + y = 7$ $A = (0,7)$ $E = (7,0)$ $C = (3,4)$

At A, $f = 2(0) + 7 = 7$; at B, $f = 2(4) + 0 = 8$; and at C, $f = 2(3) + 4 = 10$, so the maximum value of f is 10 and it occurs at the point C.

5. $2x + y = 4$ $A = (0,4)$ $E = (2,0)$

     $6x + y = 8$ $F = (0,8)$ $B = (4/3,0)$ $C = (1,2)$

At A, $f = 4(0) + 4 = 4$; at B, $f = 4(4/3) + 0 = 16/3$; and at C, $f = 4(1) + 2 = 6$, so the maximum value of f is 6 and it occurs at the point C.

6. $x + y = 150$ $A = (0,150)$ $E = (150,0)$

     $4x + y = 450$ $F = (0,450)$ $B = (112.5,0)$ $C = (100,50)$

At A, $-f = 3(0) + 150 = 20$; at B, $-f = 3(112.5) + 0 = 337.5$; and at C, $-f = 3(100) + 50 = 350$, so the maximum value of $-f$ is 350 and it occurs at the point C. Thus the minimum value of f is -350 and it occurs at the point C.

7. $2x + y = 440$ $A = (0,440)$ $E = (220,0)$

     $4x + y = 680$ $F = (0,680)$ $B = (170,0)$ $C = (120,200)$

At A, $-f = 2(0) - 440 = -440$; at B, $-f = 2(170) - 0 = 340$; and at C, $-f = 2(120) - 200 = 40$, so the maximum value of $-f$ is 340 and it occurs at the point B. Thus the minimum value of f is -340 and it occurs at the point B.

8. $x + 2y = 4$ $F = (0,2)$ $B = (4,0)$

     $x + 4y = 6$ $A = (0,3/2)$ $E = (6,0)$ $C = (2,1)$

Section 10.1

At A, –f = 0 – 2(3/2) = –3; at B, –f = 4 – 2(0) = 4; and at C,

–f = 2 – 2(1) = 0, so the maximum value of –f is 4 and it occurs at the point B.

Thus the minimum value of f is –4 and it occurs at the point B.

9.

	hours	cost($)	profit($)	number to manufacture
C1	1	30	10	x
C2	4	20	8	y
totals	≤1600	≤18000		

We are to maximize profit, f = 10x + 8y, subject to the constraints

$$x + 4y \le 1600 \qquad \text{(hours)},$$

$$30x + 20y \le 18000 \text{ (divide by 10: } 3x + 2y \le 1800) \qquad \text{(cost)},$$

$$x \ge 0, \text{ and } y \ge 0.$$

x + 4y = 1600   A = (0,400)   E = (1600,0)

3x + 2y = 1800   F = (0,900)   B = (600,0)   C = (400,300)

At A, f = 10(0) + 8(400) = 3200; at B, f = 10(600) + 8(0) = 6000; and at C,

f = 10(400) + 8(300) = 6400, so the maximum value of f is 6400 and it occurs at the point C. To ensure the maximum profit of $6400, the company should manufacture 400 of model C1 and 300 of model C2.

10.

	machine I time(min)	machine II time(min)	profit($)	number
X	3	3	15	x
Y	1	2	7	y
totals	≤3000	≤4500		

We are to maximize profit, f = 15x + 7y, subject to the constraints

$$3x + y \le 3000 \qquad \text{(machine I time)},$$

$$3x + 2y \le 4500 \qquad \text{(machine II time)},$$

$$x \ge 0, \text{ and } y \ge 0.$$

Section 10.1

$3x + y = 3000$ $F = (0,3000)$ $B = (1000,0)$

$3x + 2y = 4500$ $A = (0,2250)$ $E = (1500,0)$ $C = (500,1500)$

At A, $f = 15(0) + 7(2250) = 15750$; at B, $f = 15(1000) + 7(0) = 15000$; and at C, $f = 15(500) + 7(1500) = 18000$, so the maximum value of f is 18000 and it occurs at the point C. To ensure the maximum profit of $18000, the company should manufacture 500 of product X and 1500 of product Y.

11.

	hours	cost($)	profit($)	number to ship
X	20	60	40	x
Y	10	10	20	y
totals	≤1200	≤2400		

We are to maximize profit, $f = 40x + 20y$, subject to the constraints

$20x + 10y \le 1200$ (divide by 10: $2x + y \le 120$)   (hours),

$60x + 10y \le 2400$ (divide by 10: $6x + y \le 240$)   (cost),

$x \ge 0$, and $y \ge 0$.

$2x + y = 120$ $A = (0,120)$ $E = (60,0)$

$6x + y = 240$ $F = (0,240)$ $B = (40,0)$ $C = (30,60)$

At A, $f = 40(0) + 20(120) = 2400$; at B, $f = 40(40) + 20(0) = 1600$; and at C, $f = 40(30) + 20(60) = 2400$, so the maximum value of f is 2400 and it occurs at all points on the line segment joining points A and C. To ensure the maximum profit of $2400, the company should ship x refrigerators from the plant at town X and 120 − 2x refrigerators from the plant at town Y, where $0 \le x \le 30$.

12.

	cost($)	profit($)	quantity(barrels)
gasoline	6	3.50	x
heating oil	8	4.00	y
totals	≤9600		≤1400

We are to maximize profit, $f = 3.5x + 4y$, subject to the constraints

370

Section 10.1

$x + y \leq 1400$ (quantity),

$6x + 8y \leq 9600$ (divide by 2: $3x + 4y \leq 4800$) (cost),

$x \geq 0$, and $y \geq 0$.

$x + y = 1400$ $F = (0, 1400)$ $B = (1400, 0)$

$3x + 4y = 4800$ $A = (0, 1200)$ $E = (1600, 0)$ $C = (800, 600)$

At A, $f = 3.5(0) + 4(1200) = 4800$; at B, $f = 3.5(1400) + 4(0) = 4900$; and at C, $f = 3.5(800) + 4(600) = 5200$, so the maximum value of f is 5200 and it occurs at the point C. To ensure the maximum profit of $5200, the company should produce 800 barrels of gasoline and 600 barrels of heating oil.

13.

	cotton(sq. yd.)	wool(sq. yd.)	number to make	income($)
suit	2	1	x	90
dress	1	3	y	90
totals	≤80	≤120		

We are to maximize income, $f = 90x + 90y$, subject to the constraints

$2x + y \leq 80$ (sq. yd. cotton),

$x + 3y \leq 120$ (sq. yd. wool),

$x \geq 0$, and $y \geq 0$.

$2x + y = 80$ $F = (0, 80)$ $B = (40, 0)$

$x + 3y = 120$ $A = (0, 40)$ $E = (120, 0)$ $C = (24, 32)$

At A, $f = 90(0) + 90(40) = 3600$; at B, $f = 90(40) + 90(0) = 3600$; and at C, $f = 90(24) + 90(32) = 5040$, so the maximum value of f is 5040 and it occurs at the point C. To ensure the maximum income of $5040, the tailor should make 24 suits and 32 dresses.

14.

	purchase cost($)	maintenance cost($)	number
Arrow	4000	400	x
Gazelle	5000	300	y
totals	≤600000	≤40000	

We are to maximize $f = 24x + 20y$, subject to the constraints

$$4000x + 5000y \le 600000 \text{ (divide by 1000: } 4x + 5y \le 600\text{)},$$

$$400x + 300y \le 40000 \text{ (divide by 100: } 4x + 3y \le 400\text{)},$$

$$x \ge 0, \text{ and } y \ge 0.$$

$4x + 5y = 600$ $\quad A = (0,120) \quad E = (150,0)$

$4x + 3y = 400$ $\quad F = (0,400/3) \quad B = (100,0) \quad C = (25,100)$

At A, $f = 24(0) + 20(120) = 2400$; at B, $f = 24(100) + 20(0) = 2400$; and at C,

$f = 24(25) + 20(100) = 2600$, so the maximum value of f is 2600 and it occurs at the point

C. To ensure the maximum gasoline efficiency number of 2600, the company should

purchase 25 Arrows and 100 Gazelles.

15.

	from A	from B	time
cars to C	x	120 − x	2x + 6(120 − x)
cars to D	y	180 − y	4y + 3(180 − y)
totals	≤100	≤200	≤1030

We are to maximize $f = 120 - x$, subject to the constraints

$$x + y \le 100,$$

$$120 - x + 180 - y \le 200 \text{ (simplified: } x + y \ge 100\text{)}$$

(the first two constraints imply $x + y = 0$),

$$2x + 6(120 - x) + 4y + 3(180 - y) \le 1030 \text{ (simplified: } 4x - y \ge 230\text{)},$$

$$x \ge 0, \text{ and } y \ge 0.$$

Section 10.1

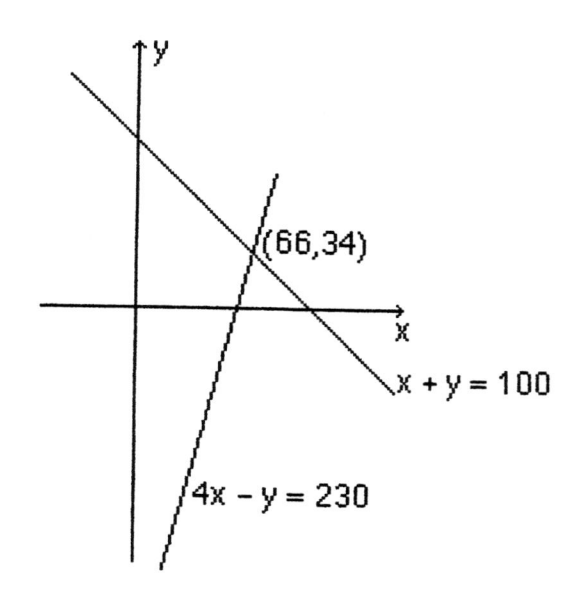

The only point in the feasible region is the point (66,34). Thus from port A, 66 cars should be moved to city C and 34 cars should be moved to city D. From port B, 54 cars should be moved to city C and 146 cars should be moved to city D.

16.

	cost($)	passengers	number to buy
Torro	18000	25	x
Sprite	22000	30	y
totals	≤572000	≤30	

We are to maximize $f = 25x + 30y$, subject to the constraints

$18000x + 22000y \le 572000$ (divide by 2000: $9x + 11y \le 286$),

$x + y \le 30$,

$x \ge 0$, and $y \ge 17$.

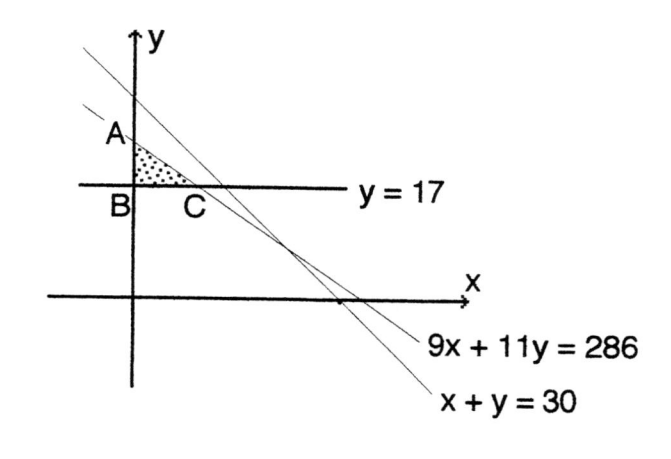

$A = (0,26)$, $B = (0,17)$, and $C = (11,17)$. The values of f at these points are 780, 510, and 785, so to maximize the number of students that can be transported, the school district should purchase 11 Torros and 17 Sprites.

17.

	A (tons)	B (tons)	tons of fertilizer
X	.8	.2	x
Y	.6	.4	y
totals	≤100	≤50	

We are to maximize the amount of fertilizer, $f = x + y$, subject to the constraints

$.8x + .6y \leq 100$ (multiply by 5: $4x + 3y \leq 500$) (amount of A),

$.2x + .4y \leq 50$ (multiply by 5: $x + 2y \leq 250$) (amount of B),

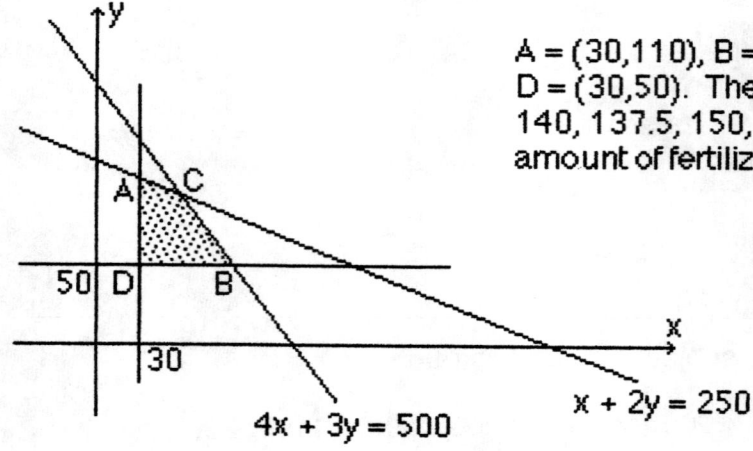

$x \geq 30$, and $y \geq 50$.

$A = (30, 110)$, $B = (87.5, 50)$, $C = (50, 100)$, and $D = (30, 50)$. The values of f at these points are 140, 137.5, 150, and 80. Thus the maximum amount of fertilizer that can be made is 150 tons. To achieve this maximum, the manufacturer should make 50 tons of X and 100 tons of Y.

18.

	units of A per oz.	units of B per oz.	cost(cents per oz.)	oz.
M	1	2	8	x
N	1	1	12	y
totals	≥7	≥10		

We are to minimize the cost, $f = 8x + 12y$, subject to the constraints

$x + y \geq 7$ (units of A),
$2x + y \geq 10$ (units of B),

$x \geq 0$, and $y \geq 0$.

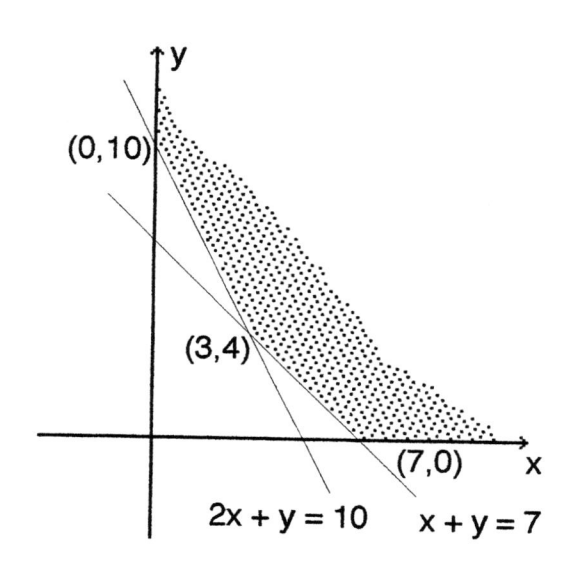

The values of f are 8(0) + 12(10) = 120, 8(3) + 12(4) = 72, and 8(7) + 12(0) = 56. Thus the hospital can achieve the minimum cost of 56 cents by serving 7 oz. of item M and no item N.

19.

	cotton	wool	silk	profit($)	number to make
slacks	1	2	1	3	x
skirts	2	1	1	4	y
totals	≤10	≤10	6		

We are to maximize profit, f = 3x + 4y, subject to the constraints

$x + 2y \leq 10$ (sq. yd. of cotton),

$2x + y \leq 10$ (sq. yd. of wool),

$x + y \leq 6$ (sq. yd. of silk),

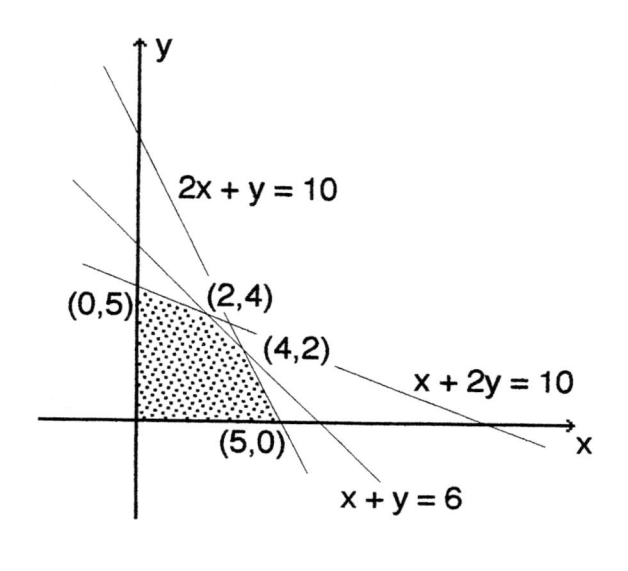

$x \geq 0$, and $y \geq 0$.

The values of f at (0,5), (2,4), (4,2), and (5,0) are 20, 22, 20, and 15. Thus the maximum profit of $22 can be achieved by making 2 pairs of slacks and 4 skirts.

Section 10.2

20. (a)

	pounds	cu. ft.	profit($)	number of packages
Pringle	5	5	.30	x
Williams	6	3	.40	y
totals	≤12000	≤9000		

We are to maximize profit, $f = .3x + .4y$, subject to the constraints

$$5x + 6y \le 12000 \quad \text{(weight in pounds)},$$

$$5x + 3y \le 9000 \quad \text{(volume in cu. ft.)},$$

$$x \ge 0, \text{ and } y \ge 0.$$

At A, B, and C, $f = 800, 540,$ and 760. Thus the profit is maximized if the shipper carries no packages for Pringle and 2000 packages for Williams.

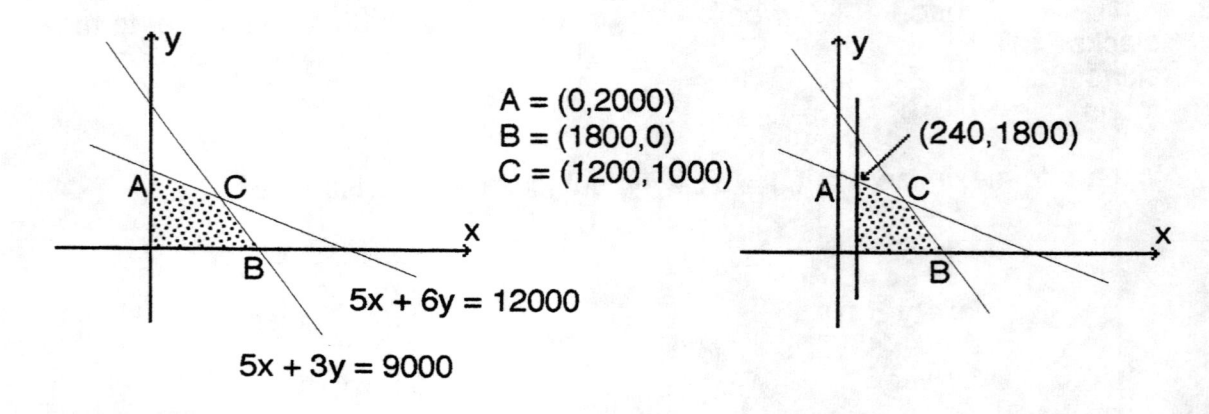

A = (0,2000)
B = (1800,0)
C = (1200,1000)

$5x + 6y = 12000$
$5x + 3y = 9000$

(240, 1800)

(b) The condition that the shipper must carry at least 240 packages for Pringle is the constraint $x \ge 240$. Thus the point A (where the maximum value occurred) is no longer in the feasible region. The maximum value of f now occurs at the point (240, 1800) and is $792. The clause in the fine print cost the shipper $8.

Exercise Set 10.2

1. We indicate the row and column of the initial pivot element with arrows.

$$\begin{array}{c} \begin{array}{ccccc} x & y & u & v & f \end{array} \\ \to \begin{bmatrix} 4 & 1 & 1 & 0 & 0 & 36 \\ 4 & 3 & 0 & 1 & 0 & 60 \\ -2 & -1 & 0 & 0 & 1 & 0 \end{bmatrix} \end{array} \approx \begin{bmatrix} 1 & 1/4 & 1/4 & 0 & 0 & 9 \\ 4 & 3 & 0 & 1 & 0 & 60 \\ -2 & -1 & 0 & 0 & 1 & 0 \end{bmatrix} \approx \begin{bmatrix} 1 & 1/4 & 1/4 & 0 & 0 & 9 \\ 0 & 2 & -1 & 1 & 0 & 24 \\ 0 & -1/2 & 1/2 & 0 & 1 & 18 \end{bmatrix}$$

$$\approx \begin{bmatrix} 1 & 1/4 & 1/4 & 0 & 0 & 9 \\ 0 & 1 & -1/2 & 1/2 & 0 & 12 \\ 0 & -1/2 & 1/2 & 0 & 1 & 18 \end{bmatrix} \approx \begin{bmatrix} 1 & 0 & 3/8 & -1/8 & 0 & 6 \\ 0 & 1 & -1/2 & 1/2 & 0 & 12 \\ 0 & 0 & 1/4 & 1/4 & 1 & 24 \end{bmatrix}.$$ The elements in the last row are all positive, so this is the final tableau. The maximum value of f is 24 when u and v are both zero. With $u = v = 0$, the first two rows give $x = 6$ and $y = 12$.

2. $\quad\quad\quad x\ y\ u\ v\ f$
$$\rightarrow \begin{bmatrix} 1 & 2 & 1 & 0 & 0 & 4 \\ 1 & 6 & 0 & 1 & 0 & 8 \\ -1 & 4 & 0 & 0 & 1 & 0 \end{bmatrix} \approx \begin{bmatrix} 1 & 2 & 1 & 0 & 0 & 4 \\ 0 & 4 & -1 & 1 & 0 & 4 \\ 0 & 6 & 1 & 0 & 1 & 4 \end{bmatrix}.$$ The maximum value of f is 4 when $y = u = 0$. The first row gives $x = 4$.

3. $\quad\quad\quad x\ y\ u\ v\ f$
$$\rightarrow \begin{bmatrix} 1 & 3 & 1 & 0 & 0 & 6 \\ 3 & 1 & 0 & 1 & 0 & 8 \\ -4 & -6 & 0 & 0 & 1 & 0 \end{bmatrix} \approx \begin{bmatrix} 1/3 & 1 & 1/3 & 0 & 0 & 2 \\ 3 & 1 & 0 & 1 & 0 & 8 \\ -4 & -6 & 0 & 0 & 1 & 0 \end{bmatrix} \approx \begin{bmatrix} 1/3 & 1 & 1/3 & 0 & 0 & 2 \\ 8/3 & 0 & -1/3 & 1 & 0 & 6 \\ -2 & 0 & 2 & 0 & 1 & 12 \end{bmatrix}$$

$$\approx \begin{bmatrix} 1/3 & 1 & 1/3 & 0 & 0 & 2 \\ 1 & 0 & -1/8 & 3/8 & 0 & 9/4 \\ -2 & 0 & 2 & 0 & 1 & 12 \end{bmatrix} \approx \begin{bmatrix} 0 & 1 & 3/8 & -1/8 & 0 & 5/4 \\ 1 & 0 & -1/8 & 3/8 & 0 & 9/4 \\ 0 & 0 & 7/4 & 3/4 & 1 & 33/2 \end{bmatrix}.$$ The maximum value of f is 33/2 when $u = v = 0$, $y = 5/4$, and $x = 9/4$.

4. $\quad\quad\quad x\ y\ u\ v\ f$
$$\rightarrow \begin{bmatrix} 1 & 1 & 1 & 0 & 0 & 180 \\ 3 & 2 & 0 & 1 & 0 & 480 \\ -10 & -5 & 0 & 0 & 1 & 0 \end{bmatrix} \approx \begin{bmatrix} 1 & 1 & 1 & 0 & 0 & 180 \\ 1 & 2/3 & 0 & 1/3 & 0 & 160 \\ -10 & -5 & 0 & 0 & 1 & 0 \end{bmatrix}$$

$$\approx \begin{bmatrix} 0 & 1/3 & 1 & -1/3 & 0 & 20 \\ 1 & 2/3 & 0 & 1/3 & 0 & 160 \\ 0 & 5/3 & 0 & 10/3 & 1 & 1600 \end{bmatrix}.$$ The maximum value of f is 1600 when $y = v = 0$. From the second row $x = 160$.

5.
$$\begin{array}{c} \begin{array}{cccccc} x & y & z & u & v & f \end{array} \\ \to \begin{bmatrix} 3 & 1 & 1 & 1 & 0 & 0 & 3 \\ 1 & -10 & -4 & 0 & 1 & 0 & 20 \\ -1 & -2 & -1 & 0 & 0 & 1 & 0 \end{bmatrix} \\ \uparrow \end{array} \approx \begin{bmatrix} 3 & 1 & 1 & 1 & 0 & 0 & 3 \\ 31 & 0 & 6 & 10 & 1 & 0 & 50 \\ 5 & 0 & 1 & 2 & 0 & 1 & 6 \end{bmatrix}.$$ The maximum value of f is 6 when $x = z = u = 0$. The first row then gives $y = 3$.

6.
$$\begin{array}{c} \begin{array}{ccccccc} x & y & z & u & v & w & f \end{array} \\ \to \begin{bmatrix} 5 & 5 & 10 & 1 & 0 & 0 & 0 & 1000 \\ 10 & 8 & 5 & 0 & 1 & 0 & 0 & 2000 \\ 10 & 5 & 0 & 0 & 0 & 1 & 0 & 500 \\ -100 & -200 & -50 & 0 & 0 & 0 & 1 & 0 \end{bmatrix} \\ \uparrow \end{array} \approx \begin{bmatrix} 5 & 5 & 10 & 1 & 0 & 0 & 0 & 1000 \\ 10 & 8 & 5 & 0 & 1 & 0 & 0 & 2000 \\ 2 & 1 & 0 & 0 & 0 & 1/5 & 0 & 100 \\ -100 & -200 & -50 & 0 & 0 & 0 & 1 & 0 \end{bmatrix}$$

$$\approx \begin{bmatrix} -5 & 0 & 10 & 1 & 0 & -1 & 0 & 500 \\ -6 & 0 & 5 & 0 & 1 & -8/5 & 0 & 1200 \\ 2 & 1 & 0 & 0 & 0 & 1/5 & 0 & 100 \\ 300 & 0 & -50 & 0 & 0 & 40 & 1 & 20000 \end{bmatrix}$$

$$\approx \begin{bmatrix} -1/2 & 0 & 1 & 1/10 & 0 & -1/10 & 0 & 50 \\ -6 & 0 & 5 & 0 & 1 & -8/5 & 0 & 1200 \\ 2 & 1 & 0 & 0 & 0 & 1/5 & 0 & 100 \\ 300 & 0 & -50 & 0 & 0 & 40 & 1 & 20000 \end{bmatrix}$$

$$\approx \begin{bmatrix} -1/2 & 0 & 1 & 1/10 & 0 & -1/10 & 0 & 50 \\ -7/2 & 0 & 0 & -1/2 & 1 & -11/10 & 0 & 950 \\ 2 & 1 & 0 & 0 & 0 & 1/5 & 0 & 100 \\ 275 & 0 & 0 & 5 & 0 & 35 & 1 & 22500 \end{bmatrix}.$$ The maximum value of f is 22500 when $x = u = w = 0$. Row 3 gives $y = 100$ and row 1 gives $z = 50$.

7.

$$\begin{array}{c} \begin{array}{cccccccc}x & y & z & u & v & w & f\end{array} \\ \to \begin{bmatrix} -1 & 2 & 3 & 1 & 0 & 0 & 0 & 6 \\ -1 & 4 & 5 & 0 & 1 & 0 & 0 & 5 \\ -1 & 5 & 7 & 0 & 0 & 1 & 0 & 7 \\ -2 & -4 & -1 & 0 & 0 & 0 & 1 & 0 \end{bmatrix} \end{array} \approx \begin{bmatrix} -1 & 2 & 3 & 1 & 0 & 0 & 0 & 6 \\ -1/4 & 1 & 5/4 & 0 & 1/4 & 0 & 0 & 5/4 \\ -1 & 5 & 7 & 0 & 0 & 1 & 0 & 7 \\ -2 & -4 & -1 & 0 & 0 & 0 & 1 & 0 \end{bmatrix}$$

$$\approx \begin{bmatrix} -1/2 & 0 & 1/2 & 1 & -1/2 & 0 & 0 & 7/2 \\ -1/4 & 1 & 5/4 & 0 & 1/4 & 0 & 0 & 5/4 \\ 1/4 & 0 & 3/4 & 0 & -5/4 & 1 & 0 & 3/4 \\ -3 & 0 & 4 & 0 & 1 & 0 & 1 & 5 \end{bmatrix} \approx \begin{bmatrix} -1/2 & 0 & 1/2 & 1 & -1/2 & 0 & 0 & 7/2 \\ -1/4 & 1 & 5/4 & 0 & 1/4 & 0 & 0 & 5/4 \\ 1 & 0 & 3 & 0 & -5 & 4 & 0 & 3 \\ -3 & 0 & 4 & 0 & 1 & 0 & 1 & 5 \end{bmatrix}$$

$$\approx \begin{bmatrix} 0 & 0 & 2 & 1 & -3 & 2 & 0 & 5 \\ 0 & 1 & 2 & 0 & -1 & 1 & 0 & 2 \\ 1 & 0 & 3 & 0 & -5 & 4 & 0 & 3 \\ 0 & 0 & 13 & 0 & -14 & 12 & 1 & 14 \end{bmatrix}. \text{ We can do no more.}$$

There are no positive terms in the pivot column. This means that the feasible region is unbounded and the objective function is unbounded. That is, there is no maximum.

8.

$$\begin{array}{c} \begin{array}{ccccccc}x & y & z & w & u & v & f\end{array} \\ \to \begin{bmatrix} 5 & 0 & 4 & 6 & 1 & 0 & 0 & 20 \\ 4 & 2 & 2 & 8 & 0 & 1 & 0 & 40 \\ -1 & -2 & -4 & 1 & 0 & 0 & 1 & 0 \end{bmatrix} \end{array} \approx \begin{bmatrix} 5/4 & 0 & 1 & 3/2 & 1/4 & 0 & 0 & 5 \\ 4 & 2 & 2 & 8 & 0 & 1 & 0 & 40 \\ -1 & -2 & -4 & 1 & 0 & 0 & 1 & 0 \end{bmatrix}$$

$$\approx \begin{bmatrix} 5/4 & 0 & 1 & 3/2 & 1/4 & 0 & 0 & 5 \\ 3/2 & 2 & 0 & 5 & -1/2 & 1 & 0 & 30 \\ 4 & -2 & 0 & 7 & 1 & 0 & 1 & 20 \end{bmatrix} \approx \begin{bmatrix} 5/4 & 0 & 1 & 3/2 & 1/4 & 0 & 0 & 5 \\ 3/4 & 1 & 0 & 5/2 & -1/4 & 1/2 & 0 & 15 \\ 4 & -2 & 0 & 7 & 1 & 0 & 1 & 20 \end{bmatrix}$$

$$\approx \begin{bmatrix} 5/4 & 0 & 1 & 3/2 & 1/4 & 0 & 0 & 5 \\ 3/4 & 1 & 0 & 5/2 & -1/4 & 1/2 & 0 & 15 \\ 11/2 & 0 & 0 & 12 & 1/2 & 1 & 1 & 50 \end{bmatrix}. \text{ The maximum value of } f \text{ is 50 when}$$

$x = w = u = v = 0$. The first row then yields $z = 5$ and the second row gives $y = 15$.

9.
$$\rightarrow \begin{bmatrix} x & y & z & w & u & v & f & \\ 2 & 4 & 5 & 6 & 1 & 0 & 0 & 24 \\ 4 & 4 & 2 & 2 & 0 & 1 & 0 & 4 \\ -1 & -2 & 1 & -3 & 0 & 0 & 1 & 0 \end{bmatrix} \approx \begin{bmatrix} 2 & 4 & 5 & 6 & 1 & 0 & 0 & 24 \\ 2 & 2 & 1 & 1 & 0 & 1/2 & 0 & 2 \\ -1 & -2 & 1 & -3 & 0 & 0 & 1 & 0 \end{bmatrix}$$
$$\uparrow$$

$$\approx \begin{bmatrix} -10 & -8 & -1 & 0 & 1 & -3 & 0 & 12 \\ 2 & 2 & 1 & 1 & 0 & 1/2 & 0 & 2 \\ 5 & 4 & 4 & 0 & 0 & 3/2 & 1 & 6 \end{bmatrix}.$$ The maximum value of f is 6 when

$x = y = z = v = 0$. Row 2 gives $w = 2$.

10.

	I	II	III	profit($)	number to make
X	2	4		10	x
Y	3		6	8	y
Z	1	2	3	12	z
totals	≤360	≤360	≤360		

We are to maximize profit, $f = 10x + 8y + 12z$, subject to the constraints

$2x + 3y + z \leq 360$, $4x + 2z \leq 360$, $6y + 3z \leq 360$, $x \geq 0, y \geq 0, z \geq 0$.

$$\rightarrow \begin{bmatrix} x & y & z & u & v & w & f & \\ 2 & 3 & 1 & 1 & 0 & 0 & 0 & 360 \\ 4 & 0 & 2 & 0 & 1 & 0 & 0 & 360 \\ 0 & 6 & 3 & 0 & 0 & 1 & 0 & 360 \\ -10 & -8 & -12 & 0 & 0 & 0 & 1 & 0 \end{bmatrix} \approx \begin{bmatrix} 2 & 3 & 1 & 1 & 0 & 0 & 0 & 360 \\ 4 & 0 & 2 & 0 & 1 & 0 & 0 & 360 \\ 0 & 2 & 1 & 0 & 0 & 1/3 & 0 & 120 \\ -10 & -8 & -12 & 0 & 0 & 0 & 1 & 0 \end{bmatrix}$$
$$\uparrow$$

$$\approx \begin{bmatrix} 2 & 1 & 0 & 1 & 0 & -1/3 & 0 & 240 \\ 4 & -4 & 0 & 0 & 1 & -2/3 & 0 & 120 \\ 0 & 2 & 1 & 0 & 0 & 1/3 & 0 & 120 \\ -10 & 16 & 0 & 0 & 0 & 4 & 1 & 1440 \end{bmatrix} \approx \begin{bmatrix} 2 & 1 & 0 & 1 & 0 & -1/3 & 0 & 240 \\ 1 & -1 & 0 & 0 & 1/4 & -1/6 & 0 & 30 \\ 0 & 2 & 1 & 0 & 0 & 1/3 & 0 & 120 \\ -10 & 16 & 0 & 0 & 0 & 4 & 1 & 1440 \end{bmatrix}$$

$$\approx \begin{bmatrix} 0 & 3 & 0 & 1 & -1/2 & 0 & 0 & 180 \\ 1 & -1 & 0 & 0 & 1/4 & -1/6 & 0 & 30 \\ 0 & 2 & 1 & 0 & 0 & 1/3 & 0 & 120 \\ 0 & 6 & 0 & 0 & 5/2 & 7/3 & 1 & 1740 \end{bmatrix}.$$ The maximum value of f is 1740 when

$y = v = w = 0$. Row 2 gives $x = 30$ and row 3 gives $z = 120$. So the company should produce 30 of item X, no item Y, and 120 of item Z to make maximum profit of $1740 per day.

Section 10.2

11. Let x be the number of desks, y be the number of cabinets, and z be the number of chairs. We are to maximize profit, f = 16x + 12y + 6z, subject to the constraints

$3x + 6y + z \leq 800$, $4x + y + 2z \leq 400$, $2x + y + 2z \leq 100$, $x \geq 0, y \geq 0, z \geq 0$.

$$\rightarrow \begin{bmatrix} x & y & z & u & v & w & f & \\ 3 & 6 & 1 & 1 & 0 & 0 & 0 & 800 \\ 4 & 1 & 2 & 0 & 1 & 0 & 0 & 400 \\ 2 & 1 & 2 & 0 & 0 & 1 & 0 & 100 \\ -16 & -12 & -6 & 0 & 0 & 0 & 1 & 0 \end{bmatrix} \approx \begin{bmatrix} 3 & 6 & 1 & 1 & 0 & 0 & 0 & 800 \\ 4 & 1 & 2 & 0 & 1 & 0 & 0 & 400 \\ 1 & 1/2 & 1 & 0 & 0 & 1/2 & 0 & 50 \\ -16 & -12 & -6 & 0 & 0 & 0 & 1 & 0 \end{bmatrix}$$

$$\approx \begin{bmatrix} 0 & 9/2 & -2 & 1 & 0 & -3/2 & 0 & 650 \\ 0 & -1 & -2 & 0 & 1 & -2 & 0 & 200 \\ 1 & 1/2 & 1 & 0 & 0 & 1/2 & 0 & 50 \\ 0 & -4 & 10 & 0 & 0 & 8 & 1 & 800 \end{bmatrix} \approx \begin{bmatrix} 0 & 9/2 & -2 & 1 & 0 & -3/2 & 0 & 650 \\ 0 & -1 & -2 & 0 & 1 & -2 & 0 & 200 \\ 2 & 1 & 2 & 0 & 0 & 1 & 0 & 100 \\ 0 & -4 & 10 & 0 & 0 & 8 & 1 & 800 \end{bmatrix}$$

$$\approx \begin{bmatrix} -9 & 0 & -11 & 1 & 0 & -6 & 0 & 200 \\ 2 & 0 & 0 & 0 & 1 & -1 & 0 & 300 \\ 2 & 1 & 2 & 0 & 0 & 1 & 0 & 100 \\ 8 & 0 & 18 & 0 & 0 & 12 & 1 & 1200 \end{bmatrix}.$$ The maximum value of f is 1200 when

x = z = w = 0. Row 3 then gives y = 100. Thus the company should manufacture 100 cabinets and no desks or chairs to make the maximum profit of $1200.

12.
	cost($)	time(hrs)	profit($)	number to transport
A	10	6	12	x
B	20	4	20	y
C	40	2	16	z
totals	≤6000	≤4000		

We are to maximize profit, f = 12x + 20y + 16z, subject to the constraints

$10x + 20y + 40z \leq 6000$, $6x + 4y + 2z \leq 4000$, $x \geq 0, y \geq 0, z \geq 0$.

$$\rightarrow \begin{bmatrix} x & y & z & u & v & f & \\ 10 & 20 & 40 & 1 & 0 & 0 & 6000 \\ 6 & 4 & 2 & 0 & 1 & 0 & 4000 \\ -12 & -20 & -16 & 0 & 0 & 1 & 0 \end{bmatrix} \approx \begin{bmatrix} 1/2 & 1 & 2 & 1/20 & 0 & 0 & 300 \\ 6 & 4 & 2 & 0 & 1 & 0 & 4000 \\ -12 & -20 & -16 & 0 & 0 & 1 & 0 \end{bmatrix}$$

381

$$\approx \begin{bmatrix} 1/2 & 1 & 2 & 1/20 & 0 & 0 & 300 \\ 4 & 0 & -6 & -1/5 & 1 & 0 & 2800 \\ -2 & 0 & 24 & 1 & 0 & 1 & 6000 \end{bmatrix} \approx \begin{bmatrix} 1 & 2 & 4 & 1/10 & 0 & 0 & 600 \\ 4 & 0 & -6 & -1/5 & 1 & 0 & 2800 \\ -2 & 0 & 24 & 1 & 0 & 1 & 6000 \end{bmatrix}$$

$$\approx \begin{bmatrix} 1 & 2 & 4 & 1/10 & 0 & 0 & 600 \\ 0 & -8 & -22 & -3/5 & 1 & 0 & 400 \\ 0 & 4 & 32 & 6/5 & 0 & 1 & 7200 \end{bmatrix}.$$ The maximum value of f is 7200 when

$y = z = u = 0$. Row 1 then gives $x = 600$. The maximum profit of \$7200 is attained if 600 washing machines are transported from A to P and none is transported from B or C to P.

13.

	material (sq.yd.)	cost($)	profit	number to make
Aspen	60	32	12	x
Alpine	30	20	8	y
Cub	15	12	4	z
totals	≤7800	≤8320		

We are to maximize profit, $f = 12x + 8y + 4z$, subject to the constraints

$60x + 30y + 15z \leq 7800$, $32x + 20y + 12z \leq 8320$, $x \geq 0$, $y \geq 0$, $z \geq 0$.

$$\begin{array}{c} x \ y \ z \ u \ v \ f \\ \rightarrow \begin{bmatrix} 60 & 30 & 15 & 1 & 0 & 0 & 7800 \\ 32 & 20 & 12 & 0 & 1 & 0 & 8320 \\ -12 & -8 & -4 & 0 & 0 & 1 & 0 \end{bmatrix} \approx \begin{bmatrix} 1 & 1/2 & 1/4 & 1/60 & 0 & 0 & 130 \\ 32 & 20 & 12 & 0 & 1 & 0 & 8320 \\ -12 & -8 & -4 & 0 & 0 & 1 & 0 \end{bmatrix} \\ \uparrow \end{array}$$

$$\approx \rightarrow \begin{bmatrix} 1 & 1/2 & 1/4 & 1/60 & 0 & 0 & 130 \\ 0 & 4 & 4 & -8/15 & 1 & 0 & 4160 \\ 0 & -2 & -1 & 1/5 & 0 & 1 & 1560 \end{bmatrix} \approx \begin{bmatrix} 2 & 1 & 1/2 & 1/30 & 0 & 0 & 260 \\ 0 & 4 & 4 & -8/15 & 1 & 0 & 4160 \\ 0 & -2 & -1 & 1/5 & 0 & 1 & 1560 \end{bmatrix}$$
$\uparrow$

$$\approx \begin{bmatrix} 2 & 1 & 1/2 & 1/30 & 0 & 0 & 260 \\ -8 & 0 & 2 & -2/3 & 1 & 0 & 3120 \\ 4 & 0 & 0 & 4/15 & 0 & 1 & 2080 \end{bmatrix}.$$ The maximum value of f is 2080 when

$x = u = 0$. From row 1, $y + z/2 = 260$. Thus the maximum profit of \$2080 will be achieved if no Aspens are manufactured and the number of Alpines manufactured plus one-half the number of Cubs manufactured is 260.

Section 10.2

14. To minimize $f = -x + 2y$, maximize $-f = x - 2y$.

$$\begin{array}{c} x\ y\ u\ v\ f \\ \begin{bmatrix} 1 & 2 & 1 & 0 & 0 & 4 \\ 1 & 4 & 0 & 1 & 0 & 6 \\ -1 & 2 & 0 & 0 & 1 & 0 \end{bmatrix} \approx \begin{bmatrix} 1 & 2 & 1 & 0 & 0 & 4 \\ 0 & 2 & -1 & 1 & 0 & 2 \\ 0 & 4 & 1 & 0 & 1 & 4 \end{bmatrix} \end{array}$$. The maximum value of $-f$ is 4 when

$y = u = 0$. Row 1 gives $x = 4$. Thus the minimum value of f is -4 at $x = 4$, $y = 0$.

15. To minimize $f = -2x + y$, maximize $-f = 2x - y$.

$$\begin{array}{c} x\ y\ u\ v\ f \\ \begin{bmatrix} 2 & 2 & 1 & 0 & 0 & 8 \\ 1 & -1 & 0 & 1 & 0 & 2 \\ -2 & 1 & 0 & 0 & 1 & 0 \end{bmatrix} \approx \begin{bmatrix} 0 & 4 & 1 & -2 & 0 & 4 \\ 1 & -1 & 0 & 1 & 0 & 2 \\ 0 & -1 & 0 & 2 & 1 & 4 \end{bmatrix} \approx \begin{bmatrix} 0 & 1 & 1/4 & -1/2 & 0 & 1 \\ 1 & -1 & 0 & 1 & 0 & 2 \\ 0 & -1 & 0 & 2 & 1 & 4 \end{bmatrix} \end{array}$$

$$\approx \begin{bmatrix} 0 & 1 & 1/4 & -1/2 & 0 & 1 \\ 1 & 0 & 1/4 & 1/2 & 0 & 3 \\ 0 & 0 & 1/4 & 3/2 & 1 & 5 \end{bmatrix}$$. The maximum value of $-f$ is 5 when $u = v = 0$.

Row 2 gives $x = 3$ and row 1 gives $y = 1$. Thus the minimum value of f is -5 at $x = 3$, $y = 1$.

16. To minimize $f = 2x + y - z$, maximize $-f = -2x - y + z$.

$$\begin{array}{c} x\ y\ z\ u\ v\ w\ f \\ \begin{bmatrix} 1 & 2 & -2 & 1 & 0 & 0 & 0 & 20 \\ 2 & 1 & 0 & 0 & 1 & 0 & 0 & 10 \\ 1 & 3 & 4 & 0 & 0 & 1 & 0 & 15 \\ 2 & 1 & -1 & 0 & 0 & 0 & 1 & 0 \end{bmatrix} \approx \begin{bmatrix} 1 & 2 & -2 & 1 & 0 & 0 & 0 & 20 \\ 2 & 1 & 0 & 0 & 1 & 0 & 0 & 10 \\ 1/4 & 3/4 & 1 & 0 & 0 & 1/4 & 0 & 15/4 \\ 2 & 1 & -1 & 0 & 0 & 0 & 1 & 0 \end{bmatrix} \end{array}$$

$$\approx \begin{bmatrix} 3/2 & 7/2 & 0 & 1 & 0 & 1/2 & 0 & 55/2 \\ 2 & 1 & 0 & 0 & 1 & 0 & 0 & 10 \\ 1/4 & 3/4 & 1 & 0 & 0 & 1/4 & 0 & 15/4 \\ 9/4 & 7/4 & 0 & 0 & 0 & 1/4 & 1 & 15/4 \end{bmatrix}$$. The maximum value of $-f$ is 15/4 when

$x = y = w = 0$. Row 3 gives $z = 15/4$. Thus the minimum value of f is $-15/4$ at $x = 0$, $y = 0$, $z = 15/4$.

Exercise Set 10.3

1.
$$\begin{matrix} & x & y & u & v & f & \\ \rightarrow & \begin{bmatrix} 4 & 1 & 1 & 0 & 0 & 36 \\ 4 & 3 & 0 & 1 & 0 & 60 \\ -2 & -1 & 0 & 0 & 1 & 0 \end{bmatrix} & \approx & \begin{bmatrix} 1 & 1/4 & 1/4 & 0 & 0 & 9 \\ 4 & 3 & 0 & 1 & 0 & 60 \\ -2 & -1 & 0 & 0 & 1 & 0 \end{bmatrix} & \approx & \begin{bmatrix} 1 & 1/4 & 1/4 & 0 & 0 & 9 \\ 0 & 2 & -1 & 1 & 0 & 24 \\ 0 & -1/2 & 1/2 & 0 & 1 & 18 \end{bmatrix} \end{matrix}$$

basic:	u,v	x,v
nonbasic:	x,y	y,u
entering:	x	y
departing:	u	v

$$\approx \begin{bmatrix} 1 & 1/4 & 1/4 & 0 & 0 & 9 \\ 0 & 1 & -1/2 & 1/2 & 0 & 12 \\ 0 & -1/2 & 1/2 & 0 & 1 & 18 \end{bmatrix} \approx \begin{bmatrix} 1 & 0 & 3/8 & -1/8 & 0 & 6 \\ 0 & 1 & -1/2 & 1/2 & 0 & 12 \\ 0 & 0 & 1/4 & 1/4 & 1 & 24 \end{bmatrix}.$$

basic:	x,y
nonbasic:	u,v

The maximum value of f is 24 when u and v are both zero. With u = v = 0, the first two rows give x = 6 and y = 12. Thus the optimal solution is f = 24 at x = 6, y = 12.

2.
$$\begin{matrix} & x & y & u & v & f & \\ \rightarrow & \begin{bmatrix} 1 & 2 & 1 & 0 & 0 & 4 \\ 1 & 6 & 0 & 1 & 0 & 8 \\ -1 & 4 & 0 & 0 & 1 & 0 \end{bmatrix} & \approx & \begin{bmatrix} 1 & 2 & 1 & 0 & 0 & 4 \\ 0 & 4 & -1 & 1 & 0 & 4 \\ 0 & 6 & 1 & 0 & 1 & 4 \end{bmatrix}. \end{matrix}$$

basic:	u,v	x,v
nonbasic:	x,y	y,u
entering:	x	
departing:	u	

The maximum value of f is 4 when y = u = 0. The first two rows give x = 4 and v = 4. Thus the optimal solution is f = 4 at x = 4, y = 0. (y is a nonbasic variable.)

3.
$$\begin{array}{c} \begin{array}{cccccc} x & y & z & u & v & f \end{array} \\ \rightarrow \begin{bmatrix} 3 & 1 & 1 & 1 & 0 & 0 & 3 \\ 1 & -10 & -4 & 0 & 1 & 0 & 20 \\ -1 & -2 & -1 & 0 & 0 & 1 & 0 \end{bmatrix} \approx \begin{bmatrix} 3 & 1 & 1 & 1 & 0 & 0 & 3 \\ 31 & 0 & 6 & 10 & 1 & 0 & 50 \\ 5 & 0 & 1 & 2 & 0 & 1 & 6 \end{bmatrix}. \end{array}$$

            ↑

basic: u,v   y,v
nonbasic: x,y,z   x,z,u
entering: y
departing: u

The maximum value of f is 6 when $x = z = u = 0$. The first two rows give $y = 3$ and $v = 50$. Thus the optimal solution is $f = 6$ at $x = 0, y = 3, z = 0$. (x and z are nonbasic variables.)

4.
$$\begin{array}{c} \begin{array}{ccccccc} x & y & z & u & v & w & f \end{array} \\ \rightarrow \begin{bmatrix} 5 & 5 & 10 & 1 & 0 & 0 & 0 & 1000 \\ 10 & 8 & 5 & 0 & 1 & 0 & 0 & 2000 \\ 10 & 5 & 0 & 0 & 0 & 1 & 0 & 500 \\ -100 & -200 & -50 & 0 & 0 & 0 & 1 & 0 \end{bmatrix} \approx \begin{bmatrix} 5 & 5 & 10 & 1 & 0 & 0 & 0 & 1000 \\ 10 & 8 & 5 & 0 & 1 & 0 & 0 & 2000 \\ 2 & 1 & 0 & 0 & 0 & 1/5 & 0 & 100 \\ -100 & -200 & -50 & 0 & 0 & 0 & 1 & 0 \end{bmatrix} \end{array}$$

           ↑

basic: u,v,w
nonbasic: x,y,z
entering: y
departing: w

$$\approx \begin{bmatrix} -5 & 0 & 10 & 1 & 0 & -1 & 0 & 500 \\ -6 & 0 & 5 & 0 & 1 & -8/5 & 0 & 1200 \\ 2 & 1 & 0 & 0 & 0 & 1/5 & 0 & 100 \\ 300 & 0 & -50 & 0 & 0 & 40 & 1 & 20000 \end{bmatrix}$$

basic: u,v,y
nonbasic: x,z,w
entering: z
departing: u

$$\approx \begin{bmatrix} -1/2 & 0 & 1 & 1/10 & 0 & -1/10 & 0 & 50 \\ -6 & 0 & 5 & 0 & 1 & -8/5 & 0 & 1200 \\ 2 & 1 & 0 & 0 & 0 & 1/5 & 0 & 100 \\ 300 & 0 & -50 & 0 & 0 & 40 & 1 & 20000 \end{bmatrix}$$

Section 10.3

$$\approx \begin{bmatrix} -1/2 & 0 & 1 & 1/10 & 0 & -1/10 & 0 & 50 \\ -7/2 & 0 & 0 & -1/2 & 1 & -11/10 & 0 & 950 \\ 2 & 1 & 0 & 0 & 0 & 1/5 & 0 & 100 \\ 275 & 0 & 0 & 5 & 0 & 35 & 1 & 22500 \end{bmatrix}.$$

basic: z, v, y
nonbasic: x, u, w

The maximum value of f is 22500 when $x = u = w = 0$. Row 1 gives $z = 50$, row 2 gives $v = 950$, and row 3 gives $y = 100$. Thus the optimal solution is $f = 22500$ at $x = 0$, $y = 100$, $z = 50$. (x is a nonbasic variable.)

5.
$$\quad\;\; x \;\;\; y \;\; z \;\; u \;\; v \;\; w \;\; f$$
$$\rightarrow \begin{bmatrix} -1 & 2 & 3 & 1 & 0 & 0 & 0 & 6 \\ -1 & 4 & 5 & 0 & 1 & 0 & 0 & 5 \\ -1 & 5 & 7 & 0 & 0 & 1 & 0 & 7 \\ -2 & -4 & -1 & 0 & 0 & 0 & 1 & 0 \end{bmatrix} \approx \begin{bmatrix} -1 & 2 & 3 & 1 & 0 & 0 & 0 & 6 \\ -1/4 & 1 & 5/4 & 0 & 1/4 & 0 & 0 & 5/4 \\ -1 & 5 & 7 & 0 & 0 & 1 & 0 & 7 \\ -2 & -4 & -1 & 0 & 0 & 0 & 1 & 0 \end{bmatrix}$$
$$\qquad\;\;\uparrow$$

basic: u, v, w
nonbasic: x, y, z
entering: y
departing: v

$$\approx \begin{bmatrix} -1/2 & 0 & 1/2 & 1 & -1/2 & 0 & 0 & 7/2 \\ -1/4 & 1 & 5/4 & 0 & 1/4 & 0 & 0 & 5/4 \\ 1/4 & 0 & 3/4 & 0 & -5/4 & 1 & 0 & 3/4 \\ -3 & 0 & 4 & 0 & 1 & 0 & 1 & 5 \end{bmatrix} \approx \begin{bmatrix} -1/2 & 0 & 1/2 & 1 & -1/2 & 0 & 0 & 7/2 \\ -1/4 & 1 & 5/4 & 0 & 1/4 & 0 & 0 & 5/4 \\ 1 & 0 & 3 & 0 & -5 & 4 & 0 & 3 \\ -3 & 0 & 4 & 0 & 1 & 0 & 1 & 5 \end{bmatrix}$$

basic: u, y, w
nonbasic: x, z, v
entering: x
departing: v

$$\approx \begin{bmatrix} 0 & 0 & 2 & 1 & -3 & 2 & 0 & 5 \\ 0 & 1 & 2 & 0 & -1 & 1 & 0 & 2 \\ 1 & 0 & 3 & 0 & -5 & 4 & 0 & 3 \\ 0 & 0 & 13 & 0 & -14 & 12 & 1 & 14 \end{bmatrix}.$$

basic: u, y, x
nonbasic: z, v, w

386

Section 10.3

We can do no more. There are no positive terms in the pivot column. This means that the feasible region is unbounded and the objective function is unbounded. That is, there is no optimal solution.

6.
$$\begin{array}{c} \begin{array}{ccccccc} x & y & z & w & u & v & f \end{array} \\ \to \begin{bmatrix} 5 & 0 & 4 & 6 & 1 & 0 & 0 & 20 \\ 4 & 2 & 2 & 8 & 0 & 1 & 0 & 40 \\ -1 & -2 & -4 & 1 & 0 & 0 & 1 & 0 \end{bmatrix} \approx \begin{bmatrix} 5/4 & 0 & 1 & 3/2 & 1/4 & 0 & 0 & 5 \\ 4 & 2 & 2 & 8 & 0 & 1 & 0 & 40 \\ -1 & -2 & -4 & 1 & 0 & 0 & 1 & 0 \end{bmatrix} \end{array}$$

basic: u,v
nonbasic: x,y,z,w
entering: z
departing: u

$$\approx \begin{bmatrix} 5/4 & 0 & 1 & 3/2 & 1/4 & 0 & 0 & 5 \\ 3/2 & 2 & 0 & 5 & -1/2 & 1 & 0 & 30 \\ 4 & -2 & 0 & 7 & 1 & 0 & 1 & 20 \end{bmatrix} \approx \begin{bmatrix} 5/4 & 0 & 1 & 3/2 & 1/4 & 0 & 0 & 5 \\ 3/4 & 1 & 0 & 5/2 & -1/4 & 1/2 & 0 & 15 \\ 4 & -2 & 0 & 7 & 1 & 0 & 1 & 20 \end{bmatrix}$$

basic: z,v
nonbasic: x,y,w,v
entering: y
departing: v

$$\approx \begin{bmatrix} 5/4 & 0 & 1 & 3/2 & 1/4 & 0 & 0 & 5 \\ 3/4 & 1 & 0 & 5/2 & -1/4 & 1/2 & 0 & 15 \\ 11/2 & 0 & 0 & 12 & 1/2 & 1 & 1 & 50 \end{bmatrix}.$$

basic: z,y
nonbasic: x,w,u,v

The maximum value of f is 50 when $x = w = u = v = 0$. The first row then yields $z = 5$ and the second row gives $y = 15$. Thus the optimal solution is $f = 50$ at $x = 0$, $y = 15$, $z = 5$, $w = 0$. (x and w are nonbasic variables.)

Chapter 10 Review Exercises

In Exercises 1 - 4, we refer to the drawing at the right. The points A, B, C, E, and F will be identified, and the objective function will be evaluated at the three nonzero vertices A, B, and C of the feasible region.

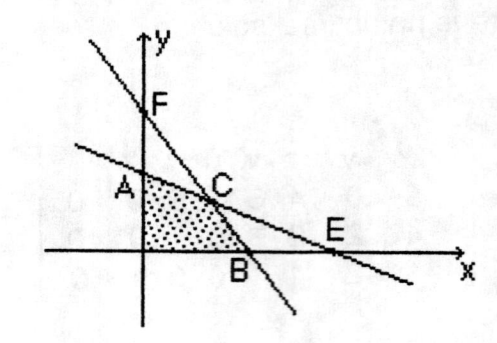

1. $2x + 4y = 16$ A = (0,4)   E = (8,0)

    $3x + 2y = 12$ F = (0,6)   B = (4,0)   C = (2,3)

    At A, f = 2(0) + 3(4) = 12; at B, f = 2(4) + 3(0) = 8; and at C, f = 2(2) + 3(3) = 13, so the maximum value of f is 13 and it occurs at the point C.

2. $x + 2y = 16$ A = (0,8)   E = (16,0)

    $3x + 2y = 24$ F = (0,12)   B = (8,0)   C = (4,6)

    At A, f = 6(0) + 4(8) = 32; at B, f = 6(8) + 4(0) = 48; and at C, f = 6(4) + 4(6) = 48, so the maximum value of f is 48 and it occurs at every point on the line segment joining B and C.

3. $3x + 2y = 21$ F = (0,21/2)   B = (7,0)

    $x + 5y = 20$ A = (0,4)   E = (20,0)   C = (5,3)

    At A, –f = –4(0) – 4 = –3; at B, –f = –4(7) – 0 = –28; and at C, –f = –4(5) – 3 = –23, so the maximum value of –f is zero and it occurs at the origin. Thus the minimum value of f is zero and it occurs at the origin.

Chapter 10 Review Exercises

4.

	acres	picking time(hrs)	profit($)
strawberries	x	8	700
tomatoes	y	6	600
total	≤40	≤300	

We are to maximize profit, f = 700x + 600y, subject to the constraints

$\quad$ x + y ≤ 40,

$\quad$ 8x + 6y ≤ 300 (divide by 2: 4x + 3y ≤ 150),

$\quad$ x ≥ 0, and y ≥ 0.

x + y = 40 $\qquad$ A = (0,40) $\qquad$ E = (40,0)

4x + 3y = 150 $\qquad$ F = (0,50) $\qquad$ B = (75/2,0) $\qquad$ C = (30,10)

At A, f = 700(0) + 600(40) = 24000; at B, f = 700(75/2) + 600(0) = 26250; and at C, f = 700(30) + 600(10) = 27000, so the maximum value of f is 27000 and it occurs at the point C. To ensure the maximum profit of $27000, the farmer should plant 30 acres of strawberries and 10 acres of tomatoes.

5.

	cu. ft.	sq. ft.	cost($)	number
X	36	6	54	x
Y	44	8	60	y
totals		≤256	≤2100	

We are to maximize volume, f = 36x + 44y, subject to the constraints

$\quad$ 6x + 8y ≤ 256 (divide by 2: 3x + 4y ≤ 128),

$\quad$ 54x + 60y ≤ 2100 (divide by 6: 9x + 10y ≤ 350),

$\quad$ x ≥ 0, and y ≥ 20.

Chapter 10 Review Exercises

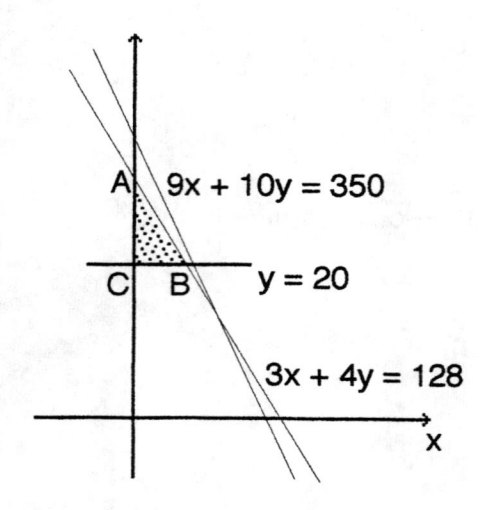

A = (0,32), B = (16,20), and C = (0,20). The values of f at these three points are 1408, 1456, and 880. Thus volume will be maximized if the company purchases 16 lockers of type X and 20 of type Y.

6.
$$\begin{array}{c} x \; y \; z \; u \; v \; w \; f \\ \begin{bmatrix} 1 & 2 & 4 & 1 & 0 & 0 & 0 & 20 \\ 2 & 4 & 4 & 0 & 1 & 0 & 0 & 60 \\ 3 & 4 & 1 & 0 & 0 & 1 & 0 & 90 \\ -2 & -1 & -1 & 0 & 0 & 0 & 1 & 0 \end{bmatrix} \approx \begin{bmatrix} 1 & 2 & 4 & 1 & 0 & 0 & 0 & 20 \\ 0 & 0 & -4 & -2 & 1 & 0 & 0 & 20 \\ 0 & -2 & -11 & -3 & 0 & 1 & 0 & 30 \\ 0 & 3 & 7 & 2 & 0 & 0 & 1 & 40 \end{bmatrix} \end{array}.$$

The maximum value of f is 40 when $y = z = u = 0$. Row 1 gives $x = 20$.

7. Let x be the number of X tables and y be the number of Y tables.

We are to maximize profit, $f = 8x + 4y$, subject to the constraints

$10x + 5y \leq 300$, $8x + 4y \leq 300$, $4x + 8y \leq 300$, $x \geq 0$, and $y \geq 0$.

$$\begin{array}{c} x \; y \; u \; v \; w \; f \\ \begin{bmatrix} 10 & 5 & 1 & 0 & 0 & 0 & 300 \\ 8 & 4 & 0 & 1 & 0 & 0 & 300 \\ 4 & 8 & 0 & 0 & 1 & 0 & 300 \\ -8 & -4 & 0 & 0 & 0 & 1 & 0 \end{bmatrix} \approx \begin{bmatrix} 1 & 1/2 & 1/10 & 0 & 0 & 0 & 30 \\ 8 & 4 & 0 & 1 & 0 & 0 & 300 \\ 4 & 8 & 0 & 0 & 1 & 0 & 300 \\ -8 & -4 & 0 & 0 & 0 & 1 & 0 \end{bmatrix} \end{array}$$

$$\approx \begin{bmatrix} 1 & 1/2 & 1/10 & 0 & 0 & 0 & 30 \\ 0 & 0 & -4/5 & 1 & 0 & 0 & 60 \\ 0 & 6 & -2/5 & 0 & 1 & 0 & 180 \\ 0 & 0 & 4/5 & 0 & 0 & 1 & 240 \end{bmatrix}.$$ The maximum value of f is 240 when $u = 0$.

Chapter 10 Review Exercises

Row 1 gives $x + y/2 = 30$. Thus the maximum daily profit of $240 is realized when the number of X tables plus one-half the number of Y tables finished is 30.

8. To minimize $f = x - 2y + 4z$, maximize $-f = -x + 2y - 4z$.

$$\begin{array}{cccccccc} x & y & z & u & v & w & f \end{array}$$

$$\rightarrow \begin{bmatrix} 1 & -1 & 3 & 1 & 0 & 0 & 0 & 4 \\ 2 & 2 & -3 & 0 & 1 & 0 & 0 & 6 \\ -1 & 2 & 3 & 0 & 0 & 1 & 0 & 2 \\ 1 & -2 & 4 & 0 & 0 & 0 & 1 & 0 \end{bmatrix} \approx \begin{bmatrix} 1 & -1 & 3 & 1 & 0 & 0 & 0 & 4 \\ 2 & 2 & -3 & 0 & 1 & 0 & 0 & 6 \\ -1/2 & 1 & 3/2 & 0 & 0 & 1/2 & 0 & 1 \\ 1 & -2 & 4 & 0 & 0 & 0 & 1 & 0 \end{bmatrix}$$

$$\approx \begin{bmatrix} 1/2 & 0 & 9/2 & 1 & 0 & 1/2 & 0 & 5 \\ 3 & 0 & -6 & 0 & 1 & -1 & 0 & 4 \\ -1/2 & 1 & 3/2 & 0 & 0 & 1/2 & 0 & 1 \\ 0 & 0 & 7 & 0 & 0 & 1 & 1 & 2 \end{bmatrix}$$. The maximum value of $-f$ is 2 when $z = w = 0$.

Row 3 then gives $-x/2 + y = 1$, and rows 1 and 2 give $x/2 + u = 5$ and $3x + v = 4$. Thus the minimum value of f is -2 for all x,y,z where $z = 0$ and $y = 1 + x/2$, with $0 \le x \le 4/3$.

Appendix A

We use the symbol x to denote cross product of vectors.

1. (a) $\mathbf{u} \times \mathbf{v} = (2 \times 4 - 3 \times 0, 3 \times -1 - 1 \times 4, 1 \times 0 - 2 \times -1) = (8, -7, 2)$.

   (b) $\mathbf{v} \times \mathbf{u} = (0 \times 3 - 4 \times 2, 4 \times 1 - (-1) \times 3, -1 \times 2 - 0 \times 1) = (-8, 7, -2)$.

   (c) $\mathbf{u} \times \mathbf{w} = (2 \times -1 - 3 \times 2, 3 \times 1 - 1 \times -1, 1 \times 2 - 2 \times 1) = (-8, 4, 0)$.

   (d) $\mathbf{v} \times \mathbf{w} = (0 \times -1 - 4 \times 2, 4 \times 1 - (-1) \times -1, (-1) \times 2 - 0 \times 1) = (-8, 3, -2)$.

   (e) $(\mathbf{u} \times \mathbf{v}) \times \mathbf{w} = (8, -7, 2) \times \mathbf{w} = (-7 \times -1 - 2 \times 2, 2 \times 1 - 8 \times -1, 8 \times 2 - (-7) \times 1) = (3, 10, 23)$.

2. (a) $\mathbf{u} \times \mathbf{v} = \begin{vmatrix} \mathbf{i} & \mathbf{j} & \mathbf{k} \\ -2 & 2 & 4 \\ 3 & 0 & 5 \end{vmatrix} = (10, 22, -6)$.   (b) $\mathbf{u} \times \mathbf{w} = \begin{vmatrix} \mathbf{i} & \mathbf{j} & \mathbf{k} \\ -2 & 2 & 4 \\ 4 & -2 & 1 \end{vmatrix} = (10, 18, -4)$.

   (c) $\mathbf{w} \times \mathbf{v} = \begin{vmatrix} \mathbf{i} & \mathbf{j} & \mathbf{k} \\ 4 & -2 & 1 \\ 3 & 0 & 5 \end{vmatrix} = (-10, -17, 6)$.   (d) $\mathbf{v} \times \mathbf{w} = \begin{vmatrix} \mathbf{i} & \mathbf{j} & \mathbf{k} \\ 3 & 0 & 5 \\ 4 & -2 & 1 \end{vmatrix} = (10, 17, -6)$.

   (e) $(\mathbf{w} \times \mathbf{v}) \times \mathbf{u} = \begin{vmatrix} \mathbf{i} & \mathbf{j} & \mathbf{k} \\ -10 & -17 & 6 \\ -2 & 2 & 4 \end{vmatrix} = (-80, 28, -54)$.

3. (a) $\mathbf{u} \times \mathbf{v} = \begin{vmatrix} \mathbf{i} & \mathbf{j} & \mathbf{k} \\ 2 & 3 & 1 \\ -1 & 2 & 4 \end{vmatrix} = 10\mathbf{i} - 9\mathbf{j} + 7\mathbf{k}$.

   (b) $\mathbf{u} \times \mathbf{w} = \begin{vmatrix} \mathbf{i} & \mathbf{j} & \mathbf{k} \\ 2 & 3 & 1 \\ 3 & 0 & -7 \end{vmatrix} = -21\mathbf{i} + 17\mathbf{j} - 9\mathbf{k}$.

   (c) $\mathbf{w} \times \mathbf{u} = \begin{vmatrix} \mathbf{i} & \mathbf{j} & \mathbf{k} \\ 3 & 0 & -7 \\ 2 & 3 & 1 \end{vmatrix} = 21\mathbf{i} - 17\mathbf{j} + 9\mathbf{k}$.

Answers to Appendix A

(d) $\mathbf{v} \times \mathbf{w} = \begin{vmatrix} \mathbf{i} & \mathbf{j} & \mathbf{k} \\ -1 & 2 & 4 \\ 3 & 0 & -7 \end{vmatrix} = -14\mathbf{i} + 5\mathbf{j} - 6\mathbf{k}.$

(e) $(\mathbf{w} \times \mathbf{u}) \times \mathbf{v} = \begin{vmatrix} \mathbf{i} & \mathbf{j} & \mathbf{k} \\ 21 & -17 & 9 \\ -1 & 2 & 4 \end{vmatrix} = -86\mathbf{i} - 93\mathbf{j} + 25\mathbf{k}.$

4. (a) $\mathbf{u} \times \mathbf{v} = \begin{vmatrix} \mathbf{i} & \mathbf{j} & \mathbf{k} \\ 3 & 1 & -2 \\ 4 & -1 & 2 \end{vmatrix} = (0,-14,-7).$

(b) $3\mathbf{v} \times 2\mathbf{w} = \begin{vmatrix} \mathbf{i} & \mathbf{j} & \mathbf{k} \\ 12 & -3 & 6 \\ 0 & 6 & -4 \end{vmatrix} = (-24, 48, 72).$

(c) $(\mathbf{w} \times \mathbf{u}) \cdot \mathbf{v} = \begin{vmatrix} \mathbf{i} & \mathbf{j} & \mathbf{k} \\ 0 & 3 & -2 \\ 3 & 1 & -2 \end{vmatrix} \cdot \mathbf{v} = (-4,-6,-9) \cdot (4,-1,2) = -28.$

(d) $(\mathbf{w}+2\mathbf{u}) \times \mathbf{v} = \begin{vmatrix} \mathbf{i} & \mathbf{j} & \mathbf{k} \\ 6 & 5 & -6 \\ 4 & -1 & 2 \end{vmatrix} = (4,-36,-26).$

(e) $\mathbf{u} \cdot (\mathbf{v} \times \mathbf{w}) = \mathbf{u} \cdot \begin{vmatrix} \mathbf{i} & \mathbf{j} & \mathbf{k} \\ 4 & -1 & 2 \\ 0 & 3 & -2 \end{vmatrix} = (3,1,-2) \cdot (-4,8,12) = -28.$

(f) $(\mathbf{v} \times \mathbf{u}) \cdot (\mathbf{w} \times \mathbf{v}) = (0,14,7) \cdot (4,-8,-12) = -196.$

5. (a) $\vec{AB} = (-3,4,6) - (1,2,1) = (-4,2,5)$ and $\vec{AC} = (1,8,3) - (1,2,1) = (0,6,2).$

$\vec{AB} \times \vec{AC} = \begin{vmatrix} \mathbf{i} & \mathbf{j} & \mathbf{k} \\ -4 & 2 & 5 \\ 0 & 6 & 2 \end{vmatrix} = (-26,8,-24)$, so area $= \frac{1}{2} \|(-26,8,-24)\| = \sqrt{329}.$

(b) $\vec{AB} = (0,2,6) - (3,-1,2) = (-3,3,4)$ and $\vec{AC} = (7,1,5) - (3,-1,2) = (4,2,3).$

Answers to Appendix A

$$\vec{AB} \times \vec{AC} = \begin{vmatrix} \mathbf{i} & \mathbf{j} & \mathbf{k} \\ -3 & 3 & 4 \\ 4 & 2 & 3 \end{vmatrix} = (1,25,-18), \text{ so area} = \frac{1}{2}\|(1,25,-18)\| = \frac{5}{2}\sqrt{38}.$$

(c) $\vec{AB} = (0,5,2) - (1,0,0) = (-1,5,2)$ and $\vec{AC} = (3,-4,8) - (1,0,0) = (2,-4,8)$.

$$\vec{AB} \times \vec{AC} = \begin{vmatrix} \mathbf{i} & \mathbf{j} & \mathbf{k} \\ -1 & 5 & 2 \\ 2 & -4 & 8 \end{vmatrix} = (48,12,-6), \text{ so area} = \frac{1}{2}\|(48,12,-6)\| = 3\sqrt{69}.$$

6. (a) $\mathbf{u} = \vec{AB} = (4,8,1) - (1,2,5) = (3,6,-4)$, $\mathbf{v} = \vec{AC} = (-3,2,3) - (1,2,5) = (-4,0,-2)$, and

$\mathbf{w} = \vec{AD} = (0,3,9) - (1,2,5) = (-1,1,4)$.

$$\mathbf{w} \cdot (\mathbf{u} \times \mathbf{v}) = \begin{vmatrix} 3 & 6 & -4 \\ -4 & 0 & -2 \\ -1 & 1 & 4 \end{vmatrix} = 130, \text{ so the volume} = |130| = 130.$$

(b) $\mathbf{u} = \vec{AB} = (-2,3,4) - (3,1,6) = (-5,2,-2)$, $\mathbf{v} = \vec{AC} = (0,2,-5) - (3,1,6) = (-3,1,-11)$,

and $\mathbf{w} = \vec{AD} = (3,-1,4) - (3,1,6) = (0,-2,-2)$.

$$\mathbf{w} \cdot (\mathbf{u} \times \mathbf{v}) = \begin{vmatrix} -5 & 2 & -2 \\ -3 & 1 & -11 \\ 0 & -2 & -2 \end{vmatrix} = 96, \text{ so the volume} = |96| = 96.$$

(c) $\mathbf{u} = \vec{AB} = (-3,1,4) - (0,1,2) = (-3,0,2)$, $\mathbf{v} = \vec{AC} = (5,2,3) - (0,1,2) = (5,1,1)$,

and $\mathbf{w} = \vec{AD} = (-3,-2,1) - (0,1,2) = (-3,-3,-1)$.

$$\mathbf{w} \cdot (\mathbf{u} \times \mathbf{v}) = \begin{vmatrix} -3 & 0 & 2 \\ 5 & 1 & 1 \\ -3 & -3 & -1 \end{vmatrix} = -30, \text{ so the volume} = |-30| = 30.$$

Answers to Appendix A

7. (a) $\mathbf{i}\cdot\mathbf{i} = (1,0,0)\cdot(1,0,0) = 1+0+0 = 1$, $\mathbf{j}\cdot\mathbf{j} = (0,1,0)\cdot(0,1,0) = 0+1+0 = 1$,
    $\mathbf{k}\cdot\mathbf{k} = (0,0,1)\cdot(0,0,1) = 0+0+1 = 1$.

    (b) $\mathbf{i}\cdot\mathbf{j} = (1,0,0)\cdot(0,1,0) = 0+0+0 = 0$, $\mathbf{i}\cdot\mathbf{k} = (1,0,0)\cdot(0,0,1) = 0+0+0 = 0$,
    $\mathbf{j}\cdot\mathbf{k} = (0,1,0)\cdot(0,0,1) = 0+0+0 = 0$.

8. $(\mathbf{i} \times \mathbf{j})\cdot\mathbf{k} = (0\times 0 - 0\times 1, 0\times 0 - 1\times 0, 1\times 1 - 0\times 0)\cdot(0,0,1) = (0,0,1)\cdot(0,0,1) = 0+0+1 = 1$.

9. $\mathbf{u} = (u_1, u_2, u_3)$ and $\mathbf{0} = (0,0,0)$.

    $\mathbf{u} \times \mathbf{0} = (u_2\times 0 - u_3\times 0, u_3\times 0 - u_1\times 0, u_1\times 0 - u_2\times 0) = (0,0,0) = \mathbf{0}$.

    $\mathbf{0} \times \mathbf{u} = (0\times u_3 - 0\times u_2, 0\times u_1 - 0\times u_3, 0\times u_2 - 0\times u_1) = (0,0,0) = \mathbf{0}$.

10. $\mathbf{u} = (u_1, u_2, u_3)$, $\mathbf{v} = (v_1, v_2, v_3)$, and $\mathbf{w} = (w_1, w_2, w_3)$.

    $(\mathbf{u} \times \mathbf{v})\cdot\mathbf{w} = (u_2 v_3 - u_3 v_2, u_3 v_1 - u_1 v_3, u_1 v_2 - u_2 v_1)\cdot(w_1, w_2, w_3)$
    $= u_2 v_3 w_1 - u_3 v_2 w_1 + u_3 v_1 w_2 - u_1 v_3 w_2 + u_1 v_2 w_3 - u_2 v_1 w_3$
    $= u_1 v_2 w_3 - u_1 v_3 w_2 + u_2 v_3 w_1 - u_2 v_1 w_3 + u_3 v_1 w_2 - u_3 v_2 w_1$
    $= (u_1, u_2, u_3)\cdot(v_2 w_3 - v_3 w_2, v_3 w_1 - v_1 w_3, v_1 w_2 - v_2 w_1) = \mathbf{u}\cdot(\mathbf{v} \times \mathbf{w})$.

11. $\mathbf{u} = (u_1, u_2, u_3)$, $\mathbf{v} = (v_1, v_2, v_3)$, and $\mathbf{w} = (w_1, w_2, w_3)$.

    The coordinates of $\mathbf{u} \times (\mathbf{v} \times \mathbf{w}) = (u_1, u_2, u_3) \times (v_2 w_3 - v_3 w_2, v_3 w_1 - v_1 w_3, v_1 w_2 - v_2 w_1)$
    are $u_2(v_1 w_2 - v_2 w_1) - u_3(v_3 w_1 - v_1 w_3)$,
    $u_3(v_2 w_3 - v_3 w_2) - u_1(v_1 w_2 - v_2 w_1)$, and $u_1(v_3 w_1 - v_1 w_3) - u_2(v_2 w_3 - v_3 w_2)$.
    Rearranging the terms, the coordinates are $(u_2 w_2 + u_3 w_3)v_1 - (u_2 v_2 + u_3 v_3)w_1$,
    $(u_1 w_1 + u_3 w_3)v_2 - (u_1 v_1 + u_3 v_3)w_2$, and $(u_1 w_1 + u_2 w_2)v_3 - (u_1 v_1 + u_2 v_2)w_3$.

    Now add $0 = u_1 v_1 w_1 - u_1 v_1 w_1$ to the first coordinate, $0 = u_2 v_2 w_2 - u_2 v_2 w_2$ to the second coordinate, and $0 = u_3 v_3 w_3 - u_3 v_3 w_3$ to the third coordinate. The coordinates become $(u_1 w_1 + u_2 w_2 + u_3 w_3)v_1 - (u_1 v_1 + u_2 v_2 + u_3 v_3)w_1$,

Answers to Appendix A

$(u_1w_1 + u_2w_2 + u_3w_3)v_2 - (u_1v_1 + u_2v_2 + u_3v_3)w_2$, and
$(u_1w_1 + u_2w_2 + u_3w_3)v_3 - (u_1v_1 + u_2v_2 + u_3v_3)w_3$. These are the coordinates of
$(\mathbf{u}\cdot\mathbf{w})\mathbf{v} - (\mathbf{u}\cdot\mathbf{v})\mathbf{w}$, so $\mathbf{u} \times (\mathbf{v} \times \mathbf{w}) = (\mathbf{u}\cdot\mathbf{w})\mathbf{v} - (\mathbf{u}\cdot\mathbf{v})\mathbf{w}$.

12. $c(\mathbf{u} \times \mathbf{v}) = c(u_2v_3 - u_3v_2, u_3v_1 - u_1v_3, u_1v_2 - u_2v_1)$
 $= (cu_2v_3 - cu_3v_2, cu_3v_1 - cu_1v_3, cu_1v_2 - cu_2v_1) = c\mathbf{u} \times \mathbf{v}$
 $= (u_2cv_3 - u_3cv_2, u_3cv_1 - u_1cv_3, u_1cv_2 - u_2cv_1) = \mathbf{u} \times c\mathbf{v}$.

13. $\mathbf{u}$ and $\mathbf{v}$ are parallel if and only if the angle between them is zero. If $0 \leq \theta < \pi$ then $\sin \theta = 0$ if and only if $\theta = 0$, so $\mathbf{u}$ and $\mathbf{v}$ are parallel if and only if $\sin \theta = 0$, where θ is the angle between the vectors. If neither $\mathbf{u}$ nor $\mathbf{v}$ is the zero vector, then $\|\mathbf{u}\| \neq 0$ and $\|\mathbf{v}\| \neq 0$, so $\sin \theta = \dfrac{\|\mathbf{u} \times \mathbf{v}\|}{\|\mathbf{u}\| \|\mathbf{v}\|} = 0$ if and only if $\|\mathbf{u} \times \mathbf{v}\| = 0$ if and only if $\mathbf{u} \times \mathbf{v} = \mathbf{0}$.

14. $\mathbf{u} = (u_1,u_2,u_3)$, $\mathbf{v} = (v_1,v_2,v_3)$, and $\mathbf{w} = (w_1,w_2,w_3)$. From Exercise 11, $\mathbf{u} \times (\mathbf{v} \times \mathbf{w}) = (\mathbf{u}\cdot\mathbf{w})\mathbf{v} - (\mathbf{u}\cdot\mathbf{v})\mathbf{w}$. Therefore $\mathbf{v} \times (\mathbf{w} \times \mathbf{u}) = (\mathbf{v}\cdot\mathbf{u})\mathbf{w} - (\mathbf{v}\cdot\mathbf{w})\mathbf{u}$ and $\mathbf{w} \times (\mathbf{u} \times \mathbf{v}) = (\mathbf{w}\cdot\mathbf{v})\mathbf{u} - (\mathbf{w}\cdot\mathbf{u})\mathbf{v}$. Adding gives $\mathbf{u} \times (\mathbf{v} \times \mathbf{w}) + \mathbf{v} \times (\mathbf{w} \times \mathbf{u}) + \mathbf{w} \times (\mathbf{u} \times \mathbf{v})$
 $= (\mathbf{u}\cdot\mathbf{w})\mathbf{v} - (\mathbf{u}\cdot\mathbf{v})\mathbf{w} + (\mathbf{v}\cdot\mathbf{u})\mathbf{w} - (\mathbf{v}\cdot\mathbf{w})\mathbf{u} + (\mathbf{w}\cdot\mathbf{v})\mathbf{u} - (\mathbf{w}\cdot\mathbf{u})\mathbf{v} = \mathbf{0}$.

15. $\mathbf{u} = (u_1,u_2,u_3)$, $\mathbf{v} = (v_1,v_2,v_3)$, and $\mathbf{w} = (w_1,w_2,w_3)$.

$$\mathbf{u}\cdot(\mathbf{v} \times \mathbf{w}) = \begin{vmatrix} u_1 & u_2 & u_3 \\ v_1 & v_2 & v_3 \\ w_1 & w_2 & w_3 \end{vmatrix} = 0$$ if and only if the row vectors $\mathbf{u}$, $\mathbf{v}$ and $\mathbf{w}$ are linearly dependent, i.e., if and only if one is a linear combination of the other two. So $\mathbf{u}\cdot(\mathbf{v} \times \mathbf{w}) = 0$ if and only if one vector lies in the plane of the other two.

16. $(\mathbf{t} \times \mathbf{u})\cdot(\mathbf{v} \times \mathbf{w})$
 $= (t_2u_3 - t_3u_2, t_3u_1 - t_1u_3, t_1u_2 - t_2u_1)\cdot(v_2w_3 - v_3w_2, v_3w_1 - v_1w_3, v_1w_2 - v_2w_1)$
 $= (t_2u_3 - t_3u_2)(v_2w_3 - v_3w_2) + (t_3u_1 - t_1u_3)(v_3w_1 - v_1w_3) + (t_1u_2 - t_2u_1)(v_1w_2 - v_2w_1)$
 $= t_2u_3v_2w_3 + t_3u_2v_3w_2 - t_2u_3v_3w_2 - t_3u_2v_2w_3 + t_3u_1v_3w_1 + t_1u_3v_1w_3 - t_3u_1v_1w_3 - t_1u_3v_3w_1$
 $+ t_1u_2v_1w_2 + t_2u_1v_2w_1 - t_1u_2v_2w_1 - t_2u_1v_1w_2$. Now add zero in the form
 $t_1u_1v_1w_1 - t_1u_1v_1w_1 + t_2u_2v_2w_2 - t_2u_2v_2w_2 + t_3u_3v_3w_3 - t_3u_3v_3w_3$ and rearrange terms:

396

Answers to Appendix B

$$(\mathbf{t} \times \mathbf{u}) \cdot (\mathbf{v} \times \mathbf{w}) = (t_1 v_1 + t_2 v_2 + t_3 v_3)(u_1 w_1 + u_2 w_2 + u_3 w_3)$$
$$- (u_1 v_1 + u_2 v_2 + u_3 v_3)(t_1 w_1 + t_2 w_2 + t_3 w_3) = \begin{vmatrix} \mathbf{t} \cdot \mathbf{v} & \mathbf{u} \cdot \mathbf{v} \\ \mathbf{t} \cdot \mathbf{w} & \mathbf{u} \cdot \mathbf{w} \end{vmatrix}.$$

Appendix B

1. (a) $P_0 = (x_0, y_0, z_0) = (1,-2,4)$ and $(a,b,c) = (1,1,1)$.

   point-normal form: $(x - 1) + (y + 2) + (z - 4) = 0$
   general form: $x + y + z - 3 = 0$

   (b) $P_0 = (x_0, y_0, z_0) = (-3,5,6)$ and $(a,b,c) = (-2,4,5)$.

   point-normal form: $-2(x + 3) + 4(y - 5) + 5(z - 6) = 0$
   general form: $-2x + 4y + 5z - 56 = 0$

   (c) $P_0 = (x_0, y_0, z_0) = (0,0,0)$ and $(a,b,c) = (1,2,3)$.

   point-normal form: $x + 2y + 3z = 0$
   general form: $x + 2y + 3z = 0$

   (d) $P_0 = (x_0, y_0, z_0) = (4,5,-2)$ and $(a,b,c) = (-1,4,3)$.

   point-normal form: $-(x - 4) + 4(y - 5) + 3(z + 2) = 0$
   general form: $-x + 4y + 3z - 10 = 0$

2. (a) $\overrightarrow{P_1 P_2} = (1,0,2) - (1,-2,3) = (0,2,-1)$ and $\overrightarrow{P_1 P_3} = (-1,4,6) - (1,-2,3) = (-2,6,3)$.

   $$\overrightarrow{P_1 P_2} \times \overrightarrow{P_1 P_3} = \begin{vmatrix} \mathbf{i} & \mathbf{j} & \mathbf{k} \\ 0 & 2 & -1 \\ -2 & 6 & 3 \end{vmatrix} = (12,2,4) \text{ is normal to the plane.}$$

   $P_0 = (x_0, y_0, z_0) = (1,-2,3)$ and $(a,b,c) = (12,2,4)$.

   point-normal form: $12(x - 1) + 2(y + 2) + 4(z - 3) = 0$
   general form: $12x + 2y + 4z - 20 = 0$ or $6x + y + 2z - 10 = 0$

397

Answers to Appendix B

(b) $\vec{P_1P_2} = (1,2,4) - (0,0,0) = (1,2,4)$ and $\vec{P_1P_3} = (-3,5,1) - (0,0,0) = (-3,5,1)$.

$$\vec{P_1P_2} \times \vec{P_1P_3} = \begin{vmatrix} i & j & k \\ 1 & 2 & 4 \\ -3 & 5 & 1 \end{vmatrix} = (-18,-13,11) \text{ is normal to the plane.}$$

$P_0 = (x_0, y_0, z_0) = (0,0,0)$ and $(a,b,c) = (-18,-13,11)$.

point-normal form: $-18x - 13y + 11z = 0$
general form: $-18x - 13y + 11z = 0$

(c) $\vec{P_1P_2} = (3,5,4) - (-1,-1,2) = (4,6,2)$ and $\vec{P_1P_3} = (1,2,5) - (-1,-1,2) = (2,3,3)$.

$$\vec{P_1P_2} \times \vec{P_1P_3} = \begin{vmatrix} i & j & k \\ 4 & 6 & 2 \\ 2 & 3 & 3 \end{vmatrix} = (12,-8,0) \text{ is normal to the plane.}$$

$P_0 = (x_0, y_0, z_0) = (-1,-1,2)$ and $(a,b,c) = (12,-8,0)$.

point-normal form: $12(x + 1) - 8(y + 1) = 0$
general form: $12x - 8y + 4 = 0$ or $3x - 2y + 1 = 0$

(d) $\vec{P_1P_2} = (-2,4,-3) - (7,1,3) = (-9,3,-6)$ and $\vec{P_1P_3} = (5,4,1) - (7,1,3) = (-2,3,-2)$.

$$\vec{P_1P_2} \times \vec{P_1P_3} = \begin{vmatrix} i & j & k \\ -9 & 3 & -6 \\ -2 & 3 & -2 \end{vmatrix} = (12,-6,-21) \text{ is normal to the plane.}$$

$P_0 = (x_0, y_0, z_0) = (7,1,3)$ and $(a,b,c) = (12,-6,-21)$.

point-normal form: $12(x - 7) - 6(y - 1) - 21(z - 3) = 0$
general form: $12x - 6y - 21z - 15 = 0$ or $4x - 2y - 7z - 5 = 0$

3. $(3,-2,4)$ is normal to the plane $3x - 2y + 4z - 3 = 0$ and $(-6,4,-8)$ is normal to the plane $-6x + 4y - 8z + 7 = 0$. $(-6,4,-8) = -2(3,-2,4)$, so the normals are parallel and therefore the planes are parallel.

4. $(3,-6,9)$ is normal to the plane $3x - 6y + 9z - 6 = 0$ and $(-1,2,-3)$ is normal to the plane $-x + 2y - 3z - 2 = 0$. $-3(-1,2,-3) = (3,-6,9)$, so the normals are parallel and therefore the planes are parallel. The planes are distinct so they have no points in common.

Answers to Appendix B

5. (2,–3,1) is normal to the plane $2x - 3y + z + 4 = 0$ and therefore to the parallel plane passing through (1,2,–3).

   $P_0 = (x_0, y_0, z_0) = (1,2,-3)$ and $(a,b,c) = (2,-3,1)$.

   point-normal form: $2(x - 1) - 3(y - 2) + (z + 3) = 0$
   general form: $2x - 3y + z + 7 = 0$

6. (a) $P_0 = (x_0, y_0, z_0) = (1,2,3)$ and $(a,b,c) = (-1,2,4)$.

   parametric equations: $x = 1 - t$, $y = 2 + 2t$, $z = 3 + 4t$, $-\infty < t < \infty$

   symmetric equations: $-x + 1 = \dfrac{y-2}{2} = \dfrac{z-3}{4}$

   (b) $P_0 = (x_0, y_0, z_0) = (-3,1,2)$ and $(a,b,c) = (1,1,1)$.

   parametric equations: $x = -3 + t$, $y = 1 + t$, $z = 2 + t$, $-\infty < t < \infty$

   symmetric equations: $x + 3 = y - 1 = z - 2$

   (c) $P_0 = (x_0, y_0, z_0) = (0,0,0)$ and $(a,b,c) = (-2,-3,5)$.

   parametric equations: $x = -2t$, $y = -3t$, $z = 5t$, $-\infty < t < \infty$

   symmetric equations: $-x/2 = -y/3 = z/5$

   (d) $P_0 = (x_0, y_0, z_0) = (-2,-4,1)$ and $(a,b,c) = (2,-2,4)$.

   parametric equations: $x = -2 + 2t$, $y = -4 - 2t$, $z = 1 + 4t$, $-\infty < t < \infty$

   symmetric equations: $\dfrac{x+2}{2} = -\dfrac{y+4}{2} = \dfrac{z-1}{4}$

7. $P_0 = (x_0, y_0, z_0) = (1,2,-4)$ and $(a,b,c) = (2,3,1)$.

   $(x - 1, y - 2, z + 4) = t(2,3,1)$, thus the parametric equations are

   $x = 1 + 2t$, $y = 2 + 3t$, $z = -4 + t$, $-\infty < t < \infty$.

Answers to Appendix B

8. The direction of the line must be orthogonal to (3,2,5). Let it be (a,b,c). Then (3,2,5)·(a,b,c) = 3a + 2b + 5c = 0. There are many solutions; one solution is (1,−4,1). The parametric equations of the line through (2,−3,1) with direction (1,−4,1) are
x = 2 + t, y = −3 − 4t, z = 1 + t, −∞ < t < ∞.

9. $P_0 = (x_0, y_0, z_0) = (4,-1,3)$ and (a,b,c) = (2,−1,4).

   (x − 4, y + 1, z − 3) = t(2,−1,4), thus the parametric equations are

   x = 4 + 2t, y = −1 − t, z = 3 + 4t, −∞ < t < ∞.

10. xy plane: z = 0   xz plane: y = 0   yz plane: x = 0

11. $\overrightarrow{P_1P_2}$ = (−1,1,3) − (3,−5,5) = (−4,6,−2) and $\overrightarrow{P_1P_3}$ = (5,−8,6) − (3,−5,5) = (2,−3,1).

    $\overrightarrow{P_1P_2} = -2\overrightarrow{P_1P_3}$, so the points P_1, P_2, and P_3 are colinear. There are many planes through the line containing these three points.

12. (−1,2,−4) is normal to the plane perpendicular to the given line.

    $P_0 = (x_0, y_0, z_0) = (4,-1,5)$ and (a,b,c) = (−1,2,−4).

    point-normal form: −(x − 4) + 2(y + 1) − 4(z − 5) = 0
    general form: −x + 2y − 4z + 26 = 0

13. The points on the line are of the form (1 + t, 14 − t, 2 − t). If the line lies in the plane, the points will satisfy the equation of the plane. 2(1 + t) − (14 − t) + 3(2 − t) + 6 = 2 − 14 + 6 + 6 + 2t + t − 3t = 0, so the line lies in the plane.

14. 3(4 + 2t) + 2(5 + t) − 4(7 + 2t) + 7 = 12 + 10 − 28 + 6t + 2t − 8t = −6 for all possible values of t, so there is no point on the line which lies in the plane.

15. The direction of the line must be perpendicular to (−2,3,2). Let it be (a,b,c). Then (−2,3,2)·(a,b,c) = −2a + 3b + 2c = 0. There are many solutions, all of the form $\left(\frac{3b+2c}{2}, b, c\right)$; one solution is (1,0,1). The parametric equations of the line through (5,−1,2) with direction (1,0,1) are x = 5 + t, y = −1, z = 2 + t, −∞ < t < ∞.

400

Answers to Appendix B

16. The directions of the two lines are $(-4,4,8)$ and $(3,-1,2)$. $(-4,4,8) \cdot (3,-1,2) = 0$, so the lines are orthogonal. To find the point of intersection, equate the expressions for x, y, and z: $-1 - 4t = 4 + 3h$, $4 + 4t = 1 - h$, $7 + 8t = 5 + 2h$. This system of equations has the unique solution $t = -1/2$, $h = -1$, so the point of intersection is $x = 1$, $y = 2$, $z = 3$.

17. The directions of the two lines are $(4,5,3)$ and $(1,-3,-2)$. $(4,5,3) \neq c(1,-3,-2)$, so the lines are not parallel. If the lines intersect there will be a solution to the system of equations $1 + 4t = 2 + h$, $2 + 5t = 1 - 3h$, $3 + 3t = -1 - 2h$. However the first equation gives $h = 4t - 1$ and substituting this into the other two equations gives two different values of t. Thus the system of equations has no solution and the lines do not intersect.

18. (a) Let (e,f,g) and (r,s,t) be two points on the plane and let h be any scalar. Then $2e + 3f - 4g = 0$ and $2r + 3s - 4t = 0$, so that $2(e+r) + 3(f+s) - 4(g+t) = 0$ and $2he + 3hf - 4hg = 0$; i.e., $(e,f,g) + (r,s,t) = (e+r, f+s, g+t)$ and $h(e,f,g) = (he,hf,hg)$ are also on the plane. Thus the set of all points on the plane is a subspace of $\mathbf{R}^3$.

    A basis for the space must consist of two linearly independent vectors that are orthogonal to $(2,3,-4)$. Two such vectors are $(0,4,3)$ and $(2,0,1)$.

    (b) The point $(0,0,0)$ is not on the plane so the set of all points on the plane is not a subspace of $\mathbf{R}^3$.

    (c) If $d = 0$, then the proof of part (a) with 2, 3, and -4 replaced with a, b, and c shows that the plane is a subspace of $\mathbf{R}^3$. If the plane is a subspace of $\mathbf{R}^3$, then $(0,0,0)$ is a point on the plane. Thus $ax0 + bx0 + cx0 + d = 0$, so that $d = 0$.

19. Let (a,b,c) be the direction of a line. The line can be written in symmetric form if and only if a, b, and c are all nonzero. But the line is orthogonal to the x axis if and only if $a = 0$, the line is orthogonal to the y axis if and only if $b = 0$, and the line is orthogonal to the z axis if and only if $c = 0$, so the line can be written in symmetric form if and only if it is not orthogonal to any axis.

Appendix C Calculator Exercises

C1

1. $x=1.5, y=-0.5$

2. $x=0.8, y=-1.2$

3. no solution

4. $x=r, y=-0.5r+2$

5. $x=11, y=-4$

6. $x=0.8571, y=0.1429$

7. no solution

8. $x=4, y=2$

9. $x=3, y=-1$

10. $x=r, y=r/3 + 2/3$

C2

3. $\begin{bmatrix} 1 & 3 & -2 & 0 \\ 1 & 2 & -3 & 6 \\ 8 & 3 & 2 & 5 \end{bmatrix}$

4. $\begin{bmatrix} 2 & 7 & 5 & 1 \\ 0 & -8 & 4 & 3 \\ 3 & -5 & 8 & 9 \end{bmatrix}$

5. $\begin{bmatrix} 1 & 2 & 3 & -1 \\ 0 & 3 & 10 & 0 \\ 0 & -8 & -1 & -1 \end{bmatrix}$

6. $\begin{bmatrix} 1 & 0 & -1 & -6 \\ 0 & 1 & 2 & 1 \\ 0 & 0 & 11 & -1 \end{bmatrix}$

7. $\begin{bmatrix} 1 & 0 & 0 & -23 \\ 0 & 1 & 0 & 17 \\ 0 & 0 & 1 & 5 \end{bmatrix}$

8. $\begin{bmatrix} 1 & 0 & 2 & 7 \\ 0 & 1 & 5 & -3 \\ 0 & 0 & -1 & 4 \end{bmatrix}$

9 - 13. See Section 1.1 of main text, Exercises 11 (a) - 11 (e).

14. $\begin{bmatrix} 1 & 2 & -1 & 3 \\ 1 & 3 & -2 & -6 \\ -1 & -1 & 3 & 6 \end{bmatrix} \underset{R3+R1}{\overset{R2+(-1)R1}{\approx}} \begin{bmatrix} 1 & 2 & -1 & 3 \\ 0 & 1 & -1 & -9 \\ 0 & 1 & 2 & 9 \end{bmatrix} \underset{R3+(-1)R2}{\overset{R1+(-2)R2}{\approx}} \begin{bmatrix} 1 & 0 & 1 & 21 \\ 0 & 1 & -1 & -9 \\ 0 & 0 & 3 & 18 \end{bmatrix}$

$\underset{(1/3)R3}{\approx} \begin{bmatrix} 1 & 0 & 1 & 21 \\ 0 & 1 & -1 & -9 \\ 0 & 0 & 1 & 6 \end{bmatrix} \underset{R2+R3}{\overset{R1+(-1)R3}{\approx}} \begin{bmatrix} 1 & 0 & 0 & 15 \\ 0 & 1 & 0 & -3 \\ 0 & 0 & 1 & 6 \end{bmatrix}$.

so the solution is $x_1 = 15, x_2 = -3, x_3 = 6$.

15. $x_1 = 6, x_2 = -3, x_3 = 2$

16. $x_1 = -2.8333, x_2 = 2.0556, x_3 = 0.0556$

Answers to Appendix C, Calculator Exercises

17. Start with an appropriate REF and work backwards. For example,

$$\begin{bmatrix} 1 & 0 & 0 & 1 \\ 0 & 1 & 0 & 2 \\ 0 & 0 & 1 & 3 \end{bmatrix} \underset{R2+(-2)R3}{\overset{R1+(2)R3}{\sim}} \begin{bmatrix} 1 & 0 & 2 & 7 \\ 0 & 1 & -1 & -1 \\ 0 & 0 & 1 & 3 \end{bmatrix} \underset{R1+(-2)R2}{\sim} \begin{bmatrix} 1 & -2 & 4 & 9 \\ 0 & 1 & -1 & -1 \\ 0 & 0 & 1 & 3 \end{bmatrix}$$

$$\underset{(2)R2}{\sim} \begin{bmatrix} 1 & -2 & 4 & 9 \\ 0 & 2 & -2 & -2 \\ 0 & 0 & 1 & 3 \end{bmatrix} \underset{R2<->R3}{\sim} \begin{bmatrix} 1 & -2 & 4 & 9 \\ 0 & 0 & 1 & 3 \\ 0 & 2 & -2 & -2 \end{bmatrix} \underset{R3+(3)R1}{\overset{R2+(-1)R1}{\sim}} \begin{bmatrix} 1 & -2 & 4 & 9 \\ 1 & -2 & 5 & 12 \\ -1 & 4 & -6 & -11 \end{bmatrix}$$

The following system uses all three types of row operations

$$\begin{aligned} x - 2y + 4z &= 9 \\ x - 2y + 5z &= 12 \\ -x + 4y - 6z &= -11 \end{aligned}$$

18. A 3x3 matrix represents the equations of three lines in a plane. In order for there to be a unique solution, the three lines would have to meet in a point. For there to be many solutions, the three lines would all have to be the same. It is far more likely that the lines will meet in pairs (or that one pair will be parallel), i.e., that there will be no solution, the situation represented by the reduced echelon form I_3.

C3

1. $\begin{bmatrix} 4 & 12 \\ -3 & 7 \end{bmatrix}$
2. $\begin{bmatrix} 2 & -2 \\ 3 & 11 \end{bmatrix}$
3. $\begin{bmatrix} 9 & 15 \\ 0 & -6 \end{bmatrix}$

4. $\begin{bmatrix} -2 & -14 \\ 6 & -18 \end{bmatrix}$
5. $\begin{bmatrix} 5 & 19 \\ -6 & 16 \end{bmatrix}$
6. $\begin{bmatrix} 4 & -20 \\ 15 & -51 \end{bmatrix}$

7. $\begin{bmatrix} 9.2 & 24.4 \\ -5.1 & 10.3 \end{bmatrix}$
8. $\begin{bmatrix} -17.51 & -46.57 \\ 9.78 & -19.84 \end{bmatrix}$
9. $\begin{bmatrix} 1 & 0 \\ -1 & 13 \end{bmatrix}$

10. $\begin{bmatrix} -1 & -4 \\ 9 & 1 \end{bmatrix}$
11. $\begin{bmatrix} 0 & -5 \\ 10 & 17.5 \end{bmatrix}$
12. $\begin{bmatrix} 3.76 & 7.52 \\ -18.8 & 22.56 \end{bmatrix}$

13. $\begin{bmatrix} 5 & 20 & -5 \\ 10 & 15 & -25 \end{bmatrix}$
14. Does not exist.
15. Does not exist.

C4

1. $\begin{bmatrix} -12 & 66 \\ 6 & -18 \end{bmatrix}$
2. $\begin{bmatrix} 3 & -9 \\ -9 & -33 \end{bmatrix}$
3. $\begin{bmatrix} -36 & 1198 \\ 18 & -54 \end{bmatrix}$

Answers to Appendix C, Calculator Exercises

4. $\begin{bmatrix} -6 & 76 \\ 6 & -22 \end{bmatrix}$

5. $\begin{bmatrix} -15 & 75 \\ 15 & 15 \end{bmatrix}$

6. $\begin{bmatrix} 81 & 65 \\ 0 & 16 \end{bmatrix}$

7. $\begin{bmatrix} -58800 & 65400 \\ 11400 & -10200 \end{bmatrix}$

8. $\begin{bmatrix} 6204 & -12312 \\ 1698 & -2574 \end{bmatrix}$

9. $\begin{bmatrix} 10 & -12 \\ -31 & 50 \end{bmatrix}$

10. $\begin{bmatrix} -14 & 4 & 50 \\ 69 & 26 & -219 \end{bmatrix}$

11. Does not exist.

12. $\begin{bmatrix} -7 & -8 & 19 \\ 61 & 109 & -142 \end{bmatrix}$

13. Does not exist.

14. $\begin{bmatrix} -195 & 280 & 831 \\ 89 & 1441 & 562 \end{bmatrix}$

C5

1. $\begin{bmatrix} 1 & 2 \\ 0 & 1 \end{bmatrix}$

2. $\begin{bmatrix} 1 & 9 \\ 2 & 4 \end{bmatrix}$

3. $\begin{bmatrix} 2 & 5 \\ 1 & 4 \\ 0 & 3 \end{bmatrix}$

4. $\begin{bmatrix} 0 & 7 \\ 1 & 9 \\ 1 & -1 \\ 3 & 2 \end{bmatrix}$

5. $\begin{bmatrix} 1 & 3 & 0 \\ 2 & 6 & 8 \end{bmatrix}$

6. $\begin{bmatrix} 2 & -7 & 4 \\ -3 & 0 & -6 \\ 6 & 2 & -1 \end{bmatrix}$

7. $\begin{bmatrix} -3 & -19 \\ 8 & 28 \end{bmatrix}$

8. $\begin{bmatrix} -143 & -475 \\ 200 & 632 \end{bmatrix}$

9. $\begin{bmatrix} 3 & 4 & -11 \\ 0 & 12 & -24 \\ -2 & -4 & 10 \end{bmatrix}$

10. $\begin{bmatrix} 9 & -21 \\ 22 & 70 \end{bmatrix}$

11. $\begin{bmatrix} 2 & 1 & 1 & 0 \\ 1 & 1 & 0 & 0 \\ 1 & 0 & 2 & 1 \\ 0 & 0 & 1 & 1 \end{bmatrix}$

12. $A + A^t = \begin{bmatrix} 4 & 11 & 4 \\ 11 & 6 & -11 \\ 4 & -11 & 18 \end{bmatrix}$. $A + A^t$ is symmetric for all square matrices A.

13. $A - A^t = \begin{bmatrix} 0 & -1 & 4 \\ 1 & 0 & 5 \\ -4 & -5 & 0 \end{bmatrix}$. $A - A^t = -(A - A^t)^t$ for all square matrices A.

A matrix P such that $P = -P^t$ is called antisymmetric.

Answers to Appendix C, Calculator Exercises

C6

1. $\begin{bmatrix} 1 & 0 \\ -2 & 1 \end{bmatrix}$

2. $\begin{bmatrix} -0.2857 & 0.1429 \\ 0.6429 & -0.0714 \end{bmatrix}$

3. $\begin{bmatrix} 1.5 & -0.5 \\ -2.0 & 1.0 \end{bmatrix}$

4. $\begin{bmatrix} -3 & 1 \\ 1 & 0 \end{bmatrix}$

5. Does not exist.

6. $\begin{bmatrix} -1.75 & 0.75 \\ -1.5 & 0.5 \end{bmatrix}$

7. $\begin{bmatrix} 2.3333 & -3.0000 & -0.3333 \\ -2.6667 & 3.0000 & 0.6667 \\ 1.3333 & -1.0000 & -0.3333 \end{bmatrix}$

8. $\begin{bmatrix} 0.8333 & 0.6667 & -2.0000 \\ 0.3333 & 0.6667 & -1.0000 \\ -0.1667 & -0.3333 & 1.0000 \end{bmatrix}$

9. $\begin{bmatrix} 0.1667 & 0 & 0.1667 \\ -0.0833 & -0.7500 & 0.1667 \\ -0.3333 & -0.5000 & 0.1667 \end{bmatrix}$

10. Does not exist.

11. $\begin{bmatrix} -0.4286 & 0.7143 & -1.5714 \\ 0.2857 & -0.1429 & -0.2857 \\ 0.2857 & -0.1429 & 0.7143 \end{bmatrix}$

12. $\begin{bmatrix} 1.5 & -0.5 & 0.5 \\ 0.75 & -0.25 & -0.25 \\ 2 & -1 & 0 \end{bmatrix}$

13. $x_1 = -2, x_2 = 2$

14. $x_1 = 24, x_2 = -5$

15. $x_1 = 5, x_2 = 0$

16. $x_1 = 3, x_2 = -2$

17. $x_1 = 11, x_2 = -4$

18. $x_1 = 9, x_2 = -2$

19. $x_1 = 0.7778, x_2 = 0.3333, x_3 = -0.5556$

20. $x_1 = 0.25, x_2 = -0.75, x_3 = 1.25$

21. $x_1 = 143, x_2 = -47, x_3 = -16$

22. $\begin{bmatrix} 2 & 0 \\ 0 & 1 \end{bmatrix}$

23. $\begin{bmatrix} 7 & 13 \\ -2 & -3 \end{bmatrix}$

24. $\begin{bmatrix} 6 & 14 & -7 \\ -5 & -11 & 6 \\ -3 & -6 & 5 \end{bmatrix}$

25. $\begin{bmatrix} 42.2 & -11.6 & 18.2 \\ 91.6 & -24.8 & 39.6 \\ -15.4 & 4.2 & -5.4 \end{bmatrix}$

C7

1. -2
2. 5
3. 2.13
4. 57.3432
5. -460
6. -178
7. 438
8. 3876
9. 0, Row3 = 3Row1
10. 0, Row3 = Row1 - Row 2
11. 0, Col2 = Col3
12. 0, Row3 = -3Row1

Appendix D MATLAB Exercises

D2 Solving Systems of Linear Equations (Sections 1.1, 1.2, 1.3)

1. (a) $x = 2$, $y = -1$, $z = 1$;   (b) $x = 1 - 2r$, $y = r$, $z = -2$;

   (c) $x = 1/2$, $y = 2/5$, $z = 3/4$ (d) $x = 7/10$, $y = -.2/5$, $z = 7/20$

2. (a) $x = 2$, $y = 5$ (b) no solution   (c) many solutions $(4-2r, r)$ (d) $x = 6/7$, $y = 1/7$

3. (a) $x = 4$, $y = -3$, $z = -1$ (b) $x = -1$, $y = 1$, $z = 2$ (c) $x = 2$, $y = 3$, $x = -1$

4. Start with an appropriate REF and work backwards. For example,

$$\begin{bmatrix} 1 & 0 & 0 & 1 \\ 0 & 1 & 0 & 2 \\ 0 & 0 & 1 & 3 \end{bmatrix} \underset{\substack{R1+(2)R3 \\ R2+(-2)R3}}{\approx} \begin{bmatrix} 1 & 0 & 2 & 7 \\ 0 & 1 & -1 & -1 \\ 0 & 0 & 1 & 3 \end{bmatrix} \underset{R1+(-2)R2}{\approx} \begin{bmatrix} 1 & -2 & 4 & 9 \\ 0 & 1 & -1 & -1 \\ 0 & 0 & 1 & 3 \end{bmatrix}$$

$$\underset{(2)R2}{\approx} \begin{bmatrix} 1 & -2 & 4 & 9 \\ 0 & 2 & -2 & -2 \\ 0 & 0 & 1 & 3 \end{bmatrix} \underset{R2<->R3}{\approx} \begin{bmatrix} 1 & -2 & 4 & 9 \\ 0 & 0 & 1 & 3 \\ 0 & 2 & -2 & -2 \end{bmatrix} \underset{\substack{R2+(-1)R1 \\ R3+(3)R1}}{\approx} \begin{bmatrix} 1 & -2 & 4 & 9 \\ 1 & -2 & 5 & 12 \\ -1 & 4 & -6 & -11 \end{bmatrix}$$

The following system uses all three types of row operations

$$\begin{aligned} x - 2y + 4z &= 9 \\ x - 2y + 5z &= 12 \\ -x + 4y - 6z &= -11 \end{aligned}$$

6. The matrix corresponds to a system of three equations in two variables. Probability is that such a system written down at random has no solution; the three lines do not cross in a point. Thus REF will be the identity matrix. To get another form of REF, take any three lines that cross in a point. For example, construct a system that crosses at $x = 1$, $y = 2$. Let system be

$$\begin{aligned} x + y &= 3 \\ -x + y &= 1 \\ 2x + y &= 4 \end{aligned} \quad \text{Get} \quad \begin{bmatrix} 1 & 1 & 3 \\ -1 & 1 & 1 \\ 2 & 1 & 4 \end{bmatrix} \approx \begin{bmatrix} 1 & 0 & 1 \\ 0 & 1 & 2 \\ 0 & 0 & 0 \end{bmatrix}$$

7. REF for first matrix is correct, second incorrect due to roundoff error. Crucial error takes place when a small element that should have been 0, in the (3,3) location, is normalized. This can be clearly seen in all steps in Format Long E.

8. See Section 9.1 for discussion of Gaussian elimination and echelon form.
  [GJalg and Galg are useful for class demonstration of difference in algorithms.]

Answers to Appendix D, MATLAB Exercises

Get

$$\begin{bmatrix} 1 & 1 & 1 & 3 \\ 2 & 3 & 1 & 5 \\ 1 & -1 & -2 & -5 \end{bmatrix} \underset{\substack{R2+(-2)R1 \\ R3+(-1)R1}}{\approx} \begin{bmatrix} 1 & 1 & 1 & 3 \\ 0 & 1 & -1 & -1 \\ 0 & -2 & -3 & -8 \end{bmatrix} \underset{R3+(2)R2}{\approx} \begin{bmatrix} 1 & 1 & 1 & 3 \\ 0 & 1 & -1 & -1 \\ 0 & 0 & -5 & -10 \end{bmatrix}$$

$$\underset{(1/5)R2}{\approx} \begin{bmatrix} 1 & 1 & 1 & 3 \\ 0 & 1 & -1 & -1 \\ 0 & 0 & 1 & 2 \end{bmatrix} \underset{\substack{R1+(-1)R3 \\ R2+R3}}{\approx} \begin{bmatrix} 1 & 1 & 0 & 1 \\ 0 & 1 & 0 & 1 \\ 0 & 0 & 1 & 2 \end{bmatrix} \underset{R1+(-1)R2}{\approx} \begin{bmatrix} 1 & 0 & 0 & 0 \\ 0 & 1 & 0 & 1 \\ 0 & 0 & 1 & 2 \end{bmatrix}$$

9. (a) 8 mult, 9 add, 0 swap (b) 11 mult, 12 add, 0 swap.
Creation of zeros and leading 1s are not counted as operations - they are substitutions. Gauss gains one multiplication and one addition in the back substitution. No operations take place on the (1,3) location when a zero is created in the (1,2) location.

10. Let m denote multiplications, a additions. We get

	Gauss	G/J
n=2	6m,3a	6m,3a
n=3	17m,11a	18m,12a
n=4	36m,26a	40m,30a
n=5	65m,50a	75m,60a

Thus Gauss mult: pts (2,6), (3,17), (4,36), (5,65)
          additions: pts (2,3), (3,11), (4,26), (5,50)
G/J multiplications: pts (2,6), (3,18), (4,40), (5,75)
          additions: pts (2,3), (3,12), (4,30), (5,60)

Substituting points into the polynomial $a + bn + cn^2 + dn^3$ and solving the systems of linear equations gives:

Gauss mults $\frac{n^3}{3} + n^2 - \frac{n}{3}$, adds $\frac{n^3}{3} + \frac{n^2}{2} - \frac{5n}{6}$ G/J mults $\frac{n^3}{2} + \frac{n^2}{2}$, adds $\frac{n^3}{2} - \frac{n}{2}$

For system with 20x20 matrix of coeffs, Gauss 5910 flops, G/J 8190 flops.

11. (a) $y = 2x^2 + x + 5$ (b) $y = x^3 - 2x^2 + x - 3$.

D3 Matrix Operations (Sections 2.1, 2.2, 2.3)

1. (a) $\begin{bmatrix} 4 & 5 \\ 3 & 1 \end{bmatrix}$ (b) $\begin{bmatrix} 1 & 0 \\ 0 & 1 \end{bmatrix}$ (Explore various powers of matrix B!)

   (c) $\begin{bmatrix} 1 & 0 \\ 2 & -1 \end{bmatrix}$, that is B   (d) $\begin{bmatrix} 3 & 5 \\ 5 & 8 \end{bmatrix}$

Answers to Appendix D, MATLAB Exercises

(e) $(P + P^t)$ will be symmetric. $(P + P^t)^t = P^t + (P^t)^t = P^t + P = P + P^t$.

(f) PP^t will be symmetric. $(PP^t)^t = (P^t)^t P^t = PP^t$.

2. (a) $\begin{bmatrix} -9 & 35 \\ 1 & 10 \end{bmatrix}$ (b) $\begin{bmatrix} 265 & 480 \\ 88 & 171 \end{bmatrix}$

4. 29

5.
(a) »X = A(3 , :);
    »Y = A(: , 4);
    » X * A' * Y

    ans = 821

(b) »B = A(1:3 , 2:4);
    »B^7

    ans = $\begin{matrix} 1752000 & 5792960 & 8249536 \\ 1628992 & 5367168 & 7639744 \\ 1768384 & 5792960 & 8233152 \end{matrix}$

(c) »C= [A(1, :) ; A(3, :) ; A(4, :)];
    »C*A

    ans = $\begin{matrix} 9 & 0 & 25 & 62 \\ -23 & 18 & 6 & 22 \\ 13 & -6 & -11 & 9 \end{matrix}$

D4 Computational Considerations (Section 2.2)

1. Have A 2x2, B 2x3, C 3x1.
   Let m stand for a multiplication. AB is a 2x3 matrix. It has 6 elements. Each element is computed by multiplying a row of A by a column of B. Each element requires 2m. Thus <u>12m</u> for AB. (AB)xC is 2x1 matrix, has 2 elements. Each element involves 3m. Therefore <u>6m</u> to compute (AB)xC. Thus total multiplications for ABxC is 12+6 = 18.
   BC has 2 elements, and requires 3m per element. Thus <u>6m</u> for BC. Ax(BC) has 2 elements, and takes 2m per element. Thus <u>4m</u> for Ax(BC). Total for AxBC is 6+4 = 10.

2. There are mn elements in AB. Each is computed by multiplying a row of A times a column of B, i.e., by doing r multiplications (and r-1 additions). Thus the total number of multiplications is (mn)r or mrn.
   The number of multiplications required to compute AB is mrn and the number required to compute BC is rns. (AB)C is the product of an mxn matrix and an nxs matrix, so the number of multiplications required is mns + mrn. A(BC) is the product of an mxr matrix and an rxs matrix, so the number of multiplications is mrs + rns.

3. (a) A(BC) 3,864; (AB)C 7,395   (b) (A(BC))D 4,494; A(B(CD)) 65,058

4. (a) (AB)C 4,320; A(BC) 6,345   (b) (AB)B 8,100; $A(B^2)$ 95,175
   (c) ((AB)B)B 12,150; $A(B^3)$ 186,300   (d) ((AB)B)C 8,370; $A((B^2))C$ 97,470

Answers to Appendix D, MATLAB Exercises

5. Each of the mn elements of AB is computed by multiplying a row of A times a column of B. This requires adding r terms, i.e., doing r-1 additions. Thus the total number of additions required to compute AB is mn(r-1).

To calculate (AB)C requires mn(r-1) additions for AB and ms(n-1) additions for the second product. Thus the total is mn(r-1) + ms(n-1). To calculate A(BC) requires rs(n-1) additions for BC and ms(r-1) for the second product. Thus the total is rs(n-1) + ms(r-1).

If A, B, and C are 2x2, 2x3, and 3x1, respectively, the number of additions is 2x3x1 + 2x1x2 = 10 for (AB)C and 2x1x2 + 2x1x1 = 6 for A(BC).

D5 Inverse of a Matrix (Section 2.4)

2. $\begin{bmatrix} 1 & 0 \\ 2 & -1 \end{bmatrix}^2 = \begin{bmatrix} 1 & 0 \\ 0 & 1 \end{bmatrix}$, $\begin{bmatrix} 1 & 0 \\ 2 & -1 \end{bmatrix}^3 = \begin{bmatrix} 1 & 0 \\ 2 & -1 \end{bmatrix}$, $\begin{bmatrix} 1 & 0 \\ 2 & -1 \end{bmatrix}^4 = \begin{bmatrix} 1 & 0 \\ 0 & 1 \end{bmatrix}$.

Further, $\begin{bmatrix} 1 & 0 \\ 2 & -1 \end{bmatrix}^n = \begin{bmatrix} 1 & 0 \\ 2 & -1 \end{bmatrix}$ if n is odd, $\begin{bmatrix} 1 & 0 \\ 2 & -1 \end{bmatrix}^n = \begin{bmatrix} 1 & 0 \\ 0 & 1 \end{bmatrix}$ if n is even.

(a) $\begin{bmatrix} 1 & 0 \\ 2 & -1 \end{bmatrix}^2 = \begin{bmatrix} 1 & 0 \\ 0 & 1 \end{bmatrix}$, thus $B^2 B^2 = I$, implying that $(B^2)^{-1} = B^2$.

(b) $(B^{194})^{-1} = I^{-1} = I$.

3. (a) $A^{-1} = \begin{bmatrix} 1 & 0 \\ -2 & 1 \end{bmatrix}$

(b) $B^{-1} = \begin{bmatrix} 0.15116279069767 & 0.10465116279070 \\ 0.12790697674419 & -0.01162790697674 \end{bmatrix} = \begin{bmatrix} -13/86 & 9/86 \\ 11/86 & -1/86 \end{bmatrix}$

4. (a) $B = \begin{bmatrix} 2 & 0 \\ 0 & 1 \end{bmatrix}$ (b) $B = \begin{bmatrix} 7 & 13 \\ -2 & -3 \end{bmatrix}$ (c) $B = \begin{bmatrix} 6 & 14 & -7 \\ -5 & -11 & 6 \\ -3 & -6 & 5 \end{bmatrix}$

(d) $B = \begin{bmatrix} 42.2000 & -11.6000 & 18.2000 \\ 91.6000 & -24.8000 & 39.6000 \\ -15.4000 & 4.2000 & -5.4000 \end{bmatrix}$

5. ```
function R=sim(P,Q)
%Performs similarity transf inv(B)*A*B
%Calling format: sim(A,B)
R=inv(Q)*P*Q;
```
```

Answers to Appendix D, MATLAB Exercises

D6 Solving Systems of Equations using Matrix Inverse (Section 2.4)

1. $x = 10, y = -2$
2. $x = -1, y = 2.5, z = -1.5$
3. $x = 9, y = -14$
4. $x = 7/9, y = 3/9, z = -5/9$
5. $x=1, y=2, z=3$; $x=-1, y=2, z=0$; $x=0, y=1, y=2$

D7 Cryptography (Section 2.4)

1. (a) 1 27 6 15 18 27 20 8 9 19 27 5 24 5 18 3 9 19 27
     A - F O R - T H I S - E X E R C I S E

   (b) 19 21 14 27 4 5 3 11 27 4 1 25 20 15 14 1 27 2 5 1 3 8
     S U N - D E C K - D A Y T O N A - B E A C H

D8 Leontief I/O Model (Section 2.5)

1. $X = \begin{bmatrix} 165 \\ 480 \\ 250 \end{bmatrix}$ 2. $X_1 = \begin{bmatrix} 60 \\ 40 \end{bmatrix}$, $X_2 = \begin{bmatrix} 22.5 \\ 16.333 \end{bmatrix}$, $X_3 = \begin{bmatrix} 15 \\ 20 \end{bmatrix}$ 3. $D = \begin{bmatrix} 4 \\ .9 \\ .65 \end{bmatrix}$

4. Each column of A adds to 1. A is in fact a stochastic matrix, $0 \le a_{ij} \le 1$, and $\sum_j a_{ij} = 1$. If A is stochastic, $|A - I| = 0$ [Add all columns to 1st column]. Thus $(A - I)^{-1}$ does not exist. In this situation, $X = AX$. The industrial ouput is all taken up by industrial demand, and there is nothing left to meet the demand of the open sector.
 [Note - prerequisite for this exercise is determinants and the fact that the inverse does not exist if the determinant is zero - this exercise is best left until these topics have been covered. X is in fact an eigenvector of A corresponding to eigenvector 1 here.]

D9 Markov Chains (Sections 2.6, 6.2)

1. (a) $X_{01} = \begin{bmatrix} 5199.2 \\ 60.8 \end{bmatrix}$, $X_{02} = \begin{bmatrix} 198.424 \\ 61.576 \end{bmatrix}$, $X_{03} = \begin{bmatrix} 197.6713 \\ 62.3287 \end{bmatrix}$, $X_{04} = \begin{bmatrix} 196.9411 \\ 63.0589 \end{bmatrix}$, $X_{05} = \begin{bmatrix} 196.2329 \\ 63.7671 \end{bmatrix}$ City / Suburb

   (b) $\begin{bmatrix} 0.99 & 0.02 \\ 0.01 & 0.98 \end{bmatrix}^5 = \begin{bmatrix} .9529 & .0942 \\ .0471 & .9058 \end{bmatrix}$. $p_{11}^{(5)} = 0.9529$.

2. (a) Using mathematics, solve $PX = X$. Get $X = \begin{bmatrix} 40 \\ 160 \end{bmatrix}$.

   (b) $P = \begin{bmatrix} 0.2 & 0.2 \\ 0.8 & 0.8 \end{bmatrix}$. $p_{11} = p_{12} = 0.2$, long term probabilities of living in state 1, city, is 0.2.

   $p_{22} = p_{21} = 0.8$, long term probabilities of living in state 2, suburbs, is 0.8.

Answers to Appendix D, MATLAB Exercises

3. (a) Use $1.01\begin{bmatrix} 0.96 & 0.01 \\ 0.04 & 0.99 \end{bmatrix} = \begin{bmatrix} .9696 & .0101 \\ .0404 & .9999 \end{bmatrix}$ as transition matrix, to get

$X_{01} = \begin{bmatrix} 57.671 \\ 144.329 \end{bmatrix}$, $X_{02} = \begin{bmatrix} 57.3755 \\ 146.6445 \end{bmatrix}$, $X_{03} = \begin{bmatrix} 57.1124 \\ 148.9478 \end{bmatrix}$ City Suburb,

$X_{04} = \begin{bmatrix} 56.8806 \\ 151.2402 \end{bmatrix}$ City Suburb, $X_{05} = \begin{bmatrix} 56.6789 \\ 153.5231 \end{bmatrix}$ City Suburb

(b) Use $\begin{bmatrix} 1.012 & 0 \\ 0 & 1.008 \end{bmatrix}\begin{bmatrix} 0.96 & 0.01 \\ 0.04 & 0.99 \end{bmatrix} = \begin{bmatrix} .9715 & .0101 \\ .0403 & .9980 \end{bmatrix}$ as transition matrix, to

get $X_{01} = \begin{bmatrix} 57.7812 \\ 144.0534 \end{bmatrix}$, $X_{02} = \begin{bmatrix} 57.5894 \\ 146.0939 \end{bmatrix}$, $X_{03} = \begin{bmatrix} 57.4236 \\ 148.1225 \end{bmatrix}$ City Suburb,

$X_{04} = \begin{bmatrix} 57.2831 \\ 150.1405 \end{bmatrix}$, $X_{05} = \begin{bmatrix} 57.1669 \\ 152.1487 \end{bmatrix}$ City Suburb

4. (a) With P, there is a steady decrease in city population, and steady increase in suburban population. With Q, bar graphs show fluctuations. The P model is most realistic. Diagonal elements of P are dominant, non-diagonal elements of Q are dominant. P is more realistic since it reflects the larger probability of person staying in same state.

   (b) Result: Let $A = \begin{bmatrix} a & b \\ 1-a & 1-b \end{bmatrix}$. If a>b, steady trends. If a<b, fluctuating trends.

5. $X_{n-1} = A^{-1} X_n$. $A^{-1} = \begin{bmatrix} 1.0421 & -0.0105 \\ -0.0421 & 1.0105 \end{bmatrix}$. $X_{99} = \begin{bmatrix} 58.9474 \\ 141.0526 \end{bmatrix}$, $X_{98} = \begin{bmatrix} 59.9446 \\ 140.0554 \end{bmatrix}$,

$X_{97} = \begin{bmatrix} 60.9943 \\ 139.0057 \end{bmatrix}$, $X_{96} = \begin{bmatrix} 62.0993 \\ 137.9007 \end{bmatrix}$ City Suburb, $X_{95} = \begin{bmatrix} 63.2624 \\ 136.7376 \end{bmatrix}$ City Suburb

Not stochastic, A^{-1} has negative numbers, but sum of a column is 1.

6. (a) $X_{01} = \begin{bmatrix} 58 \\ 141.2 \\ 60.8 \end{bmatrix}$, $X_{02} = \begin{bmatrix} 58.0040 \\ 140.4200 \\ 61.5760 \end{bmatrix}$, $X_{03} = \begin{bmatrix} 58.0117 \\ 139.6596 \\ 62.3287 \end{bmatrix}$ City Suburb Nonmetro

$X_{04} = \begin{bmatrix} 58.0227 \\ 138.9184 \\ 63.0589 \end{bmatrix}$, $X_{05} = \begin{bmatrix} 58.0369 \\ 138.1960 \\ 63.7671 \end{bmatrix}$ City Suburb Nonmetro

(b) $\begin{bmatrix} 60.6667 \\ 112.6667 \\ 86.6667 \end{bmatrix}$ City Suburb Nonmetro   (c) $\begin{bmatrix} .2333 \\ .4333 \\ .3333 \end{bmatrix}$ City Suburb Nonmetro

D10 Digraphs (Section 2.7)

1. Distance = 3. Two paths 2 -> 4 -> 5 -> 1 and 2 -> 4 -> 3 -> 1.

Answers to Appendix D, MATLAB Exercises

2.

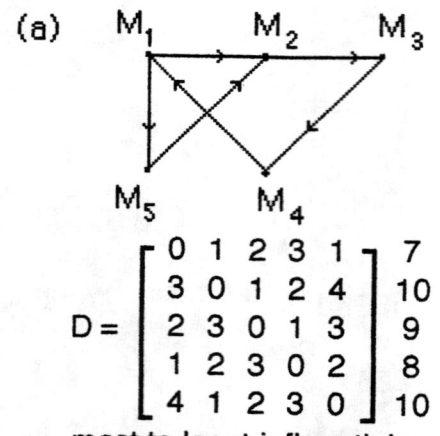

(a) M_1, M_2, M_3, M_5, M_4

$$D = \begin{bmatrix} 0 & 1 & 2 & 3 & 1 \\ 3 & 0 & 1 & 2 & 4 \\ 2 & 3 & 0 & 1 & 3 \\ 1 & 2 & 3 & 0 & 2 \\ 4 & 1 & 2 & 3 & 0 \end{bmatrix} \begin{matrix} 7 \\ 10 \\ 9 \\ 8 \\ 10 \end{matrix}$$

most to least influential:
M_1, M_4, M_3, with M_2 and M_5 equally uninfluential.

(b) M_2, M_1, M_3, M_5, M_4

$$D = \begin{bmatrix} 0 & 1 & 1 & 3 & 2 \\ x & 0 & x & x & x \\ 2 & 3 & 0 & 2 & 1 \\ 3 & 4 & 1 & 0 & 2 \\ 1 & 2 & 2 & 1 & 0 \end{bmatrix} \begin{matrix} 7 \\ 4x \\ 8 \\ 9 \\ 6 \end{matrix}$$

most to least influential:
M_5, M_1, M_3, M_4, M_2

M_3 becomes most influencial person by influencing M_1 in both groups.

D11 Determinants   (Section 3.1, 3.2, 3.3)

1. (a) $|A| = 7$, $M(a_{22}) = -5$, $M(a_{31}) = 5$.   (b) $|B| = 12$, $M(b_{21}) = 29$, $M(b_{33}) = 15$.

2. All determinants are zero.
   (a) Row 3 is 3 times row 1.
   (b) Column 3 is −3/2 times column 1.
   (c) Columns 2 and 3 are equal.
   (d) Row 3 is −3 times row 1.

3. (a) $\begin{vmatrix} 1 & -1 & 0 & 2 \\ -1 & 1 & 0 & 0 \\ 2 & -2 & 0 & 1 \\ 3 & 1 & 5 & -1 \end{vmatrix} \underset{R4+(-3)R1}{\overset{R2+R1}{\underset{R3+(-2)R1}{=}}} \begin{vmatrix} 1 & -1 & 0 & 2 \\ 0 & 0 & 0 & 2 \\ 0 & 0 & 0 & -3 \\ 0 & 4 & 5 & -7 \end{vmatrix} \overset{=}{\underset{R2 \leftrightarrow R4}{}} - \begin{vmatrix} 1 & -1 & 0 & 2 \\ 0 & 4 & 5 & -7 \\ 0 & 0 & 0 & -3 \\ 0 & 0 & 0 & 2 \end{vmatrix} = 0.$

   (b) $\begin{vmatrix} 2 & 1 & 3 & 1 \\ -2 & 3 & -1 & 2 \\ 2 & 1 & 2 & 3 \\ -4 & -2 & 0 & -1 \end{vmatrix} \underset{R4+(2)R1}{\overset{R2+R1}{\underset{R3+(-1)R1}{=}}} \begin{vmatrix} 2 & 1 & 3 & 1 \\ 0 & 4 & 2 & 3 \\ 0 & 0 & -1 & 2 \\ 0 & 0 & 6 & 1 \end{vmatrix} \overset{=}{\underset{R4+(6)R3}{}} \begin{vmatrix} 2 & 1 & 3 & 1 \\ 0 & 4 & 2 & 3 \\ 0 & 0 & -1 & 2 \\ 0 & 0 & 0 & 13 \end{vmatrix} = -104.$

D12 Cramer's Rule   (Section 3.4)

1. (a) $\dfrac{-1}{3} \begin{bmatrix} -7 & 9 & 1 \\ 8 & -9 & -2 \\ -4 & 3 & 1 \end{bmatrix}$ (b) $\dfrac{-1}{3} \begin{bmatrix} 0 & -6 & 3 \\ -3 & -3 & 3 \\ 2 & 3 & -3 \end{bmatrix}$ (c) $\dfrac{-1}{4} \begin{bmatrix} -6 & 2 & -2 \\ -3 & 1 & 1 \\ -8 & 4 & 0 \end{bmatrix}$

2. (a) 1, 2, −1   (b) −1, 3, 4   (c) 0.5, 0.25, 0.25

Answers to Appendix D, MATLAB Exercises

D13 Dot Product, Norm, Angle, Distance, Projection (Section 4.2)

1. $u \cdot v = 2$, $\|u\| = 6.3246$, angle between u and v = 87.2031°, $d(X,Y) = 5.7446$.

2. $u \cdot v = -26$, $\|u\| = 6.245$, angle between u and v = 117.0171°, $d(X,Y) = 10.3923$.

3. $\|u\| = 3.7417$, $\|v\| = 11.0454$, $\|u + v\| = 11.5758 < 3.7417 + 11.0454$.

4. A is an orthogonal matrix. Rows and columns of A form orthonormal sets. A is invertible with $A^{-1} = A^t$. $|A| = 1$ (in general $|A| = \pm 1$).

D14 Transformations defined by Matrices (Section 4.3, 4.4)

1. »map([cos(2 * pi / 3) -sin(2 * pi / 3); sin(2 * pi / 3) cos(2 * pi / 3)])

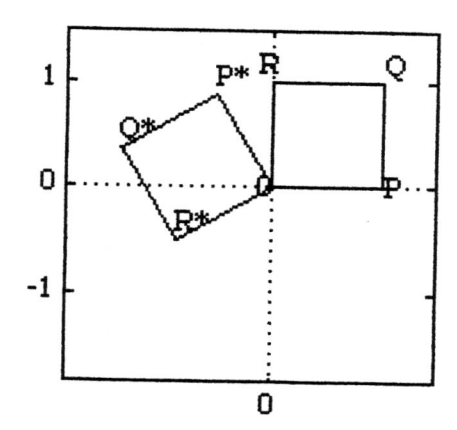

   $O^* = (0, 0)$, $P^* = (-0.5000, 0.8600)$, $Q^* = (-1.3660, -1.5000)$
   $R^* = (-0.8660, -1.5000)$

2. $O^* = (0, 0)$, $P^* = (0.8660, -0.5000)$, $Q^* = (1.3660, 0.3660)$,
   $R^* = (0.5000, 0.8660)$

3. A = [3 0;0 3]
   $O^* = (0, 0)$, $P^* = (3, 0)$, $Q^* = (3, 3)$, $R^* = (0, 3)$

4. A = [-1 0;0 1]
   $O^* = (0, 0)$, $P^* = (-1, 0)$, $Q^* = (-1, 1)$, $R^* = (0, 1)$

5. (a) $O^* = (0, 0)$, $P^* = (3, 1)$, $Q^* = (3, 5)$, $R^* = (0, 4)$
   (b) $O^* = (0, 0)$, $P^* = (4, 1)$, $Q^* = (3, 6)$, $R^* = (-1, 5)$
   (c) $O^* = (0, 0)$, $P^* = (0, -3)$, $Q^* = (3, -3)$, $R^* = (3, 0)$

6. (a) A = [cos(pi/2) -sin(pi/2); sin(pi/2) cos(pi/2)], B = [2 0; 0 2]
   $O^\# = (0, 0)$, $P^\# = (0, 2)$, $Q^\# = (-2, 2)$, $R^\# = (-2, 0)$

   (b) A = [4 0,0 4], B = [1 0;0 -1]
   $O^\# = (0, 0)$, $P^\# = (4, 0)$, $Q^\# = (4, -4)$, $R^\# = (0, -4)$

Answers to Appendix D, MATLAB Exercises

(c) A = [cos(pi/3) -sin(pi/3); sin(pi/3) cos(pi/3)], B = [0 1;1 0]
$O^\# = (0, 0)$, $P^\# = (0.8660, 0.5000)$, $Q^\# = (1.3660, -0.3600)$,
$R^\# = (0.5000, -0.8660)$

7. (a) A = [3 0;0 3], B = [1 2;0 1]
$O^\# = (0, 0)$, $P^\# = (3, 0)$, $Q^\# = (9, 3)$, $R^\# = (6, 3)$

(b) A = 3 0;0 2], B = 0 1;1 0]
$O^\# = (0, 0)$, $P^\# = (0, 3)$, $Q^\# = (2, 3)$, $R^\# = (2, 0)$

(c) A = [2 0;0 2], B = [1 0, 3 1];
C = [cos(pi/3) -sin(pi/3); sin(pi/3) cos(pi/3)]
compmap can have two arguments only
Let D = B*A the use compmap(D, C) to get
$O^\# = (0, 0)$, $P^\# = (-4.1962, 4.7321)$, $Q^\# = (-5.9282, 5.7321)$,
$R^\# = (-1.7321, 1.0000)$

8. $O^* = (0, 0)$, $P^* = (6, 2)$, $Q^* = (8, 3)$, $R^* = (5, 2)$

9. (a) (2,2)   (b) (3, 2)

D15 Fractals   (Section 4.4)

1. T_2 gives bottom left leaf. T_1 then gives all the leaves above it. T_3 gives bottom right leaf, T_1 then gives all the leaves above it. T_4 gives bottom of main stem, T_1 then gives the main stem above it.

4. (a) Let $\begin{bmatrix} 0.86 & 0.03 \\ -0.03 & 0.86 \end{bmatrix} = \begin{bmatrix} k & 0 \\ 0 & k \end{bmatrix}\begin{bmatrix} \cos\alpha & -\sin\alpha \\ \sin\alpha & \cos\alpha \end{bmatrix}$. $k\cos\alpha = 0.86$, $k\sin\alpha = -0.03$.
$k^2 = 0.86^2 + 0.03^2$. $k = 0.8605230967$. $\alpha = -1.998°$, to three decimal places.
dilation factor = 0.8605230967, angle of rotation = $-1.998°$.

(b) $1.5\alpha = -2.997°$.
$T_1 = \begin{bmatrix} 0.8605230967 & 0 \\ 0 & 0.8605230967 \end{bmatrix}\begin{bmatrix} \cos(-2.997) & -\sin(-2.997) \\ \sin(-2.997) & \cos(-2.997) \end{bmatrix}\begin{bmatrix} x \\ y \end{bmatrix} + \begin{bmatrix} 0 \\ 1.5 \end{bmatrix}$
$= \begin{bmatrix} 0.8593461366 & 0.0449913039 \\ 0.0449913039 & 0.8593461366 \end{bmatrix}\begin{bmatrix} x \\ y \end{bmatrix} + \begin{bmatrix} 0 \\ 1.5 \end{bmatrix}$. T_2, T_3, T_4 remain same.

(c) Write T_1 in the following form for rotation through angle α.
$T_1 = \begin{bmatrix} 0.8605230967 & 0 \\ 0 & 0.8605230967 \end{bmatrix}\begin{bmatrix} \cos(\alpha) & \sin(\alpha) \\ -\sin(\alpha) & \cos(\alpha) \end{bmatrix}\begin{bmatrix} x \\ y \end{bmatrix} + \begin{bmatrix} 0 \\ 1.5 \end{bmatrix}$

(d) e.g. let $\alpha>0$ for a rotation through angle to right to the left in formula for (c).
let $\alpha<0$ for a rotation through angle to right to the left.

Answers to Appendix D, MATLAB Exercises

let $\alpha = 0$ for vertical fern.

5. Tip of fern will be point where $T_1(x,y) = (x,y)$.

$$\begin{bmatrix} 0.86 & 0.03 \\ -0.03 & 0.86 \end{bmatrix}\begin{bmatrix} x \\ y \end{bmatrix} + \begin{bmatrix} 0 \\ 1.5 \end{bmatrix} = \begin{bmatrix} x \\ y \end{bmatrix}. \quad \begin{array}{l} -0.14x + 0.03y = 0 \\ 0.03x + 0.14y = 1.5 \end{array}$$

$x = 2.1951, y = 10.2439$

D16 Linear Combinations, Linear Dependence, Basis (Sections 5.1 - 5.5)

1. (a) $(-3, 3, 7) = 2(1, -1, 2) - (2, 1, 0) + 3(-1, 2, 1)$ (b) not a combination
   (c) $(0, 10, 8) = (2-r)(-1, 2, 3) + (2-2r)(1, 3, 1) + r(1, 8, 5)$

2. (a) $2(-1, 3, 2) - 3(1, -1, -3) - (-5, 9, 13) = 0$ (b) Linearly independent

3. (a) linearly independent, $\det(A) = 2268 \ne 0$. $\text{rank}(A) = 4$
   (b) linearly dependent, $\det(A) = 0$, $\text{rank}(A) = 3 < 4$

4. Rank is 2. R2-R1 gives 5 5 5 5 5, R3-R2 gives 5 5 5 5 5 etc.

   Any matrix of the form $\begin{bmatrix} 1 & \cdot & \cdot & \cdot \\ \cdot & \cdot & \cdot & \cdot \\ \cdot & \cdot & \cdot & n^2 \end{bmatrix}$ is of rank 2.

5. (a) $\{(1, 0, 1), (0, 1, 1)\}$ (b) $\{(1, 0, 0.6, -0.2, 1.6), (0, 1, 1.8, 0.4, 0.8)\}$
   (c) $\{(1, 0, 0, 0, 0, 0), (0, 1, 0, 0.4, 2.6, 1.4), (0, 0, 1, 0.8, 1.2, 1.8)\}$

D17 Gram - Schmidt Orthogonalization (Section 5.6)

1. $(0.1905, -0.2381)$.   2. $(0, -0.2857, -0.1429, 0.5714)$

3. (a) $\{(0.5345, 0.8018, 0.2673), (0.3841, -0.5121, 0.7682)\}$
   (b) $\{(0.8729, 0.4364, 0.2182, 0), (-0.2264, 0.3843, 0.1369, -0.8845),$
       $(-0.4309, 0.6424, 0.4387, 0.4573)\}$
   (c) $\{(0.1826, 0.3651, 0.5477, 0.7303), (0.5477, -0.7303, -0.1826, 0.3651)\}$
       The vectors are linearly dependent. They span a 2D subspace.

D18 Eigenvalues and Eigenvectors and Applications (Sections 6.1 - 6.4)

1. $\lambda_1 = 1, v_1 = \begin{bmatrix} -.6017 \\ .7453 \\ -.2872 \end{bmatrix}, \lambda_2 = 1, v_2 = \begin{bmatrix} .4399 \\ .0091 \\ -.8980 \end{bmatrix}, \lambda_3 = 10, v_3 = \begin{bmatrix} .6667 \\ .6667 \\ .3333 \end{bmatrix}.$

Answers to Appendix D, MATLAB Exercises

2. $A = \begin{bmatrix} 1 & 2 \\ 1 & 0 \end{bmatrix}$, $\lambda_1 = 2$, $v_1 = \begin{bmatrix} 0.8944 \\ 0.4472 \end{bmatrix}$, $\lambda_2 = -1$, $v_2 = \begin{bmatrix} -.7071 \\ .7071 \end{bmatrix}$.

$A = \begin{bmatrix} 2 & 3 \\ 1 & 0 \end{bmatrix}$, $\lambda_1 = 3$, $v_1 = \begin{bmatrix} 0.9487 \\ 0.3162 \end{bmatrix}$, $\lambda_2 = -1$, $v_2 = \begin{bmatrix} -.7071 \\ .7071 \end{bmatrix}$.

$A = \begin{bmatrix} 3 & 4 \\ 1 & 0 \end{bmatrix}$, $\lambda_1 = 4$, $v_1 = \begin{bmatrix} 0.9701 \\ 0.2425 \end{bmatrix}$, $\lambda_2 = -1$, $v_2 = \begin{bmatrix} -.7071 \\ .7071 \end{bmatrix}$.

Conjecture: $A = \begin{bmatrix} a & a+1 \\ 1 & 0 \end{bmatrix}$, $\lambda_1 = a+1$, $v_1 = \begin{bmatrix} a+1 \\ 1 \end{bmatrix}$, $\lambda_2 = -1$, $v_2 = \begin{bmatrix} -1 \\ 1 \end{bmatrix}$.

Proof: $\begin{vmatrix} a-\lambda & a+1 \\ 1 & -\lambda \end{vmatrix} = 0$ gives $\lambda^2 - \lambda a - (a+1) = 0$. $\lambda = a+1, -1$.

Corresponding eigenvectors are $\begin{bmatrix} a+1 \\ 1 \end{bmatrix}$, $\begin{bmatrix} -1 \\ 1 \end{bmatrix}$.

4. The eigenvectors of $\lambda = 1$ are vectors of the form $r \begin{bmatrix} 2 \\ 1 \end{bmatrix}$. If there is no change in total population $2r + r = 200 + 60 = 260$, so $r = 86.6667$. Thus the long-term prediction is that population in metropolitan areas will be 173.3333 million and population in nonmetropolitan areas will be 86.6667 million.

5. $P = \begin{bmatrix} 0 & 1/3 & 0 & 1/4 \\ 1/2 & 0 & 1/3 & 1/4 \\ 0 & 1/3 & 0 & 1/2 \\ 1/2 & 1/3 & 2/3 & 0 \end{bmatrix}$, P is regular since P^2 is positive. Eigenvectors of P for $\lambda = 1$ are vectors of the form $r \begin{bmatrix} 2 & 3 & 3 & 4 \end{bmatrix}^t$; distribution of rats in rooms 1, 2, 3, 4 is 2:3:3:4.

Powers of P approach the stochastic matrix $Q = \begin{bmatrix} 2s & 2s & 2s & 2s \\ 3s & 3s & 3s & 3s \\ 3s & 3s & 3s & 3s \\ 4s & 4s & 4s & 4s \end{bmatrix}$, so $2s + 3s + 3s + 4s = 1$

and $s = 1/12$. The long-term probability that a given rat will be in room 4 is $4/12 = 1/3$.

D19 Kernel and Range (Section 7.1)

1. (a) kernel $\{(-2, -3, 1)^t\}$, range $\{(1, 0, 2)^t, (0, 1, -1)^t\}$
   (b) kernel $\{(.5, -1.75, 1)^t\}$, range $\{(1, 0, 2)^t, (0, 1, 1)^t\}$
   (c) kernel is zero vector, range $\{(1, 0, 0)^t, (0, 1, 0)^t, (0, 0, 1)^t\}$
   (d) kernel $\{(3, 1, 0)^t, (6, 0, 1)^t\}$, range $\{(1, -2, -1)^t\}$

Answers to Appendix D, MATLAB Exercises

D20 Inner Product, Non-Euclidean Geometry (Sections 8.1, 8.2)

1. (a) $\|(0, 1)\| = 2$. (b) $90°$. The vectors $(1,1)$, and $(-4, 1)$ are at right angles.

   (c) $\text{dist}((1, 0), (0, 1)) = \sqrt{5}$.

   (d) $\text{dist}((x, y), (0, 0)) = 1, 2, 3$; $\langle (x, y), (x, y) \rangle = 1, 4, 9$; $[x\ y]A[x\ y]^t = 1, 4, 9$; $x^2 + 4y^2 = 1, 4, 9$.

2. Matrix for dot product is $\begin{bmatrix} 1 & 0 \\ 0 & 1 \end{bmatrix}$.

3. Equations of the circles are $[x\ y]A[x\ y]^t = r^2$; $6x^2 + 4xy + 9y^2 = 1, 4, 9$.

4. (a), (b) All vectors of form $a(1, 1)$ or $b(1, -1)$. i.e. lie on the cone through the origin.

   (c) Any two points of the form (x, y) and $(x+1, y+1)$; or (x, y) and $(x-1), (y+1)$.

   (d) $-x^2 + y^2 = 1, 4, 9$.

5. (a) $y = \pm 1, \pm\sqrt{2}, \pm\sqrt{3}$. (b) $x = \pm 1, \pm\sqrt{2}, \pm\sqrt{3}$.
   (c) $x^2 + 2xy + y^2 = 1, 2$ or 3; $(x+y)^2 = 1, 2$ or 3; $y = -x + 1, \sqrt{2}$ or $\sqrt{3}$.

6. (a) $\{(x,y,z): z^2 - x^2 - y^2 = 0\}$. (b) Surfaces $z^2 - x^2 - y^2 = 1, 4, 9$.

D21 Space-Time Travel (Section 8.2)

1. Earth time 120 yrs, spaceship time 79.3725 yrs.

2. Earth time 820.0820 yrs, spaceship time 11.5974 yrs.

D22 Pseudoinverse and Least Squares Curves (Section 8.4)

2. (a) $\begin{bmatrix} 0.3333 & 0 & 0.6667 \\ 0 & 0.1667 & -0.1667 \end{bmatrix}$ (b) $\begin{bmatrix} 0.5325 & 0.3896 & -0.3117 \\ -0.1688 & -0.0260 & 0.2208 \end{bmatrix}$

3. Pseudoinverse of an invertible matrix is that matrix. $\text{pinv}(A) = (A^tA)^{-1}A^t = A^{-1}(A^t)^{-1}A^t = A^{-1}$.

4. $(A^tA)^{-1}A^t$ does not exist since rank (A^tA) = rank(A) =2. Thus $\det(A^tA) = 0$ and $(A^tA)^{-1}$ does not exist. Same applies whenever A has more columns than rows. Note that pinv(A) exists. pinv(A) is based on the more general Moore-Penrose definition of pseudoinverse, which agrees with $(A^tA)^{-1}A^t$ when A^tA is invertible. Get

Answers to Appendix D, MATLAB Exercises

$$\text{pinv}(A) = \begin{bmatrix} -0.9444 & 0.4444 \\ -0.1111 & 0.1111 \\ 0.7222 & -0.2222 \end{bmatrix}$$

5. (a) $x = 1.7333$, $y = 0.6$ (b) $x = 0.4857$, $y = 1.0857$ (c) $x = 2.2521$, $y = -0.8$

6. $y = 4x - 3$.

7. $y = 15.25 - 10.05x + 1.75x^2$.

8. U.K $N = 57.33673469Y + 1275.897959$
   U.S.A. $N = 5214.785714Y + 6.357142857$
   Japan $N = 2032.989796Y - 0.8163265306$
   year 2000 is $y = 25$. Predictions: U.K. 2709, U.S.A. 5374, Japan 2013.

9. Using long format: $y = 13.594830131858504 + .24460060963367x - 0.01641543655465x^2 - 0.000021513018277x^3 - 0.00000078817010x^4$.
   In the year 2010, $x=110$. $y(110)=12.81640381$. Thus the prediction is that 12.8% will be foreign born in the year 2010.

11. $y > 9.4$

D23 LU Decomposition (Section 9.2)

1. (a) $L = \begin{bmatrix} 1 & 0 \\ .3333 & 1 \end{bmatrix}$, $U = \begin{bmatrix} 3 & 5 \\ 0 & .3333 \end{bmatrix}$

   (b) $L = \begin{bmatrix} 1.0000 & 0 & 0 \\ -0.5000 & 1.0000 & 0 \\ 0.2500 & 0.7308 & 1.0000 \end{bmatrix}$, $U = \begin{bmatrix} 4.0000 & 1.0000 & 3.0000 \\ 0 & 6.5000 & 1.5000 \\ 0 & 0 & -0.8462 \end{bmatrix}$

   (c) $L = \begin{bmatrix} 1.0000 & 0 & 0 \\ 0.5000 & 1.0000 & 0 \\ 1.0000 & 1.2500 & 1.0000 \end{bmatrix}$, $U = \begin{bmatrix} 2 & 0 & 2 \\ 0 & 4 & 0 \\ 0 & 0 & 3 \end{bmatrix}$

2. (a) $X = \begin{bmatrix} 1.0000 & 1.0000 \\ 1.0000 & -1.0000 \\ 2.0000 & 1.0000 \end{bmatrix}$ (b) $X = \begin{bmatrix} 1.0000 & -2.0000 \\ 1.0000 & 1.0000 \end{bmatrix}$

D24 Condition Number of a Matrix (Section 9.3)

1. (a) 21 (b) 225 (c) 65 (d) 493.5   2. 4,606

3. Very often, if determinant is very large or small, then the condition number will be large, but not always. For example,

Answers to Appendix D, MATLAB Exercises

$A = \begin{bmatrix} 2 & 2 \\ 2.0000000001 & 2 \end{bmatrix}$, to make |A| almost zero, columns are almost the same. Have
|A| = -2.000000165490742e-11, c(A) = 7.999999338077087e+11.
If condition number is high, the determinant will be very large or very small. e.g. consider the Hilbert matrix A of Exercise 2. $c(A) = c(A^{-1}) = 4{,}606$. |A| = 2.6455e-06, and
$|A^{-1}| = 1/|A| = 3.7800\text{e}+05$. However, the converse is not true, since

$$c(kA) = \|kA\| \|(kA)^{-1}\| = \|kA\| \|k^{-1}A^{-1}\| = |k| \, |k^{-1}| \|A\| \|A^{-1}\| = c(A)$$
$$|kA| = k|A|$$

Thus condition number is invariant under scalar multiplication of matrix, but determinant is not.

4. $y = 25 - 26\tfrac{1}{3}x + 9x^2 - \tfrac{2}{3}x^3$.

$A = \begin{bmatrix} 1 & 1 & 1 & 1 \\ 1 & 2 & 4 & 16 \\ 1 & 3 & 9 & 27 \\ 1 & 4 & 16 & 64 \end{bmatrix}$, a Vandermonde matrix. c(A) = 2000, system is ill-conditioned.

D25 Jacobi and Gauss-Seidel Iterative Methods (Section 9.4)

1. (a) Solution x = 1, y = 0.5, z = 0. With tolerance of .0001, Jacobi takes 11 iterations to converge; Gauss - Seidel takes 4 iterations.   (b) Does not converge

D26 The Simplex Method in Linear Programming (Section 10.2)

1. f = 24 at x = 6, y = 12.   2. f = 4 at x = 4, y = 0.   3. f = 16.5, at x = 1.25, y = 2.25

D27 Cross Product (Appendix A)

1. (a) (36, -27, -26)   (b) (-19, 3, 30)   2. (a) 108   (b) 0   (c) 0